Chiral Separations

METHODS IN MOLECULAR BIOLOGY™

John M. Walker, SERIES EDITOR

METHODS IN MOLECULAR BIOLOGY™

Chiral Separations

Methods and Protocols

Edited by

Gerald Gübitz

and

Martin G. Schmid

Institute of Pharmaceutical Chemistry and Pharmaceutical Technology,
Karl-Franzens University, Graz, Austria

Humana Press ⋇ Totowa, New Jersey

Production Editor: Tracy Catanese

Cover Illustration: Figure 1 from Chapter 25, "Chiral Separation by Capillary Electrochromatography Using Cyclodextrin Phases" by Dorothee Wistuba, Jingwu Kang, and Volker Schurig.

Cover design by Patricia F. Cleary.

For additional copies, pricing for bulk purchases, and/or information about other Humana titles, contact Humana at the above address or at any of the following numbers: Tel.: 973-256-1699; Fax: 973-256-8341; E-mail: humana@humanapr.com; or visit our Website: www.humanapress.com

Printed in the United States of America. 10 9 8 7 6 5 4 3 2 1

1-59259-648-7 (e-book)

Library of Congress Cataloging in Publication Data

Chiral separations : methods and protocols / edited by Gerald Gübitz and Martin G. Schmid.
      p. cm. -- (Methods in molecular biology ; 243)
   Includes bibliographical references and index.
   ISBN 1-58829-150-2 (alk. paper)
   ISSN 1064-3745
      1. Enantiomers--Separation--Laboratory manuals. 2. Chirality. I. Gübitz, Gerald. II. Schmid, Martin G. III. Series.

QP517.C57C46 2003
615'.19--dc21

200347771

Preface

Many compounds of biological and pharmacological interest are asymmetric and show optical activity. Approximately 40% of the drugs in use are known to be chiral and only about 25% are administered as pure enantiomers. It is well established that the pharmacological activity is mostly restricted to one of the enantiomers (eutomer). In several cases, unwanted side effects or even toxic effects may occur with the inactive enantiomer (distomer). Even if the side effects are not that drastic, the inactive enantiomer has to be metabolized, which represents an unnecessary burden for the organism. The administration of pure, pharmacologically active enantiomers is therefore of great importance. The ideal way to get to pure enantiomers would be by enantioselective synthesis. However, this approach is usually expensive and not often practicable. Usually, the racemates are obtained in a synthesis, and the separation of the enantiomers on a preparative scale is necessary. On the other hand, there is also a great demand for methods of enantiomer separation on an analytical scale for controlling synthesis, checking for racemization processes, controlling enantiomeric purity, and for pharmacokinetic studies. Conventional methods for enantiomer separation on a preparative scale are fractionated crystallization, the formation of diastereomeric pairs followed by repeated recrystallization, and enzymatic procedures. In recent years, chromatographic methods such as gas chromatography and, especially, liquid chromatography have attracted increasing interest for chiral separation, both on analytical and preparative scales. More recently, capillary electrophoresis and electrochromatography have also proven useful for chiral separation on an analytical scale.

Chiral Separations: Methods and Protocols focuses on chromatographic and electroseparation techniques for chiral separation on an analytical scale. It is not the aim of this book to give a comprehensive overview of all applications of chiral separation principles. Because there are several thousand publications on this topic, this would require a series of books. For comprehensive overviews the reader is referred to specialized review articles.

Chiral Separations: Methods and Protocols begins with an introduction to the different techniques, principles, and mechanisms of chiral separation, and includes a historical background (Chapter 1). Chapters 2–4 review some special techniques and include practical advice for users. The remainder of the book is devoted to articles describing typical procedures for enantiomer separation by chromatographic and electromigration techniques applying different chiral separation principles. These procedures may be of general character, or are otherwise presented by means of applications to substance classes or special compounds. These chapters differ from conventional articles, because primary emphasis is set on giving reliable procedures for users. Special attention is given to important experimental data, and practical hints in the "Notes" section enable the reader to adapt these procedures to one's separation problems.

Forty-three authors from twenty-four research laboratories all over the world have contributed to *Chiral Separations: Methods and Protocols.* We want to express our thanks to all of our authors and coauthors for making their expertise and knowledge available to those who are not already versed in this area.

This book should be helpful to biochemists, pharmaceutical chemists, clinical chemists, molecular biologists, and pharmacologists, both in research institutions and in industry.

Gerald Gübitz
Martin G. Schmid

Contents

Contents

Contents

Contributors

HASSAN Y. ABOUL-ENEIN • *Pharmaceutical Analysis Laboratory, Biological and Medical Research Department (MBC-03), King Faisal Specialist Hospital and Research Center, Riyadh, Saudi Arabia*

IMRAN ALI • *Pharmaceutical Analysis Laboratory, Biological and Medical Research Department (MBC-03), King Faisal Specialist Hospital and Research Center, Riyadh, Saudi Arabia*

LARS I. ANDERSSON • *DMPK and Bioanalytical Chemistry, AstraZeneca Research and Development, Södertälje, Sweden*

DANIEL W. ARMSTRONG • *Department of Chemistry, Iowa State University, Ames, IA*

BEZHAN CHANKVETADZE • *Molecular Recognition and Separation Science Laboratory, School of Chemistry, Tbilisi State University, Tbilisi, Georgia*

YOON JAE CHO • *Department of Chemistry, Pusan National University, Pusan, South Korea*

VADIM A. DAVANKOV • *Institute of Organoelement Compounds (INEOS), Russian Academy of Sciences, Moscow, Russia*

ELZBIETA EKIERT • *Department of Chemistry, Warsaw University, Warsaw, Poland*

SALVATORE FANALI • *Istituto di Metodologie Chimiche, C. N. R., Area della Ricerca di Roma, Monterotondo Scalo (Roma) Italy*

GERALD GÜBITZ • *Institute of Pharmaceutical Chemistry and Pharmaceutical Technology, Karl-Franzens University, Graz, Austria*

KURT GÜNTHER • *Degussa AG, Industriepark Wolfgang GmbH, Hanau, Germany*

JUN HAGINAKA • *Faculty of Pharmaceutical Sciences, Mukogawa Women's University, Nishinomiya, Japan*

LEO HSU • *Research and Development, GlaxoSmithKline, King of Prussia, PA*

MYUNG HO HYUN • *Department of Chemistry, Pusan National University, Pusan, South Korea*

ROLAND ISAKSSON • *Department of Chemistry and Biomedical Sciences, University of Kalmar, Kalmar, Sweden*

GUNNAR JOHANSSON • *Department of Biochemistry, Uppsala University, Uppsala, Sweden*

JINGWU KANG • *Institute of Organic Chemistry, University of Tübingen, Tübingen, Germany*

GENEVIEVE KENNEDY • *Research and Development, GlaxoSmithKline, King of Prussia, PA*

MICHAEL LÄMMERHOFER • *Christian Doppler Laboratory for Molecular Recognition Materials, Institute of Analytical Chemistry, University of Vienna, Vienna, Austria*

WOLFGANG LINDNER • *Christian Doppler Laboratory for Molecular Recognition Materials, Institute of Analytical Chemistry, University of Vienna, Vienna, Austria*

WIOLETA MARUSZAK • *Pharmaceutical Research Institute, Warsaw, Poland*

CLIFFORD R. MITCHELL • *Department of Chemistry, Iowa State University, Ames, IA*

KLAUS MÖLLER • *Macherey-Nagel, Düren, Germany*

JAKOB NILSSON • *Department of Technical Analytical Chemistry, Lund University, Lund, Sweden*

STAFFAN NILSSON • *Department of Technical Analytical Chemistry, Lund University, Lund, Sweden*

HIROYUKI NISHI • *Analytical Chemistry Department, CMC Research Laboratory, Tanabe Seiyaku Co., Ltd., Yodogawa-ku, Osaka, Japan*

YOSHIO OKAMOTO • *Department of Applied Chemistry, Graduate School of Engineering, Nagoya University, Furo-cho, Chikusa-ku, Nagoya, Japan*

KOJI OTSUKA • *Department of Material Chemistry, Graduate School of Engineering, Kyoto University, Nishikyo-ku, Kyoto, Japan*

GÖRAN PETTERSSON • *Department of Biochemistry, Uppsala University, Uppsala, Sweden*

COLETTE M. RABAI • *Department of Chemistry, Millsaps College, Jackson, MS*

PETER RICHTER • *Degussa AG, Industriepark Wolfgang GmbH, Hanau, Germany*

MARJA-LIISA RIEKKOLA • *Laboratory of Analytical Chemistry, University of Helsinki, Finland*

MARTIN G. SCHMID • *Institute of Pharmaceutical Chemistry and Pharmaceutical Technology, Karl-Franzens University, Graz, Austria*

VOLKER SCHURIG • *Institute of Organic Chemistry, University of Tübingen, Tübingen, Germany*

HELI SIRÉN • *Laboratory of Analytical Chemistry, University of Helsinki, Finland*

PETER SPÉGEL • *Department of Technical Analytical Chemistry, Lund University, Lund, Sweden*

SHIGERU TERABE • *Department of Material Science, Graduate School of Science, Himeji Institute of Technology, Kamigori, Hyogo, Japan*

GERALD TERFLOTH • *Research and Development, GlaxoSmithKline, King of Prussia, PA*

TOSHIMASA TOYO'OKA • *School of Pharmaceutical Sciences, University of Shizuoka, Shizuoka, Japan*

MAREK TROJANOWICZ • *Department of Chemistry, Warsaw University, Warsaw, Poland*

TIMOTHY J. WARD • *Department of Chemistry, Millsaps College, Jackson, MS*

DOROTHEE WISTUBA • *Institute of Organic Chemistry, University of Tübingen, Tübingen, Germany*

TOM LING XIAO • *Department of Chemistry, Iowa State University, Ames, IA*

CHIYO YAMAMOTO • *Department of Applied Chemistry, Graduate School of Engineering, Nagoya University, Furo-cho, Chikusa-ku, Nagoya, Japan*

1

Chiral Separation Principles

An Introduction

Gerald Gübitz and Martin G. Schmid

1. Introduction

The development of methods for chiral separation on an analytical as well as on a preparative scale has attracted great attention during the past two decades. Chromatographic methods such as gas chromatography (GC) *(1)*, high-performance liquid chromatography (HPLC) *(2–6)*, supercritical fluid chromatography (SFC) *(7–9)*, and thin-layer chromatography (TLC) *(10–13)* have been developed using different chiral separation principles. More recently, capillary electrophoresis (CE) *(14–21)* and capillary electrochromatography (CEC) *(22–25)* have been shown to be powerful alternatives to chromatographic methods. Several separation principles successfully used in HPLC have been transferred to CE and CEC. For the separation of enantiomers on a preparative scale, LC has become increasingly attractive.

The main domain of chromatographic and electromigration techniques is obviously the separation on an analytical scale for enantiomer purity control in synthesis, check for racemization processes, pharmaceutical quality control, pharmacokinetic studies, etc. Chromatographic enantiomer separations can be carried out either indirectly by using chiral derivatization reagents to form diastereomeric derivatives or directly using chiral selectors, which can be incorporated either in the stationary phase or the mobile phase. Similarly, in CE, indirect and direct ways are possible, thereby, in the latter approach, the chiral selector is simply added to the electrolyte.

CEC represent a new hybrid method between HPLC and CE. Accordingly, the chiral selector can be present in the mobile phase or in the stationary phase. Open tubular capillaries containing the stationary phase coated to the wall and packed capillaries are used. A new trend is to move away from packed capillaries.

From: *Methods in Molecular Biology, Vol. 243: Chiral Separations: Methods and Protocols*
Edited by: G. Gübitz and M. G. Schmid © Humana Press Inc., Totowa, NJ

Since packing of capillaries with silica-based materials is not easy, and the preparation of frits by sintering a zone of the packing is a rather sophisticated procedure, a new technique, the preparation of monolithic phases, was introduced. Such monolithic phases were prepared either on silica bases or by *in situ* polymerization of monomers, including the chiral selector directly in the capillary (continuous beds). The latter technique was introduced by Hjertén et al. *(26)*. Monolithic phases also found application in micro- and nano-HPLC. General overviews of the application of chromatographic and electromigration techniques for chiral separation are given in comprehensive overviews *(3)* and books *(27,28)*.

2. Indirect Separation

A broad spectrum of chiral derivatization reagents have been developed for GC, HPLC, and CE. Specialized reviews report on the application of chiral derivatization reagents for various substance classes *(29–34)*. For HPLC and CE, fluorescence reagents are of particular interest with respect to enhancing detection sensitivity *(35)*.

A certain disadvantage of this approach is the additional step. Furthermore, the chiral derivatization reagent has to be optically pure, and one must ensure that no racemization takes place during the reaction. On the other hand, many problems cannot be solved by direct separation approaches.

3. Direct Separation

The easiest way to perform direct separation is to add a chiral selector to the mobile phase in the case of HPLC, TLC, and CE. This simple approach gives good results in many cases, but is not always practicable and is cost-intensive with expensive reagents. More elegant and convenient is the use of chiral stationary phases (CSPs), where the chiral selector is adsorbed or chemically bonded to the stationary phase.

Several models for the requirements to obtain chiral recognition have been discussed. The most reliable model is the three-point contact model, proposed by Dalgliesh *(36)*, which postulates that three interactions have to take effect and at least one of them has to be stereoselective (**Fig. 1**). This model can be applied to most of the chiral separation principles. A detailed discussion of theoretical aspects of different chiral separation principles on atomic-level molecular modeling is given by Lipkowitz *(37)*. An overview and a description of various chiral separation principles will be presented in the following.

3.1. Formation of Multiple Hydrogen Bonds

Pioneering work in the field of chromatographic chiral separation was done by Gil-Av et al. *(38)*. This group developed chiral GC phases based on *N*-trifluoroacetyl-L-amino acid esters and resolved *N*-trifluoroacetyl amino acids.

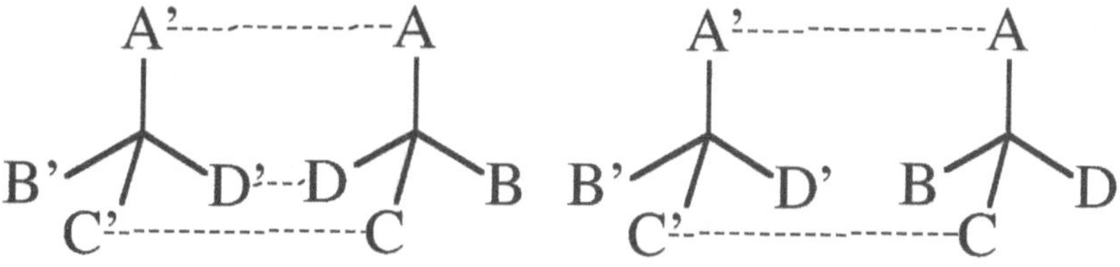

Fig. 1. Three-point interaction model.

The separation is based on the formation of multiple hydrogen bonds. Later, Bayerís group prepared a GC phase based on valine diamide linked to polysiloxanes, which was commercialized under the name Chirasil-Val *(39)*. Subsequently, several other chiral GC phases have been developed *(40)*.

HPLC phases using amino acid amides as chiral selectors were prepared by Dobashi and Hara *(41–43)*. The authors resolved on these phases derivatives of amino acids, hydroxy acids, and amino alcohols based on the formation of multiple hydrogen bonds.

3.2. Chiral π-Donor and π-Acceptor Phases

This principle had already been introduced by Pirkle's group at the end of the 1970s *(44,45)*. An (R)-*N*-(3,5-dinitrobezoyl)phenylglycine phase, having π-acceptor properties, showed chiral recognition ability for a broad spectrum of compounds with π-donor groups. In addition to π-π-interactions, dipole stacking and hydrogen bonds are assumed to be the interactions responsible for chiral recognition *(46)*.

An article by Welch *(47)* gives an overview of the large series of π-acceptor and π-donor phases prepared in Pirkle's group and their application to various compound classes. Several of these phases are commercially available (Regis Technologies, Morton Groove, IL, USA).

Subsequently, numerous π-acceptor and π-donor phases were developed by different groups *(48–51)*. Recently, it has been shown that phases of this type can also be used in CEC *(52,53)*.

3.3. Ionic Interactions

Ionic interactions exclusively are not sufficient to provide chiral recognition according to the three-point interaction model *(36)*. Additional supporting interactions such as hydrogen bonds, dipole-dipole interactions, or π-π-interactions have to take effect.

Lindner's group prepared cation-exchange-based CSPs using cinchona alkaloids as chiral selectors, which were used in HPLC *(54)* and CEC *(55,56)*. In this case π-π-interactions and hydrogen bonds are additonal interactions.

The formation of ion pairs using chiral counter ions such as (+)-S-10-camphor sulfonic acid *(57,58)*, N-benzoylcarbonyl glycyl-L-proline *(59)*, (-)2,3,4,6-di-O-isopropylidene-2-keto-L-gulonic acid *(60)*, and quinine *(59,61,62)* was utilized for the HPLC separation of various basic and acidic drugs, respectively. Also, with this principle, lateral binding forces have to support chiral recognition. The use of ion-pairing reagents in CE was successful only in nonaqueous medium. (+)-S-10-camphoric acid *(63)* was used for the chiral separation of bases and quinine *(64)* and quinine derivatives *(65)* for acidic compounds using nonaqueous electrolytes.

3.4. Chiral Surfactants

Surfactants are amphiphilic molecules containing a polar head group and a hydrophobic tail, which form micelles above the critical micelle concentration (CMC). The use of surfactants in CE was introduced by Terabe et al. *(66)* and called "micellar electrokinetic chromatography" (MEKC), since the hydrophobic micelles act as pseudostationary phases.

The analytes distribute between the electrolyte bulk phase and the chiral micelle phase. As chiral surfactants, bile salts, saponines, long chain N-alkyl-L-amino acids, N-alkanoyl-L-amino acids, alkylglycosides and polymeric amino acid, and dipetide derivatives were used. Overviews of the use of chiral surfactants are given in recent reviews *(67–70)*.

3.5. Chiral Metal Complexes: Ligand Exchange

The principle of ligand-exchange chromatography was introduced by Davankov and Rogozhin *(71)* in the early seventies. Chiral recognition is based on the formation of ternary mixed metal complexes between a chiral selector ligand and the analyte ligand. The different complex stability constants of the mixed complexes with D- and L-enantiomers are responsible for separation.

Mobile phase A_m

$$\updownarrow$$

Stationary phase $A_S + MS_S \rightleftharpoons AMS_S$

in which A represents the analyte; M represents the metal; and S represents the selector.

Generally, the chiral selector can be fixed to the stationary phase or added to the mobile phase. The first chiral liquid-exchange chromatography (LEC)-phases were prepared by Davankov for classical column chromatography and were based on polystyrene-divinylbenzene polymers containing amino acid residues complexed with metal ions. This basic principle was adapted by Gübitz et al. *(72–75)* to HPLC preparing chemically bonded phases on silica gel basis. These phases showed enantioselectivity for underivatized amino acids *(72–75)*, α-alkyl- and *N*-alkyl amino acids *(75,76)*, dipeptides *(75)*, hydroxy acids *(77)*, and thyroid hormones *(78)*. Phases of this type have been commercialized by Serva, Heidelberg, Germany (Chiral=Si-L-Pro, L-Hypro, L-Val) and Daicel, Tokyo Japan (Chiralpak WH). Subsequently, a considerably high number of chiral LEC-phases has been published *(79–85)*. Instead of chemically binding of ligands to silica gel, LEC-phases were also prepared by coating ligands with hydrophobic chains to reversed phases *(86–91)*. Addition of the selector ligand to the mobile phase was also found to be a successful alternative in several cases *(92,93)*. The following equilibria are to be taken into account in this approach:

Mobile phase $\qquad\qquad A_m + MS_m \rightleftharpoons AMS_m$
$$\qquad\qquad\qquad\qquad \updownarrow \quad \updownarrow \qquad\quad \updownarrow$$
Stationary phase $\qquad\qquad A_S + MS_S \rightleftharpoons AMS_S$

in which A represents the analyte; M represents the metal; and S represents the selector.

TLC plates containing the copper(II)complex of (2S,4R,2,RS)-4-hydroxy-1-(2,-hydroxydodecyl)proline as selector coated on a C-18 layer were developed by Günther et al. *(94)*.

Plates of this type have been commercialized by Macherey-Nagel (Düren, Germany) (Chiralplate®) and Merck (Darmstadt, Germany) (HPTLC-CHIR®). Chapter 2 is devoted to the use of TLC for chiral separations focusing on ligand-exchange thin-layer chromatography (LE-TLC).

The principle of LE has also been shown to be applicable in CE. In this case the selector complex is simply added to the electrolyte. A recent review gives an overview of developments and applications of this technique *(95)*.

More recently, LE was also successfully applied in CEC. Schmid et al. *(96)* prepared an LE-continuous bed by *in situ* co-polymerization of methacrylamide and *N*-(2-hydroxy-3-allyloxypropyl)-L-4-hydroxyproline as a chiral selector in the presence of piperazine diacrylamide as a crosslinker and vinylsulfonic acid as a charge providing agent. The applicability of this phase for chiral separation was demonstrated by the separation of amino acids *(96)* and hydroxy acids *(97)*. An alternative technique for preparing monolithic phases was published by Chen and Hobo *(98)*. A silica-based monolithic phase was prepared by a sol-gel procedure starting from tetramethoxysilane. The monolith was subsequently derivatized with L-prolineamide as chiral selector via 3-glycidoxypropyltrimethoxysilane. This CSP was applied to the chiral separation of dansyl amino acids and hydroxy acids.

The use of metal complexes, such as rhodium and nickel camphorates and 1,3-diketonate-bis-chelates of manganese(II), cobalt(II), and nickel(II) derived from perfluoroacetylated terpene-ketones in GC and their application to the chiral separation of pheromones, flavors, and oxiranes was described by Schurig et al. *(99–102)*.

3.6. Cyclodextrins

Cyclodextrins (CDs) are the most frequently used chiral selectors to have found application in HPLC, GC, SFC, TLC, CE, and CEC. CDs are cyclic oligo-saccharides consisting of six (α-CD), seven (β-CD), or eight (γ-CD) glucopyran-ose units. They form a truncated cone with a hydrophobic cavity. The outer surface is hydrophilic. The hydroxyl groups at the rim of the CD at positions 2, 3, and 6 are available for derivatization. Thereby the solubility of the CDs can be increased, and and the depth of the cavity modified. The chiral recognition mechanism is based on inclusion of a bulky hydrophobic group of the analyte, preferably aromatic groups, into the hydrophobic cavity of the CD. A second prerequisite for chiral recognition is the possibility of the formation of hydro-gen bonds or dipole-dipole interactions between the hydroxyl groups at the mouth of the CD and polar substituents close to the chiral center of the analyte.

In HPLC, CDs can be used either in CSPs or as chiral mobile phase additives. The first CSPs containing CDs chemically bonded to silica gel were developed by Armstrong et al. *(103)*. An overview of the application of CDs in HPLC and CE has recently been given by Bresolle et al. *(104)*. Chapter 3 in this book gives detailed information about CD-CSPs and their applications.

CDs were also used as CSPs for GC *(105)*. Permethylated *(106)* or perpen-tylated CDs *(106)* or other derivatives with varying polarity *(107)* were used as chiral selectors for the preparation of GC phases. These CSPs found also appli-cation for SFC *(7–9)*. The use CDs in TLC has been summarized in several reviews *(10,12,13)*.

The broadest spectrum of application of CDs was certainly found in CE *(17, 19,20,108)*. In addition to the native CDs, several neutral *(109)* and charged derivatives *(110,111)* were used. The most frequently used neutral CD derivatives are heptakis-O-methyl-CD, heptakis (2,6-di-O-methyl)-CD, heptakis (2,3,6-tri-O-methyl)-CD, hydroxyethyl-CD, and hydroxypropyl-CD. Since neutral CDs migrate with the same velocity as the electroosmotic flow (EOF), they cannot be used for neutral analytes. Negatively charged CDs, such as sulfated CDs, sulfobutyl- and sulfoethyl-β-CD, carboxymethyl-β-CD, and succinyl-β-CD, were applied to the chiral separation of neutral and basic compounds, since they show a counter-current mobility. Positively charged CDs, which contain amine or quaternary ammonium functions, on the other hand, found application to the chiral resolution of neutral and acidic analytes. Recently, also amphoteric CDs were developed *(112)*. It has been found that the combination of neutral and charged CDs often improves or even enables separation *(113,114)*.

Also, the combination of CDs with other chiral or nonchiral reagents was described. One example is the addition of sodium dodecyl sulfate (SDS), which forms negatively charged micelles *(115)*. These micelles migrate in the direction opposite to the EOF, while neutral CDs migrate with the same velocity as the EOF. Partition of the analyte takes place beween the bulk solution, the CD, and the micelle. Thereby, a neutral analyte is retarded and can be resolved using a neutral CD. This principle, named CD-mediated micellar electrokinetic chromatography (CD-MEKC) *(115)* can be also used as a means for reversing the enantiomer migration order *(116)*. The combination of CDs with nonchiral crown ethers *(117,118)* or ion-pairing reagents *(119,120)* were found to support or enable chiral resolution in may cases. Compounds containing diol structure can be resolved by using a mixture of a CD and borate *(121–123)*. The formation of mixed CD-borate-diol complexes is assumed.

The first application of CEC using CDs was described by Schurig's group *(124, 125)* using open tubular capillaries. The capillary wall was coated with permethylated β-CD, which was attached to dimethylpolysiloxane via an octamethylene spacer. The same capillary was used for nano-HPLC, GC, SFC, and CEC *(126)*. Later, the same group prepared packed capillaries containing permethylated β-CD chemically bonded to silica gel *(127,128)*. An overview of the applications of CDs in chiral CEC is given by Schurig and Wistuba *(129)*. Phases based on continuous bed technology, prepared by *in situ* polymerization directly in the capillary were described by Koide and Ueno *(130)* and Végvári et al. *(131)*. Recently, Wistuba and Schurig *(132)* prepared a monolithic phase by sintering the silica bed of a packed capillary at 380°C and binding a permethylated β-CD onto the surface.

3.7. Carbohydrates

Native polysaccharides showed only weak chiral recognition ability. Microcrystalline cellulose triacetate (CTA-I) was found to be able to include stereoselectively compounds with aromatic moieties into cavities formed by swelling *(133)*. Phases containing cellulose triacetate, prepared by a different way (CTA-II), coated onto macroporous silica gel, showed distinct enantioselectivity *(134)*. In this case, hydrogen bondings and dipole-dipole interactions were assumed to be the main interactions *(135)*. Okamoto's group prepared a broad spectrum of cellulose ester and cellulose carbamate-based phases. These phases were commercialized by Daicel (Tokyo, Japan). Several polysaccharide-based phases can be used in addition to the normal phase mode also in the polar organic- and reversed-phase mode *(136)*. Specialized reviews give an overview of the development and application of various polysaccharide-based CSPs *(137–141)*. X-ray, nuclear magnetic resonance (NMR) studies, and computer simulations brought some insight into the chiral recognition mechanism of phases based on the cellulose trisphenyl carbamate type (CTPC).

CTPC has a left-handed 3/2 helical conformation, and the glucose residues are regularly arranged along the helical axis. A chiral groove exists with polar carbamate groups inside the groove and hydrophobic aromatic groups outside of the groove. Polar groups of the analytes may interact with the carbamate residues inside the groove via hydrogen-bonds. π-π-interactions might be additional contributions for chiral recognition *(140)*. When cellulose was substituted by amylose, different enantioselectivity was observed *(142)*.

Other polysaccharides described for the preparation of CSPs are chitosan *(143)*, chitin *(144)* and amylopectin *(145)*. Detailed information about polysaccharide-based phases and their applications are given in Chapters 5 and 6 in this book. Several polysaccharide phases used in HPLC also found application in SFC *(7–9)*. Native cellulose and cellulose derivatives were also described as stationary phases for TLC *(10,12)*.

Maltodextrins and dextrans were found to be useful chiral selectors in CE. Also in this case, the formation of a helical structure supported by additional interactions, such as hydrogen bonds and dipole-dipole interactions, is assumed to be responsible for chiral recognition *(146,147)*. Other polysaccharides such as amyloses, laminaran, pullulan, methylcellulose and carboxymethyl cellulose *(148)*, and even some monosaccharides *(149)* were found to exhibit some limited chiral recognition ability. Several negatively charged polysaccharides, such as heparin, various sulfated glycoseaminoglycans, and polygalacturonic acid, were tested in CE and found application for the chiral separation of basic compounds *(21,146)*. Furthermore, some positively charged polysaccharides, such as diethylaminoethyl dextran, and the aminoglycoside antibiotics streptomycin sulfate, kanamycin sulfate, and fradiomycin sulfate were investigated *(150)*.

3.8. Macrocyclic Antibiotics

Macrocyclic antibiotics were introduced as chiral selectors by Armstrong *(151)*. These selectors found application in HPLC *(152–156)*, TLC *(157,158)*, CE *(156,159–161)*, and recently in CEC *(22,25,162–168)*. Two main groups of macrocyclic antibiotics, the ansamycins rifamycin B and rifamycin SV, and the glycopeptides vancomycin, ristocetin, teicoplanin, and avoparcin are the most frequently used selectors. CSPs on this basis have been commercialized by Astec (Whippany, NJ, USA).

Recently, a series of other glycopeptide antibiotics were also investigated for their chiral recognition ability. The glycopeptides consist of an aglycon portion of fused macrocyclic rings that form a hydrophobic basket shape, which can include hydrophobic parts of an analyte and a carbohydrate moiety. There are pendant polar arms, which form hydrogen bonds and dipole-dipole interactions with polar groups of the analyte. Furthermore, ionic interactions and π-π-interactions might support the separation. While rifamycin B was found to be superior for basic compounds, rifamycin SV and the glycopeptide antibiotics are more suitable for acidic analytes in CE separations. Since these selectors may cause

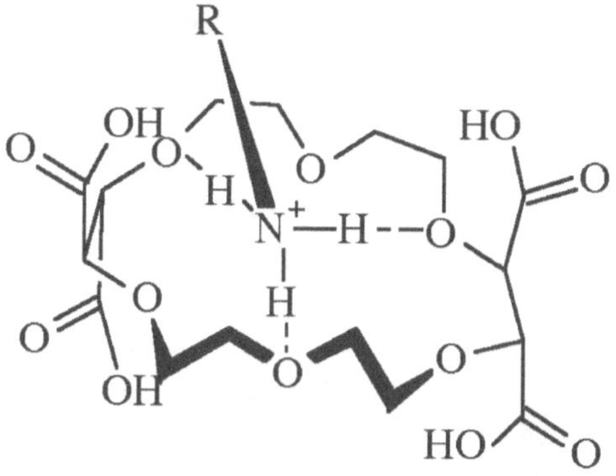

Fig. 2. Stereoselective inclusion of an amine into a chiral crown ether.

detection problems in CE due to their UV-absorption, a partial filling method and a counter-current process was applied to overcome these problems *(169)*. Interestingly, the teicoplanin aglycon showed distinct stereoselectivity compared to the intact molecule *(168,170)*. Chapter 6 gives an overview of CSPs based on macrocyclic antibiotics and their application.

3.9. Chiral Crown Ethers

Crown ethers are macrocyclic polyethers that form host-guest complexes with alkali-, earth-alkali metal ions, and ammonium cations. Sousa, Cram, and coworkers *(171)* found that chiral crown ethers can include enantioselectively primary amines and developed the first chiral crown ether phases for LC *(172)*. As a chiral recognition mechanism, the formation of hydrogen bonds beween the three hydrogens attached to the amine nitrogen and the dipoles of the oxygens of the macrocyclic ether is postulated (**Fig. 2**). Furthermore, the substituents of the crown ether are arranged perpendicular to the plane of the macrocyclic ring, forming a kind of chiral barrier, which divides the space available for the substituents at the chiral centers of the analyte into two domains. Thus, two different diastereomeric inclusion complexes are formed.

Shinbo et al. *(173)* developed an HPLC phase containing a polymeric crown ether derivative adsorbed on silica gel and demonstrated the applicability of this phase for chiral separations by means of amino acids. HPLC columns of this type are commercially available under the name Crownpack CRr from Daicel. Recently, several chemically bonded chiral crown ether phases and their application to the chiral separation of amino acids aand other compounds with primary amino groups were published *(174–177)*. Such a phase is now commercially available under the name Oticrown from (Usmac, Thousand Oaks, CA, USA).

More recently, Steffek et al. showed that contrary to original observations, such a chiral crown ether phase responds stereoselectively not only to primary amines but also to some secondary amines *(178)*.

The application of chiral crown ethers in CE was first described by Kuhn et al. *(179)*. These authors used 18-crown tetracarboxylic acid ($18C_6H_4$) in an electrolyte of low pH for the chiral separation of amino acids. In addition to the inclusion into the cavity, lateral interactions, such as hydrogen bonds, dipole-dipole interactions, and ionic interactions, between the carboxylic groups of the selector and the analyte are assumed to take effect. This chiral crown ether found application to the chiral separation of sympathomimetics *(180)*, dipeptides *(181,182)*, and various drugs containing primary amino groups *(183)*. Mori et al. *(184)* showed that CE separations with this crown ether are also possible in non-aqueous medium.

3.10. Calixarenes

Calixarenes represent an interesting new type of chiral selectors. Chiral GC phases based on calix[4]arenes have recently been published *(185,186)*. The applicability of these phases was demonstrated by means of the chiral separation of selected amino acids, amino alcohols, and amines. An inclusion mechanism supported by dipole-dipole interactions and hydrogen bonds might be assumed as the chiral recognition basis. Recently, the use of calixarenes for chiral CE *(187)* and CEC *(188)* separations was described. To date, no chiral HPLC application of calixarenes has been reported.

3.11. Other Synthetic Macrocycles

Several interesting chiral receptor-like selectors for HPLC phases were synthesized *(189–192)*, which, however, will not be discussed in detail within this frame. The synthesis of such tailor-made selectors will be without doubt an approach with future.

3.12. Chiral Synthetic Polymers

Blaschke and coworkers *(193)* developed polyacrylamides containing an L-phenylalanine moiety. HPLC phases containing such polymers bonded to silica gel are commercially available (Merck) under the name Chiraspher®. Okamoto's group synthesized helical isotactic polymethacrylamides supported on macroporous silica gel, which are commercially available (Daicel) under the name Chiralpak OT. Hjertén's group developed the "continuous bed" technology by *in situ* co-polymerization of monomers including a chiral selector with a cross-linker *(26)*. With this simple process, monolithic phases are obtained and no frits are needed. This technique found application for the preparation of chiral HPLC- *(194)* and CEC-phases *(95,130,131,195–197)*. Sinner and Buchmeiser *(198)*

Table 1
Proteins Used as Chiral Selectors

Protein	Trade name of CSP	Manufacturer
BSA	Chiral BSA	Shandon
	Resolvosil BSA-7	Nagel-Macherey
	Resolvosil BSA-7PX	Nagel-Macherey
	Ultron ES-BSA	Shinwa Chemical Ind.
HSA	Chiral-HSA	Chrom Tech AB
	Chiral HSA	Shandon
α1-Acid glycoprotein	Chiral-AGP	Chrom Tech AB
Ovomucoid		
Ovoglycoprotein	Ultron ES-OVM	Shinwa Chemical Ind.
Avidin	Bioptic AV-1	GL Sciences/Ansys Techn.
Riboflavin binding protein		
Cellobiohydrolase I	Chiral-CBH	Chrom Tech AB
Lysozyme		
Pepsin	Ultron ES-Pepsin	Shinwa Chemical Ind.
Amyloglucosidase		
Ovotransferrin		
β-Lactoglobulin		

recently published a ring-opening metathesis polymerization for the preparation of monolithic phases using a norborene derivative of β-CD as chiral monomer. An overview of the synthesis and application of chiral synthetic polymers is given by Nakano *(199)*.

3.13. Molecularly Imprinted Polymers

This principle was introduced by Wulff *(200)*. A monomer is polymerized with a crosslinker in the presence of a chiral template molecule. After removing the template molecule, a chiral imprinted cavity remains, which shows stereo-selectivity to the template or closely related molecules. This technique found application in HPLC, TLC, and CEC. Several groups prepared chiral mono-lithic phases for CEC using the imprint approach *(201–204)*. For detailed informa-tions the reader is referred to specialized reviews *(205–207)* (*see also* Chapters 9 and 25).

3.14. Use of Proteins as Chiral Selectors

Proteins are known to be able to bind drugs stereoselectively. This behavior has been utilized for chromatographic and capillary electrophoretic separations of drug enantiomers. Proteins used as chiral selectors in HPLC and CE are listed in **Table 1**. Specialized reviews summarize the use of proteins as chiral selectors in HPLC *(208)* and CE *(209–211)* (*see also* Chapters 15 and 16). Bovine serum albumin (BSA) found also some application as chiral selector in TLC and CEC.

The chiral recognition ability of proteins is related to the formation of a three-dimensional structure. Dipole-dipole interactions, hydrogen bonds, and hydrophobic interactions are assumed to be the main interactions. Dependent on pH, they can be negatively or positively charged. Ionic strength and pH, type, and concentration of organic modifiers were found to affect strongly retention and resolution. Proteins show enantioselectivity for a broad spectrum of compounds, however, predictions are hardly possible.

4. Miscellaneous

4.1. Nonaqueous CE

The use of nonaqueous solvents in CE is sometimes advantageous, for solubility reasons, to reduce interactions with the capillary wall and to avoid the interference of water in the case of weak interactions between analytes and chiral selector. Chiral ion-pairing CE, for example, is only practicable in nonaqueous medium *(63,64)*. Selectivity is often improved in nonaqueous solvents. Since Joule heating is lower in nonaqueous solvents, higher voltage can be applied resulting in shorter migration times. Last but not least, nonaqueous solvents show less interferences when coupling CE with mass spectrometry (MS). Many chiral separation principles used in aqueous systems were successfully transferred to nonaqueous systems *(212)*.

4.2. Isotachophoresis and Isoelectric Focusing

There are only a few papers dealing with chiral separation by isotachophoresis (ITP) *(213)*. Coupled isotachophoresis capillary zone electrophoresis (ITP-CZE) systems for sample clean-up and preconcentration were developed by Dankova et al. *(214)*, Fanali et al. *(215)*, and Tousaint *(216)*. ITP systems for preparative isolation and purification of enantiomers were designed by Kaniansky et al. *(217)*, and Hoffmann et al. *(218)*. Glukhovsky and Vigh *(219)* used preparative isoelectric focusing (IEF) for the chiral separation of Dns-amino acids on a mg/h scale.

4.3. Reversal of Enantiomeric Elution (Migration) Order

Reversal of the enantiomeric elution order (EEO) or enantiomeric migration order (EMO), respectively, is sometimes necessary, for example for checking the enantiomeric purity of drugs. It is important to be able to detect traces of the inactive enantiomer, which can exhibit side effects, beside a high excess of the active enantiomer. To avoid overlapping with the tailing of the large peak of the active enantiomer, the inactive enantiomer should appear always as first peak.

The simplest way would be to change the chirality of the selector. This is, however, not always possible. Other tools in CE for achieving reversal of the EMO are to change from a neutral to a charged selector, to change the mobility of the analyte or the selector by varying the pH or by reversing the direction of the EOF.

An excellent survey of different possibilities for reversing the EMO in CE has been given by Chankvetadze et al. *(116)*. The possibilities for changing the EEO in HPLC are restricted, since only few chiral phases exist in both enantiomeric forms.

4.4. Chiral Analysis of Compounds in Biological Samples

The chiral separation of compounds of biological or pharmacological interest in biological samples is required, for example, in connection with pharmacodynamic studies, metabolism studies, and toxicological analysis. This requires usually intensive sample pretreatment and preconcentration steps. Column coupling and column switching methods have widely been used for analysis of biological samples *(220–222)*. Another important point is the detection sensitivity. The use of sensitive detection systems such as laser-induced fluorescence (LIF) detection or coupling of HPLC or CE with MS is often a requirement. Specialized reviews on chiral drug analysis in biological samples using chromatographic or capillary electrophoretic methods give more insight into these problems *(33,34,103,223–225)*.

4.5. Future Trends

Miniaturization of the systems is a recent trend. Increasing research is being done using nano-HPLC systems or developing microfabricated chips for CE-separation. CEC is becoming more and more popular. The use of monolithic phases in CEC and nano-HPLC will certainly make these techniques more convenient. A recent interesting technique, with which several millions of plates can be achieved, represents synchronous cyclic CE introduced by Zhao and Jorgenson *(226)*. On-line coupling of flow-injection analysis (FIA) systems with CE enable sample pretreatment steps and enhancement in sample throughput *(227–229)*. Another challenging approach will be the application of stereoselective antibodies used for enantioselective enzyme-linked immunosorbent assay (ELISA) *(230)*, immunosensors *(231)*, and flow-injection immunoassay (FIIAs) *(232,233)* as chiral selectors.

4.6. Selection of the Chiral Separation Principle

According to the nature of stereoselectivity there will never be a universally applicable chiral selector or CSP, respectively. The separation principle has always to be selected according to structure of the analytes. There are some chiral selectors that respond to a broad spectrum of compound classes, however predictions are possible only in few cases. Application guides from reagent and column suppliers are often very helpful.

References

1. Schurig, V. (2001) Separation of enantiomers by gas chromatography. *J. Chromatogr. A* **906,** 275–299.
2. Gübitz, G. (1990) Separation of drug enantiomers by HPLC using chiral stationary phases—a selective review. *Chromatographia* **30,** 555–564.
3. Bojarski, J. (1997) Recent progress in chromatographic enantioseparations. *Chem. Anal.* **42,** 139–185.
4. Gasparrini, F., Misiti, D., and Villani, C. (2001) HPLC chiral stationary phases based on low-molecular-mass selectors. *J. Chromatogr. A* **906,** 35–50.
5. Subramanian, G. (ed.) (1994) *A Practical Approach to Chiral Separations by Liquid Chromatography.* Wiley-VCH, Weinheim, Germany.
6. Ahuja, S. (ed.) (1997) *Chiral Separations—Applications and Technology.* American Chemical Society, Washington, DC.
7. Terfloth, G. (2001) Enantioseparations in super- and subcritical fluid chromatography. *J. Chromatogr. A* **906,** 301–307.
8. Williams, K. L. and Sander, L. C. (1997) Enantiomer separations on chiral stationary phases in supercritical fluid chromatography. *J. Chromatogr. A* **785,** 149–158.
9. Petersson, P. and Markides, K. E. (1994) Chiral separations performed by supercritical fluid chromatography. *J. Chromatogr. A* **666,** 381–394.
10. Günther, K. and Möller, K. (eds.) (1996) *Handbook of Thin-Layer Chromatography, 2nd Ed.* (Sherma, J. and Fried, B., ed.), Marcel Dekker, New York, pp. 621–682.
11. Duncan, J. D. (1990) Chiral separations—a comparison of HPLC and TLC. *J. Liq. Chromatogr.* **13,** 2737–2755.
12. Lepri, L. (1997) Enantiomer separation by thin-layer chromatography. *J. Planar. Chromatogr.-Modern TLC* **10,** 320–331.
13. Aboul-Enein, H. Y., El-Awady, M. I., Heard, C. M., and Nicholls, P. J. (1999) Application of thin-layer chromatography in enantiomeric chiral analysis—an overview. *Biomed. Chromatogr.* **13,** 531–537.
14. Nishi, H. and Terabe, S. (1995) Optical resolution drugs by capillary electrophoretic techniques. *J. Chromatogr. A* **694,** 245–276.
15. Fanali, S. (1996) Identification of chiral drug isomers by capillary electrophoresis. *J. Chromatogr. A* **735,** 77–121.
16. Chankvetadze, B. (1997) Separation selectivity in chiral capillary electrophoresis with charged selectors. *J. Chromatogr. A* **792,** 269–295.
17. Fanali, S. (1997) Controlling enantioselectivity in chiral capillary electrophoresis with inclusion-complexation. *J. Chromatogr. A* **792,** 227–267.
18. Gübitz, G. and Schmid, M. G. (1997) Chiral separation principles in capillary electrophoresis. *J. Chromatogr. A* **792,** 179–225.

19. Fanali, S. (2000) Enantioselective determination by capillary electrophoresis with cyclodextrins as chiral selectors. *J. Chromatogr. A* **875,** 89–122.
20. Verleysen, K. and Sandra, P. (1998) Separation of chiral compounds by capillary-electrophoresis. *Electrophoresis* **19,** 2798–2833.
21. Gübitz, G. and Schmid, M. G. (2000) Recent progress in chiral separation principles in capillary electrophoresis. *Electrophoresis* **21,** 4112–4135.
22. Gübitz, G. and Schmid, M. G. (2000) Chiral separation by capillary electrochromatography (minireview). *Enantiomer* **5,** 5–11.
23. Wistuba, D. and Schurig, V. (2000) Enantiomer separation of chiral pharmaceuticals by capillary electrochromatography. *J. Chromatogr. A* **875,** 255–276.
24. Dermaux, A. and Sandra, P. (1999) Applications of capillary electrochromatography. *Electrophoresis* **20,** 3027–3065.
25. Wistuba, D. and Schurig, V. (2000) Recent progress in enantiomer separation by CEC. *Electrophoresis* **21,** 4036–4058.
26. Hjertén, S., Liao, J.-L., and Zhang, R. (1989) High-performance liquid chromatography on continuous polymer beds. *J. Chromatogr. A* **473,** 273–275.
27. Subramanian, G. (ed.) (2000) *Chiral Separation Techniques: A Practical Approach.* Wiley-VCH, Weinheim, Germany.
28. Chankvetadze, B. (ed.) (2001) *Chiral Separations.* Elsevier Science, Amsterdam.
29. Bhushan, R. and Joshi, S. (1993) Resolution of enantiomers of amino-acids by HPLC. *Biomed. Chromatogr.* **7,** 235–250.
30. Zhou, Y., Luan, P., Liu, L., and Sun, Z. P. (1994) Chiral derivatizing reagents for drug enantiomers bearing hydroxyl-groups. *J. Chromatogr. B* **659,** 109–126.
31. Bovingdon, M. E. and Webster, R. A. (1994) Derivatization reactions for neurotransmitters and their automation. *J. Chromatogr. B* **659,** 157–183.
32. Campíns-Falcó, P., Sevillano-Cabeza, A., and Molina-Legua, C. (1994) Amphetamine and methamphetamine determinations in biological samples by high-performance liquid-chromatography. *J. Liq. Chromatogr.* **17,** 731–747.
33. Görög, S. and Gazdag, M. (1994) Enantiomeric derivatization for biomedical chromatography. *J. Chromatogr. B* **659,** 51–84.
34. Srinivas, N. R., Shyu, W. C., and Barbhaiya, R. H. (1995) Gaschromatographic determination of enantiomers as diastereomers following pre-column derivatization and applications to pharmacokinetic studies: a review. *Biomed. Chromatogr.* **9,** 1–9
35. Toyo'oka, T. (1996) Recent progress in liquid chromatographic enantioseparation based upon diastereomer formation with fluorescent chiral derivatization reagents. *Biomed. Chromatogr.* **10,** 265–277.
36. Dalgliesh, C. E. (1952) The optical resolution of aromatic amino-acids on paper chromatograms. *J. Chem. Soc.* **137,** 3940–3942.
37. Lipkowitz, K. B. (2001) Atomistic modelling of enantioselection in chromatography. *J. Chromatogr. A* **906,** 417–442.
38. Gil-Av, E., Feibush, B., and Charles-Sigler, R. (1966) Separation of enantiomers by gas liquid chromatography with an optically active stationary phase. *Tetrahedron Lett.* 1009–1015.

39. Frank, H., Nicholson, G. J., and Bayer, E. (1978) Chiral polysiloxanes for resolution of optical antipodes. *Angew. Chem. Int. Ed. Engl.* **17**, 363–365.
40. Schurig, V. (2001) Separation of enantiomers by gas chromatography. *J. Chromatogr. A* **906**, 275–299.
41. Dobashi, A., Dobashi, Y., and Hara, S. (1986) Enantioselectivity of hydrogen-bond association in liquid-solid chromatography. *J. Liq. Chromatogr.* **9**, 243–267.
42. Dobashi, Y. and Hara, S. (1985) Direct resolution of enantiomers by liquid-chromatography with the novel chiral stationary phase derived from (R,R)-tartramide. *Tetrahedron Lett.* **26**, 4217–4220.
43. Dobashi, Y. and Hara, S. (1987) A chiral stationary phase derived from (R,R)-tartramide with broadened scope of application to the liquid-chromatographic resolution of enantiomers. *J. Org. Chem.* **52**, 2490–2496.
44. Pirkle, W. H., House, D. W., and Finn, J. M. (1980) Broad-spectrum resolution of optical isomers using chiral high-performance liquid-chromatographic bonded phases. *J. Chromatogr.* **192**, 143–158.
45. Pirkle, W. H., Finn, J. M., Schreiner, J. L., and Hamper, B. C. J. (1981) A widely useful chiral stationary phase for the high-performance liquid-chromatography separation of enantiomers. *J. Am. Chem. Soc.* **103**, 3964–3966.
46. Pirkle, W. H., Welch, C. J., and Hyun, M. H. (1983) A chiral recognition model for the chromatographic resolution of n-acylated 1-aryl-1-aminoalkanes. *J. Org. Chem.* **48**, 5022–5026.
47. Welch, C. J. (1994) Evolution of chiral stationary phase design in the Pirkle laboratories. *J. Chromatogr. A* **666**, 3–26.
48. Hyun, M. H. and Min, C. S. (1998) Chiral recognition mechnism for the resolution of enantiomers on a highly effective HPLC chiral stationary phase derived from (R)-4-hydroxyphenylglycine. *Chirality* **10**, 592–599.
49. Lin, C.-E. and Lin, C.-H. (1994) Enantiomer separation of amino-acids on a chiral stationary-phase derived from 1-alanyl-disubstituted and pyrrolidinyl-disubstituted cyanuric chloride. *J. Chromatogr. A* **676**, 303–309.
50. Gasparrini, F., Misiti, D., Pierini, M., and Villani, C. (1996) Enantioselective chromatography on brush-type chiral stationary phases containing totally synthetic selectors. Theoretical aspects and practical applications. *J. Chromatogr. A* **724**, 79–90.
51. Uray, G., Maier, N. M., Niederreiter, K. S., and Spitaler, M. M. (1998) Diphenylethanediamine derivatives as chiral selectors VIII. Influence of the second amido function on the high-performance liquid chromatographic enantioseparation characteristics of (N-3,5-dinitrobenzoyl)-diphenylethanediamine based chiral stationary phases. *J. Chromatogr. A* **799**, 67–81.
52. Wolf, C., Spence, P. L., Pirkle, W. H., Derrico, E. M., Cavender, D. M., and Rozing, G. P. (1997) Enantioseparations by electrochromatography with packed capillaries. *J. Chromatogr. A* **782**, 175–179.
53. Wolf, C., Spence, P. L., Pirkle, W. H., Cavender, D. M., and Derrico, E. M. (2000) Investigation of capillary electrochromatography with brush-type chiral stationary phases. *Electrophoresis* **21**, 917–924.

54. Lämmerhofer, M. and Lindner, W. (1996) Quinine and quinidine derivatives as chiral selectors I. Brush type chiral stationary phases for high-performance liquid chromatography based on cinchonan carbamates and their application as chiral anion exchangers. *J. Chromatogr. A* **741,** 33–48.

55. Lämmerhofer, M. and Lindner, W. (1998) High-efficiency chiral separations of N-derivatized amino acids by packed-capillary electrochromatography with a quinine-based chiral anion-exchange type stationary phase. *J. Chromatogr. A* **829,** 115–125.

56. Tobler, E. M., Lämmerhofer, M., and Lindner, W. (2000) Investigation of an enantioselective non-aqueous capillary electrochromatography system applied to the separation of chiral acids. *J. Chromatogr. A* **875,** 341–352.

57. Pettersson, C. and Schill, G. (1981) Separation of enantiomeric amines by ion-pair chromatography. *J. Chromatogr.* **204,** 179–183.

58. Salva, P. S., Hite, J. G., and Henkel, J. G. (1982) The preparative scale reverse phase HPLC separation of epimeric alkaloids using camphorsulfonic acid as an ion pairing reagent. *J. Liq. Chromatogr.* **5,** 305–312.

59. Pettersson, C. and Karlsson, A. (1992) Separation of enantiomeric amines and acids using chiral ion-pair chromatography on porous graphitic carbon. *Chirality* **4,** 323–332.

60. Pettersson, C. and Gioeli, C. (1993) Chiral separation of amines using reversed-phased ion-pair chromatography. *Chirality* **5,** 241–245.

61. Pettersson, C. and No, K. (1983) Chiral resolution of carboxylic and sulfonic acids by ion-pair chromatography. *J. Chromatogr.* **282,** 671–684.

62. Pettersson, C. (1984) Chromatographic separation of enantiomers of acids with quinine as chiral counter ion. *J. Chromatogr.* **316,** 553–567.

63. Bjornsdottir, I., Hansen, S. H., and Terabe, S. (1996) Chiral separation in non-aqueous media by capillary electrophoresis using the ion-pair principle. *J. Chromatogr. A* **745,** 37–44.

64. Stalcup, A. M. and Gahm, K. H. (1996) Quinine as a chiral additive in nonaqueous capillary zone electrophoresis. *J. Microcol. Separ.* **8,** 145–150.

65. Piette, V., Lämmerhofer, M., Lindner, W., and Crommen, J. (1999) Enantiomeric separation of N-protected amino acids by non-aqueous capillary electrophoresis using quinine or tert-butyl carbamoylated quinine as chiral additive. *Chirality* **11,** 622–630.

66. Terabe, S., Ichikawa, K. T., Otsuka, K., and Tsuchiya, A. (1984) Electrokinetic separations with micellar solutions and open-tubular capillaries. *Anal. Chem.* **56,** 111–113.

67. Cammileri, P. (1997) Chiral surfactants in micellar electrokinetic capillary chromatography. *Electrophoresis* **18,** 2322–2330.

68. Palmer, C. P. and Tanaka, N. (1997) Selectivity of polymeric and polymer-supported pseudo-stationary phases in micellar electrokinetic chromatography. *J. Chromatogr. A* **792,** 105–124.

69. Otsuka, K. and Terabe, S. (2000) Enantiomer separation of drugs by micellar electrokinetic chromatography using chiral surfactants. *J. Chromatogr. A* **875,** 163–178.

70. Shamsi, S. A. and Warner, I. M. (1997) Monomeric and polymeric chiral surfactants as pseudo-stationary phases for chiral separations. *Electrophoresis* **18**, 853–872.
71. Davankov, V. A. and Rogozhin, S. V. (1971) Ligand chromatography as a novel method for the investigation of mixed complexes: stereoselective effects in α-amino acid copper(II) complexes. *J. Chromatogr.* **60**, 280–283.
72. Gübitz, G., Jellenz, W., Löffler, G., and Santi, W. (1979) Chemically bonded chiral stationary phases for the separation of racemates by HPLC. *J. High Resol. Chromatogr. Chromatogr. Commun.* **2**, 145–146.
73. Gübitz, G., Jellenz, W., and Santi, W. (1981) Separation of the optical isomers of amino acids by ligand-exchange chromatography using chemically bonded phases. *J. Chromatogr.* **203**, 377–384.
74. Gübitz, G., Juffmann, W., and Jellenz, W. (1982) Direct separation of amino acid enantiomers by high performance ligand-exchange chromatography on chemically bonded chiral phases. *Chromatographia* **16**, 103–106.
75. Gübitz, G. (1986) Direct separation of enantiomers by high performance ligand-exchange chromatography on chemically bonded chiral phases. *J. Liq. Chromatogr.* **9**, 519–535.
76. Brückner, H. (1987) Enantiomeric resolution of N-methyl-α-amino acids by ligand-exchange chromatography. *Chromatographia* **24**, 725–738.
77. Gübitz, G. and Mihellyes, S. (1984) Direct separation of 2-hydroxy acids enantiomers by high-performance liquid chromatography on chemically bonded chiral phases. *Chromatographia* **19**, 257–259.
78. Gübitz, G. and Juffmann, F. (1987) Resolution of the enantiomers of thyroid hormones by high performance ligand-exchange chromatography using a chemically bonded chiral stationary phase. *J. Chromatogr.* **404**, 391–393.
79. Davankov, V. A., Navratil, J. D., and Walton, H. F. (eds.) (1988) *Ligand Exchange Chromatography.* CRC Press, Boca Raton.
80. Davankov, V. A. (1994) Chiral selectors with chelating properties in liquid chromatography: fundamental reflections and selective review of recent developments. *J. Chromatogr. A* **666**, 55–76.
81. Kurganov, A. (2001) Chiral chromatographic separations based on ligand exchange. *J. Chromatogr. A* **906**, 51–71.
82. Davankov, V. A. (2000). 30 years of chiral ligand exchange. *Enantiomer* **5**, 209–223.
83. Marchelli, R., Corradini, R., Bertuzzi, T., et al. (1996) Chiral discrimination by ligand-exchange chromatography: a comparison between phenylalaninamide-based stationary and mobile phases. *Chirality* **8**, 452–461.
84. Gübitz, G., Mihellyes, S., Kobinger, G., and Wutte, A. (1994) New chemically bonded chiral ligand-exchange chromatographic stationary phases. *J. Chromatogr. A* **666**, 91–97.
85. Wachsmann, M. and Brückner, H. (1998) Ligand-exchange chromatographic separation of DL-amino acids on aminopropylsilica-bonded chiral s-triazines. *Chromatographia* **47**, 637–642.

86. Davankov, V. A., Bochkov, A. S., Kurganov, A. A., Roumeliotis, P., and Unger, K. K. (1980) Dealing with the ligand-exchange chromatography. 13. Separation of unmodified alpha-amino-acid enantiomers by reverse phase HPLC. *Chromatographia* **13,** 677–685.

87. Remelli, M., Fornasari, P., Dondi, F., and Pulidori, F. (1993) Dynamic column-coating procedure for chiral ligand-exchange chromatography. *Chromatographia* **37,** 23–30.

88. Yamazaki, S., Takeuchi, T., and Tanimura, T. (1989) Direct enantiomeric separation of norephedrine and its analogs by high-performance liquid-chromatography. *J. Liq. Chromatogr.* **12,** 2239–2248.

89. Ôi, N., Kitahara, H., and Aoki, F. (1993) Enantiomer separation by high-performance liquid-chromatography with copper(ii) complexes of Schiff-bases as chiral stationary phases. *J. Chromatogr* **631,** 177–182.

90. Ôi, N., Kitahara, H., and Kira, R. (1992) Direct separation of enantiomers by high-performance liquid-chromatography on a new chiral ligand-exchange phase. *J. Chromatogr.* **592,** 291–296.

91. Wan, Q. H., Shaw, P. N., Davies, M. C., and Barrett, D. A. (1997) Role of alkyl and aryl substituents in chiral ligand exchange chromatography of amino acids study using porous graphitic carbon coated with N-substituted-L-proline selectors. *J. Chromatogr. A* **786,** 249–257.

92. Gil-Av, E., Tishbee, A., and Hare, P. E. (1980) Resolution of underivatized amino-acids by reversed-phase chromatography. *J. Am. Chem. Soc.* **102,** 5115–5117.

93. Galaverna, G., Pantó, F., Dossena, A., Marchelli, R., and Bigi, F. (1985) Chiral separation of unmodified alpha-hydroxy acids by ligand exchange HPLC using chiral copper(II) complexes of (S)-phenylalaninamide as additives to the eluent. *Chirality* **7,** 331–336.

94. Günther, K., Martens, J., and Schickedanz, M. (1984) Dünnschichtchromatographische Enantiomerentrennung mittels Ligandenaustausch. *Angew. Chem.* **96,** 514–515; (1984) Thin-layer chromatographic enantiomeric resolution via ligand exchange. *Angew. Chem. Int. Ed. Engl.* **23,** 506.

95. Schmid, M. G., Grobuschek, N., Lecnik, O., and Gübitz, G. (2001) Chiral ligand-exchange capillary electrophoresis. *J. Biochem. Biophys. Methods* **48,** 143–154.

96. Schmid, M. G., Grobuschek, N., Tuscher, C., et al. (2000) Chiral separation of amino acids by ligand-exchange capillary electrochromatography using continuous beds. *Electrophoresis* **21,** 3141–3144.

97. Schmid, M. G., Grobuschek, N., Lecnik, O., Gübitz, G., Végvári, Á., and Hjertens, S. (2001) Enantioseparation of hydroxy acids on easy-to-prepare continuous beds for capillary electrochromatography. *Electrophoresis* **22,** 2616–2619.

98. Chen, Z. and Hobo, T. (2001) Chemically L-prolinamide-modified monolithic silica column for enantiomeric separation of dansyl amino acids and hydroxy acids by capillary electrochromatography and high-performance liquid chromatography. *Electrophoresis* **22,** 3339–3346.

99. Schurig, V. (1977) Resolution of a chiral olefin by complexation chromatography on an optically active rhodium(I) complex. *Angew. Chem. Int. Ed. Engl.* **16,** 110.

100. Schurig, V., Burkle, W., Hintzer, K., and Weber, R. (1989) Evaluation of nickel(II) bis(alpha-(heptafluorobutanoyl)-terpeneketonates) as chiral stationary phases for the enantiomer separation of alkyl-substituted cyclic ethers by complexation gas-chromatography. *J. Chromatogr.* **475**, 23–44.

101. Schurig, V., Schmalzing, D., and Schleimer, M. (1991) Enantiomer separation on immobilized Chirasil-Metal and Chirasil-Dex by gas-chromatography and super-critical fluid chromatography. *Angew. Chem. Int. Ed. Engl.* **30**, 987–989.

102. Jung, M., Schmalzing, D., and Schurig, V. (1991) Theoretical approach to the gas-chromatographic separation of enantiomers on dissolved cyclodextrin derivatives. *J. Chromatogr.* **552**, 43–57.

103. Armstrong, D. W. and DeMond, W. (1984) Cyclodextrin bonded phases for the liquid-chromatographic separation of optical, geometrical, and structural isomers. *J. Chromatogr. Science* **22**, 411–415.

104. Bressolle, F., Audran, M., Pham, T. N., and Vallon, J. J. (1996) Cyclodextrins and enantiomeric separations of drugs by liquid chromatography and capillary electro-phoresis: basic principles and new developments. *J. Chromatogr. B* **687**, 303–336.

105. Schurig, V. (2001) Separation of enantiomers by gas chromatography. *J. Chromatogr. A* **906**, 275–299.

106. König, W. A., Lutz, S., Mischnick-Lubbecke, P., Brassat, B., and Wenz, G. (1988) Cyclodextrins as chiral stationary phases in capillary gas-chromatography. 1. Pen-tylated alpha-cyclodextrin. *J. Chromatogr.* **447**, 193–197.

107. Armstrong, D. W., Li, W. Y., and Pitha, J. (1990) Reversing enantioselectivity in capillary gas-chromatography with polar and nonpolar cyclodextrin derivative phases. *Anal. Chem.* **62**, 214–217.

108. Vigh, Gy. and Sokolowski, A. D. (1997) Capillary electrophoretic separations of enantiomers using cyclodextrin-containing background electrolytes. *Electrophoresis* **18**, 2305–2310.

109. Koppenhoefer, B., Zhu, X., Jakob, A., Wuerthner, S., and Lin, B. (2000) Separa-tion of drug enantiomers by capillary electrophoresis in the presence of neutral cyclodextrins. *J. Chromatogr. A* **875**, 135–161.

110. Chankvetadze, B. (1997) Separation selectivity in chiral capillary electrophore-sis with charged selectors. *J. Chromatogr. A* **792**, 269–295.

111. De Boer, T., De Zeeuw, R. A., De Jong, G. J., and Ensing, K. (2000) Recent innovations in the use of charged cyclodextrins in capillary electrophoresis for chiral separations in pharmaceutical analysis. *Electrophoresis* **21**, 3220–3239.

112. Tanaka, Y. and Terabe, S. (1997) Enantiomer separation of acidic racemates by capillary electrophoresis using cationic and amphoteric beta-cyclodextrins as chiral selectors. *J. Chromatogr. A* **781**, 151–160.

113. Lurie, I. S. (1997) Separation selectivity in chiral and achiral capillary electro-phoresis with mixed cyclodextrins. *J. Chromatogr. A* **792**, 297–307.

114. Fillet, M., Hubert, P., and Crommen, J. (2000) Enantiomeric separations of drugs using mixtures of charged and neutral cyclodextrins. *J. Chromatogr. A* **875**, 123–134.

115. Terabe, S., Miyashita, Y., Shibata, O., et al. (1990) Separation of highly hydrophobic compounds by cyclodextrin-modified micellar electrokinetic chromatography. *J. Chromatogr.* **516,** 23–31.

116. Chankvetadze, B., Schulte, G., and Blaschke, G. (1997) Nature and design of enantiomer migration order in chiral capillary electrophoresis. *Enantiomer* **2,** 157–179.

117. Huang, W. X., Xu, H., Fazio, S. D., and Vivilecchia, R. V. (2000) Enhancement of chiral recognition by formation of a sandwiched complex in capillary electrophoresis. *J. Chromatogr. A* **875,** 361–369.

118. Armstrong, D. W., Chang, L. W., and Chang, S. S. C. (1998) Mechanism of capillary electrophoresis enantioseparations using a combination of an achiral crown-ether plus cyclodextrins. *J. Chromatogr. A* **793,** 115–134.

119. Bunke, A., Jira, T., and Gübitz, G. (1995) Chiral separation of cyclodrine by means of capillary electrophoresis. *Pharmazie* **50,** 570–571.

120. Jira, T., Bunke, A., and Karbaum, A. (1998) Use of chiral and achiral ion-pairing reagents in combination with cyclodextrins in capillary electrophoresis. *J. Chromatogr. A* **798,** 281–288.

121. Schmid, M. G., Wirnsberger, K., Jira, T., Bunke, A., and Gübitz, G. (1997) Capillary electrophoretic chiral resolution of vicinal diols by complexation with borate and cyclodextrin—comparative studies on different cyclodextrin derivatives. *Chirality* **9,** 153–156.

122. Stefansson, M. and Novotny, M. (1993) Electrophoretic resolution of monosaccharide enantiomers in borate oligosaccharide complexation media. *J. Am. Chem. Soc.* **115,** 11573–11580.

123. Jira, T., Bunke, A., Schmid, M. G., and Gübitz, G. (1997) Chiral resolution of diols by capillary electrophoresis using borate-cyclodextrin complexation. *J. Chromatogr. A* **761,** 269–276.

124. Mayer, S. and Schurig, V. (1993) Enantiomer separation by electrochromatography in open tubular columns coated with Chirasil-Dex. *J. Liq. Chromatogr.* **16,** 915–931.

125. Mayer, S. and Schurig, V. (1994) Enantiomer separation using mobile and immobile cyclodextrin derivatives with electromigration. *Electrophoresis* **15,** 835–841.

126. Schurig, V., Jung, M., Mayer, S., Fluck, M., Negura, S., and Jakubetz, H. (1995) Unified enantioselective capillary chromatography on a Chirasil-DEX stationary phas. Advantages of column miniaturization. *J. Chromatogr. A* **694,** 119–128.

127. Wistuba, D., Czesla, H., Roeder, M., and Schurig, V. (1998) Enantiomer separation by pressure-supported electrochromatography using capillaries packed with a permethyl-beta-cyclodextrin stationary-phase. *J. Chromatogr. A* **815,** 183–188.

128. Wistuba, D. and Schurig, V. (1999) Enantiomer separation by pressure-supported electrochromatography using capillaries packed with Chirasil-Dex polymer-coated silica. *Electrophoresis* **20,** 2779–2785.

129. Schurig, V. and Wistuba, D. (1999) Recent innovations in enantiomer separation by electrochromatography utilizing modified cyclodextrins as stationary phases. *Electrophoresis* **20,** 2313–2328.

130. Koide, T. and Ueno, K. (1998) Enantiomeric separations of cationic and neutral compounds by capillary electrochromatography with charged polyacrylamide gels incorporating chiral selectors. *Anal. Sci.* **14,** 1021–1023.

131. Végvári, Á., Földesi, A., Hetényi, C. S., et al. (2000) A new easy-to-prepare homogeneous continuous electrochromatographic bed for enantiomer recognition. *Electrophoresis* **21,** 3116–3125.

132. Wistuba, D. and Schurig, V. (2000) Enantiomer separation by capillary electrochromatography on a cyclodextrin-modified monolith. *Electrophoresis* **21,** 3152–3159.

133. Hesse, G. and Hagel, R. (1973) A complete separation of a racemic mixture by elution chromatography on cellulose triacetate. *Chromatographia* **6,** 277–280.

134. Okamoto, Y., Hatada, K., Kawashima, M., and Yamamoto, K. (1984) Chromatographic resolution. 6. Useful chiral packing materials for high-performance liquid-chromatographic resolution-cellulose triacetate and tribenzoate coated on macroporous silica-gel. *Chem. Lett.* **5,** 739–742.

135. Wainer, I. W. and Alembik, M. C. (1986) Resolution of enantiomeric amides on a cellulose-based chiral stationary phase—steric and electronic effects. *J. Chromatogr.* **358,** 85–93.

136. Tachibana, K. and Ohnishi, A. (2001) Reversed-phase liquid chromatographic separations of enantiomers on polysaccharide type chiral stationary phases. *J. Chromatogr. A* **906,** 127–154.

137. Okamoto, Y. and Kaida, Y. (1994) Resolution by high-performance liquid-chromatography using polysaccharide carbamates and benzoates as chiral stationary phases. *J. Chromatogr. A* **666,** 403–419.

138. Oguni, K., Oda, H., and Ichida, A. (1995) Development of chiral stationary phases consisting of polysaccharide derivatives. *J. Chromatogr. A* **694,** 91–100.

139. Yashima, E. and Okamoto, Y. (1995) Chiral discrimination on polysaccharide derivatives. *Bull. Chem. Soc. Jpn.* **68,** 3289–3307.

140. Okamoto, Y. and Yashima, E. (1998) Polysaccharide derivatives for chromatographic separation of enantiomers. *Angew. Chem. Int. Ed.* **37,** 1020–1043.

141. Yashima, E. (2001) Polysaccharide-based chiral stationary phases for high-performance liquid chromatographic enantioseparation. *J. Chromatogr. A* **906,** 105–125.

142. Okamoto, Y., Aburatani, R., Hatano, K., and Hatada, K. (1988) Optical resolution of racemic drugs by chiral HPLC on cellulose and amylose tris(phenylcarbamate) derivatives. *J. Liq. Chromatogr.* **11,** 2147–2163.

143. Senso, A., Oliveros, L., and Minguillón, C. (1999) Chitosan derivatives as chiral selectors bonded on allyl silica gel: preparation, characterisation and study of the resulting high-performance liquid chromatography chiral stationary phases. *J. Chromatogr. A* **839,** 15–21.

144. Cass, Q. B., Bassi, A. I., and Matlin, S. A. (1996) Chiral discrimination by HPLC on aryl carbamate derivatives of chitin coated onto microporous aminopropyl silica. *Chirality* **8,** 131–135.

145. Felix, G. and Zhang, T. (1993) Chiral packing materials for high-performance liquid-chromatographic resolution of enantiomers based on substituted branched polysaccharides coated on silica-gel. *J. Chromatogr.* **639,** 141–149.

146. Nishi, H. (1997) Enantioselectivity in chiral capillary electrophoresis with poly-saccharides. *J. Chromatogr. A* **792,** 327–347.

147. Soini, H., Stefansson, M., Riekkola, M. L., and Novotny, M. V. (1994) Maltooligo-saccharides as chiral selectors for the separation of pharmaceuticals by capillary electrophoresis. *Anal. Chem.* **66,** 3477–3484.

148. Chankvetadze, B., Saito, M., Yashima, E., and Okamoto, Y. (1997) Enantio-sepa-ration using selected polysaccharides as chiral buffer additives in capillary elec-trophoresis. *J. Chromatogr. A* **773,** 331–338.

149. Nakamura, H., Sano, A., and Sumii, H. (1998) Chiral separation of (R,S)-1,1'-binaphthyl-2,2'-diyl hydrogenphosphate by capillary electrophoresis using monosaccharides as chiral selectors. *Anal. Sci.* **14,** 375–378.

150. Nishi, H., Nakamura, K., Nakai, H., and Sato, T. (1996) Enantiomer separation by capillary electrophoresis using DEAE-dextran and aminoglycosidic antibiot-ics. *Chromatographia* **43,** 426–430.

151. Armstrong, D. W., Rundlett, K. L., and Chen, J. R. (1994) Evaluation of the macro-cyclic antibiotic vancomycin as a chiral selector for capillary electrophoresis. *Chirality* **6,** 496–509.

152. Armstrong, D. W., Tang, Y. B., Chen, S. S., Zhou, Y. W., Bagwill, C., and Chen, J. R. (1994) Macrocyclic antibiotics as a new class of chiral selectors for liquid-chromatography. *Anal. Chem.* **66,** 1473–1484.

153. Armstrong, D. W., Liu, Y., and Ekborg-Ott, K. H. (1995) Covalently bonded teicoplanin chiral stationary-phase for HPLC enantioseparations. *Chirality* **7,** 474–497.

154. Ekborg-Ott, K. H., Liu, Y., and Armstrong, D. W. (1998) Highly enantioselective HPLC separations using the covalently bonded macrocyclic antibiotic, ristocetin A, chiral stationary phase. *Chirality* **10,** 434–483.

155. Ekborg-Ott, K. H., Zientara, G. A., Schneiderheinze, J. M., Gahm, K., and Armstrong, D. W. (1999) Avoparcin, a new macrocyclic antibiotic chiral run buffer additive for capillary electrophoresis. *Electrophoresis* **20,** 2438–2457.

156. Ward, T. J. and Farris, A. B. III. (2001) Chiral separations using the macrocyclic antibiotics: a review. *J. Chromatogr. A* **906,** 73–89.

157. Armstrong, D. W. and Zhou, Y. W. (1994) Use of a macrocyclic antibiotic as a chiral selector for the enantiomeric separation by TLC. *J. Liq. Chromatogr.* **17,** 1695–1707.

158. Bhushan, R. and Parshad, V. (1996) Thin-layer chromatographic-separation of enantiomeric dansylamino acids using a macrocyclic antibiotic as a chiral selec-tor. *J. Chromatogr. A* **736,** 235–238.

159. Ward, T. J. and Oswald T. M. (1997) Enantioselectivity in capillary electrophore-sis using the macrocyclic antibiotics. *J. Chromatogr. A* **792,** 309–325.

160. Desiderio, C. and Fanali, S. (1998) Chiral analysis by capillary electrophoresis using antibiotics as chiral selector. *J. Chromatogr. A* **807,** 37–56.

161. Armstrong, D. W. and Nair, U. B. (1997) Capillary electrophoretic enantio-separations using macrocyclic antibiotics as chiral selectors. *Electrophoresis* **18,** 2331–2342.

162. Dermaux, A., Lynen, P., and Sandra, P. (1998) Chiral capillary electrochromatography on a vancomycin stationary phase. *J. High Resol. Chromatogr.* **21,** 575–576.
163. Wikström, H., Svensson, L. A., Torstensson, A., and Owens, P. K. (2000) Immobilisation and evaluation of a vancomycin chiral stationary phase for capillary electrochromatography. *J. Chromatogr. A* **869,** 395–409.
164. Carter-Finch, A. S. and Smith, N. W. (1999) Enantiomeric separations by capillary electrochromatography using a macrocyclic antibiotic chiral stationary phase. *J. Chromatogr. A* **848,** 375–385.
165. Karlsson, C., Wikström, H., Armstrong, D. W., and Owens, P. K. (2000) Enantioselective reversed-phase and non-aqueous capillary electrochromatography using a teicoplanin chiral stationary phase. *J. Chromatogr. A* **897,** 349–363.
166. Karlsson, C., Karlsson, K., Armstrong, D. W., and Owens, P. K. (2000) Evaluation of a vancomycin chiral stationary phase in capillary electrochromatography using polar organic and reversed-phase modes. *Anal. Chem.* **72,** 4394–4401.
167. Desiderio, C., Aturki, Z., and Fanali, S. (2001) Use of vancomycin silica stationary phase in packed capillary electrochromatography I. Enantiomer separation of basic compounds. *Electrophoresis* **22,** 535–543.
168. Grobuschek, N., Schmid, M. G., Koidl, J., and Gübitz, G. (2002) Enantio-separation of amino acids and drugs by CEC, pressure supported CEC and micro-HPLC using a teicoplanin aglycone stationary phase. *J. Sep. Sci.* **25,** 1297–1302.
169. Desiderio, C., Polcaro, C. M., Padiglioni, P., and Fanali, S. (1997) Enantiomeric separation of acidic herbicides by capillary electrophoresis using vancomycin as chiral selector. *J. Chromatogr. A* **781,** 503–513.
170. Berthod, A., Chen, X., Kullman, J. P., et al. (2000) Role of the carbohydrate moieties in chiral recognition on teicoplanin-based LC stationary phase *Anal. Chem.* **72,** 1767–1780.
171. Sousa, L. R., Sogah, G. D. Y., Hoffmann, D. H., and Cram, D. J. (1978) Host-guest complexation. 12. Optical resolution of amine and amino ester salts by chromatography. *J. Am. Chem. Soc.* **100,** 4569–4576.
172. Sogah, G. D. Y. and Cram, D. J. (1979) Host-guest complexation. 14. Host covalently bound to polystyrene resin for chromatographic resolution of enantiomers of amino acids and ester salts. *J. Am. Chem. Soc.* **101,** 3035–3042.
173. Shinbo, T., Nishimura, K., Sugiura, M., and Yamaguchi, T. (1987) Chromatographic-separation of racemic amino-acids by use of chiral crown ether-coated reversed-phase packings. *J. Chromatogr.* **405,** 145–153.
174. Machida, Y., Nishi, H., Nakamura, K., Nakai, H., and Sato, T. (1998) Enantiomer separation of amino compounds by a novel chiral stationary phase derived from crown ether. *J. Chromatogr. A* **805,** 85–92.
175. Hyun, M. H., Jin, J. S., and Lee, W. (2002) Liquid chromatographic resolution of racemic amino acids and their derivatives on a new chiral stationary phase based on crown ether. *J. Chromatogr. A* **822,** 155–161.
176. Hyun, M. H., Jin, J. S., Koo, H. J., and Lee, W. (1999) Liquid chromatographic resolution of racemic amines and amino alcohols on a chiral stationary phase derived from crown ether. *J. Chromatogr. A* **837,** 75–82.

177. Hyun, M. H., Han, S. C., Lipshutz, B. H., Shin, Y. J., and Welch, C. J. (2001) New chiral crown ether stationary phase for the liquid chromatographic resolution of α-amino acid enantiomers. *J. Chromatogr. A* **910**, 359–365.

178. Steffeck, R. J., Zelechonok, Y., and Gahm, K. H. (2002) Enantioselective separation of racemic secondary amines on a chiral crown ether-based liquid chromatography stationary phase. *J. Chromatogr. A* **947**, 301–305.

179. Kuhn, R., Erni, F., Bereuter, T., and Häusler, J. (1992) Chiral recognition and enantiomeric resolution based on host guest complexation with crown ethers in capillary zone electrophoresis. *Anal. Chem.* **64**, 2815–2820.

180. Höhne, E., Krauss, G.-J., and Gübitz, G. (1992) Capillary zone electrophoresis of the enantiomers of aminoalcohols based on host-guest complexation with a chiral crown-ether. *J. High Resol. Chromatogr.* **15**, 698–700.

181. Kuhn, R., Riester, D., Fleckenstein, B., and Wiesmüller, K.-H. (1995) Evaluation of an optically-active crown-ether for the chiral separation of dipeptides and tripeptides. *J. Chromatogr. A* **716**, 371–379.

182. Schmid, M. G. and Gübitz, G. (1995) Capillary zone electrophoretic separation of the enantiomers of dipeptides based on host-guest complexation with a chiral crown-ether. *J. Chromatogr. A* **709**, 81–88.

183. Nishi, H., Nakamura, K., Nakai, H., and Sato, T. (1997) Separation of enantiomers and isomers of amino-compounds by capillary electrophoresis and high-performance liquid-chromatography utilizing crown-ethers. *J. Chromatogr. A* **757**, 225–235.

184. Mori, Y., Ueno, K., and Umeda, T. (1997) Enantiomeric separations of primary amino-compounds by nonaqueous capillary zone electrophoresis with a chiral crown-ether. *J. Chromatogr. A* **757**, 328–332.

185. Pfeiffer, J. and Schurig, V. (1999) Enantiomer separation of amino acid derivatives on a new polymeric chiral resorc[4]arene stationary phase by capillary gas chromatography. *J. Chromatogr. A* **840**, 145–150.

186. Narumi, F., Iki, N., Suzuki, T., Onodera, T., and Miyano, S. (2000) Syntheses of chirally modified thiacalix[4]arenes with enantiomeric amines and their application to chiral stationary phases for gas chromatography. *Enantiomer* **5**, 83–93.

187. Peña, M. S., Zhang, Y. L., and Warner, I. M. (1997) Enantiomeric separations by use of calixarene electrokinetic chromatography. *Anal. Chem.* **69**, 3239–3242.

188. Grady, T., Joyce, T., Smyth, M. R., Harris, S. J., and Diamond, D. (1998) Chiral resolution of the enantiomers of phenylglycinol using (S)-di-naphthylprolinol calix[4]arene by capillary electrophoresis and fluorescence spectroscopy. *Anal. Commun.* **35**, 123–125.

189. Gasparrini, F., Misiti, D., Villani, C., Borchardt, A., Burger, M. T., and Still, W. C. (1995) Enantioselective recognition by a new chiral stationary-phase at receptorial level. *J. Org. Chem.* **60**, 4314–4315.

190. Gasparrini, F., Misiti, D., Still, W. C., Villani, C., and Wennemers, H. (1997) Enantioselective and diastereoselective binding study of silica bound macrobicyclic receptors by HPLC. *J. Org. Chem.* **62**, 8221–8224.

191. Pieters, R. J., Cuntze, J., Bonnet, M., and Diederich, F. (1995) Enantioselective recognition with C3-symmetric cage-like receptors in solution and on a stationary phase. *J. Chem. Soc. Perkin. Trans.* **2**, 1891–1900.
192. Hu, K. J., Bradshaw, J. S., Dalley, N. K., Krakowiak, K. E., Wu, N. J., and Lee, M. L. (1999) Synthesis of a chiral macrocyclic dibenzodicyclohexanotetraamide-containing stationary-phase for liquid-chromatography. *J. Heterocycl. Chem.* **36**, 381–387.
193. Blaschke, G. (1986) Chromatographic resolution of chiral drugs on polyamides and cellulose triacetate. *J. Liq. Chromatogr.* **9**, 341–368.
194. Mohammad, J., Li, Y. M., El-Ahmad, M., Nakazato, K., Pettersson, G., and Hjertén, S. (1993) Chiral recognition chromatography of β-blockers on continuous polymer beds with immobilized cellulase as enantioselective protein. *Chirality* **5**, 464–470.
195. Koide, T. and Ueno, K. (2001) Enantiomeric separations of primary amino compounds by capillary electrochromatography with monolithic chiral stationary phases of chiral crown ether-bonded negatively charged polyacrylamide gels. *J. Chromatogr. A* **909**, 305–315.
196. Peters, E. C., Lewandowski, K., Petro, M., Svec, F., and Frechet, J. H. J. (1998) Chiral electrochromatography with a moulded rigid monolithic capillary column. *Anal. Commun.* **35**, 83–86.
197. Lämmerhofer, M., Svec, F., Fréchet, J. M. J., and Lindner, W. (2000) Chiral monolithic columns for enantioselective capillary electrochromatography prepared by copoly-merization of a monomer with quinidine functionality. 2. Effect of chromatographic conditions on the chiral separations. *Anal. Chem.* **72**, 4623–4628.
198. Sinner, F. and Buchmeiser, M. R. (2000) Ringöffnende Metathesepolymerisation: Zugang zu einer neuen Klasse funktionalisierter, monolithischer stationärer Phasen für die Flüssigkeitschromatographie. *Angew. Chem.* **112**, 1491–1494.
199. Nakano, T. (2001) Optically active synthetic polymers as chiral stationary phases in HPLC. *J. Chromatogr. A* **906**, 205–225.
200. Wulff, G. and Vesper, W. (1978) Preparation of chromatographic sorbents with chiral cavities for racemic resolution. *J. Chromatogr.* **167**, 171–186.
201. Schweitz, L., Andersson, L. I., and Nilsson, S. (1997) Capillary electrochromatography with predetermined selectivity obtained through molecular imprinting. *Anal. Chem.* **69**, 1179–1183.
202. Schweitz, L., Andersson, L. I., and Nilsson, S. (1999) Molecular imprinting for chiral separations and drug screening purposes using monolithic stationary phases in CEC. *Chromatographia* **49**, S93–S94.
203. Lin, J.-M., Nakagama, T., Wu, X. Z., Uchiyama, K., and Hobo, T. (1997) Capillary electrochromatographic separation of amino acid enantiomers with molecularly imprinted polymers as chiral recognition agents. *Fresenius J. Anal. Chem.* **357**, 130–132.
204. Chirica, G. and Remcho, V. T. (1999) Silicate entrapped columns—new columns designed for capillary electrochromatography. *Electrophoresis* **20**, 50–56.
205. Sellegren, B. (2001) Imprinted chiral stationary phases in high-performance liquid chromatography. *J. Chromatogr. A* **906**, 227–252.

206. Takeuchi, T. and Haginaka, J. (1999) Separation and sensing based on molecular recognition using molecularly imprinted polymers. *J. Chromatogr. B* **728,** 1–20.
207. Remcho, V. T. and Tan, Z. J. (1999) MIPs as chromatographic stationary phases for molecular recognition. *Anal. Chem. News. Features* 248A–255A.
208. Haginaka, J. (2001) Protein based chiral stationary phases for HPLC enantio-separations. *J. Chromatogr. A* **906,** 253–273.
209. Nilsson, S., Schweitz, L., and Petersson, M. (1997) Three approaches to enantiomer separation of beta-adrenergic antagonists by capillary electrochromatography. *Electrophoresis* **18,** 884–890.
210. Valtcheva, L., Mohammad, J., Pettersson, G., and Hjertén, S. (1993) Chiral separation of beta-blockers by high-performance capillary electrophoresis based on non-immobilized cellulase as enantioselective protein. *J. Chromatogr.* **638,** 263–267.
211. Hedeland, M., Isaksson, R., and Pettersson, C. (1998) Cellobiohydrolase-I as a chiral additive in capillary electrophoresis and liquid-chromatography. *J. Chromatogr. A* **807,** 297–305.
212. Wang, F. and Khaledi, M. G. (2000) Enantiomeric separations by nonaqueous capillary electrophoresis. *J. Chromatogr. A* **875,** 277–293.
213. Snopek, J., Jelinek, I., and Smolkova-Keulemansova, E. (1988) Use of cyclodextrins in isotachophoresis. 4. The influence of cyclodextrins on the chiral resolution of ephedrine alkaloid enantiomers. *J. Chromatogr.* **438,** 211–218.
214. Danková, M., Kaniansky, D., Fanali, S., and Iványi, F. (1999) Capillary zone electrophoresis separations of enantiomers present in complex ionic matrices with on-line isotachophoretic sample pretreatment. *J. Chromatogr. A* **838,** 31–43.
215. Fanali, S., Desiderio, C., Ölvecka, E., Kaniansky, D., Vojtek, M ., and Ferancova, A. (2000) Separation of enantiomers by on-line capillary isotachophoresis-capillary zone electrophoresis. *J. High Resolut. Chromatogr.* **23,** 531–538.
216. Toussaint, B., Hubert, Ph., Tjaden, U. R., van der Greef, J., and Crommen, J. (2000) Enantiomeric separation of clenbuterol by transient isotachophoresis capillary zone electrophoresis-UV detection. New optimization technique for transient isotachophoresis. *J. Chromatogr. A* **871,** 173–180.
217. Kaniansky, D., Simunicova, E., Ölvecka, E., and Ferancova, A. (1999) Separations of enantiomers by preparative capillary isotachophoresis. *Electrophoresis* **20,** 2786–2793.
218. Hoffmann, P., Wagner, H., Weber, G., Lanz, M., Caslavska, J., and Thormann, W. (1999) Separation and purification of methadone enantiomersby continuous- and interval-flow electrophoresis. *Anal. Chem.* **71,** 1840–1850.
219. Glukhovsky, P. and Vigh, Gy. (1999) Analytical- and preparative-scale isoelectric focusing separation of enantiomers. *Anal. Chem.* **71,** 3814–3820.
220. Fried, K. and Wainer, I. W. (1997) Column-switching techniques in the biomedical analysis of stereoisomeric drugs: why, how and when. *J. Chromatogr. B* **689,** 91–104.
221. Ba, B., Eckert, G., and Leube, J. (1991) Use of dabsylation column switching and chiral separation for the determination of a renin inhibitor in rat marmoset and human plasma. *J. Chromatogr.* **572,** 277–289.

222. Eto, S., Noda, H., and Noda, A. (1994) Simultaneous determination of antiepileptic drugs and their metabolites including chiral compounds via β-cyclodextrin inclusion complexes by a column-switching chromatographic technique. *J. Chromatogr. B* **658,** 385–390.
223. Ducharme, J., Fernandez, C., Gimenez, F., and Farinotti, R. (1996) Critical issues in chiral drug analysis in biological fluids by high-performance liquid-chromatography. *J. Chromatogr. B* **686,** 65–75.
224. Bojarski, J. and Aboul-Enein, H. Y. (1997) Application of capillary electrophoresis for the analysis of chiral drugs in biological fluids. *Electrophoresis* **18,** 965–969.
225. Zaugg, S. and Thormann, W. (2000) Enantioselective determination of drugs in body fluids by capillary electrophoresis. *J. Chromatogr. A* **875,** 27–41.
226. Zhao, J. and Jorgenson, J. W. (1999) Application of synchronous cyclic capillary electrophoresis: isotopic and chiral separations. *J. Microcolumn Separations* **11,** 439–449.
227. Arce, L., Tena, M. T., Rios, A., and Valcáreel, M. (1998) Determination of trans-resveratrol and other polyphenols in wines by a continuous flow sample clean-up system followed by capillary electrophoresis separation. *Anal. Chim. Acta* **359,** 27–38.
228. Fang, Z.-L., Liu, Z.-S., and Shen, Q. (1997) Combination of flow injection with capillary electrophoresis. Part I. The basic system. *Anal. Chim. Acta* **346,** 135–143.
229. Kuban, P., Pirmohammadi, R., and Karlberg, B. (1999) Flow injection analysis-capillary electrophoresis system with hydrodynamic injection. *Anal. Chim. Acta* **378,** 55–62.
230. Hofstetter, O., Hofstetter, H., Wilchek, M., Schurig, V., and Green, B. S. (1998) Antibodies can recognize the chiral center of free α-amino acids. *J. Am. Chem. Soc.* **120,** 3251–3252.
231. Hofstetter, O., Hofstetter, H., Wilchek, M., Schurig, V., and Green, B. S. (1999) Chiral discrimination using an immunosensor. *Nat. Biotechnol.* **17,** 371–374.
232. Silvaieh, H., Schmid, M. G., Hofstetter, O., Schurig, V., and Gübitz, G. (2002) Development of enantioselective chemiluminescence flow- and sequential-injection immunoassays for α-amino acids. *J. Biochem. Biophys. Methods* **53,** 1–14.
233. Silvaieh, H., Wintersteiger, R., Schmid, M. G., Hofstetter, O., Schurig, V., and Gübitz, G. (2002) Enantioselective sequential injection chemiluminescence immunoassays for 3, 3', 5-triiodothyronine (T_3) and thyroxine (T_4) *Anal. Chim. Acta* **463,** 5–14.

2

Separation of Enantiomers by Thin-Layer Chromatography

An Overview

Kurt Günther, Peter Richter, and Klaus Möller

1. Introduction

The present review discusses the versatile applicability and separation mechanisms of thin-layer chromatographic enantiomeric separations. More detailed descriptions will be given for practical applications—separations of underivatized samples—on commercially available, ready-to-use plates, focusing on the thin-layer chromatographic racemate separation based on ligand exchange, which was introduced in 1983 by Günther et al. *(1)* and on use of a densitometer for determination of antipode distributions at trace level. This chapter will not discuss the numerous and interesting diastereomeric separations by paper and thin-layer chromatography (TLC). We refer to the literature on amino acids *(2–8)*, peptides, diketopiperazines *(9–18)*, and other classes of compounds *(19–31)*.

1.1. Chromatographic Methods of Configurational Analysis

In TLC one may utilize one of three techniques for separation of enantiomeric compounds:

1. Direct separation by using chiral stationary phases, effected by the formation of diastereomeric association complexes.
   a. Paper chromatography (chiral cavities) *(32–61)*.
   b. Molecular imprinting technique *(62–65)*.
   c. Poly(meth)acrylic acid amides TLC-plates *(66)*.
   d. Cellulose thin-layer plates (see **Subheadings 1.2.** and **1.3.**).
   e. Cyclodextrin thin-layer plates (*see* **Subheading 1.4.**).
   f. TLC plates coated with chiral compounds (*see* **Subheading 1.5.**).
   g. Ligand-exchange TLC plates (*see* **Subheading 1.6.**).

From: *Methods in Molecular Biology, Vol. 243: Chiral Separations: Methods and Protocols*
Edited by: G. Gübitz and M. G. Schmid © Humana Press Inc., Totowa, NJ

2. Separation on ordinary stationary phases by means of chiral additives in the eluent, which form diastereomeric complexes with the substrate (*see* also **Subheadings 1.4.–1.6.**).
3. Separation on achiral stationary phases via diastereomeric derivatives formed by reaction of sample with a chiral reagent (*see* **Subheading 1.7.**).

A detailed review is given in ref. *67*.

1.2. Enantiomeric Separations on Cellulose Thin-Layer Plates

1.2.1. Resolution Mechanism

Cellulose is a linear macromolecule composed of optically active D-glucose units, with its chains arranged on a partially crystalline fiber structure with helical cavities. Separation of enantiomers is effected by their different fit into the lamellar chiral layer structure of the support.

1.2.2. Survey of Applications of Racemic Separations

TLC on cellulose can be considered a continuation of classical paper chromatography. Consequently, the first investigations during the mid-1970s concentrated on transfer of paper chromatographic racemate separations to cellulose layers with the aim of shortening development times and improving separation efficiencies. The research (*68–85*) deals mainly with separation problems concerning acids, amino acid derivatives, and dipeptides, focusing on the influence of the structure of the chiral support and the eluent temperature on the separation behavior of the racemates. Separation of aromatic amino acids phenylalanine, β-2-thienylalanine, 4-fluorophenylalanine, and tyrosine could not be achieved on microcrystalline or amorphous cellulose; tryptophan isomers, however, could be reproducibly be resolved on microcrystalline cellulose layers (*70*). Lowering the eluent temperature from 30° to 0°C enhances enantiomeric resolution and lengthens developing times up to 10 h. Hydrophobic eluent combinations further enhance separation, because they improve formation of the helical cellulose conformation (*74*). Separation of racemic 3,4-dihydroxyphenylalanine, tryptophan, and 5-hydroxytryptophan can be achieved in only 2 h, on a cellulose high-performance thin-layer chromatography (HPTLC) plate (*76*) (these experiments are described in detail in **Subheading 3.1.**).

Lederer et al. (*77,78,81–85*) investigated the influence of various salt concentrations in mobile phase on the separation of tryptophan, methyltryptophan, and fluorotryptophan. In experiments, lithium chloride, sodium chloride, and ammonium sulfate solutions were used for the separation on native and microcrystalline cellulose and, also, separations with aqueous copper sulfate and sodium chloride solutions containing α-cyclodextrin (CD) were described.

Mixtures of microcrystalline cellulose and cellulose derivatives were used by Suedee and Heard *(86)*. The best resolution of propranolol was obtained on cellulose tris (3,5-dimethylphenylcarbamate), of bupranolol on cellulose tris (3,4-dichlorophenylcarbamate), with mobile phase hexane/propan-2-ol (80:20, v/v).

Cyclohexylcarbamates of cellulose and amylose were prepared by Kubota, Okamoto, et al. *(87)* and were tested also as pure chiral stationary phase for TLC showing good separation factors for different compounds such as 1-(9-anthryl)-2.2.2-trifluoroethanol, Troeger's base or benzoin.

1.3. Enantiomeric Separations on Microcrystalline Triacetylcellulose Thin-Layer Plates

1.3.1. Resolution Mechanism

The resolving capability of this polysaccharide derivative is based on its morphological structure. Peracetylation of the cellulose has to be performed such that the conformation and relative position of the carbohydrate bands in their crystalline domains remain intact. In this state, cellulose triacetate includes enantioselectivity; i.e., antipode separations are possible *(88)*.

1.3.2. Survey of Applications of Racemic Separations

In 1973, Hesse and Hagel *(89)* for the first time described the thin-layer chromatographic racemate separation of Troeger's base on cellulose triacetate. Systematic investigations of this chiral support by Faupel *(90)* resulted in commercialization of a microcrystalline triacetylcellulose plate by Antec, Bennwil. These plates are stable with aqueous eluent systems and resistant toward dilute acids and bases. They are stable in alcoholic and phenolic eluents, but are attacked by glacial acetic acid and ketonic solvents. Enantiomeric separation of racemic oxindanac was first described by Faupel. Using this racemate as "pilot substance" and transferring the separation conditions *(90)*, other separation examples were published *(91)* and are described in detail in **Subheading 3.2.** Günther et al. were successful in separating the pesticide (±)-2-(4-chloro-6-ethylamino-[1,3,5]-triazin-2-ylamino)-2-methyl-butyronitrile on microcrystalline triacetyl cellulose plate OPTI-T.A.C. *(92)*. Further separations of microcrystalline triacetylcellulose plates were done by Lepri et al. *(93,94,96–100)* and Huang *(95)*.

1.4. Separation of Enantiomers Using Chiral β-CD

1.4.1. Resolution Mechanism

β-CD is a chiral toroidal-shaped molecule consisting of seven glucose units connected via α-1,4-linkages. The enantiomers are selectively retained as they fit differently into the cavity of the oligomer.

1.4.2. Survey of Applications of Racemic Separations

Armstrong et al. *(101–104)* investigated the influence of different silicas and binders on the separation behavior of β-CD TLC plates. Besides nine racemates, three diastereomeric compounds and six structural isomers were separated. Wilson *(105)* impregnated silica plates with a 1% solution of β-CD in ethanol-dimethylsulfoxide (80:20 by vol); racemic mandelic acid was barely separated, and the antipode separation of β-blockers was not possible. Bhushan and Martens presented a paper regarding impregnation of thin layer materials with a variety of reagents and the role of impregnation in enantiomeric separation *(106)*. Armstrong et al. *(107)* were the first to describe application of β-CD as a chiral eluent additive for separations on reversed-phase TLC plates. The success of separation was strongly dependent on type and quantity of modifier applied, but above all on the concentration of β-CD. The low solubility of β-CD in water (0.017 M, 25°C) can be improved by addition of urea; sodium chloride stabilizes the binder of the reversed-phase plates. Compared to β-CD bonded phases, a reversed retention behavior was noticed, the D-enantiomer eluting above the L-isomer. The separation of steroid epimers and other diastereomeric classes of compounds is also possible with this technique. Hydroxypropyl and hydroxyethyl β-CDs are also suitable as chiral mobile-phase additives for thin-layer chromatographic enantiomer separations *(108)*. Their better solubility in water and aqueous-organic eluents (compared to β-CD) enhances enantioselectivity; 0.6 M substituted β-CD has proven especially active for separation. Duncan and Armstrong *(104,109)* also described the separation of amino acids and alkaloids on different types of reversed-phase plates using the mobile phase acetonitrile-water containing maltosyl-β-CD. The preferred TLC plate was the ethyl-modified, because a greater number of compounds were separated using this plate type.

Lepri et al. *(110–113)* investigated the chromatographic behavior of dansyl-, dinitrophenyl-, and β-naphthyl-substituted amino acids, and alkaloids on layers of partially C18-modified silica with aqueous-organic solutions containing β-CD as chiral agent. Also, the influence of the concentration of urea in the eluent was studied.

As mentioned before, CDs are often used as mobile phase additives *(114–121)*, and here are interesting results using microcrystalline cellulose as thin layer *(114,115)*.

Bhushan et al. *(122,123)* achieved the resolution of (±)-atenolol, (±)-propranolol, and (±)-metoprolol into their enantiomers on silica gel plates impregnated with optically pure L-lysine (0.5%) and L-arginine (0.5%) as chiral selector. He also performed good separations of 2-arylpropionic acids on (–)-brucine-impregnated silica gel plates *(124)*.

Also, ammonium-D-10-camphorsulfonates were used for the enantiomeric separation. Huang et al. showed separations of propranolol, propafenone, pindolol, and atenolol with good separation factors *(125,126)*. Here, methylene chloride/methanol in different ratios, with 8.8 m*M* ammonium-D-10-camphorsulfonate as chiral ion-interaction agent, were used as mobile phase.

The other strategy with CD as chiral support is the use of β-CD-bonded stationary phases. Deng et al. *(127–129)* prepared a phenylcarbamate-substituted β-CD-bonded stationary phase and separated a great number of binaphthalene derivatives on this layer using petroleum ether/ethyl acetate/methanol mixtures as mobile phase.

1.5. Direct Separation of Enantiomers on TLC Plates Coated With Chiral Compounds

Standardized commercial TLC plates are essential for routine handling of large sample volumes. "Homemade" layers usually do not meet the quality requirements of modern analysis. However, they often contribute substantially to the understanding of chiral separation principles *(130–153)*. It is not the purpose of this chapter to present a detailed description of layer preparations; we refer to the separation examples listed in ref. *67*. In this context, the published works of Lepri et al. *(144–150)* and Armstrong and Zhou *(153)* are worth mentioning.

Lepri et al. investigated the chromatographic behavior of racemic dinitropyridyl, dinitrophenyl, dinitrobenzoyl, 9-fluorenylmethoxycarbonyl amino acids, tryptophanamides, lactic acid derivatives, and unusual enantiomers, such as binaphthols on reversed-phase TLC plates developed with aqueous-organic mobile phase containing bovine serum albumin (BSA) as chiral agent. More than 75 racemates have been separated in these experiments with planar chromatography using BSA in mobile phase. BSA showed enantioselectivity towards racemates with structures completely different from amino acids, their derivatives, and similar compounds such as hydroxy acids.

Armstrong and Zhou *(153)* published a work focusing on the use of the macrocyclic antibiotic vancomycin as a chiral mobile phase additive. In this work, the separations of carbamates, derivatized amino acids, racemic drugs, and dansyl-amino acids were performed on diphenyl-modified stationary phases with the eluent systems acetonitrile, 0.6 *M* NaCl, 1% triethylammonium acetate buffer (pH 4.1).

Also, another working group *(154)* used the macrocyclic antibiotic vancomycine as a chiral selector on silica gel layers. The mobile phase enabling successful resolutions of the most racemic dansyl amino acids were acetonitrile/0.5 *M* aqueous NaCl (5:2 and 14:3, v/v). The same group prepared a chiral stationary phase using a slurry of silica gel in 0.05% erythromycin solution, which was spread on glass plates *(155)*. Spots of the dansyl derivatives of DL- and L-amino acids were applied and detected under 254 nm radiation. The best mobile phases were 0.5 M NaCl/acetonitrile (1.5–25:1, v/v), in some instances with small addition of methanol. Results were reported with the development distance of 10 cm. Separation factors ranged from 1.06 to 1.36, with the D-form having the higher mobility.

1.6. Thin-Layer Enantiomeric Resolution Via Ligand Exchange

1.6.1. Resolution Mechanism

The results prove that the separation models developed for ligand exchange by high-performance liquid chromatography (HPLC) *(156–158)* are also valid for TLC; the diastereomeric complexes formed with the metal ion (e.g., Cu^{2+}) and the chiral adsorbent have different stabilities for the different antipodes, and thus, chromatographic separation is achieved.

1.6.2. Survey of Applications of Racemic Separations

Thin-layer chromatographic enantiomeric separations based on ligand exchange were published independently by Günther et al. *(159)* and Weinstein *(160)* in 1984. Though very similar in their technique, the procedures differ in their choice of chiral selector and, consequently, in their range of applicability. Using commercially available reversed-phase TLC plates, Weinstein *(160)* impregnated the layers with the optically active copper complex of N,N-di-n-propyl-L-alanine after preconditioning the ready-to-use plate with buffer A (0.3 M sodium acetate in 40% acetonitrile and 60% water, adjusted to pH 7 with acetic acid). With the exception of proline, all proteinogenic amino acids are resolved, as dansyl derivatives, into L- and D-enantiomers. A detailed description of this procedure for some selected separation examples is given in Chapter 3. Another paper from this group *(161)* describes a two-dimensional reversed-phase thin-layer chromatographic procedure for simultaneous separation of racemic dansyl amino acid mixtures. In the first direction, the dansyl amino acids were separated on RP-18 TLC plates with eluents without chiral additives using, e.g., a convex gradient with increasing acetonitrile content (2–30%) in 0.3 M sodium acetate (pH 6.3). In the second direction, the plate was treated with the above-mentioned chiral selector and then again developed with aqueous acetonitrile/sodium acetate buffer. The separation was further improved by using a temperature gradient (6.2°C/cm). The influence of the temperature on enantiomeric separation behavior is detailed in ref. *162*. Chiral diaminodiamide copper(II) complexes are also suited as chiral selectors for thin-layer chromatographic enantiomeric separations of racemic dansyl amino acids *(163)*. In these ligands, two L-amino acids are joined by an amide bond by ethylene and trimethylene bridges and are endowed with varying degrees of lipophilicity and bulkiness, depending on the nature of the amino acid side chain. The coating procedure in general corresponds to that of Weinstein *(160)*. The authors also work with one- or two-dimensional techniques with or without chiral additive in the eluent (acetonitrile/water, 33:67, adjusted to pH 6.8 with acetic acid).

Based on the work of Davankov et al. *(157,164)*, who modified commercial HPLC columns for distribution chromatography with alkyl derivatives of L-amino acids, such as *n*-decyl-L-histidine or *n*-hexadecyl-L-proline, Günther et al. *(159)* used (2S,4R,2'RS)-*N*-(2'-hydroxydodecyl)-4-hydroxy-proline, which is easier to prepare, as a chiral selector *(165)*. The following impregnation procedure proved to be most efficient. A glass plate coated with hydrophobic silica gel (RP-18 TLC) was dipped into a 0.25% copper(II) acetate solution (methanol/water, 1:9, v/v) and dried. Then the plate was immersed in a 0.8% methanolic solution of the chiral selector for 1 min. After air drying, the plate was ready for enantiomeric separations. Contrary to the procedures described above, in this case, antipode separation of amino acids was possible without derivatization. Because the commercially available chiral TLC/HPTLC plates are based on this ligand-exchange chromatography (LEC) technique, a detailed description of chromatographic conditions will be given in Chapter 3.

In the last 2 yr, efforts were made to illuminate the structure of the complex of the 4-hydroxyproline selector and to find new selectors for the enantiomeric separation based on LEC. Lübben and Martens et al. *(166)* tried to do X-ray investigations of the 4-hydroxyproline-copper^{2+}-complex, but it was not possible to get a crystalline complex of this selector. Therefore, they synthesized a model compound with a methyl group instead of the $C_{10}H_{21}$-group. With this short alkyl group as a modified selector, the chelate complex crystallized in the ortho rhombic crystal system, and X-ray data are available. The same group also mentioned that the configuration in the 2'-position of the side chain of the 4-hydroxyproline selector has no influence on the stereoselectivity of its copper complex in the enantiomeric separation of amino acids *(167)*. Only recently, new attempts on crystallization and structure determination of the copper(II)-complex were successful *(168)*. The results show that coordination at the copper center of the selector complex is fundamentally different as compared to that of the short alkyl chain model compound mentioned previously.

However, a simple method is performed by Bhushan et al. *(169)*. Here, L-proline was used as a chiral selector on normal phase silica gel, and amino acids were resolved with the eluent systems *n*-butanol/acetonitrile/water (6:2:3, v/v/v), chloroform/methanol/propionic acid (15:6:4, v/v/v), and acetonitrile/methanol/water (2:2:1, v/v/v).

Until now however, these selectors showed no eminent advantage compared with the 4-hydroxyproline selector. Therefore, until today, these layers using the 4-hydroproline selector are the only commercial available ready-to-use plates (ChiralPlate® [Macherey-Nagel, Düren, Germany] and Chir® [Merck, Darmstadt, Germany]).

1.6.3. Selected Examples of Separation

Under license from Degussa (1), ChiralPlate, the first chiral TLC ready-to-use plate based on LEC was developed and commercialized in 1985 in cooperation with Macherey-Nagel (170). In 1988, again under license from Degussa, the commercialization of the chiral HPTLC ready-to-use plate CHIR with concentrating zone by Merck followed. The following separation examples focus on elaborations with ChiralPlate; however, since they are based on the same separation principle, they can be easily transferred to the HPTLC-CHIR plates. A comparison of separation results on both plates will be given for the thin-layer chromatographic separation of α-hydroxycarboxylic acids (171). Other applications from external groups have been published previously (172–192). This chapter will not discuss the successful application of ChiralPlate in forced-flow planar chromatographic techniques such as overpressured layer chromatography (OPLC) and analytical rotation planar chromatography (RPC); we refer to the literature (193,194).

With the technique described (in **Subheading 3.3.2.**), more than 100 racemate separations have been accomplished by Günther, most of which have been published (67,159,170,171,195–201). We will not describe all separations accomplished so far, but rather will demonstrate the versatile applicability of this method for some selected classes of compounds. A selection of separation examples is reproduced in **Figs. 1** and **2**.

Amino acids (see Fig. 1). Thus far, 12 proteinogenic amino acids have been separated without derivatization; cysteine can be determined as thiazolidine-4-carboxylic acid, which is formed from cysteine by a simple reaction with formaldehyde. The separation of nonproteinogenic amino acids is shown in ref. *67*.

Dipeptides. For the enantiomeric separation of dipeptides, it is remarkable that the enantiomer with the C-terminal L-configuration always has a lower R_f value than the one with the C-terminal D-configuration. This method can also resolve diastereomeric dipeptides (198). Wang et al. (75) compared the migration and separation characteristics of dipeptides on ChiralPlate with those on cellulose. Marseigne (189) separated D,L-asp-acc-OPr (dipeptide 56410 RP), a dipeptide with sweetening properties, whereas another group (188) investigated the separation of D,L-asp-D,L-phe-OCH$_3$ (aspartame).

α-Methylamino acids. α-Methylamino acids are very important as specific enzyme inhibitors. Furthermore, they can be directly inserted into numerous biologically active peptides to modify their range of activity.

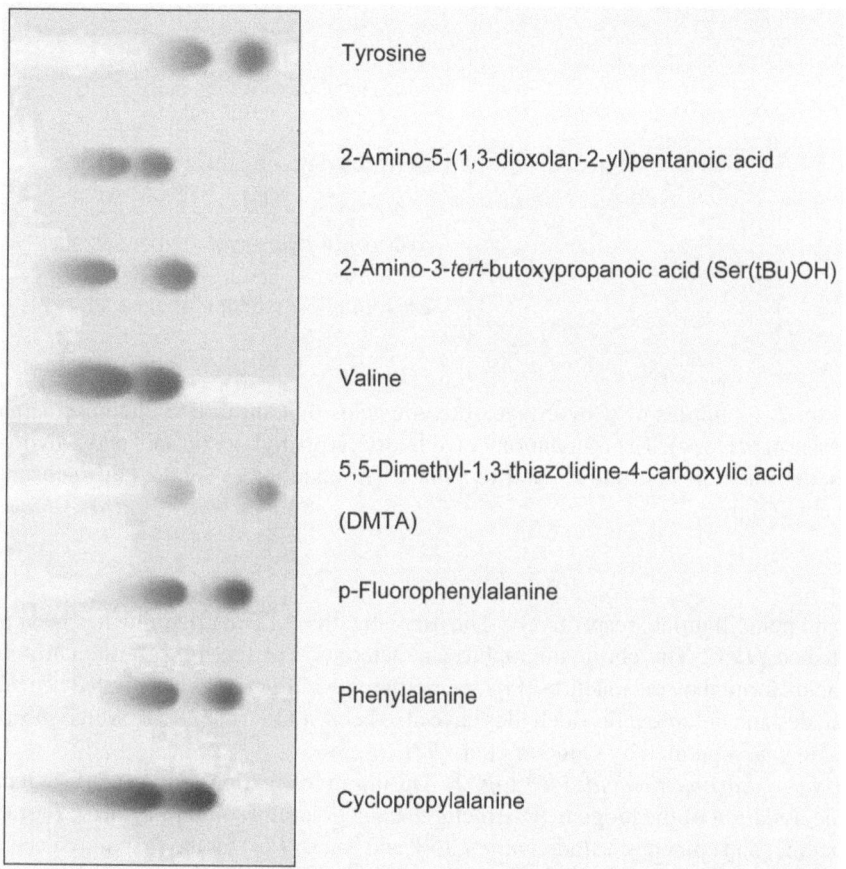

Tyrosine

2-Amino-5-(1,3-dioxolan-2-yl)pentanoic acid

2-Amino-3-*tert*-butoxypropanoic acid (Ser(tBu)OH)

Valine

5,5-Dimethyl-1,3-thiazolidine-4-carboxylic acid

(DMTA)

p-Fluorophenylalanine

Phenylalanine

Cyclopropylalanine

Fig. 1. Examples of different amino acids and derivatives on ChiralPlate (acetonitrile/methanol/water, 4/1/1 v/v/v). In these cases, the D-enantiomers show lower R_f-values.

Separations in this field with different eluent systems have been published independently *(174,175,200)*. D,L-methyldopa can also be separated without problem *(199)*. N-*Alkylamino acids*. Examples have been published recently *(67,92,170,195)*. In contrast to the examples described above, the detection of *N,N*-dimethylphenylalanine was achieved with iodine. The enantiomeric separation of *N*-carbamoyltryptophan has also been described *(176)*.

Halogenated amino acids. Another class of compounds that shows good enantiomeric resolution is the halogenated amino acids. However, a differentiation between 4-chloro-, 4-bromo-, and 4-iodophenylalanines is not possible *(170,195)*.

Heterocyclic compounds. Thiazolidine-4-carboxylic acid and 5,6-dimethylthiazolidine-4-carboxylic acid are formed by formaldehyde condensation from cysteine

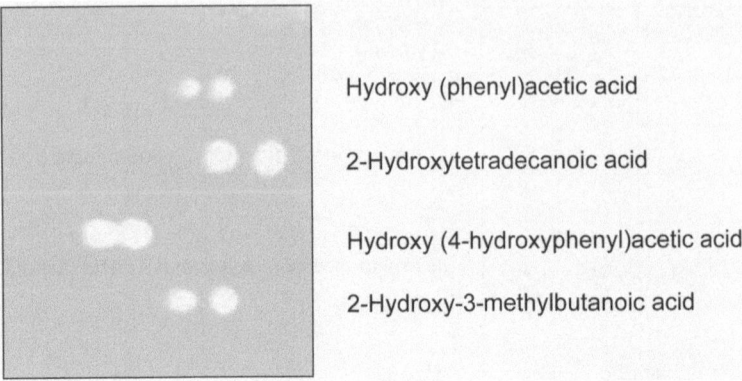

Hydroxy (phenyl)acetic acid

2-Hydroxytetradecanoic acid

Hydroxy (4-hydroxyphenyl)acetic acid

2-Hydroxy-3-methylbutanoic acid

Fig. 2. Examples of α-hydroxycarboxylic acids on ChiralPlate (dichloromethane/methanol; 9/1, v/v). The D-enantiomers of hydroxy(phenyl)acetic acid and 2-hydroxy-3-methylbutanoic acid show lower R_f values. The enantiomers of the other substances are not assigned.

and penicillamine, respectively. The derivatization of penicillamine has been published *(195)*. The chromatographic characteristics of the thiazolidine carboxylic acids formed by the reaction of D,L-penicillamine with various substituted benzaldehydes and heterocyclic aldehydes have also been studied *(177)*. 3-Carboxymorpholine was separated by Günther et al. *(92)*.

α-*Hydroxycarboxylic acids* (*see* **Fig. 2**). During investigation of the enantioselective degradation of the biogenic R-structured catecholamines norepinephrine (noradrenaline) and epinephrine (adrenaline), Jork and Kany *(173)* for the first time succeeded in the enantiomeric separation of the resulting 3,4-dihydroxymandelic acid and vanillylmandelic acid, respectively, using the lipophilic eluent mixture dichloromethane/methanol (45:5, v/v) and postchromatographic detection with 2,6-dichloroquinone-4-chloroimide (Merck; cat. no. 3037).

Vanadium pentoxide was especially useful for postchromatographic derivatization *(202)* of the aromatic and aliphatic α-hydroxycarboxylic acids. For aromatic α-hydroxycarboxylic acids, manganese chloride-sulfuric acid (for 30 min at 120°C) was also suitable *(180)*.

1.6.4. Quantitative Evaluation of TLC-Separated Enantiomers

Phenylalanine, tert-leucine, 5,5-dimethylthiazolidine-4-carboxylic acid, and α-hydroxyphenylalanine have been chosen as models for the direct quantitative evaluation of thin-layer chromatograms. Emphasis has been placed on the evaluation of detection limits for the TLC-separated enantiomers, because exact determination of trace levels of a D- or L-enantiomer in an excess of the other is increasingly important *(171,201,203–205)*.

In order to enhance specificity and sensitivity, postchromatographic derivatization with ninhydrin or vanadium pentoxide was used. Dipping the plates into the reagent solution proved most useful, because it could be automated *(206)*. Quantification of the minor enantiomer was achieved by *in situ* remission measurement with the CS-930 double-beam scanner (Shimadzu, Kyoto, Japan) or the densitometer CD 60 (Desaga, Heidelberg, Germany), and comparison with external standard solutions. Additionally, possible proportional systematic deviations were excluded by the standard addition method *(207)*.

For every substance investigated, the absorption maximum was determined independently prior to the quantification experiments. A detailed description of chromatographic conditions will be given in **Subheading 3.3.3.**

1.6.4.1. RESULTS

1. Phenylalanine. The calibration line shows that quantitative determinations of L-phenylalanine in D-phenylalanine are possible in a working range of 0.04–0.4 µg/ spot without any problem.
2. Tert-leucine. The D-enantiomer can also be determined with high sensitivity in the L-amino acid.
3. 5,5-Dimethylthiazolidine-4-carboxylic acid. Even 0.1–1% of the L-enantiomer can easily be quantitated.
4. α-Hydroxyphenylalanine. The calibration curve shows good evaluation of the amount of D-enantiomer in the working range 1–6%.

1.7. Enantiomer Separation Using Diastereomeric Derivatives

With the increasing number of commercially available, extremely pure chiral auxiliaries, thin-layer chromatographic purity control via formation of diastereomers has gained increasing importance. In contrast to direct enantiomer separations, antipode separation via diastereomers is usually not achieved with chiral adsorbents; however, enhanced "diastereomer selectivity" is also noted for asymmetric supports. The type of chiral reagent for formation of the diastereomer depends among other parameters on the structure—mono- or bifunctional —of the compound to be derivatized.

The published work *(208–222)* focuses on reactions of racemic compounds with NH$_2$(NH)-, OH-, and COOH-functionalities with the auxiliaries known from liquid chromatography, especially with commercial ready-to-use reagents *(67)*.

1.8. Summary

This review does not claim completeness. We intended to demonstrate for a few selected examples the present possibilities of thin-layer chromatographic enantiomeric separations. Emphasis was placed on racemate separations with commercial plates based on cellulose, cellulose triacetate, ChiralPlate, and HPTLC-CHIR, with detailed descriptions of the respective separation procedures and applications.

Because precise determinations of minute D- or L-concentrations in an excess of the other enantiomer become more and more important, the quantitation of TLC-separated antipodes was treated explicitly; further optimization of separation parameters and detection by fluorescence should enable improvement of the present detection limit of ≥0.1% D- or L-component. Here, it is worth mentioning that until today only the layers based on LEC with the 4-hydroxyproline selector are generally accepted, and these are the only ready-to-use plates commercially available on the market.

Compared to the classical methods of GC and HPLC, the TLC enantiomeric separation technique implies parallel (simultaneous) separations and is therefore especially suited for economical routine analyses.

2. Materials

2.1. Cellulose TLC

  1. Cellulose-precoated HPTLC plates (Merck; cat. no. 5786); size 10 × 20 cm; layer thickness 0.1 mm, without fluorescent indicator.

2.1.2. Solvents

  1. Water.
  2. Methanol.

2.1.3. Reagents

  1. Ninhydrin.

2.2. Triacetylcellulose TLC

2.2.1. Plates

  1. OPTI-T.A.C. TLC plates L.254 (Antec, Bennwil; cat. no. 4006).

2.2.2. Solvents

  1. Water.
  2. Methanol.
  3. Ethanol.

2.3. Ligand-Exchange TLC

2.3.1. Plates

  1. RP-18 TLC-precoated plate (Merck; cat. no. 15389), size 20 × 20 cm; layer thickness 0.25 mm, with fluorescent indicator.

2. TLC-precoated plates, ChiralPlate (Macherey-Nagel; cat. no. 811 055/056); size 10–20 cm; layer thickness 0.25 mm).
3. HPTLC-precoated plates CHIR with concentrating zone (Merck; cat. no. 14285); size 10 × 10 cm; concentrating zone 2.5 × 10 cm.

2.3.2. Solvents

1. Water.
2. Methanol.
3. Acetonitrile.
4. Acetone.
5. Dichloromethane.
6. 0.1 *M* Hydrochloric acid.
7. Acetic acid.
8. 2.5 *M* Sulfuric acid.

2.3.3. Reagents

1. Sodium acetate.
2. *N,N*-di-*n*-propyl-L-alanine.
3. Ninhydrin.
4. Vanadium pentoxide (Merck; cat. no. 824).
5. Cupric acetate.
6. Sodium carbonate.

3. Methods

3.1. Cellulose TLC-Plates

3.1.1. Chromatographic Conditions
for the Racemic Compounds Cited in Ref. 76

1. Method: ascending, one-dimensional development in a TLC chamber with chamber saturation.
2. Plates: cellulose-precoated HPTLC plates (Merck; cat. no. 5786).
3. Eluent: methanol/water, 3:2 (v/v).
4. Sample vol: 1 µL of a 0.05% methanolic solution (1:1) applied as a 10-mm streak.
5. Length of run: 17 cm.
6. Time of run: 2 h.
7. Detection: the dried plates were immersed for 3 s in a 0.3% ninhydrin solution in acetone (Tauchfix, Baron) and then dried in a cabinet for approx 4 min at 105°C. Blue-violet derivatives formed on the white background.
8. Spectroscopy λ equals 565 nm (reflectance) (*see* **Notes 1 and 2**).

3.2. Triacetylcellulose Plates

3.2.1. Chromatographic Conditions for the Substances Cited in Ref. 91

1. Method: ascending, one-dimensional development in a TLC chamber without chamber saturation.

2. Plates: OPTI-T.A.C. TLC plates L.254.
3. Eluent: ethanol-water, 80:20 (v/v) [for oxindanac, 85:15 (v/v)].
4. Sample vol:
   a. Oxindanac: 5 µL of a 0.2% methanolic solution applied as a 15-mm streak.
   b. 2-Phenylcyclohexanone: 10 µL of a 1% methanolic solution applied as a 15-mm streak.
   c. (R,S)-2,2,2-Trifluoro-1-(9-anthryl)-ethanol: 1 µL of a 0.2% methanolic solution applied as a 10-mm streak.
   d. Troeger's base: 2 µL of a methanolic solution applied as a 15-mm streak.
5. Length of run: 10 cm.
6. Time of run: 1.3 h.
7. Detection: UV (254 nm resp. 366 nm).
8. Spectroscopy: λ equals 254 nm (deuterium lamp) or λ_{exc} equals 366 nm, λ_{em} equals 420 nm (cut-off filter, mercury lamp) (for anthryl derivative) (*see* **Notes 1** and **2**).

3.3. Ligand-Exchange TLC

3.3.1. Chromatographic Conditions (According to Weinstein [160])

1. Method: ascending, one-dimensional development in a TLC chamber with chamber saturation.
2. Plates: RP-18 TLC-precoated plate (Merck; cat. no. 15389).
   Preparation of plates: reversed-phase TLC plates were developed (prior to application of the dansyl amino acids) in 0.3 M sodium acetate in 40% acetonitrile and 60% water, adjusted to pH 7.0 with acetic acid (buffer A). After fan-drying, the plates were immersed in a solution of 8 mM N,N-di-n-propyl-L-alanine and 4 mM cupric acetate in 97.5% acetonitrile and 2.5% water for 1 h and left to dry in the air. The plates are stable and can be stored for further use.
3. Eluent: 0.3 M sodium acetate in 40% acetonitrile and 60% water, adjusted to pH 7.0 with acetic acid (buffer A).
4. Sample vol: 0.5 µL of a 0.6% methanolic solution (1:1).
5. Length of run: 16 cm.
6. Time of run: 1.5 h.
7. Detection: UV (366 nm).
8. Spectroscopy λ_{exc} equals 366 nm, λ_{em} equals 420 nm (cut-off filter, mercury lamp) (*see* **Notes 1** and **2**).
9. Results. R_f values of selected dansyl amino acids:
   a. Dansyl-D,L-aspartic acid, 0.45 (L)/0.48 (D).
   b. Dansyl-D,L-serine, 0.32/0.34.
   c. Dansyl-D,L-glutamic acid, 0.51 (L)/0.58 (D).

3.3.2. Examples of Separations With ChiralPlate and HPTLC-CHIR

3.3.2.1. Chromatographic Conditions (ChiralPlate)

1. Method: ascending one-dimensional development in a TLC chamber with chamber saturation.

2. Plates: TLC-precoated plates, ChiralPlate.
3. Eluent: to achieve short analysis times, ternary mixtures of water-miscible alcohol, water, and acetonitrile proved useful. Most racemate separations could be accomplished using one of two eluent systems:
   a. Methanol/water/acetonitrile, 50:50:200 (v/v/v).
   b. Methanol/water/acetonitrile, 50:50:30 (v/v/v).

   For some substances, however, different eluent systems were more suitable:

   a. Methanol/water, 10:80 (v/v).
   b. Acetone/methanol/water, 10:2:2 (v/v/v).
   c. Dichloromethane/methanol, 45:5 (v/v).
4. Sample vol: with eluents A, B, and C, 2 µL of a 1% solution of the racemate (methanol or methanol/water) were applied. With eluent D, 2 µL of a 0.5% solution of the racemate [0.1 M hydrochloric acid-methanol, 1:1 (v/v)] were applied. With eluent E, 2 µL of a 0.5% solution of the racemate [methanol or methanol/dichloromethane (1:1)] were applied.
5. Length of run: 13 cm.
6. Time of run: 0.5 h (eluent A), 1 h (eluent B), 1.5 h (eluent C), 0.8 h (eluent D), and 0.3 h (eluent E).
7. Detection: different detection methods were used, depending on the type of compound.

   For proteinogenic and nonproteinogenic amino acids, the dried plates were dipped for 3 s in a 0.3% ninhydrin solution in acetone and then dried in a drying cabinet for approx 5 min at 110°C. Red derivatives formed on a white background. For α-hydroxycarboxylic acids, 1.82 g of vanadium pentoxide were weighed into a 100-mL measuring flask, 30 mL of 1 M sodium carbonate were added and completely dissolved by treatment in an ultrasonic bath. After cooling, 46 mL of 2.5 M sulfuric acid and acetonitrile to 100 mL were added. The dried plates were briefly (set 2 s on the Tauchfix) dipped into this solution and then left to stand at room temperature for approx 45 min. Blue derivatives formed on a yellow background.

3.3.2.2. Chromatographic Conditions (HPTLC-CHIR)

1. Method: ascending, one-dimensional development in a TLC chamber with chamber saturation.
2. Plates: HPTLC-precoated plates CHIR with concentrating zone.
3. Eluent: E: dichloromethane/methanol, 45:5 (v/v).
4. Sample volume: 1 µL of a 0.5% solution (methanol or methanol/dichloromethane, 1:1) applied as a 10-mm streak.
5. Length of run: 5.5 cm.
6. Time of run: 0.1 h.
7. Detection: *see* **Subheading 3.3.2.1.**
8. Spectroscopy: λ equal 595 nm (tungsten lamp) (*see* **Notes 1** and **2**).

3.3.3. Quantitative Evaluation of TLC-Separated Enantiomers

For every substance investigated (*see* **Subheading 1.6.4.**), the absorption maximum was determined independently prior to the quantification experiments.

3.3.3.2. PREPARATION OF TEST SOLUTIONS AND STANDARD SOLUTIONS

Successful separation of amino acids on the TLC plate depends, inherently, on the concentration and often on the hydrochloric acid content of the applied solution. Addition of hydrochloric acid generally improves the solubility of the amino acids and often considerably enhances the enantiomeric resolution.

1. Phenylalanine test solution (U_{Ph}). Weigh 200 mg of D-phenylalanine into a 10-mL measuring flask and fill to the mark with 50% methanolic hydrochloric acid solution (10 g of acid per liter of solution).
2. Phenylalanine standard solution (V_{Ph}). Weigh 100 mg of L-phenylalanine into a 100-mL measuring flask and fill to the mark with methanol/0.1 *M* hydrochloric acid (1:1). From this stock solution, the standard solutions are prepared for the working range required. Dilute 200, 400, 600 µL, etc., of the stock solution to 10 mL with hydrochloric acid (10 g of acid per liter of solution)/methanol (1:1). Thus 0.1–0.3% solutions of the L-enantiomer relative to the 200 mg of D-phenylalanine are obtained.
3. Tert-leucine test solution (U_L). Dissolve 200 mg of L-tert-leucine in 10.0 mL of 50% methanol.
4. Tert-leucine standard solution (V_L). Dissolve 100 mg of tert-leucine in 100 mL of 50% methanol. Dilute 200, 400, 600 µL, etc., of this stock solution to 10 mL with 50% methanol to obtain 0.1–1.3% D-enantiomer relative to 200 mg of L-tert-leucine.
5. 5,5-Dimethylthiazolidine-4-carboxylic acid test solution (U_D). Add 500 µL of concentrated hydrochloric acid to 500 mg of D-5,5-dimethylthiazolidine-4-carboxylic acid and make up to 10 mL with isopropanol.
6. 5,5-Dimethylthiazolidine-4-carboxylic acid standard solution (V_D). Add 500 µL of concentrated hydrochloric acid to 100 mg of L-5,5-dimethylthiazolidine-4-carboxylic acid and make up to 100 mL with isopropanol. Add 500 µL of concentrated hydrochloric acid to 500, 1000, 1500 µL, etc., of this stock solution, and make up to 10 mL with isopropanol. These solutions correspond to 0.1–0.3% of the L-enantiomer relative to 500 mg of D-5,5-dimethylthiazolidine-4-carboxylic acid.
7. Hydroxyphenylalanine test solution (U_H). Weigh 300 mg of L-hydroxyphenylalanine into a 10-mL measuring flask and fill to the mark with methanol.
8. Hydroxyphenylalanine standard solution (V_H). Dissolve 30 mg of D-hydroxyphenylalanine in 100 mL of methanol; 1, 2, 3 µL, etc., of this stock solution correspond to 1–3% of the D-enantiomer relative to 300 mg of L-hydroxyphenylalanine.

3.3.3.3. CHROMATOGRAPHIC CONDITIONS

In general, the separation conditions for quantitative evaluation were similar to those for qualitative enantiomer separations by TLC. Any differences will be explained below. The plates were TLC-precoated ChiralPlates, sizes, 20 × 20 cm;

layer thickness, 0.25 mm. They were activated for 15 min at 110°C in a drying cabinet prior to use. The details of the eluents and detection procedures were as given above for the qualitative separation (*see* **Subheading 3.3.2.1.**).

3.3.3.4. Spectrophotometric Conditions (*see* **Notes 1** and **2**)

For the evaluation, the absorption curve was measured in the chromatographic direction. The measured peak areas resp. peak heights, plotted against the amount of sample per spot, gave the calibration lines.

4. Notes

1. Spectrophotometric conditions: Shimadzu scanner.
   a. Instrument: double-beam TLC scanner CS 930 (Shimadzu).
   b. Measuring setup: monochromator-sample (remission).
   c. Light source: tungsten lamp.
   d. Measuring area: 1.2 × 3 mm.
   e. Feed: 0.05 mm.
2. Spectrophotometric conditions: Desaga densitometer.
   a. Instrument: densitometer CD 60 (Desaga).
   b. Measuring setup: monochromator-sample (remission).
   c. Light source: tungsten lamp.
   d. Measuring area: 6.0 × 0.4 mm.
   e. Feed: 0.1 mm.

References

1. Günther, K., Martens, J., and Schickedanz, M. (1987) Verfahren zur dünnschicht-chromatographischen Trennung von Enantiomeren. EP-PS 143147, Degussa AG, Hanau, Germany.
2. Hardy, T. L. and Holland, D. O. (1952) The separation of DL-threonine from DL-allothreonine. *Chem. and Ind.* **855**.
3. Shaw, K. N. F. and Fox, S. W. (1953) Stereochemistry of the β-phenylserines: improved preparation of allophenylserine. *J. Am. Chem. Soc.* **75**, 3421–3424.
4. Drell, W. (1955) The separation of substituted *threo-* and *erythro*-phenylserines by paper chromatography. The configuration of arterenol and epinephrine. *J. Am. Chem. Soc.* **77**, 5429–5431.
5. Piez, K. A., Irreverre, F., and Wolff, H. L. (1956)The separation and determination of cyclic imino acids. *J. Biol. Chem.* **233**, 687–697.
6. Hegarty, M. P. (1957) The isolation and identification of 5-hydroxypiperidine-2-carboxylic acid from *leucaena glauca* benth. *Aust. J. Chem.* **10**, 484–488.
7. Gray, D. O., Blake, J., Brown, D. H., and Fowden, L. (1964) Chromatographic separation of amino acid isomers. *J. Chromatogr.* **13**, 276–277.
8. Bellon, G., Berg, R., Chastang, F., Malgras, A., and Borel, J. P. (1984) Separation and evaluation of the *cis* and *trans* isomers of hydroxyprolines: effect of hydrolysis on the epimerisation. *Anal. Biochem.* **137**, 151–155.

9. Hinman, J. W., Caron, E. L., and Christensen, H. N. (1950) The isomeric dipeptides of valine including a correction. *J. Am. Chem. Soc.* **72,** 1620–1625.

10. Sachs, H. and Brand, E. (1954) Optical rotation of peptides. VIII. Glutamic acid tripeptides. *J. Am. Chem. Soc.* **76,** 1811–1814.

11. Taschner, E., Sokolowska, T., Biernat, J. F., Chimiak, A., Wasielewski, C., and Rzeszotarska, B. (1963) Aminosäuren und Peptide, XI. Nachweis der Racemisierung bei Peptidsynthesen durch Papierchromatographie. *Liebigs Ann. Chem.* **663,** 197–202.

12. Sokolowska, T. and Biernat, J. F. (1964) The chromatographic separation of diastereoisomeric dipeptides. *J. Chromatogr.* **13,** 269–270.

13. Wieland, T. and Bende, H. (1965) Chromatographische Trennung einiger diastereomerer Dipeptide und Betrachtungen zur Konformation. *Chem. Ber.* **98,** 504–515.

14. Nitecki, D. E., Halpern, B., and Westley, J. W. (1968) A simple route to sterically pure diketopiperazines. *J. Org. Chem.* **33,** 864–866.

15. Westley, J. W., Close, V. A., Nitecki, D. N., and Halpern, B. (1968) Determination of steric purity and configuration of diketopiperazines by gas-liquid chromatography, thin-layer chromatography, and nuclear magnetic resonance spectrometry. *Anal. Chem.* **40,** 1888–1890.

16. Hubert, P. and Dellacherie, E. (1973) Thin-layer chromatographic separation of dipeptide and tripeptide diastereoisomers. *J. Chromatogr.* **80,** 144–145.

17. Arendt, A., Kolodziejczyk, B., and Sokolowska, T. (1976) Separation of diastereomers of protected dipeptides by thin-layer chromatography. *Chromatographia* **9,** 123–126.

18. Lepri, L., Desideri, P. G., Heimler, D., and Giannessi, S. (1983) High-performance thin-layer chromatography of diastereomeric di- and tripeptides on ready-for-use plates of silanized silica gel and on ammonium tungstophosphate layers. *J. Chromatogr.* **265,** 328–334.

19. Sarsúnová, M., Semonsky, M., and Cerny, A. (1970) Ergot alkaloids. XXXVI. Separation of diastereomeric (+)-1-hydroxy-2-butylamides of D- and L-lysergic and D- and L-isolysergic acids by thin-layer chromatography. *J. Chromatogr.* **50,** 442–446.

20. Sondack, D. L. (1974) TLC separation and identification of diastereomers of D-ergonovine maleate. *J. Pharm. Sci.* **63,** 1141–1143.

21. Giron, D. and Groell, P. (1978) Separation of diastereoisomers by reversed phase thin-layer chromatography. *J. High Resol. Chromatogr. Chromatogr. Commun.* **1,** 67–68.

22. Kohli, J. C., Alang, N. K., and Khushminder, A. (1981) Specific separation of isopulegol from stereoisomeric isopulegols by thin layer chromatography. *Sci. Cult.* **47,** 170.

23. Iida, T., Momose, T., Shinohara, T., Goto, J., Nambara, T., and Chang, F. C. (1986) Separation of allo bile acid stereoisomers by thin-layer and high-performance liquid chromatography. *J. Chromatogr.* **366,** 396–402.

24. Souter R. W. (1987) *Chromatographic Separations of Stereoisomers*. CRC Press, Boca Raton, pp. 212–221.

25. Palamareva, M. D., Haimova, M., Stefanovsky, J., Viteva, L., and Kurtev, B. (1971) Thin-layer chromatography on silica gel as a method for assigning the relative configurations to some aliphatic diastereomeric compounds. *J. Chromatogr.* **54,** 383–391.

26. Palamareva, M. D. and Snyder, L. R. (1984) Liquid-solid chromatography of diastereomers: tetrasubstitutes ethanes. *Chromatographia* **19,** 352–354.

27. Palamareva, M. D. (1988) Chromatographic behaviour of diastereomers. IX. Application of a microcomputer program to the thin-layer chromatographic separation of some diastereomers on silica. *J. Chromatogr.* **438,** 219–224.

28. Palamareva, M. D., Kurtev, B. J., and Kavrakova, I. (1991) Chromatographic behaviour of diastereoisomers. X. Thin-layer chromatographic retentions on silica of some (E)- and (Z)-oxazolones and related cinnamates as a function of mobile phase effects or Hammett constants. *J. Chromatogr.* **545,** 161–175.

29. Palamareva, M. D. and Kozekov, I. (1996) Theoretical threatment of the adsorptivity of esters of racemic and meso-2,3-dibromobutane-1,4-dioic acids on alumina. *J. Planar Chromatogr.* **9,** 439–444.

30. Palamareva, M. D. and Kozekov, I. (1997) Chromatographic behaviour of diastereoisomers. XIII. Adsorptivity of esters of Z- and E-2,3-diphenylpropenoic acids and similar compounds on silica in terms of the Snyder theory. *J. Chromatogr. A* **758,** 135–144.

31. Lippmann, T. and Mann, G. (1995) Chromatographie diastereomerer Calix[4] arene. *Git Fachz. Lab.* **3,** 203–207.

32. Kotake, M., Sakan, T., Nakamura, N., and Senoh, S. (1951) Resolution into optical isomers of some amino acids by paper chromatography. *J. Am. Chem. Soc.* **73,** 2973–2974.

33. Bonino, G. B. and Carassiti, V. (1951) Separation of the optical antipodes of racemic β-naphthol benzylamine by paper chromatography. *Nature* **167,** 569–570.

34. Berlingozzi, S., Serchi, G., and Adembri, G. (1951) Ricerche sugli aminoacidi rotatori. -Nota XII. *Sperimentale. Sez. Chim. Biol.* **2,** 89–94.

35. Fujisawa, Y. (1951) Resolution of the racemic amino acids on the paper chromatograms. *J. Osaka City Med. Center* **1,** 7–13.

36. Mason, M. and Berg, C. P. (1952) The metabolism of D- and L-tryptophan and D- and L-kynurenine by liver and kidney preparations. *J. Biol. Chem.* **195,** 515–529.

37. Dalgliesh, C. E. (1952) The optical resolution of aromatic amino acids on paper chromatograms. *J. Chem. Soc.* 3940–3942.

38. Closs, K. and Haug, C. M. (1953) Note on the resolution of DL-tryptophan by paper chromatography. *Chem. Ind.* **103.**

39. Makino, K. and Takahashi, H. (1953) Synthesis of 5-hydroxykynurenine. *Science* **118,** 699.

40. Roberts, E. A. H. and Wood, D. J. (1953) Separation of tea polyphenols on paper chromatograms. *Biochem. J.* **53,** 332–336.

41. Lambooy, J. P. (1954) The syntheses, paper chromatography and substrate specificity for tyrosinase of 2,3-, 2,4-, 2,5-, 2,6-, and 3,5-dihydroxyphenylalanines. *J. Chem. Am. Soc.* **76,** 133–138.

42. Weichert, R. (1954) Über Molekülverbindungen zwischen α-Aminosäuren und nicht-reduzierenden Zuckern. II. *Acta Chem. Scand.* **8,** 1542–1546.
43. Berlingozzi, S., Adembri, G., and Bucci, G. (1954) Ricerche sugli aminoacidi rotatori. -Nota XVII. Sdoppiamento die D.L.forme per cromatografia su carta. *Gazz. Chim. Ital.* **84,** 393–404.
44. Rhuland, L. E., Work, B., Denman, R. F., and Hoare, D. S. (1955) The behavior of the isomers of α,ε-diaminopimelic acid on paper chromatograms. *J. Am. Chem. Soc.* **77,** 4844–4846.
45. Hoare, D. S. and Work, B. (1955) The stereoisomers of α,ε-diaminopimelic acid: their distribution in nature and behaviour towards certain enzyme preparations. *Biochem. J.* **61,** 562–568.
46. Weichert, R. (1955) Spaltung von D,L-Histidin bei Papierchromatographie. *Acta Chem. Scand.* **9,** 547.
47. Blaschko, H. and Hope, D. B. (1956) Excretion of cystathione in pyridoxine deficiency. *Biochem. J.* **63,** 7P.
48. Klingmüller, V. and Maier-Sihle, L. (1957) Papierchromatographische Spaltung der racemischen Aminosäuren Histidin und Tryptophan. *Z. Physiol. Chem.* **308,** 49–50.
49. Butenandt, A., Biekert, B., Koga, N., and Traub, P. (1960) Über Ommochrome, XXI. Konstitution und Synthese des Ommatins D. *Z. Physiol. Chem.* **321,** 258–275.
50. Mayer, W. and Merger, F. (1961) Darstellung optisch aktiver Catechine durch Racemattrennung mit Hilfe der Adsorptionschromatographie an Cellulose. *Liebigs Ann. Chem.* **644,** 65–69.
51. Kikkawa, I. (1961) Studies on the alkaloids of menispermaceous plants. CLXXXI. Optical resolution of *dl*-magnoflorine by paper chromatography. *J. Pharm. Soc. Japan* **81,** 732–735.
52. Franck, B. and Schlingloff, G. (1962) Biogeneseähnliche Alkaloidsynthesen durch oxidative Kondensation, II. Synthese eines Aporphins. *Liebigs Ann. Chem.* **659,** 123–132.
53. Tomita, M. and Sugamoto, M. (1962) Optical resolution of some alkaloids by paper chromatography and cellulose column chromatography. *J. Pharm. Soc. Japan* **82,** 1141–1144.
54. De Ligny, C. L., Nieboer, H., De Vijlder, J. J. M., and van Willigen, J. H. H. G. (1963) Separation of some DL-amino acids into their optical isomers by paper adsorption chromatography. *Recueil* **82,** 213–224.
55. Butenandt, A., Biekert, E., Kübler, H., and Linzen, B., and Traub, P. (1963) Über Ommochrome, XXII. Konstitution und Synthese des Rhodommatins. *Z. Physiol. Chem.* **334,** 71–83.
56. Contractor, S. F. and Wragg, J. (1965) Resolution of the optical isomers of DL-tryptophan, 5-hydroxy-DL-tryptophan and 6-hydroxy-DL-tryptophan by paper and thin-layer chromatography. *Nature* **208,** 71–72.
57. Franck, B. and Blaschke, G. (1966) Biogeneseähnliche Alkaloidsynthesen durch oxydative Kondensation, VII. Neue Aporphin-Synthesen. *Liebigs Ann. Chem.* **695,** 144–157.

58. Zaltzman-Nirenberg, P., Daly, J., Guroff, G., and Udenfriend, S. (1966) Separation of the enantiomers of the aminonaphthylalanines by paper chromatography. *Anal. Biochem.* **15,** 517–522.
59. Weichert, R. (1969) Zur Racemattrennung von Aminosäuren durch Papieradsorptionschromatographie. *Ark Kemi* **31,** 517–532.
60. El Din Awad, A. M. and El Din Awad, O. M. (1974) Ion-exchange paper chromatographic resolution of some racemic DL-amino acids on alginate. *J. Chromatogr.* **93,** 393–398.
61. Fu, Z., Sun, L., Cai, J., et al. (1997) Paper chromatographic separation of racemic diphenylmethyl alkohols using only pure water surfactant micellar mobile phase and host-guest chromatography. *J. Chromatogr. Sci.* **35,** 309–314.
62. Kriz, D., Kriz, C. B., Andersson, L. I., and Mosbach, K. (1994) Thin-layer chromatography based on the molecular imprinting technique. *Anal. Chem.* **66,** 2636–2639.
63. Suedee, R., Songkram, C., Petmoreekul, A., and Sangkunakup, S., Sankasa, S., and Kongyarit, N. (1998) Thin-layer chromatography using synthetic polymers imprinted with quinine as chiral stationary phase. *J. Planar Chromatogr.* **11,** 272–276.
64. Suedee, R., Srichana, T., Saelim, J., and Thavornpibulbut, T. (1999) Chiral determination of various adrenergic drugs by thin layer chromatography using molecularly imprinted chiral stationary phases prepared with α-agonists. *Analyst* **124,** 1003–1009.
65. Suedee, R., Songkram, C., Petmoreekul, A., Sangkunakup, S., Sankasa, S., and Kongyarit, N. (1999) Direct enantioseparation of adrenergic drugs via thin-layer chromatography using molecularly imprinted polymers. *J. Pharm. Biomed. Anal.* **19,** 519–527.
66. Mack, M. and Kinkel, J. (1990) Trennmaterialien. DE 4005868, Merck Patent GmbH, Darmstadt, Germany.
67. Günther, K. and Möller, K. (1996) Separation of enantiomers by thin-layer chromatography. *Handbook of Thin-Layer Chromatography, 2nd ed.,* Marcel Dekker, New York, pp. 621–682.
68. Haworth, D. T. and Hung, Y. W. (1973) Thin-layer chromatography of an optically active complex. *J. Chromatogr.* **78,** 314–315.
69. Chimiak, A. and Polonski, T. (1975) Thin-layer chromatographic separation of diaminodicarboxylic acid stereoisomers and their dansyl derivatives. *J. Chromatogr.* **115,** 635–638.
70. Munier, R. L., Drapier, A. M., Gervais, C., and Tréfouël, J. (1976) Biochimie Analytique—Comportement des isoméres optiques du tryptophane et de ses analogues structuraux inhibiteurs de la croissance bactérienne en chromatographie sur cellulose à haute cristallinité. *C. R. Acad. Sc. Paris, Ser. D* **282,** 1761–1764.
71. Bach, K. and Haas, H. J. (1977) Dünnschichtchromatographische Spaltung der Racemate einiger Aminosäuren. *J. Chromatogr.* **136,** 186–188.
72. Yuasa, S., Shimada, A., Kameyama, K., Yasui, M., and Adzuma, K. (1980) Cellulose thin-layer and column chromatography for resolution of DL-tryptophan. *J. Chromatogr. Sci.* **18,** 311–314.

73. Yuasa, S. and Shimada, A. (1982) Separation of DL-amino acids by cellulose thin-layer chromatography. *Sci. Rep. Osaka Univ.* **31,** 13–22.
74. Yuasa, S., Shimada, A., Isoyama, M., Fukuhara, T., and Itoh, M. (1986) Cellulose conformation responsible for resolution of DL-amino acids. *Chromatographia* **21,** 79–82.
75. Wang, K. T., Chen, S. T., and Lo, L. C. (1986) The thin-layer chromatographic separation of enantiomeric dipeptides. *Z. Anal. Chem.* **324,** 338–339.
76. Günther, K. and Zeller, M. Degussa AG, Hanau, Germany.
77. Kuhn, A. O., Lederer, M., and Sinibaldi, M. (1989) Adsorption chromatography on cellulose. IV. Separation of D- and L-tryptophan and D- and L-methyltryptophan on cellulose with aqueous solvents. *J. Chromatogr.* **469,** 253–260.
78. Lederer, M. (1990) Adsorption chromatography on cellulose. VI. Further studies on the separation of D- and L-tryptophan on cellulose with aqueous solvents. *J. Chromatogr.* **510,** 367–371.
79. Shimada, A., Fukuhara, T., Isoyama, M., Itoh, M., and Yuasa, S. (1989) Construction of chiral evolution model. *Viva Origino* **17,** 53–62.
80. Shimada, A., Fukuhara, T., Isoyama, M., Itoh, M., and Yuasa, S. (1990) Construction of chiral evolution model. Chromatographic resolution of amino acid racemates by use of cellulose. *Nagasaki Daigaku Kyoikugakubu Shizen Kagaka Hokoku* **42,** 53–62.
81. Lederer, M. (1992) Adsorption chromatography on cellulose. VII. Chiral separation on cellulose with aqueous solvents. *J. Chromatogr.* **604,** 55–62.
82. Xuan, H. T. K., Kuhn, A. O., and Lederer, M. (1992) Adsorption chromatography on cellulose. VIII. The salting-out behaviour of some peptides with aromatic groups. *J. Chromatogr.* **626,** 301–304.
83. Xuan, H. T. K. and Lederer, M. (1993) Adsorption chromatography on cellulose. IX. Chiral separations with aqueous solvents and liquid-liquid systems. *J. Chromatogr.* **635,** 346–348.
84. Xuan, H. T. K. and Lederer, M. (1993) Adsorption chromatography on cellulose. X. Adsorption of tryptophan and derivatives from $CuSO_4$-containing eluents. *J. Chromatogr.* **645,** 185–188.
85. Xuan, H. T. K. and Lederer, M. (1994) Adsorption chromatography on cellulose. XI. Chiral separations with aqueous solutions of cyclodextrins as eluents. *J. Chromatogr.* **659,** 191–197.
86. Suedee, R. and Heard, C. M. (1997) Direct resolution of propranolol and bupranolol by thin-layer chromatography using cellulose derivatives as stationary phase. *Chirality* **9,** 139–144.
87. Kubota, T., Yamamoto, C., and Okamoto, Y. (2000) Tris(cyclohexylcarbamate)s of cellulose and amylose as potential chiral stationary phases for high-performance liquid chromatography and thin-layer chromatography. *J. Am. Chem. Soc.* **122,** 4056–4059.
88. Hesse, G. and Hagel, R. (1976) Über Inclusionschromatographie und ein neues Retentionsprinzip für Benzolderivate. *Chromatographia* **9,** 62–68.

89. Hesse, G. and Hagel, R. (1973) Vollständige Racemattrennung durch Elutions-chromatographie an Cellulose-tri-acetat. *Chromatographia* **6,** 277–280.
90. Faupel, M. (1987) A new microcristalline triacetylcellulose thin-layer plate permitting separation of enantiomers. Separation of racemic oxindanac into its enantiomeric components. 4th Int. Symp. Instrumental TLC, Selvino/Bergamo, Italy.
91. Günther, K. and Zeller, M. Degussa AG, Hanau, Germany.
92. Günther, K. and Merget, S. Degussa AG, Hanau, Germany.
93. Lepri, L., Coas, V., Desideri, P. G., and Zocchi, A. (1994) Reversed phase planar chromatography of enantiomeric compounds on triacetylcellulose. *J. Planar Chromatogr.* **7,** 376–381.
94. Lepri, L. (1995) Reversed phase planar chromatography of enantiomeric compounds on microcristalline triacetylcellulose. *J. Planar Chromatogr.* **8,** 467–469.
95. Huang, M. B., Sun, J., Wang, J. S., et al. (1997) Separation of pharmaceutical enantiomers on column and thin-layer plate of cellulose triacetate. *Acta Pharm. Sinica* **32,** 612–616.
96. Lepri, L., Del Bubba, M., and Masi, F. (1997) Reversed-phase planar chromatography of enantiomeric compounds on microcristalline cellulose triacetate (MCTA). *J. Planar Chromatogr.* **10,** 108–113.
97. Lepri, L., Del Bubba, M., Coas, V., and Cincinelli, A. (1999) Reversed-phase planar chromatography of racemic flavanones. *J. Liq. Chrom. Rel. Technol.* **22,** 105–118.
98. Lepri, L., Cincinelli, A., and Del Bubba, M. (1999) Reversed-phase planar chromatography of optical isomers on microcristalline cellulose triacetate. *J. Planar Chromatogr.* **12,** 298–301.
99. Lepri, L., Del Bubba, M., Cincinelli, A., and Boddi, L. (2000) Inclusion planar chromatography of enantiomeric and racemic compounds. *J. Planar Chromatogr.* **13,** 384–387.
100. Lepri, L., Del Bubba, M., Cincinelli, A., and Boddi, L. (2001) Direct resolution of aromatic alcohols by planar chromatography on tribenzoylcellulose as chiral stationary phase. *J. Planar Chromatogr.* **14,** 134–136.
101. Alak, A. and Armstrong, D. W. (1986) Thin-layer chromatographic separation of optical, geometrical, and structural isomers. *Anal. Chem.* **58,** 582–584.
102. Ward, T. J. and Armstrong, D. W. (1986) Improved cyclodextrin chiral phases: a comparison and review. *J. Liq. Chromatogr.* **9,** 407–423.
103. Han, S. M. and Armstrong, D. W. (1990) Enantiomeric separation by thin-layer chromatography. *Chem. Anal.* **108,** 81–100.
104. Duncan, J. D. and Armstrong, D. W. (1990) Chiral mobile phase additives in reversed-phase TLC. *J. Planar Chromatogr.* **3,** 65–67.
105. Wilson, I. D. (1986) Towards chiral TLC plates: some preliminary studies. *Method. Surv. Biochem. Anal.* **16,** 277–281.
106. Bhushan, R. and Martens, J. (1997) Direct resolution of enantiomers by impregnated TLC. *Biomed. Chromatogr.* **11,** 280–285.

107. Armstrong, D. W., He, F. Y., and Han, S. M. (1988) Planar chromatographic separation of enantiomers and diastereomers with cyclodextrin mobile phase additives. *J. Chromatogr.* **448,** 345–354.

108. Armstrong, D. W., Faulkner, J. R., Jr., and Han, S. M. (1988) Use of hydroxypropyl- and hydroxyethyl-derivatized β-cyclodextrins for the thin-layer chromatographic separation of enantiomers and diastereomers. *J. Chromatogr.* **452,** 323–330.

109. Duncan, J. D. and Armstrong, D. W. (1991) A study of the effects of the degree of substitution of hydroxypropyl-β-cyclodextrin used as a chiral mobile phase additive in TLC. *J. Planar Chromatogr.* **4,** 204–206.

110. Lepri, L., Coas, V., Desideri, P. G., and Checchini, L. (1990) Separation of optical and structural isomers by planar chromatography with development by β-cyclodextrin solutions. *J. Planar Chromatogr.* **3,** 311–316.

111. Lepri, L., Coas, V., Desideri, P. G., et al. (1990) Planar chromatography of isomers using β-cyclodextrin solutions as mobile phase. *J. Planar Chromatogr.* **3,** 533–535.

112. Lepri, L., Coas, V., and Desideri, P. G. (1991) Reversed-phase planar chromatography of isomers using α- and β-cyclodextrin solutions as eluents. *J. Planar Chromatogr.* **4,** 338–340.

113. Lepri, L., Coas, V., and Desideri, P. G. (1994) Planar chromatography of optical and structural isomers with eluents containing modified β-cyclodextrins. *J. Planar Chromatogr.* **7,** 322–326.

114. Xuan, H. T. K. and Leipzig-Pagani, E. (1996) Quantitative thin layer chromatography on cellulose. II. Selected applications: lower alcohols, tryptophan enantiomers, gold and platinum. *J. Chromatogr. A* **746,** 261–268.

115. Huang, M. B., Li, H. K., Li, G. L., Yan, C. T., and Wang, L. P. (1996) Planar chromatographic direct separation of some aromatic amino acids and aromatic amino amino alcohols into enantiomers using cyclodextrin mobile phase additives. *J. Chromatogr. A* **742,** 289–294.

116. LeFevre, J. W. (1993) Reversed-phase thin-layer chromatographic separations of enantiomers of dansyl-amino acids using β-cyclodextrin as a mobile phase additive. *J. Chromatogr.* **653,** 293–302.

117. Hao, A. Y., Tong, L. H., Zhang, F. S., and Gao, X. M., and Inoue, Y. (1995) Direct thin-layer chromatographic separations of enantiomers of six selected amino acids using 2-O-[(R)-2-hydroxypropyl]-β-CD as a mobile phase additive. *Anal. Lett.* **28,** 2041–2048.

118. Lambroussi, V., Piperaki, S., and Tsantili-Kakoulidou, A. (1999) Formation of inclusion-complexes between cyclodextrins, as mobile phase additives in RPTLC, and fluoxetine, norfluoxetine, and promethazine. *J. Planar Chromatogr.* **12,** 124–128.

119. Aboul-Enein, H. Y., El-Awady, M. I., and Heard, C. M. (2000) Enantiomeric separation of aminoglutethimide, acetyl aminoglutethimide, and dansyl aminoglutethimide by TLC with β-cyclodextrin and derivatives as mobile phase additives. *J. Liq. Chrom. Rel. Technol.* **23,** 2715–2726.

120. LeFevre, J. W., Gublo, E. J., Botting, C., et al. (2000) Qualitative reversed-phase thin-layer chromatographic analysis of the stereochemistry of D- and L-α-amino acids in small peptides. *J. Planar Chromatogr.* **13,** 160–165.

121. LeFevre, J. W., Rogers, E. D., Pico, L. L., and Botting, C. L. (2000) The effect of structure on the resolution of dansyl amino acids using beta-cyclodextrin as a mobile phase additive in reversed-phase thin-layer chromatography. *Chromatographia* **52,** 648–652.

122. Bhushan, R. and Parshad, V. (1996) Resolution of (±)-ibuprofen using L-arginine-impregnated thin-layer chromatography. *J. Chromatogr. A* **721,** 369–372.

123. Bhushan, R. and Thiongo, G. T. (1998) Direct enantioseparation of some β-adrenergetic blocking agents using impregnated thin-layer chromatography. *J. Chromatogr. B* **708,** 330–334.

124. Bhushan, R. and Thiongo, G. T. (1999) Direct enantiomeric resolution of some 2-arylpropionic acids using (–)-brucine-impregnated thin-layer chromatography. *Biomed. Chromatogr.* **13,** 276–278.

125. Huang, M. B., Li, G. L., Yang, G. S., Shi, Y. H., Gao, J. J., and Liu, X. D. (1997) Enantiomeric separation of aromatic amino alkohol drugs by chiral ion-pair chromatography on a silica gel plate. *J. Liq. Chrom. Rel. Technol.* **20,** 1507–1514.

126. Li, G., Huang, M., Yang, G., Wu, G., and Du, A. (1999) The enantiomeric separation of aromatic amino alkohol drugs by thin-layer chromatography. *Chin. J. Chromatogr.* **17,** 215–216.

127. Deng, Q., Kang, S., Zhu, Q., and Xu, Z. (1999) Study of a new cyclodextrin bonded stationary phase for thin-layer chromatographic chiral separations. *Am. Lab.* **31,** 43–47.

128. Zhu, Q., Ma, G., Deng, Q., and Zeng, L. (2000) Preparation and application of phenylcarbamate-substituted β-cyclodextrin bonded stationary phase used for thin-layer chromatography. *Chin. J. Anal. Chem.* **28,** 349–352.

129. Zhu, Q., Yu, P., Deng, Q., and Zeng, L. (2001) β-Cyclodextrin-bonded chiral stationary phase for thin-layer chromatographic separation of enantiomers. *J. Planar Chromatogr.* **14,** 137–139.

130. Paris, R. R., Sarsunova, M., and Semonsky, M. (1967) La chromatographie en couche mince comme méthode de dédoublement des alcaloides racémiques. Application aux médicaments contenant de l'éphédrine. *Ann. Pharm. Franc.* **25,** 177–180.

131. Wainer, I. W., Brunner, C. A., and Doyle, T. D. (1983) Direct resolution of enantiomers via thin-layer chromatography using a chiral adsorbent. *J. Chromatogr.* **264,** 154.

132. Oi, N. (1985) Direct separation of optically active isomers by chromatographic methods. *Farumashia,* **21,** 747–752.

133. Bhushan, R. and Ali, I. (1987) Resolution of enantiomeric mixtures of phenylthiohydantoin amino acids on (+)-tartaric acid-impregnated silica gel plates. *J. Chromatogr.* **392,** 460–463.

134. Bhushan, R. and Ali, I. (1987) TLC Resolution of enantiomeric mixtures of amino acids. *Chromatographia* **23,** 141–142.

135. Wall, P. E. (1989) Proceedings of the International Symposium on Instrumental Thin Layer Chromatography/Planar Chromatography, Brighton, Sussex, U.K.

136. Wall, P. E. (1989) Preparation and Application of HPTLC plates for enantiomer separation. *J. Planar Chromatogr.* **2**, 228–232.

137. Brunner, C. A. and Wainer, I. (1989) Direct stereochemical resolution of enantiomeric amides via thin-layer chromatography on a covalently bonded chiral stationary phase. *J. Chromatogr.* **472**, 277–283.

138. Tivert, A. M. and Backman, A. (1989) Enantiomeric separation of aminoalcohols by TLC using chiral counter-ion in the mobile phase. *J. Planar Chromatogr.* **2**, 472–473.

139. Duncan, J. D. (1990) Chiral separations: a comparison of HPLC and TLC. *J. Liq. Chromatogr.* **13**, 2737–2755.

140. Duncan, J. D., Armstrong, D. W., and Stalcup, A. M. (1990) Normal phase TLC separation of enantiomers using chiral ion interaction agents. *J. Liq. Chromatogr.* **13**, 1091–1103.

141. Brand, K., Kinkel, J., and Nagel, J. (1988) Chromatographieplatten und Verfahren zur dünnschicht-chromatographischen Trennung von Enantiomeren. DE 3843266, Merck Patent GmbH, Darmstadt, Germany.

142. Witherow, L., Spurway, T. D., Ruane, R. J., Wilson, I. D., and Longdon, K. (1991) Problems and solutions in chiral thin-layer chromatography: a two-phase "Pirkle" modified amino-bonded plate. *J. Chromatogr.* **553**, 497–501.

143. Lohmann, D. and Däppen, R. (1991) Enantiomeric silanes, modified packing materials, and use thereof. U.S. Patent 4,997,965.

144. Lepri, L., Coas, V., and Desideri, P. G. (1992) Planar chromatography of optical isomers with bovine serum albumin in the mobile phase. *J. Planar Chromatogr.* **5**, 175–178.

145. Lepri, L., Coas, V., and Desideri, P. G. (1992) Reversed phase planar chromatography of optically active fluorenylmethoxycarbonyl amino acids with bovine serum albumin in the mobile phase. *J. Planar Chromatogr.* **5**, 294–296.

146. Lepri, L., Coas, V., Desideri, P. G., and Zocchi, A. (1992) Reversed phase planar chromatography of enantiomeric tryptophans with bovine serum albumin in the mobile phase. *J. Planar Chromatogr.* **5**, 234–238.

147. Lepri, L., Coas, V., Desideri, P. G., and Pettini, L. (1992) Reversed phase planar chromatography of enantiomeric compounds with bovine serum albumin in the mobile phase. *J. Planar Chromatogr.* **5**, 364–367.

148. Lepri, L., Coas, V., Desideri, P. G., and Santianni, D. (1993) Reversed phase planar chromatography of dansyl DL amino acids with bovine serum albumin in the mobile phase. *Chromatographia* **36**, 297–301.

149. Lepri, L., Coas, V., Desideri, P. G., and Pettini, L. (1993) Thin layer chromatographic enantioseparation of miscellaneous compounds with bovine serum albumin in the eluent. *J. Planar Chromatogr.* **6**, 100–104.

150. Lepri, L., Coas, V., Desideri, P. G., and Zocchi, A. (1994) The mechanism of retention of enantiomeric solutes on silanized silica plates eluted with albumin solutions. *J. Planar Chromatogr.* **7**, 103–107.

151. Tivert, A. M. and Backman, A. (1993) Separation of the enantiomers of β-blocking drugs with a chiral mobile phase additive. *J. Planar Chromatogr.* **6**, 216–219.

152. Bhushan, R. and Ali, I. (1993) Resolution of racemic mixtures of hyoscyamine and colchicine on impregnated silica gel layers. *Chromatographia* **35**, 679–680.

153. Armstrong, D. W. and Zhou, Y. (1994) Use of macrocyclic antibiotic as the chiral selector for enantiomeric separations by TLC. *J. Liq. Chromatogr.* **17**, 1695–1707.

154. Bhushan, R. and Thiongo, G. T. (2000) Separation of the enantiomers of dansyl-D,L-amino acids by normal-phase TLC on plates impregnated with a macrocyclic antibiotic. *J. Planar Chromatogr.* **13**, 33–36.

155. Bhushan, R. and Parshad, V. (1996) Thin-layer chromatographic separation of enantiomeric dansylamino acids using a macrocyclic antibiotic as a chiral selector. *J. Chromatogr. A* **736**, 235–238.

156. Lindner, W. (1981) Trennung von Enantiomeren mittels moderner Flüssigkeits-Chromatographie. *Chimia* **35**, 294–307.

157. Davankov, V. A., Bochkov, A. S., Kurganov, A. A., Roumeliotis, P., and Unger, K. K. (1980) Separation of unmodified α-amino acid enantiomers by reversed phase HPLC. *Chromatographia* **13**, 677–685.

158. Kurganov, A. A., Ponomaryova, T. M., and Davankov, V. A. (1984) Copper(II) complexes with optically active diamines. V. Enantioselective effects in equally-paired and mixed-ligand copper(II) complexes with diamines. *Inorg. Chim. Acta* **86**, 145–149.

159. Günther, K., Martens, J., and Schickedanz, M. (1984) Dünnschichtchromato-graphische Enantiomerentrennung mittels Ligandenaustausch. *Angew. Chem.* **96**, 514–515; (1984) Thin-layer chromatographic enantiomeric resolution via ligand exchange. *Angew. Chem. Int. Ed. Engl.* **23**, 506.

160. Weinstein, S. (1984) Resolution of optical isomers by thin layer chromatography. *Tetrahedron Lett.* **25**, 985–986.

161. Grinberg, N. and Weinstein, S. (1984) Enantiomeric separation of Dns-amino acids by reversed-phase thin-layer chromatography. *J. Chromatogr.* **303**, 251–255.

162. Grinberg, N. (1985) Thin-layer chromatographic separation using a temperature gradient. *J. Chromatogr.* **333**, 69–81.

163. Marchelli, R., Virgili, R., Armani, E., and Dossena, A. (1986) Enantiomeric separation of D,L-Dns-amino acids by one- and two-dimensional thin-layer chromatography. *J. Chromatogr.* **355**, 354–357.

164. Davankov, V. A., Bochkov, A. S., and Belov, Y. P. (1981) Ligand exchange chromatography of racemates. XV. Resolution of α-amino acids on reversed-phase silica gels coated with N-decyl-L-histidine. *J. Chromatogr.* **218**, 547–557.

165. Martens, J., Weigel, H., Busker, E., and Steigerwald, R. (1987) Optisch aktive Prolin-Derivate, Verfahren zur ihrer Herstellung und ihre Verwendung. DE-PS 3143726, Degussa AG, Hanau, Germany.

166. Lübben, S., Martens, J., Haase, D., Pohl, S., and Saak, W. (1990) Dünnschicht-chromatographische Enantiomerentrennung nach dem Ligandenaustauschprin-zip: Synthese und Struktur eines chiralen Modellkomplexes. *Tetrahedron Lett.* **31**, 7127–7128.

167. Martens, J. and Lubben, S. (1990) Synthese sterisch einheitlicher Selektoren zur dünnschichtchromatographischen Racematspaltung nach dem Ligandenaustauschprinzip. *Tetrahedron* **46**, 1231–1234.

168. Frank, W. and Günther, K., (Prof. Dr. W. Frank, Institut für Anorganische Chemie und Strukturchemie der Universität Düsseldorf, BRD), in press.

169. Bhushan, R., Reddy, G. P., and Joshi, S. (1994) TLC resolution of DL amino acids on impregnated silica gel plates. *J. Planar Chromatogr.* **7**, 126–128.

170. Günther, K. and Rausch, R. (1985) Thin-layer chromatographic enantiomeric resolution based on ligand exchange. Proc. 3rd Int. Symp. Instrumental HPTLC, Würzburg, FRG; *Inst. Chromatogr.* Bad Dürkheim, pp. 469–475.

171. Günther, K. (1988) Thin-layer chromatographic enantiomeric resolution via ligand exchange. *J. Chromatogr.* **448**, 11–30.

172. Brinkman, U. A. T. and Kamminga, D. (1985) Rapid separation of enantiomers by thin-layer chromatography on a chiral stationary phase. *J. Chromatogr.* **330**, 375–378.

173. Jork, H. and Kany, E. (1986) GDCh-Fortbildungskurs, No. 301, Saarbrücken, Germany.

174. Brückner, H., Bosch, I., Graser, T., and Fürst, P. (1987) Determination of α-alkyl-α-amino acids and α-amino alkohols by chiral-phase capillary gas chromatography and reversed-phase high-performance liquid chromatography. *J. Chromatogr.* **395**, 569–590.

175. Brückner, H. (1987) Enantiomeric resolution of N-methyl-α-amino acids and α-alkyl-α-amino acids by ligand-exchange chromatography. *Chromatographia* **24**, 725–738.

176. Gont, L. K. and Neuendorf, S. K. (1987) Enantiomeric separation of N-carbamyltryptophan by thin-layer chromatography on a chiral stationary phase. *J. Chromatogr.* **391**, 343–345.

177. Kovács-Hadady, K. and Kiss, I. T. (1987) Attempts for the chromatographic separation of D- and L-penicillamine enantiomers. *Chromatographia* **24**, 677–679.

178. Feldberg, R. S. and Reppucci, L. M. (1987) Rapid separation of anomeric purine nucleosides by thin-layer chromatography on a chiral stationary phase. *J. Chromatogr.* **410**, 226–229.

179. Detolle, S., Postaire, E., Gimenez, F., Prognon, P., and Pradeau, D. (1988) Thin layer chromatography and liquid chromatography of thiorphan enantiomers and their diastereoisomers degradation products. 1st Int. Symp. Separation of Chiral Molecules, May 31-June 2, Paris, France.

180. Mack, M., Hauck, H. E., and Herbert, H. (1988) Enantiomeric separation in TLC with the new HPTLC pre-coated plate CHIR with concentrating zone. *J. Planar Chromatogr.* **1**, 304–308.

181. Mack, M., Hauck, H. E., and Jost, W. (1988) HPTLC pre-coated plate CHIR with concentrating zone. A new layer for separation of enantiomers. 17th Int. Symp. Chromatography, September 25–30, Wien, Austria.

182. Schlauch, M. (2001) Chirale Analytik von aliphatischen und cycloaliphatischen Aminosäuren mit chromatographischen und elektrophoretischen Methoden. (Thesis, Frahm, A. W.) Albert-Ludwig Universität Freiburg.

183. Bhushan, R., Mahesh, V. K., and Mallikharjun, P. V. (1989) Thin layer chromatography of peptides and proteins: a review. *Biomed. Chromatogr.* **3**, 95–104.

184. Zydowsky, T. M., de Lara, E., and Spanton, S. G. (1990) Stereoselective synthesis of α-alkyl α-amino acids. Alkylation of 3-substituted 5*H*,10b*H*-oxazolo[3,2-c][1,3]benzoxazine-2(3*H*),5-diones. *J. Org. Chem.* **55**, 5437–5439.

185. Toth, G., Lebl, M., and Hruby, V. J. (1990) Chiral thin-layer chromatographic separation of phenylalanine and tyrosine derivatives. *J. Chromatogr.* **504**, 450–455.

186. Euerby, M. R. (1990) Resolution of the neuroexcitatory N-methylaspartic acid (2-methylaminosuccinic acid) enantiomers by ligand exchange thin layer chromatography. *J. Chromatogr.* **502**, 226–229.

187. Hauck, H. E. and Mack, M. (1990) Precoated plates with concentrating zones: a convenient tool for thin-layer chromatography. *LC-GC,* **8**, 88–96.

188. Lin, S.-L., Chen, S.-T., Wu, S.-H., and Wang, K. T. (1991) Separation of aspartame and its precursor stereoisomers by chiral chromatography. *J. Chromatogr.* **540**, 392–396.

189. Marseigne, I. (1991) Direct separation of enantiomers of amino acids and dipeptides by TLC and HPLC. Second International Symposium on Chiral Discrimination, Roma.

190. Suzuki, M., Naitho, G., and Takitani, S. (1992) Determination of optical purity of L-amino acids by thin-layer densitometry. *Iyakuhin Kenkyu* **23**, 49–52.

191. Pyka, A. (1993) A new optical topological index (I_{opt}) for predicting the separation of D and L optical isomers by TLC Part III. *J. Planar Chromatogr.* **6**, 282–288.

192. Aboul-Enein, H. Y. and Serignese, V. (1994) Optical purity of thyroxine enantiomers in bulk materials by chiral thin layer chromatography. 5th International Symposium on Chiral Discrimination, Stockholm.

193. Nyiredy, Sz., Dallenbach-Toelke, K., Günther, K., and Sticher, O. (1988) Applicability of forced-flow planarchromatographic methods for the separation of chiral molecules. 1st Int. Symp. Separation of Chiral Molecules, May 31–June 2, Paris, France.

194. Nyiredy, Sz., Dallenbach-Toelke, K., and Sticher, O. (1988) Applicability of forced-flow planar chromatographic methods for the separation of enantiomers on chiralplate. *J. Chromatogr.* **450**, 241–252.

195. Günther, K., Martens, J., and Schickedanz, M. (1985) Thin-layer chromatographic enantiomeric resolution. *Naturwissenschaften* **72**, 149–150.

196. Günther, K., Martens, J., and Schickedanz, M. (1985) Resolution of optical isomers by thin-layer chromatography (TLC). Enantiomeric Purity of L-DOPA. *Z. Anal. Chem.* **332**, 513–514.

197. Günther, K., Martens, J., and Schickedanz, M. (1986) Dünnschichtchromatographische Trennung stereoisomerer Dipeptide. *Angew. Chem.* **98**, 284–285; (1986)

Thin layer chromatographic separation of stereoisomeric dipeptides. *Angew. Chem. Int. Ed. Engl.* **25,** 278–279.

198. Günther, K., Martens, J., and Schickedanz, M. (1986) Resolution of optical isomers by thin-layer chromatography: enantiomeric purity of D-penicillamine. *Arch. Pharm. (Weinheim Ger.)* **319,** 461–465.

199. Martens, J., Günther, K., and Schickedanz, M. (1986) Resolution of optical isomers by thin-layer chromatography: enantiomeric purity of methyldopa. *Arch. Pharm. (Weinheim Ger.)* **319,** 572–574.

200. Günther, K., Schickedanz, M., Drauz, K., and Martens, J. (1986) Thin-layer chromatographic enantiomeric resolution of α-alkyl amino acids. *Z. Anal. Chem.* **325,** 298–299.

201. Günther, K. (1986) Dünnschichtchromatographische Enantiomerentrennung mittels Ligandenaustausch. *GIT; Suppl.* **3,** 6–12.

202. Klaus, R. and Fischer, W. (1987) A means of analysing glycols, especially ethylene glycol and diethylene glycol, by a method used for the determination of carbohydrates in alcoholic beverages. *Chromatographia* **23,** 137–140.

203. Günther, K. and Schickedanz, M. (1987) Quantitative Auswertung von dc-getrennten Enantiomeren *GIT; Suppl.* **3,** 27–32.

204. Günther, K. and Rausch, R. (1987) Quantitative determination of TLC-separated enantiomers of amino acides. 4th Int. Symp. Instrumental TLC, Selvino/Bergamo, Italy.

205. Günther, K. (1988) Détermination quantitative des énantiomères d'acides aminés par chromatographie sur couche mince (CCM). *Analysis* **16,** 514–518.

206. Funk, W. and Heiligenthal, M. (1984) Nachbehandlung von DC-Platten durch instrumentalisiertes Tauchen. *GIT; Suppl.* **4(5),** 49–51.

207. Funk, W., Dammann, V., Vonderheid, C., and Oehlmann, G. (1985) *Statistische Methoden in der Wasseranalytik,* VCH Weinheim.

208. Barooshian, A. V., Lautenschleger, M. J., and Harris, W. G. (1972) Thin layer chromatographic separation of optical isomers of labeled DOPA via dipeptide formation. *Anal. Biochem.* **49,** 569–571.

209. Eskes, D. (1976) A procedure for the differentiation of the optical isomers of amphetamine and methamphetamine by thin-layer chromatography. *J. Chromatogr.* **117,** 442–444.

210. Kolodziejczyk, A. M. and Arendt, A. (1979) Use of tritium labelled compounds in peptide chemistry. Determination of enantiomeric purity of amino acid derivatives by the radiochromatographic method. *Polish J. Chem.* **53,** 1017–1023.

211. Jarman, M. and Stec, W. J. (1979) Formation of diastereomeric derivatives from the enantiomers of the antitumour agent cyclophosphamide by reaction with 1-phenethyl alcohol, and their separation by thin-layer chromatography. *J. Chromatogr.* **176,** 440–443.

212. Gübitz, G. and Mihellyes, S. (1984) Optical resolution of β-blocking agents by thin-layer chromatography and high-performance liquid chromatography as diastereomeric R-(-)-1-(1-naphthyl)ethylureas. *J. Chromatogr.* **314,** 462–466.

213. Weber, H., Spahn, H., Mutschler, E., and Möhrke, W. (1984) Activated α-alkyl-α-arylacetic acid enantiomers for stereoselective thin-layer chromatographic and high-performance liquid chromatographic determination of chiral amines. *J. Chromatogr.* **307,** 145–153.

214. Rossetti, V., Lombard, A., and Buffa, M. (1986) The HPTLC resolution of the enantiomers of some 2-arylpropionic acid anti-inflammatory drugs. *J. Pharm. Biomed. Anal.* **4,** 673–676.

215. Beneytout, J. L., Tixier, M., and Rigaud, M. (1986) Capillary gas-liquid or thin-layer chromatographic resolution of 2-hydroxy-fatty acid enantiomers. *J. Chromatogr.* **351,** 363–365.

216. Slégel, P., Vereczkey-Donáth, G., Ladányi, L., and Tóth-Lauritz, M. (1987) Enantiomeric separation of chiral carboxylic acids, as their diastereomeric carboxamides, by thin layer chromatography. *J. Pharm. Biomed. Anal.* **5,** 665–673.

217. Ruterbories, K. J. and Nurok, D. (1987) Thin-layer chromatographic separation of diastereomeric amino acid derivatives prepared with Marfey's reagent. *Anal. Chem.* **59,** 2735–2736.

218. Comber, R. N. and Brouillette, W. J. (1987) Resolution of carnitine an 4-methylcarnitine via the esters formed with (*S*)-(+)-methoxyphenylacetic acid. *J. Org. Chem.* **52,** 2311–2314.

219. Pflugmann, G., Spahn, H., and Mutschler, E. (1987) Rapid determination of the enantiomers of metoprolol, oxprenolol and propranolol in urine. *J. Chromatogr.* **416,** 331–339.

220. Büyüktimkin, N. and Buschauer, A. (1988) Separation and determination of some amino acid ester enantiomers by thin-layer chromatography after derivatisation with (*S*)-(+)-naproxen. *J. Chromatogr.* **450,** 281–283.

221. Spell, J. C. and Stewart, J. T. (1997) A high-performance thin-layer chromatographic assay of pindolol enantiomers by chemical derivatization. *J. Planar Chromatogr.* **10,** 222–224.

222. Nagata, Y., Iida, T., and Sakai, M. (2001) Enantiomeric resolution of amino acids by thin-layer chromatography. *J. Mol. Catal. B Enzym.* **12,** 105–108.

3

Cyclodextrin-Based Chiral Stationary Phases for Liquid Chromatography

A Twenty-Year Overview

Clifford R. Mitchell and Daniel W. Armstrong

1. Introduction

Reversed-phase chiral stationary phases (CSPs) were important early on because pharmacokinetic and pharmocodynamic studies, which were done via reversed-phase high-performance liquid chromatography (HPLC), required a solvent-compatible CSP to separate chiral analytes and metabolites. The development of stable and effective reversed-phase CSPs eventually led to the US Food and Drug Administration's 1992 guidelines regarding the development of chiral pharmaceutical products (*1*). One of the original and more versatile reversed-phase CSPs is based on cyclodextrins and their derivatives. It has been used to separate the enantiomers of over 1000 compounds, as well as numerous diastereomers, structural isomers, homologous compounds, and structurally unrelated compounds. Over 300 articles have been published in the literature on the use of cyclodextrin stationary phases, and countless analytical methods, which utilize these stationary phases, have been developed in academia and industry.

There are two general approaches to the use of cyclodextrins in liquid chromatography, chiral mobile phase additives (CMAs) and CSPs. This chapter will outline the latter; that is, the use of cyclodextrin-based bonded stationary phases to achieve enantiomeric separations. Also, it should be noted that chromatography has played an important role in understanding chiral recognition in cyclodextrin-based systems.

1.1. History and Development

Cyclodextrins were first used in 1959 for chiral separations as a selective precipitation/crystallization agent for enantiomers (*2*). This subject, and other more

From: *Methods in Molecular Biology, Vol. 243: Chiral Separations: Methods and Protocols*
Edited by: G. Gübitz and M. G. Schmid © Humana Press Inc., Totowa, NJ

classical separations with cyclodextrins have been thoroughly reviewed *(3)*. Aqueous solutions of cyclodextrins have been used as mobile phases with achiral stationary phases, occasionally with great success. The first reported successful use of cyclodextrins in chromatography as a mobile phase additive was in thin-layer chromatography (TLC) *(4–6)*. Prior to the advent of covalently bonded cyclodextrin CSPs, polymerized and crosslinked gels composed of cyclodextrins were examined, often with mixed results *(3)*. Chiral separations were occasionally achieved, but the stationary phase was not robust, and often easily overloaded.

Bonded phase cyclodextrin CSPs for analytical column chromatography were attempted by Fujimura et al. and Kawaguchi et al. almost simultaneously *(7,8)*. β-Cyclodextrin molecules were bonded to silica via ethylenediamine linkages. While marginally effective, several disadvantages of the linkage included: (i) hydrolytic instability; (ii) low cyclodextrin loading; (iii) separation selectivity affected by amine linkage; and (iv) a tedious synthesis. For these reasons, a commercial stationary phase bonded with this linkage chain never materialized.

The first high coverage stable bonded phase cyclodextrin CSP was developed by Armstrong and DeMond *(9)*. It was composed of β-cyclodextrin bonded to silica via a "6-10" unit spacer arm free of nitrogen. This CSP was commercialized by Advanced Separations Technologies in late 1983. It was the first stable bonded cyclodextrin chiral stationary phase as well as the first commercially available reversed-phase CSP. The usefulness of this CSP was immediately realized by a number of researchers. Initial research focused on the separation of chiral aromatic compounds *(9–13)*, aromatic derivatized amino acids *(9,14)*, structural isomers *(9,13,15)*, and diasteriomeric compounds *(15)*. The use of cyclodextrins in TLC and HPLC research led directly to their use in capillary electrophoresis and gas chromatography and to their rapid success and acceptance.

2. Materials

2.1. Physical and Chemical Properties of Cyclodextrin Molecules

Cyclodextrins are toroidal shaped molecules composed of α-(1,4)-linked D-(+)-gluco-pyranose units. They are produced naturally by the digestion of starch by *Bacillus macerans* or the action of the enzyme cyclodextrin transglycosylase. The α-, β- and γ-cyclodextrins (**Fig. 1**) are produced in greatest abundance and consist of six, seven, and eight glucopyranose units, respectively. Additionally, δ-cyclodextrin and branched homologs are also produced in nature. **Table 1** lists some of the physical and chemical properties of the most common cyclodextrins.

The rims of the cyclodextrin torus are lined with hydroxyl groups. One rim is lined with secondary 2- and 3-hydroxyl groups, and the other with primary

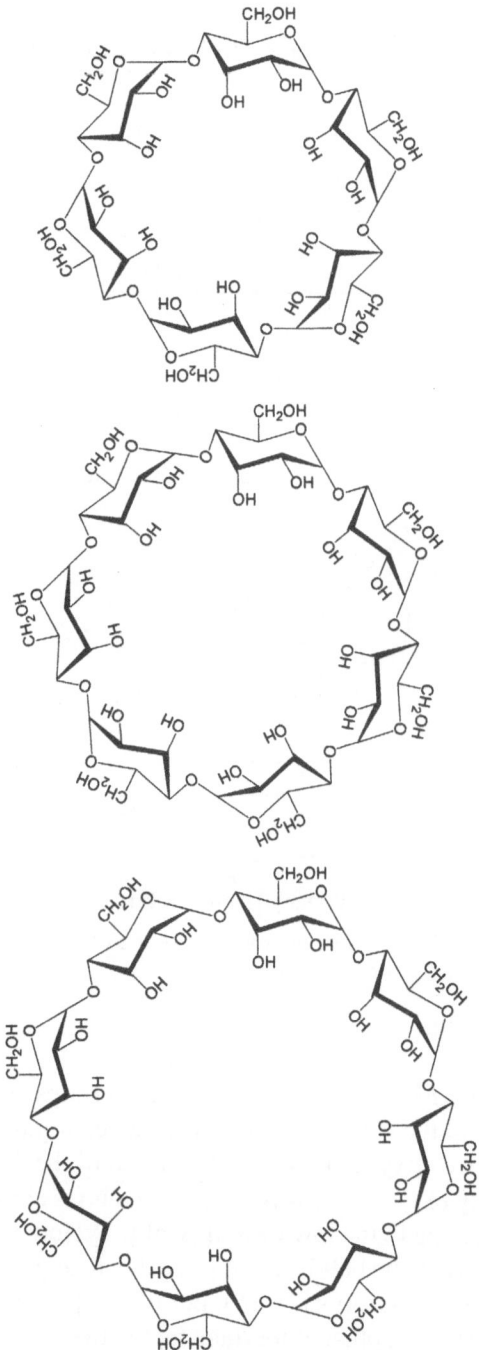

Fig. 1. Three naturally occurring cyclodextrin molecules, α, β, and γ-cyclodextrin.

Table 1
Selected Physical and Chemical Characteristics of Cyclodextrin Molecules

Cyclodextrin	No. of glucose units	Molecular mass (g/mol)	Cavity diameter (nm)	No. of stereogenic centers	Water solubility (g/100 mL)
α	6	972	0.57	30	14.5
β	7	1135	0.78	35	1.85
γ	8	1297	0.95	40	23.2

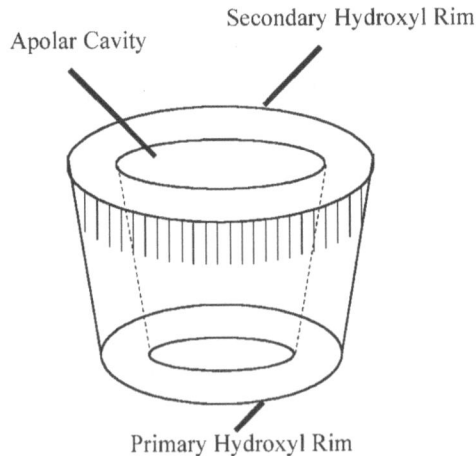

Fig. 2. Simplified depiction of the toroidal shape of a cyclodextrin molecule.

6-hydroxyl groups (**Fig. 2**). The C2 hydroxyl groups of cyclodextrins can hydrogen bond with the C3 hydroxyl on the adjacent glucose unit, creating a belt of secondary hydroxyl H-bonds at the mouth of the cavity *(16)*. This may affect the solubility of cyclodextrins in solution and is considered to be one of the factors contributing to the low solubility of β-cyclodextrin in water. On the α- and γ-cyclodextrins, the belt is incomplete due to angular strain and/or flexibility of the cyclodextrin torus *(17,18)*, but on the β-cyclodextrin, the belt of H-bonds is complete. Another important factor that affects the solubility of native cyclodextrins in water is their crystal lattice structures and their respective lattice energies.

It has been noted that many derivatized cyclodextrins (i.e., methylated, hydroxypropylated, etc.) are more water soluble than native β-cyclodextrin. The reason for this is that the derivatized cyclodextrins are rarely pure compounds. They are a mixture of related homologs and isomers, which cannot effectively crystallize *(19)*. Thus, it is the impure mixture that is more water soluble. Exceedingly pure cyclodextrin derivatives are often less soluble in water.

As all hydroxyl groups of the linked glucose units are on the rim of the cyclodextrin torus, its cavity is relatively hydrophobic in nature. Additionally, the nonbonding electrons on the glycosidic oxygen bridges in the α-(1,4) bonds are oriented toward the center of the cavity, which gives the cavity some Lewis-base character *(18)*. The hydroxyl groups are used for chemical modification of cyclodextrins. As discussed in **Subheading 2.2.**, cyclodextrins may be chemically derivatized and/or attached to silica support for use in chromatography. Cyclodextrin molecules are stable over a broad pH range. They are stable in basic solutions, but can hydrolyze in acidic solutions. The pH constraints that must be observed when using derivatized cyclodextrins and silica gel bound cyclodextrins are from pH 3.5 to 8.0. The upper pH limit is determined by the instability (dissolution) of silica at elevated pHs.

Generally, the cavity of α-cyclodextrin can accommodate a molecule the size of a six-membered aromatic ring. The cavity of the β-cyclodextrin can accommodate a naphthalene size molecule, and γ-cyclodextrin has a cavity size that can permit inclusion of three ring compounds (anthracene, phenanthrene). In this last case, often only partial inclusion of the hydrophobic aromatic group will occur if the size match is not perfect.

2.2. Commercially Available CSPs

Originally, cyclodextrin-bonded phases were proposed for both chiral separations and as an exceptionally selective reversed-phase column for separating structural isomers and geometric isomers *(9)*. It was also proposed as an alternative reversed-phase column that provided different selectivities for more routine separations *(10)*. Initially, chiral separations attracted the most attention. However, there has been a reemergence in their use as an unusual selectivity reversed-phase stationary phase.

Many types of chemistry have been used to bond cyclodextrin molecules to silica gel supports. As previously stated, amine linkages were first used to bond cyclodextrins to polyacrylamide gels *(8)* and silica gel *(7)*. When these linkages proved to be deficient, other linkages were explored (**Fig. 3**). The first successful chemistry utilized is the epoxide linkage developed by Armstrong. The linkage is free of nitrogen atoms and contains a minimal number of hydrogen bonding groups. Most commercially available cyclodextrin CSPs utilize this chemistry in bonding cyclodextrins to the silica support.

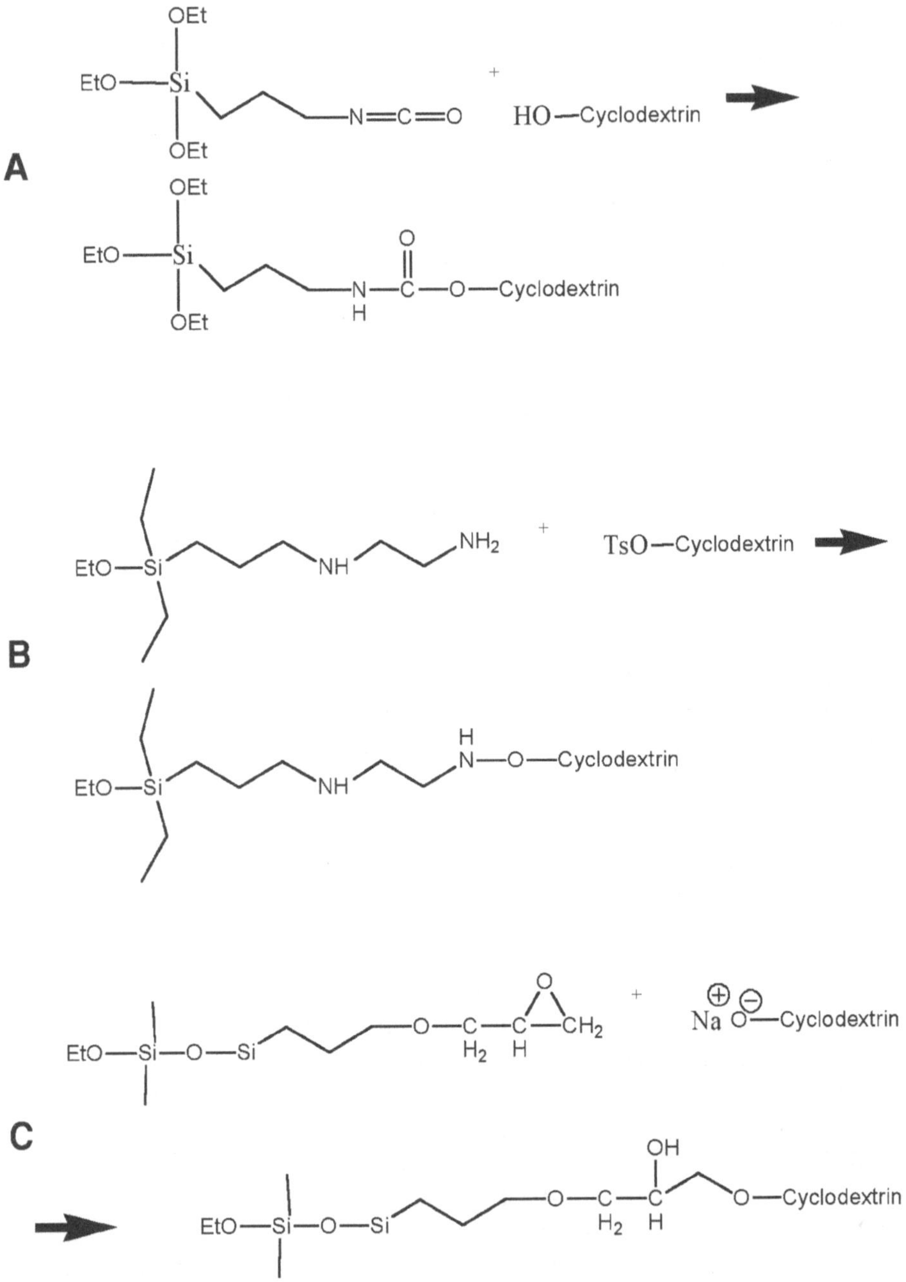

Fig. 3. The various chemistries that have been utilized to bond cyclodextrins to silica. (**A**) Isocyanate carbamate linkage. (**B**) Ethylenediamine linkage. (**C**) Epoxide ether linkage. The epoxide linkage is the most successful, as it is free of nitrogen. The other linkages have been shown to interact with analytes.

While many types of cyclodextrin CSPs are available, they may all be classified into three general categories: (i) native cyclodextrin CSPs; (ii) derivatized cyclodextrin CSPs; and (iii) aromatic derivatized cyclodextrin CSPs. **Table 2** lists all of the commercially available cyclodextrin CSPs, including the manufacturers that have produced facsimiles or analogues of the early Advanced Separation Technologies (Astec) columns.

2.2.1. Native Cyclodextrin CSPs

Native (underivatized) cyclodextrin CSPs consist of cyclodextrin molecules covalently bound to a silica gel support via a linkage chain. Easily the most popular native cyclodextrin CSP is based on β-cyclodextrin. It has been shown to be effective at resolving the enantiomers of many compounds *(9,11,12,14,15, 20–40,127)*. Subsequently, other native and derivatized cyclodextrin CSPs were developed, as will be discussed (**Table 2**).

The α- and γ-cyclodextrin CSPs, while less broadly applicable in the reversed-phase mode than the β-cyclodextrin CSP, are also useful for specific applications. The α-cyclodextrin CSP has been used to separate enantiomers of underivatized aromatic amino acids and substituted analogues *(41)*. In addition, it provides the only documented liquid chromatography (LC) chiral separation of monoterpene hydrocarbons (e.g., α-pinene, β-pinene, camphene, etc.) *(42)*. This is the only known case of a liquid chromatography separation of chiral hydrocarbon molecules. γ-Cyclodextrin CSPs have been used to separate positional isomers *(43)* and several polycyclic aromatic and bi-napthyl chiral molecules *(44)*. They also effectively separate steroid stereoisomers.

All of the native cyclodextrin CSPs are effective in the polar organic mode *(39)*. Any differences in retention, selectivity, and resolution between these CSPs in the polar organic mode arise from the difference in the size of the three different cyclodextrin molecules and how well the analyte fits the spacing and geometry of the hydroxyl groups on the cyclodextrin molecule. As will be discussed in **Subheading 3.**, the mechanism of chiral discrimination, put simply, consists of the chiral analyte interacting with the mouth of the cyclodextrin cavity and hydrogen-bonding with the secondary hydroxyls on the rim. It is intuitive that different sized cyclodextrins will accommodate analyte molecules differently *(39)*.

2.2.2. Derivatized Cyclodextrin CSPs

A few derivatized cyclodextrin CSPs are less broadly effective as chiral selectors, but excel at the separation of enantiomers of specific classes of molecules (including those not resolved on native β-cyclodextrin CSPs). Examples of this are the 2,3-dimethylated-β-cyclodextrin (Cyclobond I DM) and acetylated β-cyclodextrin (Cyclobond I AC) (**Fig. 4**). Some derivatized cyclodextrin CSPs

Table 2
Commercial Suppliers of CD CSPs[a]

Company	Trade name	Chemical description	Mode	Comments
Advanced Separation Technologies 37 Leslie Ct. P.O. Box 297 Whippany, NJ 07981 973-428-9080 astec@aol.com http://www.astecusa.com	Cyclobond I 2000	β-Cyclodextrin	RP, PO	In Europe, contact denise@asteceuro.com Cyclobond I-RSP may be the most useful RP cyclodextrin-based CSP Preparative scale stationary phases are also available for all CD CSPs.
	Cyclobond II	γ-Cyclodextrin	RP, PO	
	Cyclobond III	α-Cyclodextrin	RP, PO	
	Cyclobond AC	per-Acetylated β Cyclodextrin	RP, PO	
	Cyclobond DM	2,3-Dimethylated β Cyclodextrin	RP	
	Cyclobond RSP	R,S-Hydroxy Propyl Ether β Cyclodextrin	RP, PO	
	Cyclobond SP	S-Hydroxy Propyl Ether β Cyclodextrin	RP, PO	
	Cyclobond SN	S-Naphthyl Ethyl Carbamate β Cyclodextrin	RP, NP, PO	
	Cyclobond RN	R-Naphthyl Ethyl Carbamate β Cyclodextrin	RP, NP, PO	
	Cyclobond DMP	Dimethyl Phenyl Carbamate β Cyclodextrin	RP, NP, PO	
	Astec apHera	Carboxymethyl-β-Cyclodextrins bonded to a polymethacrylate support	RP	Can be used at basic pH to 14.0.
Macherey-Nagel P.O. Box 101352 D-52313 Duren Germany 49-2421-969-0 question@macherey-nagel.de	Nucleodex (1 type)	Native β-CD	RP, PO	
	Nucleodex PM (3 types)	Per-Methylated CDs (α, β, γ)	RP	
Merck P.O. Box 64271 Frankurter Str. 250 64293 Darmstadt Germany 49-6151-72-0 service@merck.de	ChiraDex, ChiraDex γ	Native β and γ CDs	RP, PO	

Showa Denko 5-1, Ougimach, Kawasaki-ku Kawasaki-city Kanagawa 210-0867 Japan 81-44-329-0733 shodex@hq.sdk.co.jp http://www.sdk.co.jp/ shodex/english/contents.htm	ORpak CDBS-453	Native β CD bonded to silica	RP	
	Orpak 453-HQ cyclodextrin series (3 types)	Native α, β, and γ CDs bonded to Polyhydroxymethacrylate crosslinked gel	RP	Polymer-based gel to which cyclodextrin derivatives are bonded. 100% Aqueous Mobile phases only.
Thermo Hypersil 320 Rolling Ridge Dr. Penn Eagle Industrial Park Bellefonte, PA 16823 800-292-6088 tech@keystonescientific.com http://www.thermohypersil.com	Keystone Chiral β	Native β CD and Permethylated β CD	RP, PO	
YMC Karasuma-Gojo Bldg. 284 Daigo-cho Karasuma Nisihiiru Gojo-dori Shimogyo-ku Kyoto 600-8106 Japan 81-75-342-4567 y-ohyagi@ymc.co.jp	YMC Chiral β	Native CDs	RP	

[a]Adapted from **ref. 137**.
RP, reversed-phase; PO, polar organic; NP, normal phase.

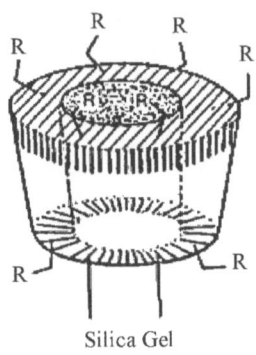

Silica Gel

—CH₃   Dimethylated

$$\overset{O}{\underset{}{\overset{\parallel}{-C}CH_3}}$$ Acetylated

$$\overset{OH}{\underset{}{-CH_2\overset{|}{\underset{*}{C}}HCH_3}}$$ Hydroxy-Propyl

Naphthyl-ethyl-carbamate

Dimethyl-phenyl-carbamate

Fig. 4. Commercially available derivatized CD CSPs. Asterisk denotes stereogenic center. *See* **Table 2** for a list of suppliers.

are more broadly useful than the native β-cyclodextrin CSP. These include the hydroxypropyl derivatized β-cyclodextrin (Cyclobond I RSP) and the naph-thyl-ethylcarbamolyated-β-cyclodextrin (Cyclobond I-RN and SN) (**Fig. 4**).

Acetylated cyclodextrin CSPs consist of cyclodextrin molecules in which the secondary hydroxyl groups in positions 2 and 3 and all available primary hydroxyl groups in position 6 are acetylated. This functionality acts as a hydrogen bond acceptor, which can interact with analyte hydroxyl or amine groups (ideally in positions α or β to the chiral center). The acetyl group can also act as a rigid π-electron system for steric interactions, and it has the effect of extending the mouth of the cyclodextrin cavity to include larger molecules. This is especially beneficial when the chiral analyte, upon binding to the cyclodextrin, protrudes from the cavity and the chiral center is not in close proximity to the rim of the cyclodextrin. For these molecules, the acetylated cyclodextrin CSP will have enhanced enantioselectivity compared to the native β-cyclodextrin CSP. Also, they tend to retain molecules less than native β-cyclodextrin when used with similar mobile phases. Additionally, this stationary phase is useful in the polar organic mode for chiral molecules that have at least two functional groups about the stereogenic center capable of participating in hydrogen bonding *(45)*. This CSP has been shown to be effective at resolving the enantiomers of alkaloids (scopolamine, homoatropine) *(45,46)* and pesticides (trihexylphenidyl, rulene) *(45)*.

The dimethylated cyclodextrin CSP consists of a cyclodextrin molecule in which the secondary 2- and 3-hydroxyls (located on the outer rim) have been derivatized with methyl groups. This eliminates the ability of the cyclodextrin molecule to hydrogen bond with other molecules. The lack of secondary hydroxyls precludes the efficacy of this CSP in the polar organic mode *(47)*. However, it has been shown to be a useful CSP for certain specific classes of compounds (usually neutral) in the reversed-phase mode, such as diaryl and aliphatic chiral sulfoxides *(48)*, substituted furo-coumarins *(49)*, coumachlor, coumafuryl, idazoxan, and warfarin. Steric interactions between the cyclodextrin molecule and chiral analytes are much more important in chiral recognition with this CSP (**Fig. 5**). Comparison of separations on both the 2,3-dimethylated cyclodextrin CSP and the native β-cyclodextrin CSP helped to elucidate the role of hydrogen bonding and steric interactions in chiral recognition. Often the dimethylated cyclodextrin CSPs will exhibit greater selectivity than the native β-cyclodextrin CSP for large neutral analytes with multiple ring structures *(50)*.

Perhaps the most broadly applicable of all cyclodextrin CSPs in the reversed-phase mode is the hydroxypropyl-β-cyclodextrin CSP (Cyclobond I RSP). Approximately seven hydroxyl groups on the β-cyclodextrin molecule are derivatized with hydroxypropyl functionalities. Since the hydroxypropyl moiety is chiral, each group adds an additional stereogenic center to the derivatized cyclodextrin molecule. These CSPs are available derivatized either with racemic or *(S)*-propylene oxide. In all but a few cases, the Cyclobond I RSP has the same selectivity as the Cyclobond I-SP. The broad applicability of this CSP is primarily owing to the nature of the derivative. The effective size of the cavity is extended by the hydroxypropyl group, while the remaining hydroxyl groups retain the ability to hydrogen bond with analytes. The pendant hydroxypropyl

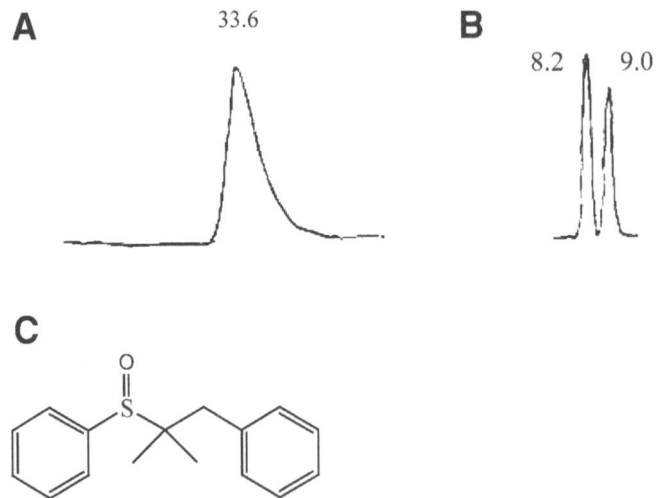

Fig. 5. Separation of a chiral sulfoxide on (**A**) native β-cyclodextrin and (**B**) dimethy-lated β-cyclodextrin. Retention times are given in minutes on each chromatogram. (**C**) Structure of sulfoxide. Steric interactions are more important for separations using dimethy-lated cyclodextrins. 40/60 methanol/water. UV/V is detection at 254 nm, 1.0 mL/min.

group allows for interactions with portions of larger molecules that project out of the cyclodextrin cavity when an inclusion complex is formed. The length and flexibility of the hydroxypropyl group allows for both hydrogen bonding and steric interactions to groups β and γ to the chiral center *(51)*. Additionally, this stationary phase has been shown to be applicable for the separation of non-aromatic racemates such as *N*-tert-buoxycarbonyl (t-BOC)-derivatized amino acids *(52)*. This CSP effectively separates enantiomers in both the reversed-phase and polar organic modes of operation, although it is much more effec-tive in the reversed-phase mode.

2.2.3. Aromatic-Derivatized Cyclodextrin CSPs

Cyclodextrins derivatized with aromatic moieties are the only effective cyclo-dextrin-based CSP in the normal phase mode *(53–55)*. They are also useful in the reversed-phase and polar organic modes *(54,56,57)*. Again, the only manu-facturer of these CSPs is Advanced Separations Technologies (Whippany, NJ, USA) (**Table 1**). The commercially available aromatic-derivatized cyclodex-trin CSPs consist of an aromatic group coupled to the secondary hydroxyl on the "mouth" of the cyclodextrin cavity by a carbamate linkage. Two types of CSP are available: dimethylphenylcarbamate-β-cyclodextrin, and naphthylethylcarbamate-β-cyclodextrin (**Fig. 4**). In the later CSP, the ethyl group, α to the naphthyl group, contains a stereogenic center. Columns are available with both the *R* and *S* configurations. Typically, there are four to six aromatic substituents per cyclodextrin molecule.

In the reversed-phase mode of operation, these CSPs are more retentive (i.e., hydrophobic) than the other categories of cyclodextrin CSPs. Generally, an additional 15–20% of organic modifier is required to elute a solute from an aromatic derivatized cyclodextrin CSP in the same amount of time compared to native or nonaromatic-derivatized cyclodextrin CSPs. The *R*- and *S*-naphthyl-ethylcarbamate cyclodextrin CSPs have proven to be more broadly applicable than most other derivatized cyclodextrin CSPs. These are able to separate the enantiomers of many classes of compounds including pesticides (fonofos, crufomate, ancymidol, and coumachlor) *(56,58)*, pharmacologically active compounds (tropicamide, indapamide, althiazide, and tolperisone) *(56)*, nonsteroidal anti-inflamitories (ibuprofen, flurbiprofen) *(54)*, various benzodiazepine (anti-anxiety agents and sedatives, lorazepam, oxazepam, temazepam) *(54)*, and several derivatized amino acids *(59)*.

As mentioned previously, these are the only cyclodextrin CSPs effective in normal phase operation. Both the naphthylethylcarbamate and dimethylphenylcarbamate derivatized CSP are π-electron donating (π-basic) in nature. Therefore, analytes that have π-electron accepting groups (π-acidic) are ideal candidates for successful chiral separations on these CSPs. If an analyte can be made to be π-acidic by chemical derivatization (with reagents such as 3,5-dinitrobenzylamine for acidic compounds [upon conversion of the acidic compound to an acid chloride] or 3,5-dinitrobenzoyl chloride for alcohols and amines), the likelihood of a separation of enantiomers on these CSPs is virtually assured *(55,60)*.

It has been demonstrated by Berthod et al. *(60)* that the potential exists to create a database of separation factors (α) for separations on the naphthylethylcarbamate-β-cyclodextrin CSPs in the normal phase mode of operation. Using data gathered from 121 chiral separations on these CSPs, both *R* and *S* configurations, they were able to estimate the separation potential for over 1.6 million chiral compounds. The separation factor of over 50 compounds in a variety of mobile phases was verified with an almost uncanny accuracy. While this work has been carried out on naphthylethylcarbamate-β-cyclodextrin CSPs, the potential to create this type of database with predictive capabilities exists for any CSP.

2.2.4. Carboxymethyl-Derivatized Cyclodextrin on a Polymer Support

Recently, carboxymethyl-derivatized β-cyclodextrin covalently bonded to a polymethylacrylate support was introduced by Advanced Separation Technologies as the Astec ApHera CSP. This CSP is stable at alkaline conditions (to pH 14.0). Furthermore, the carboxymethyl group can ionize at pHs greater than 6.0, thereby producing a anionic CSP. Consequently, electrostatic interactions can be used for chiral recognition on this CSP. Indeed, this stationary phase is particularly effective at separating amine-containing compounds *(61)*. It is available in α-, β-, and γ-cyclodextrin formats. Thus far, only reversed-phase enantiomeric separations have been reported. Its efficacy in the polar organic mode is unknown. Methods development on this CSP is similar to that for silica-based cyclodextrin stationary phases, except that there is no alkaline pH restriction for the mobile phases used.

3. Modes of Use, Chiral Recognition, and Method Development on Cyclodextrin CSPs

Cyclodextrin-based stationary phases are considered to be multimodal stationary phases. They have the ability to operate in three different modes of analytical column chromatography: reversed-phase, normal phase, and polar organic modes. When operating in these modes, these CSPs display the qualities that are essential *(54,62)* for a multimodal stationary phase, namely: (i) the stationary phase must be stable in all solvents used; (ii) any changes in the stationary phase (conformation of selector) must be reversible when changing between chromatographic modes; (iii) there must be significant selectivity and mechanistic differences in each mode; and (iv) there should be logical or systematic approaches for selecting a particular mode of separation and optimization *(54)*. Some limitations are imposed on the efficacy of cyclodextrin CSPs in the normal phase mode, as will be discussed later. As mentioned previously, cyclodextrin CSPs are also effective at achiral separations. It has been demonstrated in the literature that many structural isomers, homologs, regio-isomers, and compounds that are structurally unrelated may be separated from each other *(10,63–83)*.

The normal rules of chiral recognition apply to cyclodextrin CSPs, as they do with any CSP. A minimum of three different points of interaction about the stereogenic center or axis are required for chiral recognition to occur. There must be a small, but sufficient, difference in free energy of transfer between the mobile phase and stationary phase for the two enantiomers. Lastly, not all interactions between the stationary phase and the analyte contribute to chiral recognition *(84)*. Many interactions are not enantioselective and will contribute to chromatographic retention, but not to the separation of enantiomers. Often the retention of a molecule is only partially due to interaction with the chiral selector. For example, there are at least three general sites of interaction that do not lead to a separation of enantiomers: (i) nonenantioselective sites on the chiral selector; (ii) sites on the chain linking the selector to the silica; and (iii) sites on the silica support. Increasing the percentage of time a solute is in the

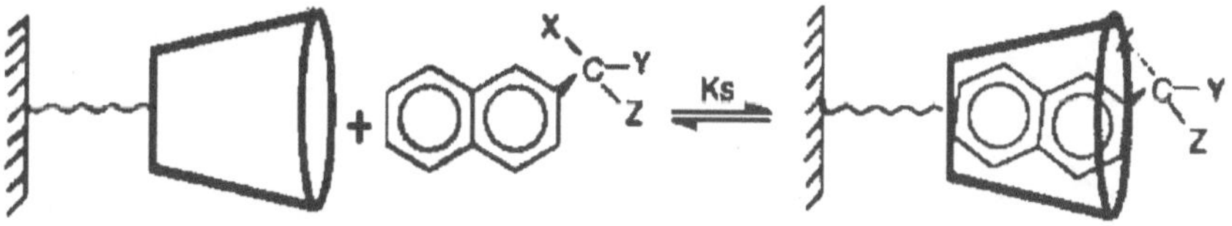

Fig. 6. Formation of an inclusion complex. From **ref.** *136*.

stationary phase may or may not enhance enantioselectivity. Furthermore, it is possible to have two different dominant interactions leading to opposite enantioselectivities that eliminate the observed separation. It has been shown that cyclodextrin CSPs derivatized with single enantiomers can be more successful at separating a given set of enantiomers, compared to a cyclodextrin CSP derivatized with a racemate *(51)*.

There are three modes of chromatographic operation in which cyclodextrin CSPs are effective, and each has a distinct mechanism of chiral recognition. In the reversed-phase mode, inclusion complexation is the primary interaction that leads to chiral recognition. The three points of interaction are completed by hydrogen bonding and steric repulsion interactions. In the polar organic mode of operation, hydrogen bonding, dipolar interactions, and steric repulsion give rise to chiral recognition. In the normal phase mode, chiral recognition is achieved via a combination of π-π complexation, hydrogen bonding, dipole-dipole stacking, and steric effects.

3.1. Reversed-Phase Mode

In reversed-phase chromatography using aqueous or hydro-organic mobile phases, the most hydrophobic portion of the molecule usually occupies the hydrophobic cavity of the cyclodextrin. This association has been termed an inclusion complex (**Fig. 6**). Often the terms "tight" and "loose" are used to describe an inclusion complex. These terms do not necessarily refer to the magnitude of the binding equilibrium between the cyclodextrin and the analyte, but rather to the fit of the analyte in the cyclodextrin cavity. For a "loose" inclusion complex, the analyte will have the ability to move and rotate in the cyclodextrin cavity, thus diminishing the possibilities for chiral recognition. A "tight" inclusion complex will impart a selective orientation to the included group with much less freedom for rotation or reorientation in the cyclodextrin cavity. With a few exceptions *(42,52)*, the presence of at least one aromatic group is beneficial for a chiral separation in the reversed-phase mode.

Chiral separations in the reversed-phase mode are achieved by the formation of enantioselective inclusion complexes. To initiate an inclusion complex, an analyte molecule in the vicinity of the chiral selector may approach the cyclodex-

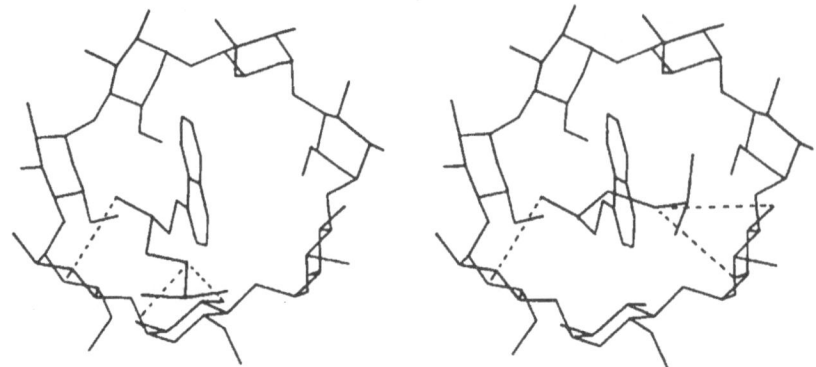

Fig. 7. Inclusion complex between β-cyclodextrin and *R*-propranolol and *S*-propra-
nolol. The aromatic portion of the molecule is included into the cyclodextrin cavity,
while hydrogen bonding is occurring to substituents near the chiral center. While the
sets of interactions are the same, the strength of hydrogen bonding is unequal for the
two enantiomers. The enantiomer that interacts more strongly with cyclodextrin will
be retained longer. From **ref. *12***.

trin molecule with any orientation. A molecule must first "dock" with the cyclodextrin with at least one point of attachment (hydrogen bonding, dipolar interaction, hydrophobic interaction, etc.). An inclusion complex is formed when a
hydrophobic portion of the analyte occupies the nonpolar cavity of the cyclodextrin. Finally, a reorientation, which produces an enantioselective inclusion
complex, must occur. During any of these steps, the analyte can dissociate from
the cyclodextrin molecule and partition into the mobile phase. For an analyte
molecule that is complexed within the cyclodextrin, both hydrogen bonding
and steric interaction can occur between the secondary hydroxyls and substituents about the stereogenic center, providing the necessary three (or more) points
of interaction that lead to chiral discrimination (**Fig. 7**). Alternatively, the secondary hydroxyl groups may be derivatized with a group that is capable of additional interactions (steric repulsion, H-bonding, dipolar, π-π complexation, etc.).

An affinity for the hydrophobic cavity will increase chromatographic retention and may or may not increase the selectivity of the inclusion complex.
Since the interior of the cyclodextrin is mildly π-basic, π-acidic groups will
have an enhanced affinity for the cavity (compared to an unmodified aromatic
system) (**Fig. 1**). Analytes with electron withdrawing groups such as nitro, sul-
fonate, phosphate, and the halogens will cause an aromatic system to become
π-acidic. These systems usually form even stronger inclusion complexes with
cyclodextrins. This phenomenon has been studied by many researchers (*3,13,85*).

Some essential characteristics for a molecule to be separated in the reversed-phase mode include the presence of a hydrophobic group (such as an aromatic ring) for the formation of an inclusion complex *(9,11,12,14,15)*. Furthermore, the aromatic group must be up a good size match to the cyclodextrin cavity. If the aromatic group is larger than the cyclodextrin cavity, it will simply not fit; if the group is much smaller than the cyclodextrin cavity, it will form a loose (conformationally mobile) inclusion complex. A second necessary characteristic is that, upon complexation, the stereogenic center, or a substituent on the stereogenic center, be located near one of the secondary hydroxyls at the mouth of the cyclodextrin cavity. Alternatively, the stereogenic center or substituent may be in close proximity to the added functional group of a derivatized cyclodextrin molecule. Interaction with the cyclodextrins hydroxyl or derivative groups is usually the only way to obtain the additional simultaneous interactions necessary for chiral recognition. It has been noted in the literature that the following structural features are beneficial to enantiomeric separations in the reversed-phase mode on cyclodextrin-based CSPs: (i) the presence of an aromatic ring system, α or β to the chiral center, occasionally in the γ position; (ii) the presence of at least one hydrogen bonding group near the chiral center; and (iii) a second π-system somewhere on the molecule *(26)*.

3.2. Polar Organic Mode

In the polar organic mode, association between analytes and cyclodextrins are achieved via a combination of hydrogen bonding and dipolar interactions, usually at the mouth of the cyclodextrin. Nonaqueous polar mobile phases are utilized, and the cyclodextrin cavity is occupied mainly by the organic solvent molecules. This effectively makes the stationary phase more polar, as analytes can only interact with the polar external surfaces (e.g., hydroxyl groups) of the cyclodextrin torus as opposed to the nonpolar cavity. Chiral recognition occurs at the "mouth" of the cavity; the analyte may form a "lid" over the larger opening of the cyclodextrin (**Fig. 8**). The primary interactions that lead to chiral recognition are hydrogen bonding between the secondary hydroxyl groups on the rim of the cyclodextrin, and the chiral analyte, dipole-dipole interactions, and steric repulsive interactions. Consequently, chiral analyte molecules must have certain structural features if they are to separate in the polar organic mode *(45)*. Specifically, an analyte must have a minimum of two separate hydrogen bonding groups. One of the hydrogen bonding groups should be on or near the stereogenic center. The other hydrogen bonding group can be anywhere in the molecule. Also, it is beneficial if the analyte has a bulky moiety (e.g., an aromatic group, branched hydrocarbon, etc.) near the stereogenic center *(45)*. The same basic rules apply to derivatized cyclodextrin CSPs that are used in the polar organic mode. However, an interaction or interactions with the derivative moiety

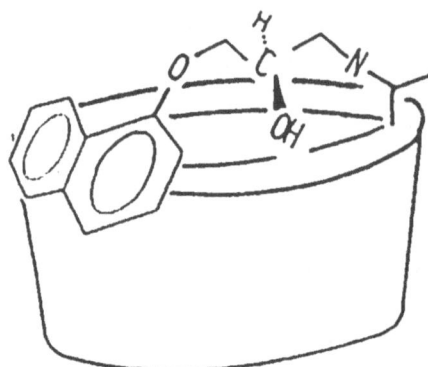

Fig. 8. Interaction between β-cyclodextrin and propranolol in the polar organic mode. Polar organic solvents occupy the cyclodextrin cavity, while the chrial analyte interacts the hydroxyl groups on the mouth of the activity.

can substitute for one or two of the interactions with the hydroxyl groups at the mouth of the cyclodextrin cavity.

It is essential to use a very high percentage of acetonitrile as the mobile phase in the polar organic mode. Acetonitrile acts as a polar but poor hydrogen-bonding solvent. In the presence of acetonitrile, hydrogen-bonding interactions between the cyclodextrin and the analyte's hydrogen-bonding groups (e.g., hydroxyl, amine, carboxyl, etc.) will be accentuated. If retention is too great in acetonitrile, methanol is added to decrease retention. Since methanol competes with the analyte for hydrogen-bonding sites on the cyclodextrin, it decreases retention, but also decreases enantioselectivity. Triethylamine and acetic acid are also used in small amounts (approx 0.1%) to modulate the degree of protonation and charge of the analyte.

The polar organic mode of operation was conceived for the separation of more polar molecules *(37,40,45)*. It can be applied equally well to chiral and achiral separations. Very polar molecules, such as amines, carboxylic acids, and their hydrochloride salts, are often poorly soluble in nonpolar organic solvents like hexane. In a mobile phase comprised of mostly acetonitrile, with small amounts of other additives, they are more soluble, but will still preferentially partition from the mobile phase to the stationary phase. The amount of additives (methanol, triethylamine, and glacial acetic acid) controls both the partition coefficient of the analyte and the selectivity of the CSP (**Fig. 9**).

3.3. Normal Phase Mode

The normal phase mode of operation is similar to the polar organic mode of operation in that hydrophobic inclusion complexation does not occur. Nonpolar

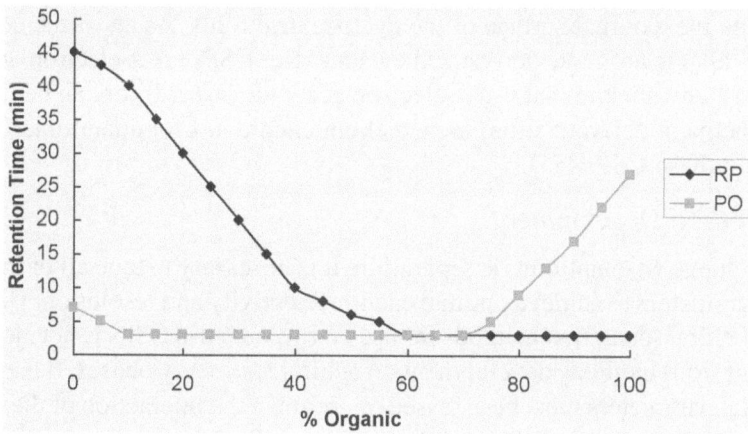

Fig. 9. Theoretical retention curves for molecules that display retention in reversed-phase and polar organic modes of operation.

organic solvents (hexane, heptane), modified by polar alcohols and ethers, are used as the mobile phase. The nonpolar solvent molecules occupy the cyclodextrin cavity. Consequently, analytes may only interact with the polar external surface of the cyclodextrin torus and any derivatizing groups that may be present. Only aromatic-derivatized CSPs have been shown to be effective in the normal phase mode. These CSPs can function as π-complex/hydrogen bonding stationary phases.

There are many types of interactions that, when combined, can give rise to a separation of enantiomers, including hydrogen bonding, steric repulsion, and π-π complexation, and dipole-dipole stacking *(53)*. These interactions occur between the analyte and the external portions of the cyclodextrin molecule (the residual cyclodextrin hydroxyl groups, aromatic moieties, and/or the linkage arm). However, it seems that for chiral recognition to occur, it is essential that the analyte contain an aromatic group. A survey of the literature does not produce any examples of enantiomeric separations of nonaromatic chiral molecules in the normal phase mode. Any observed separation must therefore occur because some π-π interactions are important. The only site available for this type of interaction in this separation mode is on the aromatic-derivatized cyclodextrins functional group. Certainly π-π interactions between the analyte and the CSP are beneficial *(53–55)*.

The interactions leading to chiral discrimination include π-electron acceptor (π-acid) and π-electron donator (π-base) interactions, plus any of the other types of interaction (mentioned above). Hydrogen bonding may occur with rim hydroxyl groups or the carbamate linkage chain. Steric repulsions may occur involving the external surface of the cyclodextrin torus. As all of the commercially available aromatic derivatized cyclodextrin CSPs are π-electron donating (π-basic), any analyte that is a π-electron acceptor (π-acid) (or can be made so upon chemical derivatization) are excellent candidates for enantiomeric separation on these CSPs (55).

3.4. Method Development

To achieve an enantiomeric separation, it is necessary to tune all of the operation parameters to achieve optimal enantioselectivity and resolution. Development of chiral separation methods for use on cyclodextrin CSPs is not altogether different from methods development on achiral stationary phases. Several fundamental parameters must be assessed to optimize the interaction of the analyte and the stationary phase. They include: (i) mobile phase composition, including the type and amount of organic modifier, pH, and buffer type; (ii) flow rate; and (iii) column temperature. When the effects of these parameters are understood, it is not difficult to determine if a separation of enantiomers can be achieved (see **Note 1**).

3.4.1. Mobile Phase Composition

As with achiral chromatography, the composition and nature of the mobile phase components has the most dramatic effect on chiral separations. The composition of the mobile phase must be tuned so that the analyte has adequate retention (retention factor, $k \approx 2\text{--}9$) and is well resolved from any other species present in the sample. It is necessary to screen different modifiers and solvent combinations to assess which will yield the optimum enantioselectivity and resolution.

3.4.1.1. REVERSED-PHASE MODE

In reversed-phase chromatography, the most commonly used organic modifiers are methanol and acetonitrile; less commonly used are tetrahydrofuran, isopropanol, and ethanol. Changes in enantioselectivity and efficiency can occur upon changing from one organic modifier to another. There are many reports in the literature showing that different selectivities and resolutions are observed when using different modifiers (14,25,30,47,52,66,86). It is difficult to predict which modifier will produce the best results. However, methanol and acetonitrile are the optimum organic modifiers in over 90% of all reported separations. Changing from a solvent that can accept and donate hydrogen bonds (methanol) to one that cannot bond hydrogen (acetonitrile, tetrahydrofuran) will often have an effect on the observed separation (whether it is an improvement or not is application specific). Ethanol and isopropanol have greater affinity for the cyclodextrin cavity and will displace solutes to a greater degree than methanol

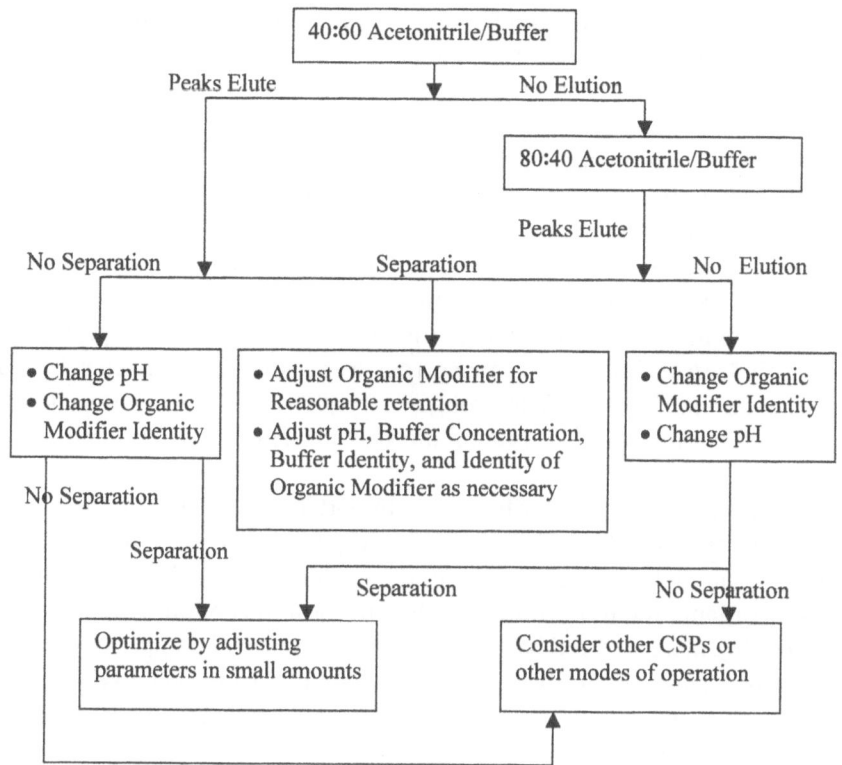

Fig. 10. Suggested method development procedure for reversed-phase chromatography on cyclodextrin-based CSPs.

or acetonitrile. Also note that acetonitrile tends to be a stronger eluent than methanol. Hence, methanol containing mobile phases usually contain approx 10–20% more organic modifier than comparable acetonitrile containing mobile phases. It is also possible to perform gradient elution on these CSPs (*see* **Note 2**). **Figure 10** is a flow chart for method development on cyclodextrin-based CSPs in the reversed-phase mode (*see* **Note 3**).

3.4.1.1.1. Buffer and pH Effects. Considerations regarding operational pH and buffers are only relevant in the reversed-phase mode. The choice of buffer and operating pH is important for several reasons. First and foremost, the polarity and charge of acidic and basic analytes are affected by pH for obvious reasons, which will in turn affect the solutes' interaction with the CSP. Clearly analytes that are ionized will interact with the CSP differently than their neutral conjugates. In some cases, charged analytes may have a higher binding constant to the cyclodextrin molecule. This is the case with the p-nitrophenolate anion *(85)*. In this instance, the ion-cyclodextrin complex is more stable than the complex with the corresponding neutral phenol. However it is more common that neutral analytes partition more strongly to the hydrophobic cyclodextrin cavity (in the reversed-phase mode) than their charged counterparts *(75)*. Ionic species are more hydrophilic and more highly solvated. Thus, they usually spend a greater percentage of time in the mobile phase (relative to their neutral analogues). Also the solvation shell may interfere with or inhibit chiral recognition.

The nature of the buffer is also important. Occasionally, different enantioselectivities and resolutions are observed upon changing the buffer. Using triethylamine-acetate can increase chromatographic efficiency by screening the analyte from the acidic sites on the silica support. This buffer is preferred for most cyclodextrin CSPs, as it is noncorrosive and enhances efficiency by masking available silanols and other strong adsorption sites. Other preferred buffers include ammonium nitrate, ammonium acetate, citrate buffers, and triethylammonium phosphate.

The concentration of the buffer is also important. Triethylamine, acetate, formate, and citrate buffers are composed of organic molecules that are capable of forming an inclusion complex with the cyclodextrin cavity. Thus, they can act as an organic modifier, in that high concentrations often reduce retention. Additionally, most acid and base modifiers are capable of competing for hydrogen-bonding sites on the cyclodextrin molecule. At high enough concentrations (>1.5%), enantioselectivity can be altered, diminished, or eliminated due to these effects *(36)*.

Cyclodextrin CSPs are stable at pHs ranging 4.0–7.0. At more basic pHs, the silica support can degrade. At more acidic pHs (especially <3.0), the cyclodextrin molecules can begin to hydrolyze. To assess the effect of pH on a chiral separation, examine the selectivity and resolution at 0.5 pH units above and below the analyte's ionization pK. If the pK is not known, examine the separation at pHs of 4.0 and 7.0 to reveal the effect of pH (*see* **Note 4**).

3.4.1.2. Polar-Organic Mode

In the polar organic mode, the typical starting mobile phase composition is 95:5:0.3:0.2 (by vol) acetonitrile/methanol/acetic acid/triethylamine. Acetonitrile acts as a polar, but poor hydrogen-bonding, solvent. Methanol will compete with the analyte for hydrogen-bonding sites on the cyclodextrin, displacing the analyte, and decreasing retention. If the solutes elute immediately, decreasing the amount of methanol and/or the acid and base (while maintaining the same molar ratio of acid to base) will cause an increase in retention. **Figure 11** is a flow chart for method development in the polar organic mode (*see* **Note 3**).

Separations in the polar organic mode may be tuned by changing the relative amounts of the acid/base modifiers. The degree of protonation and charge of the analyte (where applicable) is controlled through the relative amounts of

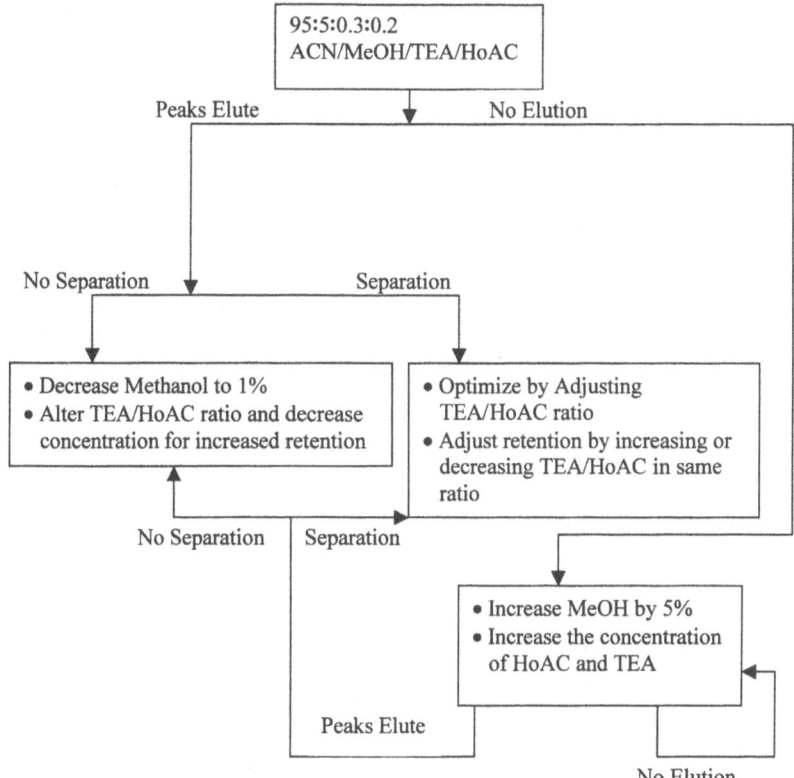

Fig. 11. Suggested method development procedure for polar organic mode on cyclo-dextrin-based CSPs.

glacial acetic acid and triethylamine. Optimizing the ratio of amine to acid modifier is essential in optimizing enantioselectivity *(37,45)*. To determine the optimal ratio of acid to base modifier, vary the concentration of each modifier while keeping the total vol of modifier constant. Start at 0.3% (3 mL) of tri-ethylamine in 1 L of 95:5 acetonitrile/methanol and assess enantioselectivity and resolution (*see* **Note 5**). Decrease the triethylamine to 0.25% (2.5 mL), add 0.05% (0.5 mL) of glacial acetic acid, and repeat the analysis. Continue this process until the triethylamine content is at 0.00% and glacial acetic acid is at 0.30%. One should observe a curve similar to **Fig. 12,** in which a maxi-mum in resolution is found at the optimal acid to base ratio.

3.4.1.3. NORMAL PHASE MODE

In the normal phase mode, the most common mobile phase is hexane/iso-propanol. Heptane may be used interchangeably with hexane with little change

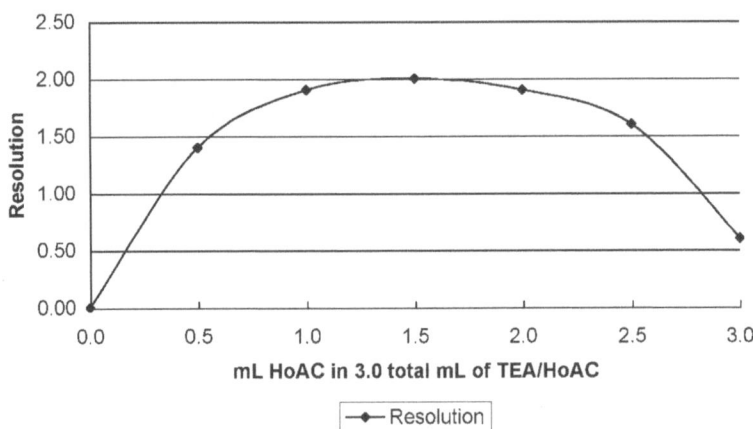

Fig. 12. Optimization of acetic acid:triethylamine ration. A maximum in resolution will occur at the optimal ratio. Total combined vol of TEA and acetic acid are constant, 3.0 mL, and the ratio is varied. Adapted from **ref. 37**.

in the separation. Acetone, ethanol, butanol, methyl t-butyl ether, and dichloromethane may be used as polar modifiers. If initial screenings with hexane/isopropanol do not produce observable enantioselectivity, change the identity of the strong eluent. Consider varying the properties of the strong eluent from a solvent that can accept and donate hydrogen bonds (i.e., the alcohols: isopropanol, ethanol, methanol, butanol) to one that can only accept hydrogen bonds (ethyl acetate, acetone) or participate in dipolar interactions (dichloromethane, chloroform). As with the reversed-phase mode, it is difficult to predict which modifier will produce the best results. **Figure 13** is a flow chart for method development in the normal phase mode (*see* **Note 3**). Recall that the only cyclodextrin-based CSPs effective in the normal phase mode (for separating enantiomers) are the aromatic-derivatized CSPs, dimethylphenylcarbamate-β-cyclodextrin and *R*- (or *S*-)naphthylethylcarbamate-β-cyclodextrin (*see* **Note 6**). However, native cyclodextrin CSPs can be used in the normal phase mode for achiral separations. In this mode, they have polarities similar to diol-type-bonded stationary phases.

3.4.2. Temperature Effects

While it is occasionally beneficial to operate at higher temperatures in achiral chromatography, the opposite trends are seen on cyclodextrin CSPs. Lowering the temperature of the column will lead to increased retention and usually increased enantiomeric resolution. At high temperatures, complexation between cyclodextrins and analytes will diminish *(3)*. This will cause an increase in the relative amount of time an analyte spends in the mobile phase versus the sta-

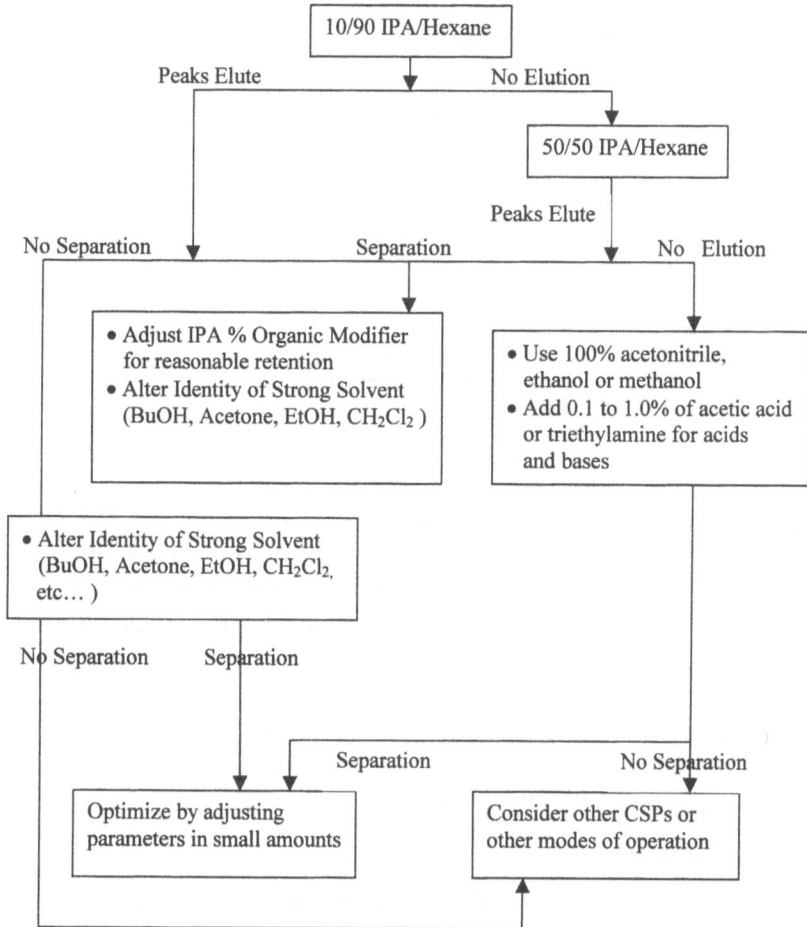

Fig. 13. Suggested method development for normal phase chromatography on aromatic-derivatized cyclodextrin-based CSPs.

tionary phase. Practically speaking, temperature is not a parameter that is often optimized in chiral separations. Generally it is only used as a last resort to improve a separation. If necessary, it is suggested that evaluating the enantioselectivity (α) and resolution (Rs) at several temperatures, bracketing standard room temperature (10°, 15°, 20°, 25°, 30°C), will reveal the effects of temperature on a particular separation (*see* **Note 7**).

3.4.3. Flow Rate Effects

The mobile phase flow rate can also affect the efficiency of chiral separations. However, in this case, the same effects can be seen in both chiral and achiral

**Table 3
The Effect of Flow Rate
on Enantioselectivity and Resolution[a]**

Flow Rate (mL/min)	α	Rs
1.00	1.20	0.85
0.80	1.24	0.92
0.60	1.24	0.97
0.40	1.25	1.12
0.30	1.26	1.22
0.20	1.24	1.20

[a]Data from **ref. 3**.
α, enantioselectivity; Rs, resolution.

separations. Most typically, analytical column chromatography on cyclodextrin CSPs is performed at flow rates of 0.5 to 1.5 mL/min. These flow rates are often selected to strike a balance between high sample throughput and reasonable column backpressure. At these flow rates, the plate height (H) is dominated by stationary phase mass transfer. It is, therefore, possible to achieve higher efficiency (N), and improved resolution (Rs), by decreasing the flow rate. **Table 3** gives an actual experimental example from Hinze et al. *(14)*. Notice that while resolution improves as the flow rate is decreased, enantioselectivity remains essentially constant (*see* **Note 8**).

3.4.4. Pressure Effects

It is also noteworthy that column backpressure can affect the chromatographic figures of merit (retention, selectivity, efficiency, and resolution) *(88–90)*. These are often modest effects at best, and they have been studied almost exclusively with cyclodextrin-based separations. The fundamental reason that pressure alters the chromatography is related to the partition equilibria between the mobile phase and stationary phase. Pressure can alter the equilibria in which a molecules' solvation shell, or partial molar volume, changes upon transfer between mobile phases and stationary phases. For enantiomers that interact differently with a cyclodextrin CSP, the change in their solvation shell will also be different upon partitioning. Therefore, pressure can affect a separation (*see* **Note 9**).

3.4.5. Column Care

Proper treatment of a stationary phase will allow for a long column lifetime. Column backpressure should be restricted to <300 bar of pressure for silica-based stationary phases. pH should be restricted to moderate values between 4 and 7. More basic conditions will degrade the silica support, and more acidic conditions will hydrolyze the cyclodextrin molecules. When use of the CSP is concluded, it is necessary to remove buffers and salts from the column and HPLC system to prevent precipitation in the fluid pathways. Remove any residual buffers by flushing first with unbuffered mobile phase, then 100% water for 10 column vol. Store the column in at least 50% hydro-organic solvents (methanol or acetonitrile) for short-term storage (overnight to 2 to 3 d) (*see* **Note 10**). Most manufacturers recommend that for long-term storage, columns be stored in 100% isopropanol. Additionally, flushing with 100% isopropanol is recommended when switching between reversed-phase and normal phase modes. Isopropanol is an excellent solvent for removing residual solutes or solvents (hexane or water).

Polymerized and crosslinked stationary phases have less mechanical strength than silica gel-based supports. Adhere to the manufacturers guidelines regarding column backpressure for these stationary phases. The advantage of these phases is the wide range of pH stability, 4.0–14.0 (again acidic conditions will hydrolyze the cyclodextrin molecules) (**Table 1**). The above mentioned storage guidelines are equally applicable here.

It is good practice to monitor a column's performance by evaluating and keeping a record of the separation figures of merit (retention, selectivity, resolution, efficiency) using a test compound. An evaluation should be done immediately upon receiving the CSP from the manufacturer, to assess the performance of the column when it is in good (new) working condition. An ideal test substance is a racemate that can just barely be baseline resolved using stringently optimized conditions. Compounds that are easily resolved are not good test substances, since they still separate even if some column degradation has occurred. If there is ever a question that the stationary phase may be damaged or degraded, analysis of the test compound will reveal the condition of the stationary phase. Manufacturers provide certificates of analysis with new columns, showing the separation of their own test compound. Evaluating a column's condition with a proper test compound is useful for all chiral stationary phases.

The quality of solvents and other additives used for chiral separations should always be HPLC grade or equivalent. This will extend column life and prevent detection issues from arising (due to high UV absorbing backgrounds or stray ions in mass spectrometry detection). Not all brands of HPLC solvents are equivalent.

3.4.6. Chiral Separations Found in the Chemical Literature

The literature regarding chiral separations using cyclodextrin-based CSPs has been reviewed many times *(91–97)*. **Table 4** is a summary of the separations reported in the literature.

Table 4
Reported Separations from the Chemical Literature

Amino acids, derivatives, and peptides

Compound	Column	Mode	Reference	
Tryptophan, phenylalanine, tyrosine, structural analogues	α	RP	41	Hydrophyllic Interaction Chromatography
Amino acids, underivatized	β	HILIC	40	
Amino acids, DANSYL derivatized	β	RP	9	
Amino acids, DANSYL derivatized	β	RP	14	
Amino acids, DANSYL substituted	β	RP	111	
Amino acids, DANSYL substituted	Methylated γ CD (experimentally prepared)	RP	132	
DL-Serine, FMOC, DANSYL derivatized	γ	RP	135	
Amino acids, t-Boc derivatized	RSP	RP	52	
Amino acids, DNP derivatized	β	RP	36	
Amino acids, AQC derivatized	α, β, γ, AC RSP, RN, SN	RP	59	
Amino acids, FMOC derivatized	β	PO	126	
Glutamate FMOC derivatized	γ	PO	125	
Phenylalanine FMOC	RN	RP	118	
Proline FMOC	RN	RP	118	
Tryptophan FMOC	RN	RP	118	
Tyrosine FMOC	RN	RP	118	
Pipecolic acid FMOC	RN	RP	118	
β-Naphthamide amino acids	β	RP	14	
β-Naphthyl amino acids	β	RP	14	
Amino phosphonic acids	AC	RP	123	

Compound	CD	Mode	Ref
DL-Glutamic acid OPA derivatized	β	RP	29
Aromatic cyclic dipeptides	β, γ	RP	34
Aromatic cyclic dipeptides	β	RP	23
Aromatic linear dipeptides	β, γ	RP	34
Aromatic linear dipeptides	β	RP	23
Glycyl dipeptides and tripeptides	α, β, γ	PO	39
di- and tri-peptides	γ	RP	109
β-Adrenergic blocking agents (β blockers)			
Alprenolol	β	PO	37
Atenolol	β	PO	45
Cateolol	β, AC	PO	45
Labetolol	β	PO	45
Nadolol	β	PO	45
Metoprolol	β	PO	45
Pindolol	β, AC	PO	45
Propranolol	γ	PO	45
Timolol	β, AC	PO	45
Metallocenes			
Ferrocene, ruthenocene, osmocene analogues (15 compounds)	β	RP	11
Calcium channel blockers			
Verapamil			
Nisolidipene			
Nimodipene	β	RP	12
Sedative-anticonvulsants			
Hexobarbital			
Mephobarbital			
Methenytoin			
Triazoline			
Phensuximide	β	RP	12

(continued)

Table 4 (Continued)

Compound	Column	Mode	Reference
Nonsteroidal anti-inflammatory			
Ibuprofen	β	RP	119
Ibuprofen and metabolites	β	RP	105
Ibuprofen	RN, SN	RP	56
Suprofen	RN, SN	PO	57
Suprofen	RN	PO	45
Suprofen and related homologues	β	RP	31
Ketoprofen	β	RP	12
Flubiprofen	SN	RP	56
Sulfoxides, diaryl and aliphatic (42 compounds)	β, AC, DM, RSP RN, SN, DMP	RP	48
Sulfoxides, diaryl and aliphatic (42 compounds)	RN, SN, DMP	NP	48
Furocoumarins, substituted (25 compounds)	β, AC, DM, RSP RN, SN, DMP	RP	49
Furocoumarins, substituted (25 compounds)	RN, SN, DMP	NP	49
Methylphenidate	β	RP	12
Scopolamine	β, AC	RP	46
Scopolamine	β	RP	101
Cocaine	β	RP	101
Homatropine	β	RP	101
Homatropine	AC	PO	45
Atropine	β	RP	101
Atropine	β, DM	RP	47
Methadone	β	RP	12
Methadone	RSP	RP	129
Methylenedioxyamphetamine	β, RSP	RP	131

Compound	CD	Mode	Ref
Methylenedioxy-N-methylamphetamine	β, RSP	RP	131
Methylenedioxy-N-ethylamphetamine	β, RSP	RP	131
N-Methyl-1-(1,3-benzodioxol-5-yl)-2-butanamine	β, RSP	RP	131
Ephedrine	β	RP	27,134
Norephedrine	β	RP	134
Pseudoephedrine	β	RP	134
N-Methylephedrine	β	RP	134
N-Methylpseuoephedrine	β	RP	134
Chlorpheniramine	β	RP	12
Chlorthalidone	β	RP	12
Aminoglutethimine	β, AC, AC-α	RP	12
Flavonones (8 compounds)	β	RP	110
Flavonone glycosides (4 compounds)	β	RP	35
2,2'-Binaphthyldiyl crown ethers and analogues (24 compounds)	β	RP	100
2,2'-bi-naphthyldiyl crown-4	RN	NP	53
bi-napthyl crown ether analogues (n = 1-6)	γ	RP	44
Hydantoins, various substitutions	β	RP	25
Hydantoins, various substitutions	β	RP	27
5-(4-methylphenyl)-5-phenylhydantoin	β, DM	RP	47
Monoterpene hydrocarbons (Pinene, camphene)	α	RP	42
Nicotine and analogues (20 compounds)	β	RP/PO	30
N'-(2-naphthyl-methyl) nomicotine	RN	NP	53
N-(α-naphthyl-methyl) nomicotine	γ	RP	44
N'-Benzyl-nomicotine	RN	NP	53
Dinitrobenzyl and dinitrophenyl derivatized compounds			
1-(1-Naphthyl)-ethylamine DMB	RN, SN	NP	55
1,2,3,4-Tetrahydro-1-naphthylamine DNB	RN, SN	NP, SFC	128

(continued)

Table 4 (Continued)

Compound	Column	Mode	Reference
1,2,3,4-Tetrahydronaphthol DNP	RN, SN	NP	55
1,3-Dimethylbutylamine DNB	RN, SN	NP	55
1,5-Dimethylhexylamine DNB	RN, SN	NP	55
1-Cyclohexylethylamine DNB	RN, SN	NP, SFC	55
1-Cyclohexylethylamine DNB	RN, SN	NP	55
1-Methylbutylamine DNB	RN, SN	NP	55
2-Amino-3,3-dimethylbutane DNB	RN, SN	NP	55
2-Aminoheptane DNB	RN, SN	NP, SFC	55,128
2-Bromobutyric acid DNP	RN, SN	NP	55
2-Bromopropionic acid DNP	RN, SN	NP	55
2-Decanol DNP	RN, SN	NP	55
2-Heptanol DNP	RN, SN	NP	55
2-Hexanol DNP	RN, SN	NP	55
2-Methoxy-2-phenyl ethanol DNP	RN, SN	NP	55
2-Octanol DNP	RN, SN	NP	55
2-Pentanol DNP	RN, SN	NP	55
2-Phenylbutyric acid DNP	RN, SN	NP	55
2-Phenylpropionic acid DNP	RN, SN	NP	55
3-Aminoheptane DNB	RN, SN	NP	55
3-Hexanol DNP	SN	NP	55
4-Chlorophenylalanine ethyl ester DNB	RN, SN	NP, SFC	128
4-Methyl-2-pentalol DNP	RN, SN	NP	55
Alanine ethyl ester DNB	RN, SN	NP, SFC	128
Alanine methyl ester DNB	RN, SN	NP, SFC	128
α-Methylbenzalamine DNB	RN, SN	NP, SFC	128
α-Methylbenzalamine DNB	RN, SN	NP	55

(continued)

Compound			
DL-2-Phenylglycine DNB	RN, SN	NP	55
DL-3(1-Naphthyl)alanine DNB	RN, SN	NP	55
DL-3-(2-Naphthyl)alanine DNB	RN, SN	NP	55
DL-Alanine DNB	SN	NP	55
DL-Alanine ethyl ester DNB	RN, SN	NP	55
DL-Alanine methyl ester DNB	RN, SN	NP	55
DL-Homophenylalanine DNB	RN, SN	NP	55
DL-Isoleucine DNB	RN, SN	NP	55
DL-Norleucine DNB	RN, SN	NP	55
DL-Norvaline DNB	RN, SN	NP	55
DL-O-Methyltryosine DNB	SN	NP	55
DL-Phenylalanine DNB	RN, SN	NP	55
DL-Phenylalanine methyl ester DNB	RN, SN	NP	55
DL-Serine methyl ester DNB	RN, SN	NP	55
DL-Threonine methyl ester DNB	RN, SN	NP	55
DL-Tryptophan butyl ester DNB	RN, SN	NP	55
DL-Tryptophan DNB	RN, SN	NP	55
DL-Tryptophan ethyl ester DNB	RN, SN	NP	55
DL-Tryptophan methyl ester DNB	RN, SN	NP	55
DL-Tryptophan octyl ester DNB	RN, SN	NP	55
DL-Tyrosine DNB	RN, SN	NP	55
DL-Tyrosine methyl ester DNB	RN, SN	NP	55
DL-Valine DNB	RN, SN	NP	55
Norleucine methyl ester DNB	RN, SN	NP, SFC	128
Phenylalanine methyl ester DNB	RN, SN	NP, SFC	128
Sec-butylamine DNB	RN, SN	NP	55
Sec-Phenethyl alcohol DNP	RN, SN	NP	55
Valine methyl ester DNB	RN, SN	NP, SFC	128

Table 4 (Continued)

Compound	Column	Mode	Reference
Miscellaneous compounds			
1-(5-Chlor-2-(Methylamino)-phenyl-1,2,3,4-tetrahydroisoquinoline	γ	RP	44
1,1'-Binaphthalene-2,2-diyl hydrogen phosphate	β	RP	130
1,2,3,4-Tetrahydro-1-naphthol	RSP, SP, DM	RP	47,51,130
1-4-Benzodiazepin-2-ones			
Various homologs	β	RP	32
1,1'-bi-2-Naphthol	γ	RP	44
1-Benzocyclobutene carbonitrile	β, DM	RP	47
1-Benzocyclobutene carboxylic acid	RSP, SP	RP	51
1-Indanol	RSP, SP, DM	RP	47,51
1-Methyl-4-tetralone	β, DM	RP	47
1-Phenyl-1-butanol	RSP	RP	131
1-Phenylethyl propionate	β	RP	26
2-(4-Chlorophenoxy) propionic acid	RN	PO	45
2,2'-Bi-2-naphthol	DMP	NP	53
2-Amino-1,2-dipehnyl-ethanol	AC	RP	133
2-Amino-9-hydroxyfluorene	DM	RP	47
2-Chloro-2-pehnylacetic acid	β	RP	26
2-Ethyldene-1,5-dimethyl-3,3-diphenylpyrrolidine	RSP	RP	129
2-Hydroxy-5,5-dimethyl-4-phenyl-1,3,2-dioxaphosphorinane 2-oxide	RSP	RP	130
2-Methoxy-2-phenylethanol	DMP	NP	53
2-Methoxy-α-methylbenzyl alcohol	β	RP	26
2-Phenyl propionaldehyde	β	RP	26
2-Phenylbutyric acid	RSP	RP	133

(continued)

2-Phenyl-butyrophenone	β	RP	27
2-Phenylpropionic acid	RSP	RP	133
3,4-Dihydroxy-phenyl-α-propyl acetamide	β	RP	26
3,5-Dinitrobenzyol phenylglycine	RN, SN	RP	56
3-Benzylphthalide	RN, SN, DM	RP	47,56
3-Hexene-1-ol, 1 hexanol, 1-heptanol (cis-trans isomers)	α, β	RP	107
3-Hydroxy-kynurenine	β	RP	26
3-Phenylphthalide	RN, SN, RSP	RP	51,56
3-Thienyl-cyclohexyl-glycolic acids	β	RP	104
4-(2-Chlorophenyl)-2-hydroxy-5,5-dimethyl-1,3,2-dioxaphosphorinane 2-oxide	β, RSP	RP	130
4-Benzyl-2-oxazolidinone	β, RSP	RP	30,130
4-Benzyl-3-propionyl-2-oxaxolidinone	SN	RP	130
4-Benzyl-5,5'-dimethyl-2-oxazolidinone	RN	RP	130
4-chloro-2-(α-methylbenzyl)-phenol	β	RP	27
4-Phenyl-2-oxazolidinone	AC	RP	130
5-Ethyl-5-(p-tolyl)-2-thiobarbituric acid	β	RP	27
α,α-Di(2-naphthyl)-2-pyrrolidinemethanol	SN	PO	133
α,α-Diphenyl-2-pyrrolidine	AC	RP	133
α-Amino-3-thiopheneacetic acid	β	RP	26
Acenaphthenol	γ	RP	44
α-Methyl-alpha-phenyl succinimide	β	RP	27
α-Methyl-tryptamine	β	RP	27
Althiazide	RN, SN	RP	56
Althiazide	β,γ	PO	45
α-Methoxy-α-(trifluoromethyl) phenylacetic acid	β	RP	130
α-Methyl benzyl acetic	β	RP	26
Amines (aromatic and aliphatic) (aromatic anhydride derivatized)	β, RN	RP	124

Table 4 (Continued)

Compound	Column	Mode	Reference
Amylase, α and β forms	β	RP	106
Ancymidol	β, DM, RN, SN	RP	47,56,58
Barbitals	β	RP	111
Barbituarates	β	RP	9
Bendroflumethiazide	RN, SN	RP	56
Bendroflumethiazide	SN	PO	45
Benzo[X]pyrene(s)	β	RP	9
Benzoin	β, RN, SN, DM	RP	30,47,56,130
Benzoin ethyl ester	β, DM, RN, SN	RP	47,56
Brompheniramine	β	RP	27
Bulan	AC	RP	58
Chlorthalidone	RSP, SP, β	RP	51,128
Ciprofibrate	SN	NP	53
cis-4,5-Diphenyl-2-oxazolidinone	AC	RP	133
Coumachlor	RN	NP	53
Coumachlor	RN, SN, DM	RP	47,56
Coumachlor	RN	PO	45
Coumafuryl	DM	RP	47
Crufomate	DMP	RP	58
Deltahedral carborane and metalloborane derivatives	β	RP	116
Di-Aryl phosphine oxides	β	SFC	24
Dinicanazole	β	RP	115
Dyfonate	RN	RP	56
E Prostaglandin and metabolites	β	RP	102
Estazolam	RSP	RP	129

(continued)

Ethyl-3-phenylglycidate	β	RP	26
Fenoxaprop-ethyl	DM	RP	47
Fonofos	RN	RP	58
Gluethimide	DMP	NP	53
Gluethimide	SN	RP	56
Homophenylalanine	SN	RP	56
Hydrobenzoin	RSP, SN	RP	56,130
Idazoxan	RSP, SP, DM	RP	47,51
Indapamide	RN, SN	RP	56
Lorazepam	SN	RP	56
Mandelic acid	β	RP	26
Mandelic acid methyl ester	β	RP	26
Methyl-5-formyl-2,4-pentadienoate	β	RP	114
Methylidazoxan	RSP, SP, DM	RP	47,51
Mianserin	β, RSP	RP	128
N-(3,5-Dinitro benzoyl)-1,2,3,4-tetrahydronaphthyl-1-amine	γ	RP	44
N-(3,5-Dinitro benzoyl)-leucine	β	RP	26
N,N'-1,2-Ethylenediyl-bis (cysteine) (Rh(V)) exo complexes	α, β, γ, DMP	RP	120
N,N'-Dibenzyl-tartramide	β	RP	27
Naphthols, nitroanalines, ethoxyanilines, chloroanalines (various substitutions)	α, β	RP	108
Naphthyl amides	β	SFC	24
Napropoamide	DMP	RP	58
Nisoldipine	SP	RP	51
N-Nitrosamine acids, syn and anti conformers	α	RP	103
O-Acetyl-mandelic acid	β	RP	26

Table 4 (Continued)

Compound	Column	Mode	Reference
Oxazepam	SN	RP	*56*
Oxazepam	SN, RN	PO	*45*
p-Bromotetramisole oxalate	β, DM	RP	*47*
p-Chlorowarfarin	RN, SN	PO	*57*
Pemoline	SN	RP	*56*
Phensuximide	DMP	NP	*53*
Phenylalanineamide	β	RP	*26*
Pilocarpine HCl and degredants	β	RP	*117*
Polyoxygenated sterols (a-chiral)	β	RP	*113*
Proglumide	β, AC, RN	PO	*45*
Rotenone and rotenoid(s) (12 compounds)	β	RP	*21*
RS-ciprofibrate	β, SN, RN	PO	*45*
Ruelene	RN	RP	*56*
Ruelene	β, AC	PO	*45*
Sobrerol	β	RP	*33*
Tamoxifen	β	RP	*22*
Temazepam	SN	RP	*56*
Terfenadine and metabolites	β	RP	*112*
Terodiline	RN, SN	RP	*56*
Terodiline	α, RSP, AC	PO	*45*
Terpenic alcohols	β	RP	*33*
Tetracyclic eudistomins	β	PO	*38*
Tetrahydrozoline	SN	RP	*56*
Tetrahydroxoline	γ, SN	PO	*45*
Tetramisole	RN, SN	RP	*56*
Thienophran	β	RP	*121*

Tocopoherols, methylated tocols (a-chiral)	β, γ	NP	122
Tolplerisone	RN, SN	RP	56
Trihexphenidyl	β, AC	PO	45
Tropicamide	SN	RP	26
Tyrosine methyl ester	β	RP	26
Warfarin	DM	RP	47
Warfarin	β, SN	PO	45

Column designations: α, α-cyclodextrin; β, β-cyclodextrin; γ, γ-cyclodextrin; AC, acetylated β-cyclodextrin; DM, dimethylated β-cyclodextrin; RSP, R,S-hydroxypropyl β-cyclodextrin; SP, S-hydroxypropyl β-cyclodextrin; RN, R-naphthylethylcarbamate β-cyclodextrin; SN, S-naphthyl-ethylcarbamate β-cyclodextrin; DMP, dimethylphenylcarbamate β-cyclodextrin.

Mode designations: RP, reversed-phase; NP, normal phase; PO, polar organic; HILIC, hydrophilic interaction liquid chromatography.

As previously mentioned, certain cyclodextrin-based CSPs are more useful than others. Generally, native β-cyclodextrin CSPs are more useful than the α- and γ-cyclodextrin homologs. The total number of separations reported in the literature for each size of cyclodextrin certainly supports this claim. However, there is always a small group of molecules or structural type for which either the α-cyclodextrin or the γ-cyclodextrin stationary phases provide superior separations. For example, enantiomers of native aromatic amino acids *(41)* and monoterpene hydrocarbons *(42)* can only be separated on the α-cyclodextrin CSP. Many steroid epimers and larger polycyclic aromatic hydrocarbons are better separated on the γ-cyclodextrin stationary phase. Additionally, the polar organic mode is more effective at separating the more polar compounds, which have two or more hydrophilic or polar functional groups. β-Adrenergic blocking agents (β-blockers) can be separated into enantiomers with limited success in the reversed-phase mode *(37)*. In the polar organic mode, virtually all enantiomers of this class of molecule are much better separated. Other polar molecules, such as sugars, are also more easily separated under polar organic conditions *(41,70,72)*.

Enantiomeric separations are more easily achieved for neutral and hydrophobic compounds in the reversed-phase mode. Many classes of compounds, including barbitals, metallocenes, polycyclic aromatic hydrocarbons, mycotoxins, binaphthyl crown ethers, furocoumarins, and flavones are all neutral and can be separated easily using β-cyclodextrin CSPs. Ionizable compounds may also separate well, but often optimal conditions require neutralization of any ionizable groups through the use of buffers adjusted to the required or optimal pH. Alkaloids, like various nicotine analogs, can be easily separated under the appropriate conditions *(30)*.

Some classes of compounds display specific trends in their interaction with cyclodextrin-based CSPs. For example, underivatized amino acids do not separate into enantiomers very well on these CSPs, but the separation of N-blocked amino acids, derivatized with 5-dimethylaminonaphthalene-1-sulfonyl (DANSYL), 9-fluorenylmethylchloroformate (FMOC), 6-aminoquinolyl-N-hydroxysuccinimidyl carbamate (AQC), o-phthaladehyde (OPA), or t-BOC moietys, are usually straight forward *(9,14,29,52,59)*. Derivatization of amino acids through the primary amine group blocks interactions with the amine moiety. It also provides the necessary hydrophobic group for inclusion complexation. These compounds still possess the ability to ionize through the carboxylic acid group as well as on the N-blocking group in some cases (e.g., DANSYL and AQC groups).

Simple molecules, such as aliphatic alcohols (2-octanol), aliphatic amines (3-amino heptane), and simple carboxylic acids, have an extremely low chance of being separated into enantiomers in any mode of operation. However, their conversion to a dinitrophenyl or dinitrobenzoate-derivative can virtually assure successful enantiomeric separation on naphthylethyl-carbamate-β-cyclodextrin CSPs, particularly in the normal phase mode *(60)*. These CSPs are incredibly effective at separating any chiral molecule that contains a π-electron acceptor (π-acid) moiety. These aromatic-derivatized cyclodextrins are among the most effective multimodal chiral stationary phases ever developed (*see* **Subheading 2.2.3.**) The variety of molecules that are separable on these CSPs may be greater than any other derivatized cyclodextrins.

4. Notes

1. It can be difficult to determine which cyclodextrin CSP to start with and when to move on to another. If the variables of mode (reversed-phase, normal phase, and polar organic mode), mobile phase composition, and pH/buffer (where applicable) have been explored and no enantiomeric separation has been achieved, it is best to move on to another CSP. It can be undesirable for certain researches to keep 30 CSPs on hand, as they are very expensive, yet at the same time no one CSP will be able to achieve every desired chiral separation. It is a difficult task to determine a ranking of each cyclodextrin CSP and say which the absolute best are. It has been shown that the most useful cyclodextrin CSPs are the hydroxypropyl-β-cyclodextrin, the *R*- (or *S*)-naphthylethylcarbamate-β-cyclodextrin, the native β-cyclodextrin, and the 2,3-dimethyl-β-cyclodextrin. Keep in mind, the efficacy of any CSP is entirely application specific.
2. The polarity of native cyclodextrin CSPs, in the reversed-phase mode, is approximately equivalent to a C4 or C8 column.
3. It is a good practice to assess the solubility of a solute in a potential mobile phase prior to injection. Take a small amount of the mobile phase (approx 1 to 2 mL) in a test tube and mix into it a small amount of the sample solution. If precipitation occurs, adjust the composition of the mobile phase solution or sample solution.
4. pH is a concept that only has meaning in a 100% aqueous solution. In hydro-organic solutions, both the acidity of the solution [H^+] from the buffer and the ionization equilibrium constant (K_a or K_b) of the analyte will change, which greatly effects the separation of acidic and basic compounds *(98)*. Despite these changes to the ionization equilibrium, one can still perform chromatography at conditions more acidic or basic than the analytes K_a or K_b. Two common methods of preparing buffered mobile phases are (i) preparing the buffer and adjusting the pH in the aqueous solution, and then mixing in the appropriate amount of organic modifier; or (ii) preparing a solution of the appropriate concentrations of organic modifier and water, and then adding the buffer components and adjusting the pH. These methods generate a solution of an apparent pH (as read from a pH meter) that is not reflective of the actual [H^+] concentration in solution. However, when the prescribed pH range (pH equals 4.0–7.0, as read from a pH meter) is adhered to, both of these methods can generate a buffered mobile phase of pH that is not deleterious to cyclodextrin

CSPs. Whether the analyte will be ionized at the apparent pH is application-specific. This is an issue that affects both chiral and achiral chromatography.

5. Alternatively, 90:10 or 100:0 acetonitrile/methanol may be used as necessary, as long as adequate retention is achieved.

6. It is possible that the *R*-naphthylethylcarbamate-β-cyclodextrin and *S*-naphthylethylcarbamate-β-cyclodextrin will have opposite enantioselectivities. If the *R* column were to retain *R* enantiomer of the analyte preferentially, the *S* column could retain the *S* enantiomer preferentially. The potential for this elution order switching exists with any of the modes of operation.

7. A more rigorous approach involves plotting the natural log of the retention factor (for both enantiomers) against the inverse of the absolute temperature, (ln k' vs 1/ T, temperature in Kelvin) *(99)*. The enthalpies and entropies of transfer (between the mobile phase and stationary phase) can be obtained from the slope and intercept respectively. When a chiral separation is achieved, two lines will be generated, one for each enantiomer. The difference between the two lines is directly proportional to the separation factor (α). Extrapolation of the lines will reveal the separation factor (α) at all temperatures. Keep in mind that at higher temperatures, the hydrolytic degradation of silica proceeds at an accelerated rate.

8. Cyclodextrin-based CSPs often have reduced chromatographic efficiency compared to a typical C18 column. The reason is due to the kinetics of association and disassociation from the stationary phase. It is believed that the kinetics associated with the formation of an inclusion complex are slower than the mass transfer kinetics associated with typical brush-type C8 or C18 stationary phases. This leads to an increase in plate height (H) and a decreased number of plates per meter at higher flow rates.

9. These effects are often small, and most often raising the pressure by using capillary restriction will lead to an overall worse separation. Theoretically, the potential exists for this type of optimization to improve separations, but an example has yet to be seen in the literature in which the overall separation is improved (when using a cyclodextrin-based CSP).

10. As with all types of HPLC, it is necessary to remove salts, buffers, and solutes from the column to prevent both precipitation of salts and microbial growth, both of which are detrimental to column performance.

References

1. FDA. (1992) FDA's policy statement for the development of new stereoisomeric drugs. *Chirality* **4,** 338–340.
2. Cramer, F. and Dietsche, W. (1959) Occlusion compounds. XV. Resolution of racemates with cyclodextrins. *Chemische Berichte* **92,** 378–384.
3. Hinze, W. L. (1981) Applications of cyclodextrins in chromatographic separations and purification methods. *Separation Purification Methods* **10,** 159–237.
4. Armstrong, D. W. (1980) Pseudophase liquid chromatography: applications to TLC. *J. Liq. Chromatogr.* **3,** 895–900.

5. Hinze, W. L. and Armstrong, D. W. (1980) Thin layer chromatographic separation of ortho, meta, and para-substituted benzoic acids and phenols with aqueous solutions of alpha-cyclodextrin. *Anal. Lett.* **13,** 1093–1104.

6. Burkert, W. G., Owensby, C. N., and Hinze, W. L. (1981) The use of an alpha-cyclodextrin mobile phase in the thin-layer chromatographic separation of ortho, meta, and para substituted phenols. *J. Liq. Chromatogr.* **4,** 1065–1085.

7. Fujimura, K., Veda, T., and Ando, T. (1983) Retention behavior of some aromatic compounds on chemically bonded cyclodextrin silica stationary phase in liquid chromatography. *Anal. Chem.* **55,** 446–450.

8. Kawaguchi, Y., Tanaka, M., Nakae, M., Funazo, K., and Shono, T. (1983) Chemically bonded cyclodextrin stationary phases for liquid chromatographic separation of aromatic compounds. *Anal. Chem.* **55,** 1852–1857.

9. Armstrong, D. W. and DeMond, W. (1984) Cyclodextrin bonded phases for the liquid chromatographic separation of optical, geometrical, and structural isomers. *J. Chromatogr. Sci.* **22,** 411–415.

10. Armstrong, D. W., Alak, A., DeMond, W., Hinze, W. L., and Riehl, T. E. (1985) Separation of mycotoxins, polycyclic aromatic hydrocarbons, quinones, and heterocyclic compounds on cyclodextrin bonded phases: an alternative LC packing. *J. Liq. Chromatogr.* **8,** 261–269.

11. Armstrong, D. W., DeMond, W., and Czech, B. P. (1985) Separation of metallocene enantiomers by liquid chromatography: chiral recognition via cyclodextrin bonded phases. *Anal. Chem.* **57,** 481–484.

12. Armstrong, D. W., Ward, T. J., Armstrong, R. D., and Beesley, T. E. (1986) Separation of drug stereoisomers by the formation of beta cyclodextrin inclusion complexes. *Science* **232,** 1132–1135.

13. Chang, C. A. and Wu, Q. (1986) Comparison of liquid chromatographic separations of geometrical isomers of substituted phenols with beta and gamma cyclodextrin bonded-phase columns. *Anal. Chem. Acta* **189,** 293–299.

14. Hinze, W. L., Riehl, T. E., Armstrong, D. W., DeMond, W., Alak, A., and Ward, T. J. (1985) Liquid chromatographic separation of enantiomers using a chiral beta-cyclodextrin bonded stationary phase and conventional aqueous-organic mobile phases. *Anal. Chem.* **57,** 237–242.

15. Armstrong, D. W., DeMond, W., Alak, A., Hinze, W. L., Riehl, T. E., and Bui, K. (1985) Liquid chromatographic separation of diastereomers and structural isomers on cyclodextrin-bonded phases. *Anal. Chem.* **57,** 234–237.

16. Gillet, B., Nicole, D. J., and Delpuech, J. J. (1982) The hydroxyl group protonation rates of alpha, beta, and gamma-cyclodextrins in dimethyl sulphoxide. *Tetrahedron Lett.* **23,** 65–68.

17. Rees, D. A. (1970) Conformational analysis of polysaccharides. Part V. The characterization of linkage conformations (chain conformations) by optical rotation at a single wavelength. Evidence for distortion of cyclohexa-amylose in aqueous solution. Optical rotation and the amylose conformation. *J. Chem. Soc.* **B5,** 877–884.

18. Szejtli, J. (1988) in *Cyclodextrin Technology.* Kulwer Academic Publishers, Boston, MA, pp. 3–4.

19. Armstrong, D. W., Li, W., Chang, C. D., and Pitha, J. (1990) Polar-liquid, derivatized cyclodextrin stationary phases for the capillary gas chromatography separation of enantiomers. *Anal. Chem.* **62,** 914–923.

20. McClanahan, J. S. and Maguire, J. H. (1986) High performance liquid chromatographic determination of the enantiomeric composition of urinary phenolic metabolites of phenytoin. *J. Chromatogr.* **381,** 438–446.

21. Abidi, S. L. (1987) Chiral phase high performance liquid chromatography of rotenoid racemates. *J. Chromatogr.* **404,** 133–143.

22. Armstrong, R. D., Ward, T. J., Pattabiraman, N., Benz, C., and Armstrong, D. W. (1987) Separation of tamoxifen geometric isomers and metabolites by bonded phase beta cyclodextrin chromatography. *J. Chromatogr.* **414,** 192–196.

23. Florance, J., Galdes, A., Kontreatis, Z., Kosarych, Z., Langer, K., and Martucci, C. (1987) High performance liquid chromatographic separation of peptides and amino acid stereoisomers. *J. Chromatogr.* **414,** 313–322.

24. MaCaudiere, P., Caude, M., Rosset, R., and Tambute, A. (1987) Resolution of racemic amides and phosphine oxides on a beta-cyclodextrin bonded stationary phase by subcritical fluid chromatography. *J. Chromatogr.* **405,** 135–143.

25. Maguire, J. H. (1987) Some structural requirements for resolution of hydantoin enantiomers with a beta-cyclodextrin liquid chromatography column. *J. Chromatogr.* **387,** 453–458.

26. Armstrong, D. W., Han, S. M., and Han, Y. I. (1988) Liquid chromatographic resolution of enantiomers containing single aromatic rings with beta-cyclodextrin-bonded phases. *Anal. Chem. Acta* **208,** 275–281.

27. Han, S. M., Han, Y. I., and Armstrong, D. W. (1988) Structural factors affecting chiral recognition and separation on beta-cyclodextrin bonded phases. *J. Chromatogr.* **441,** 376–381.

28. Krstulovic, A. M., Gianviti, J. M., Burke, J. T., and Mompon, B. (1988) Enantiomeric analysis of a new anti-inflammatory agent in rat plasma using a chiral beta cyclodextrin stationary phase. *J. Chromatogr.* **426,** 417–424.

29. Merino, I. M., Gonzalez, E. B., and Sanz-Medel, A. (1988) Liquid chromatographic enantiomeric resolution of amino acids with beta-cyclodextrin bonded phases and derivatization with o-phthalaldehyde. *Anal. Chem. Acta* **234,** 127–131.

30. Seeman, J. I., Secor, H. V., Armstrong, D. W., Timmons, K. D., and Ward, T. J. (1988) Enantiomeric resolution and chiral recognition of racemic nicotine and nicotine analogues by beta-cyclodextrin complexation. Structure-enantiomeric resolution relationships in host-guest interactions. *Anal. Chem.* **60,** 2120–2127.

31. Marziani, F. C. and Sisco, W. R. (1989) Liquid chromatographic separation of positional isomers of suprofen on a cyclodextrin bonded phase. *J. Chromatogr.* **465,** 422–428.

32. Bertucci, C., Domenici, E., Uccello-Barretta, G., and Salvadori, P. (1990) High-performance liquid chromatographic resolution of racemic 1,4-benzodiazepin-2-ones by means of a beta cyclodextrin silica bonded chiral stationary phase. *J. Chromatogr.* **506,** 617–625.

33. Italia, A., Schiavi, M., and Ventura, P. (1990) Direct liquid chromatographic separation of enantiomeric and diastereomeric terpenic alcohols as beta-cyclodex-trin inclusion complexes. *J. Chromatogr.* **503**, 266–271.
34. Florance, J. and Kontetis, Z. (1991) Chiral high performance liquid chromatography of aromatic cyclic dipeptides using cyclodextrin stationary phases. *J. Chromatogr.* **543**, 299–305.
35. Krause, M. and Galensa, R. (1991) High performance liquid chromatography of diastereomeric flavanone glycosides in citrus on a beta cyclodextrin bonded stationary phase. *J. Chromatogr.* **588**, 41–45.
36. Li, S. and Purdy, W. C. (1991) Liquid chromatographic separation of the enantiomers of dinitrophenyl amino acids using a beta-cyclodextrin-bonded stationary phase. *J. Chromatogr.* **543**, 105–112.
37. Armstrong, D. W., Chen, S., Chang, C., and Chang, S. (1992) A new approach for the direct resolution of racemic beta adrenergic blocking agents by HPLC. *J. Liq. Chromatogr.* **15**, 545–556.
38. Kuijpers, P. H., Gerding, T. K., and deJong, G. J. (1992) Improvement of the liquid chromatographic separation of the enantiomers of tetracyclic eudistomins by the combination of a beta cyclodextrin stationary phase and camphorsulphonic acid as mobile phase additive. *J. Chromatogr.* **625**, 223–230.
39. Zukowski, J., Pawlowska, M., Nagatkina, M., and Armstrong, D. W. (1993) High performance liquid chromatographic enantioseparation of glycyl di- and tripeptides on native cyclodextrin bonded phases. *J. Chromatogr.* **629**, 169–179.
40. Risley, D. S. and Strege, M. A. (2000) Chiral separations of polar compounds by hydrophilic interaction chromatography with evaporative light scattering detection. *Anal. Chem.* **72**, 1736–1739.
41. Armstrong, D. W., Yang, X., Han, S. M., and Menges, R. A. (1987) Direct liquid chromatographic separation of racemates with an alpha-cyclodextrin bonded phase. *Anal. Chem.* **59**, 2594–2596.
42. Armstrong, D. W. and Zukowski, J. (1994) Direct enantiomeric resolution of monoterpene hydrocarbons via reversed-phase high-performance liquid chromatography with an alpha-cyclodextrin bonded stationary phase. *J. Chromatogr. A* **666**, 445–448.
43. Chang, C. A. and Wu, Q. (1987) Facile liquid chromatographic separation of positional isomers with a gamma-cyclodextrin bonded phase column. *J. Liq. Chromatogr.* **10**, 1359–1368.
44. Stalcup, A. M., Jin, H. L., and Armstrong, D. W. (1990) Separation of enantiomers using an gamma-cyclodextrin liquid chromatographic bonded phase. *J. Liq. Chromatogr.* **13**, 473–484.
45. Change, S. C., Reid, G. L., III., Chen, S., Chang, C. C., and Armstrong, D. W. (1993) Evaluation of a new polar-organic high performance liquid chromatographic mobile phase for cyclodextrin bonded chiral stationary phases. *Trends Anal. Chem.* **12**, 144–153.
46. Stalcup, A. M., Faulkner, J. R., Tang, Y., Armstrong, D. W., Levy, L. W., and Regalado, E. (1991) Determination of the enantiomeric purity of scopolamine

106 *Mitchell and Armstrong*

isolated from plant extract using achiral/chiral coupled column chromatography. *Biomed. Chromatogr.* **5,** 3–7.

47. Armstrong, D. W., Chang, L. W., Chang, S. C., et al. (1997) Comparison of the enantioselectivity of beta-cyclodextrin vs. heptakis-2,3-dimethyl-beta-cyclo-dextrin lc stationary phases. *J. Liq. Chromatogr. Relat. Technol.* **20,** 3279–3295.

48. Mitchell, C. R., Desai, M., McCulla, R., Jenks, W., and Armstrong, D. W. (2002) Use of native and derivatized cyclodextrin chiral stationary phases for the enantioseparation of aromatic and aliphatic sulfoxides by high performance liquid chromatography. *Chromatographia* **56,** 127–135.

49. Mitchell, C. R., Schumacher, D. S., Rozhkov, R. V., Larock, R. C., and Armstrong, D. W. (2002) Use of native and derivatized cyclodextrin chiral stationary phases for the enantioseparation of substituted furo-coumarins by high performance liquid chromatography. *J. Chromatogr. A.,* in press.

50. Armstrong, D. W., Wang, X., Chang, L. W., Ibrahim, H., Reid, C. R., and Beesley, T. E. (1997) Comparison of the selectivity and retention of beta-cyclo-dextrin vs. heptakis-2,3-O-dimethyl-beta cyclodextrin LC stationary phases for structural and geometric isomers. *J. Liq. Chromatogr. Relat. Technol.* **20,** 3297–3308.

51. Stalcup, A. M., Chang, S. C., Armstrong, D. W., and Pitha, J. (1990) (S)-2-Hydroxy-propyl-beta-cyclodextrin, a new chiral stationary phase for reversed-phase liquid chromatography. *J. Chromatogr.* **513,** 181–194.

52. Chang, S. C., Wang, L. R., and Armstrong, D. W. (1992) Facile resolution of N-tert-butoxycarbonyl amino acids: the importance of enantiomeric purity in peptide synthesis. *J. Liq. Chromatogr.* **15,** 1411–1429.

53. Armstrong, D. W., Stalcup, A. M., Hilton, M. L., Duncan, J. D., Faulkner, J. R., and Chang, S. C. (1990) Derivatized cyclodextrins for normal-phase liquid chromatographic separation of enantiomers. *Anal. Chem.* **62,** 1610–1615.

54. Armstrong, D. W., Hilton, M., and Coffin, L. (1991) Multimodal chiral stationary phases for liquid chromatography: (R)- and (S)-naphthylethylcarbamate derivatized beta cyclodextrin. *LC GC* **9,** 646–652.

55. Stalcup, A. M., Chang, S. C., and Armstrong, D. W. (1991) Effect of the configuration of the substituents of derivatized beta-cyclodextrin bonded phases on enantioselectivity in normal phase liquid chromatography. *J. Chromatogr.* **540,** 113–128.

56. Armstrong, D. W., Chang, C. D., and Lee, S. H. (1991) (R)- and (S)-Naphthyl-ethylcarbamate substituted beta cyclodextrin bonded stationary phases for the reversed phase liquid chromatographic separation of enantiomers. *J. Chromatogr.* **539,** 83–90.

57. Richards, D. S., Davidson, S. M., and Holt, R. M. (1996) Detection of non-UV-absorbing chiral compounds by high-performance liquid chromatography-mass spectrometry. *J. Chromatogr. A* **746,** 9–15.

58. Armstrong, D. W., Reid, G. L., III., Hilton, M. L., and Chang, C. D. (1993) Relevance of enantiomeric separations in environmental science. *Environ. Pollution* **79,** 51–58.

59. Pawlowska, M., Chen, S., and Armstrong, D. W. (1993) Enantiomeric separation of fluorescent, 6-aminoquinolyl-N-hydroxysuccinimidyl carbamate, tagged amino acids. *J. Chromatogr.* **641,** 257–265.

60. Berthod, A., Chang, S. C., and Armstrong, D. W. (1992) Empirical procedure that used molecular structure to predict enantioselectivity of chiral stationary phases. *Anal. Chem.* **64,** 395–404.

61. Advanced Separation Technologies. (2000) *Astec Chromatography Product Guide.* Advanced Separation Technologies, Whippany, NJ, p. 16.

62. Issaq, H. J. (1988) The multimodal cyclodextrin bonded stationary phase for high performance liquid chromatography. *J. Liq. Chromatogr.* **11,** 2131–2146.

63. Chang, C. A., Wu, Q., Abdel-Aziz, H., Melchor, N., Pannell, K. H., and Armstrong, D. W. (1985) Liquid chromatographic retention behavior of organometallic compounds and ligands with amine, octadecyl silica, and beta cyclodextrin-bonded phase columns. *J. Chromatogr.* **347,** 51–60.

64. Abidi, S. L. (1986) Liquid chromatography of hydrocarbonaceous quaternary amines on cyclodextrin bonded silica. *J. Chromatogr.* **362,** 33–46.

65. Chang, C. A., Wu, Q., and Armstrong, D. W. (1986) Reversed-phase high-performance liquid chromatographic separation of substituted phenolic compounds with a beta-cyclodextrin bonded phase column. *J. Chromatogr.* **354,** 454–458.

66. Chang, C. A., Wu, Q., and Eastman, M. P. (1986) Mobile phase effects on the separations of substituted anilines with a beta cyclodextrin bonded column. *J. Chromatogr.* **371,** 269–282.

67. Chang, C. A., Wu, Q., and Tan, L. (1986) Normal-phase high performance liquid chromatographic separations of positional isomers of substituted benzoic acids with amine and beta-cyclodextrin bonded-phase columns. *J. Chromatogr.* **361,** 199–207.

68. Issaq, H. J., Weiss, R., Ridlon, C., Fox, S. D., and Muschik, G. M. (1986) The determination of asparatame in diet soft drinks by high performance liquid chromatography. *J. Liq. Chromatogr.* **9,** 1791–1802.

69. Snider, B. G. (1986) Separation of cis-trans isomers of prostaglandins with a cyclodextrin bonded column. *J. Chromatogr.* **351,** 548–553.

70. Jin, H. L., Stalcup, A. M., and Armstrong, D. W. (1988) Separation of cyclodextrins using cyclodextrin bonded phases. *J. Liq. Chromatogr.* **11,** 3295–3304.

71. Mathes, L. E., Muschik, G. M., Demby, L., et al. (1988) High-performance liquid chromatographic determination of 2',3'-dideocycytidine and 3'-azido-3'-deoxy-thymidine in plasma using a column switching technique. *J. Chromatogr.* **432,** 346–351.

72. Armstrong, D. W. and Jin, H. L. (1989) Evaluation of the liquid chromatographic separation of monosaccharides, disaccharides, trisaccharides, tetrasaccharides, deoxysaccharides, and sugar alcohols with stable cyclodextrin bonded phase columns. *J. Chromatogr.* **462,** 219–232.

73. Fielden, P. R. and Packham, A. J. (1989) Selective determination of benzo[a]pyrene in petroleum based products using multi column liquid chromatography. *J. Chromatogr.* **479,** 117–124.

74. Matsui, H. and Sekiyu, T. (1989) High performance liquid chromatographic separation of urinary hippuric and o-, m- and p-methylhippuric acids with a beta cyclodextrin-bonded column. *J. Chromatogr.* **496,** 189–193.

75. Seeman, J. I., Secor, H. V., Armstrong, D. W., Ward, K. D., and Ward, T. J. (1989) Separation of homologous and isomeric alkaloids related to nicotine on a beta-cyclodextrin-bonded phase. *J. Chromatogr.* **483,** 169–177.

76. Tripathi, A. M., Mhalas, J. G., and Rama Rao, N. V. (1989) Determination of 2,6- and 4,6-dinitrocresols by high-performance liquid chromatography on a beta-cyclodextrin bonded column. *J. Chromatogr.* **466,** 442–445.

77. Armstrong, D. W., Bertrand, G. L., Ward, K. D., Ward, T. J., Secor, H. V., and Seeman, J. L. (1990) Evaluation of the effect of organic modifier and ph on retention and selectivity in reversed-phase liquid chromatographic separation of alkaloids on a cyclodextrin bonded phase. *Anal. Chem.* **62,** 332–338.

78. Atamna, I. Z., Muschik, G. M., and Issaq, H. J. (1990) Effect of alcohol chain length, concentration and polarity on separations in high performance liquid chromatography using bonded cyclodextrin columns. *J. Chromatogr.* **499,** 477–488.

79. Ho, J. W. (1990) Separation of porphyrins on cyclodextrin-bonded phases with a novel mobile phase. *J. Chromatogr.* **508,** 375–381.

80. Stalcup, A. M., Jin, H. L., Armstrong, D. W., Mazur, P., Derguini, F., and Nakanishi, K. (1990) Separation of carotenes on cyclodextrin-bonded phases. *J. Chromatogr.* **499,** 627–635.

81. Packham, A. J. and Fielden, P. R. (1991) Column switching for the high-performance liquid chromatographic analysis of polynuclear aromatic hydrocarbons in petroleum products. *J. Chromatogr.* **552,** 575–582.

82. Simms, P. J., Haines, R. M., and Hicks, K. B. (1993) High performance liquid chromatography of neutral oligosaccharides on a beta-cyclodextrin bonded phase column. *J. Chromatogr.* **648,** 131–137.

83. Tsou, T. L., Wu, J. R., Young, C. D., and Wang, T. M. (1997) Simultaneous determination of amoxcillin and clavulanic acid in pharmaceutical products by HPLC with beta-cyclodextrin stationary phase. *J. Pharm. Biomed. Anal.* **15,** 1197–1205.

84. Boehm, R. E., Martire, D. E., and Armstrong, D. W. (1988) Theoretical considerations concerning the separation of enantiomeric solutes by liquid chromatography. *Anal. Chem.* **60,** 522–528.

85. Bertrand, G. L., Faulkner, J. R., Han, S. M., and Armstrong, D. W. (1989) Substituent effects on the binding of phenols to cyclodextrins in aqueous solution. *J. Phys. Chem.* **93,** 6863–6867.

86. Tarr, M. A., Nelson, G., Patonay, G., and Warner, I. M. (1988) The influence of mobile phase alcohol modifiers on HPLC of polycyclic aromatics using bonded phase cyclodextrin columns. *Anal. Lett.* **21,** 843–856.

87. Torok, G., Peter, A., Gaucher, A., Wakselman, M., Mazaleyrat, J. P., and Armstrong, D. W. (1999) High performance liquid chromatographic separation of novel atropic alpha,alpha-disubstituted beta amino acids, either on different betacyclodextrin bonded phases or as their 1-fluoro-2,4-dinitrophenyl-5-l-alanine amide derivatives. *J. Chromatogr. A* **846,** 83–91.

88. Ringo, M. C. and Evans, C. E. (1997) Role of modest pressures in chiraly selective complexation interactions. *J. Physical Chemistry B* **101,** 5525–5530.
89. Ringo, M. C. and Evans, C. E. (1997) Pressure-dependent retention and selectivity in reversed-phase liquid chromatographic separations using beta-cyclodextrin stationary phases. *Anal. Chem.* **69,** 643–649.
90. Ringo, M. C. and Evans, C. E. (1997) Pressure-induced changes in chiral separations in liquid chromatography. *Anal. Chem.* **69,** 4964–4971.
91. Armstrong, D. W. (1984) Chiral stationary phases for high performance liquid chromatographic separation of enantiomers: a mini review. *J. Liq. Chromatogr.* **7,** 353–376.
92. Dappen, R., Arm, H., and Meyer, V. R. (1986) Applications and limitations of commercially available chiral stationary phases for high performance liquid chromatography. *J. Chromatogr.* **373,** 1–20.
93. Ward, T. J. and Armstrong, D. W. (1986) Improved cyclodextrin chiral phases: a comparison and review. *J. Liq. Chromatogr.* **9,** 407–423.
94. Armstrong, D. W. (1987) Optical isomer separation by liquid chromatography. *Anal. Chem.* **59,** 84A–91A.
95. Han, S. M. and Armstrong, D. W. (1989) HPLC separation of enantiomers and other isomers with cyclodextrin-bonded phases: rules for chiral recognition, in *Chiral Separations by HPLC: Applications to Pharmaceutical Compounds* (Krstulovic, A. M., ed.), Ellis Horwood Limited, West Sussex, England, pp. 208–225.
96. Armstrong, D. W. (1997) The evolution of chiral stationary phases for liquid chromatography. *LC-GC Curr. Issues HPLC Technol.* **15,** S20–S28.
97. Ward, T. J. (2000) Chiral separations. *Anal. Chem.* **72,** 4521–4528.
98. Skora, D., Tesarová, E., and Armstrong, D. W. (2002) Practical considerations of the influence of organic modifiers on the ionization of analytes and buffers in reversed-phase LC. *LC GC* **20,** 974–981.
99. Issaq, H. J., Glennon, M. L., Weiss, D. E., and Fox, S. D. (1987) High-performance liquid chromatography using a beta-cyclodextrin-bondes silica column: effect of temperature on retention, in *Ordered Media in Chemical Separations* (Hinze, W. L. and Armstrong, D. W., ed.). American Chemical Society, Washington, DC, pp. 260–271.
100. Armstrong, D. W., Ward, T. J., Czech, A., Czech, B. P., and Bartsch, R. A. (1985) Synthesis, rapid resolution, and determination of absolute configuration of racemic 2,2'-binaphthyldiyl crown ethers and analogues via beta cyclodextrin complexiation. *J. Org. Chem.* **50,** 5556–5559.
101. Armstrong, D. W., Han, S. M., and Han, Y. I. (1987) Separation of optical isomers os scopolamine, cocaine, homatropine, and atropine. *Anal. Biochem.* **167,** 261–264.
102. Oliw, E. H. (1987) Chromatography of B prostaglandins on b-cyclodextrin silica: application to analysis of major E prostaglandins in human seminal fluid. *J. Chromatogr.* **421,** 117–122.
103. Issaq, H. J., Williams, D. G., Schults, N., and Saavedra, J. E. (1988) High performance liquid chromatography separations of nitrosamines. III. Conformers of N-nitrosamino acids. *J. Chromatogr.* **452,** 511–518.

104. Macaudiere, P., Daude, M., Rosset, R., and Tambute, A. (1988) Chiral resolution of a series of 3-thienylcyclohexylglycolic acids by liquid or subcritical fluid chromatography, a mechanistic study. *J. Chromatogr.* **450,** 255–269.

105. Geisslinger, G., Dietzel, K., Lowe, D., et al. (1989) High performance liquid chromatographic determination of ibuprofen, its metabolites and enantiomers in biological fluids. *J. Chromatogr.* **491,** 139–149.

106. Henson, C. A. and Stone, J. M. (1989) Rapid high performance liquid chromatographic separation of barley malt alpha-amylase on cyclobond columns. *J. Chromatogr.* **469,** 361–367.

107. Vigh, G., Farkas, G., and Quintero, G. (1989) Displacement chromatography on cyclodextrin-silicas. II. Separation of cis-trans isomers in the reversed phase mode on alpha-cyclodextrin silica. *J. Chromatogr.* **484,** 251–257.

108. Vigh, G., Quintero, G., and Farkas, G. (1989) Displacement chromatography on cyclodextrin-silicas. I. Separation of positional and geometrical isomers in the reversed phase mode. *J. Chromatogr.* **484,** 237–250.

109. Chang, C. A., Ji, H., and Lin, G. (1990) Effects of mobile phase composition on the reversed-phase separation of dipeptides and tripeptides with cyclodextrin bonded-phase columns. *J. Chromatogr.* **522,** 143–152.

110. Krause, M. and Galensa, R. (1990) Optical resolution of flavanones by high performance liquid chromatography on various chiral stationary phases. *J. Chromatogr.* **514,** 147–159.

111. Vigh, G., Quintero, G., and Farkas, G. (1990) Displacement chromatography on cyclodextrin-silicas. III. Enantiomer separations. *J. Chromatogr.* **506,** 481–493.

112. Chan, K. Y., George, R. C., Chen, T., and Okerholm, R. A. (1991) Direct enantiomeric separation of terfenadine and its major acid metabolite by high-performance liquid chromatography, and the lack of stereoselective terfenadine enantiomer biotransformation in man. *J. Chromatogr.* **571,** 291–297.

113. West, R. R. and Cardellina, J. H. (1991) Semi-preparative separation of polyhydroxylated sterols using a beta-cyclodextrin high-performance liquid chromatography column. *J. Chromatogr.* **539,** 15–23.

114. Xu, M. and Tran, C. D. (1991) High-performance liquid chromatographic separation of racemic and diastereomeric mixtures of 2,4-pentadienoate iron tricarbonyl derivatives. *J. Chromatogr.* **543,** 233–240.

115. Furuta, R. and Nakazawa, H. (1992) Liquid chromatographic separation of the enantiomers of diniconazole using a β-cyclodextrin-bonded column. *J. Chromatogr.* **625,** 231–235.

116. Plesek, J. and Bruner, B. (1992) Liquid chromatographic resolution of enantiomers of deltahedral carborane and metallaborane derivatives. *J. Chromatogr.* **626,** 167–206.

117. Sternitzke, K. D., Fan, T. Y., and Dunn, D. L. (1992) High-performance liquid chromatographic determination of pilocarpine hydrochloride and its degradation products using a β-cyclodextrin column. *J. Chromatogr.* **589,** 159–164.

118. Armstrong, D. W., Gasper, M., Lee, S. H., Zukowski, J., and Ercal, N. (1993) D-Amino acid levels in human physiological fluids. *Chirality* **5,** 375–378.

119. Farkas, G., Irgens, L. H., Quintero, G., Beeson, M. D., AlOSaeed, A., and Vigh, G. (1993) Displacement chromatograpy on cyclodextrin silicas, IV. Separation of the enantiomers of ibuprofen. *J. Chromatogr.* **645,** 67–74.

120. Green, J., Jones, R., Harrison, R. D., Edwards, D. S., and Glacjeh, J. L. (1993) Liquid chromatographic separation of radiopharmaceutical ligand enantiomers. *J. Chromatogr.* **635,** 203–209.

121. Shaw, C. J., Sanfilippo, P. J., McNally, J. J., Park, S. A., and Press, J. B. (1993) Analytical and preparative high-performance liquid chromatographic separation of thienopyran enantiomers. *J. Chromatogr.* **631,** 173–175.

122. Abidi, S. L. and Mounts, T. L. (1994) Separations of tocopherols and methylated tocols on cyclodextrin-bonded silica. *J. Chromatogr. A* **670,** 67–75.

123. Camilleri, P., Reid, C. A., and Manallack, D. T. (1994) Chiral recognition of structurally related aminoalkylphosphonic acid derivatives on an acetylated beta-cyclodextrin bonded phase. *Chromatographia* **38,** 771–775.

124. Pawlowska, M., Zukowski, J., and Armstrong, D. W. (1994) Sensitive enantio-meric separation of aliphatic and aromatic amines using aromatic anhydrides as non-chiral derivatizing agents. *J. Chromatogr. A* **666,** 485–491.

125. Rundlett, K. L. and Armstrong, D. W. (1994) Evaluation of free D-glutamate in processed foods. *Chirality* **6,** 277–282.

126. Ekborg-Ott, K. H. and Armstrong, D. W. (1996) Evaluation of the concentration and enantiomeric purity of selected free amino acids in fermented malt bever-ages (beers). *Chirality* **8,** 49–57.

127. Lelievre, F., Yan, C., Zare, R. N., and Gareil, P. (1996) Capillary electrochro-matography: operating characteristics and enantiomeric separations. *J. Chroma-togr. A* **723,** 145–156.

128. Williams, K. L., Sander, L. C., and Wise, S. A. (1996) Comparison of liquid and supercritical fluid chromatography using naphthylethylcarbamolated beta cyclo-dextrin chiral stationary phases. *J. Chromatogr. A* **746,** 91–101.

129. Pham-Huy, C., Chikhi-Chorfi, N., Galons, H., et al. (1997) Enantioselective high performance liquid chromatography determination of methadone enantiomers and its major metabolite in human biological fluids using a new derivatized cyclo-dextrin bonded phase. *J. Chromatogr. B* **700,** 155–163.

130. Armstrong, D. W., Lee, J. T., and Chang, L. W. (1998) Enantiomeric impurities in chiral catalysts, ausilaries, and synthons used in enantioselective synthesis. Part 1. *Tetrahedron Asymmetry* **9,** 2043–2064.

131. Sadeghipour, F. and Veuthey, J. L. (1998) Enantiomeric separation of four methylenediosylated amphetamines on beta-cyclodextrin chiral stationary phases. *Chromatographia* **47,** 285–290.

132. Araki, T., Kashiwamoto, Y., Tsunoi, S., and Tanaka, M. (1999) Preparation and enantiomer separation behavior of selectively methylated gamma-cyclodextrin bonded stationary phases for high performance liquid chromatography. *J. Chro-matogr. A* **845,** 455–462.

133. Armstrong, D. W., He, L., Yu, R., Lee, J. R., and Liu, Y. S. (1999) Enantiomeric impurities in chiral catalysts, ausilaries, synthons and resolving agents. Part 2. *Tetrahedron Asymmetry* **10,** 37–60.

134. Herraex-Hernandez, R. and Campins-Falco, P. (2001) Chiral separation of ephedrines by liquid chromatography using beta cyclodextrins. *Anal. Chem. Acta* **434,** 315–324.
135. Kim, T. Y. and Kim, H. J. (2001) Chiral separation of 9-fluorenylmethyl chloroformate and dansyl chloride-derivatized dl serine by gamma gyglocextrin bonded high performance liquid chromatography. *J. Chromatogr. A* **933,** 99–106.
136. Advanced Separation Technologies. (2002) *Cyclobond Handbook.* Advanced Separation Technologies, Whippany, NJ.
137. Armstrong, D. W. and Zhang, B. (2001) Chiral stationary phases for HPLC. *Anal. Chem.* **73,** 557A–561A.

4

Enantiomeric Separations by HPLC Using Macrocyclic Glycopeptide-Based Chiral Stationary Phases

An Overview

Tom Ling Xiao and Daniel W. Armstrong

1. Introduction

The development of effective high efficiency enantiomeric separations is a tremendous success story *(1)*. The separation of enantiomers is now accomplished by chiral chromatographic and electrophoretic methods, which includes gas chromatography (GC), high-performance liquid chromatography (HPLC), thin-layer chromatography (TLC), supercritical fluid chromatography (SFC), capillary electrochromatography (CEC), and capillary electrophoresis (CE). Enantioselective HPLC is the most widely used chromatographic method, both for analytical and preparative purposes, in most branches of science and technology including the pharmaceutical and environmental fields. The search for more effective and universal chiral stationary phases (CSPs) for HPLC is an ongoing and challenging topic for separation scientists.

The most useful and widely used CSPs are capable of separating a great variety of enantiomeric compounds while providing good efficiency, good loadability, and long-term stability. The macrocyclic glycopeptides (also known as macrocyclic antibiotics) may be the most promising chiral selectors in this respect. They are the newest class of chiral selectors, first introduced by Armstrong and coworkers in 1994 *(2)*. They are the fastest growing and one of the most useful classes of chiral selectors in the world today. High efficiency enantiomeric separations of a wide variety of biological *(2–41)*, pharmacological *(2–39,42–50)*, agrochemical *(51,52)*, and nutritional compounds *(53)*, have been achieved using this class of chiral selectors.

From: *Methods in Molecular Biology, Vol. 243: Chiral Separations: Methods and Protocols*
Edited by: G. Gübitz and M. G. Schmid © Humana Press Inc., Totowa, NJ

Prior to 1994, the use of macrocyclic glycopeptides was essentially unknown in separation science and technology. Vancomycin has been used as an antibiotic to treat severe staphylococcal infections, especially when bacterial resistance to other antibiotics has developed *(4)*. The vancomycin-related antibiotics bind to the D-alanyl-D-alanine terminal group in the bacteria cell wall, thereby stopping bacterial development *(23,54–56)*. Because the target of these antibiotics is the D-alanyl-D-alanine group, it was thought that they could be used in the separation of amino acid enantiomers *(23)*. Interestingly, CSPs based on these macrocyclic glycopeptides effectively separated a wide variety of different enantiomeric compounds. CSPs based on macrocyclic glycopeptides have many of the enantioselectivity properties of more complex proteins and other polymeric selectors, with the advantages of a smaller size, good stability, and good sample capacity *(2,7)*.

There are literally hundreds of documented macrocyclic antibiotics. However, only a few appear to be broadly effective as chiral selectors. Three classes of macrocyclic antibiotics have been used successfully as chiral selectors. The first type is made up of *ansa* compounds, which consists of a chromophore spanned by an aliphatic bridge. These have been used mainly in CE applications *(15,57–59)*. The second type consists of macrocyclic peptides such as thiostrepton. These have had limited use in liquid chromatography (LC) *(2,60, 61)*. The third and most important type is the macrocyclic glycopeptides, which have three or four fused macrocyclic rings. Various saccharide moieties are attached to the fused macrocyclic system.

The unique structure of the macrocyclic glycopeptides and their abundance of functionality (e.g., aromatic, hydroxyl, amine, and carboxylic acid moieties, amide linkages, hydrophobic pockets, etc.) give them broad selectivity for a wide variety of anionic, neutral, and cationic compounds. All of the defined molecular interactions that are necessary for chiral recognition are found within these relatively compact structures. These include possibilities for ionic interactions, hydrogen bonding, steric, dipole-dipole and π-π interactions, as well as hydrophobic interactions. This allows for a wide variety of chiral separations in all known mobile phase modes (which includes the normal phase mode, reversed-phase mode, and polar organic mode). Each separation mode provides simultaneous, but different, interactions for chiral recognition. This accounts for the large number of chiral separations and the variety of types of chiral compounds that are successfully separated with this class of CSP.

The first macrocyclic glycopeptide antibiotic introduced as a commercial CSP was vancomycin (Chirobiotic V) *(2)*, followed by teicoplanin (Chirobiotic T) *(7)*. In 1998, after having been successfully used in CE chiral separations, ristocetin A (Chirobiotic R) joined this group as the third CSP of this family *(17)*. Over 230 chiral separations achieved on Chirobiotic R column in three different modes were reported in a single paper by Ekborg-Ott et al. *(17)*. Complementary selectivity to the other two glycopeptide CSPs was found, and the column showed excellent stability. However, a drawback for the Chirobiotic R is the high cost and scarcity of the ristocetin A chiral selector. The last commercialized member of this family of CSPs is teicoplanin aglycon (Chirobiotic TAG) *(23)*, which was produced by simply removing the carbohydrate moieties from teicoplanin. The original motive in removing the pendant carbohydrate groups was to investigate the role of the sugar units in chiral recognition *(23)*. Surprisingly, the new CSP had much improved selectivity for certain analytes. Recently, separations using the vancomycin aglycon CSP (VAG) (i.e., vancomycin with its carbohydrate moieties removed) were achieved in our laboratory *(43)*. Some improved separations were observed on VAG vs V *(43)*. This CSP has not yet been commercialized. Avoparcin is another glycopeptide macrocyclic antibiotic evaluated as a CSP in HPLC by Armstrong's group in 1998 *(19)*. This CSP will not be discussed in detail since it is not widely available. However, information on this CSP can be obtained from the original reference *(19)*.

Although the macrocyclic glycopeptides have analogous structures, they have somewhat different enantioselectivities. In fact, one of the more useful features of this class of chiral selectors is their "principle of complementary separations." This means that if a partial enantiomeric separation is obtained with one glycopeptide, there is a strong probability that a baseline or better separation can be obtained with a related Chirobiotic CSP using the same or a similar mobile phase. The reason for this phenomenon is the sometimes subtle differences in the stereoselective binding sites between these related chiral stationary phases. A fast screening approach for chiral HPLC method development, based on this complementary feature, was proposed by Wang et al. *(62)*. This is discussed in detail in **Subheading 4.4.**

In a relatively short period of time after their introduction, the macrocyclic glycopeptide chiral selectors were used successfully in most chromatographic and electrophoretic methods including HPLC *(1,2,7,9,11–14,16–27,29,31–43, 45–52,62–84)*, HPLC-mass spectrometry (MS) *(44,85–89)* TLC *(3)*, CE *(4,5,8, 10,15,52,61,69,81,90–101)*, CEC *(28,30,91,102,103)*, SFC *(104–107)*, and in enhanced-fluidity liquid chromatography (EFLC) *(108)*. Over one thousand enantiomeric separations have been reported using this class of chiral selectors, including both derivatized and underivatized amino acids, peptides, β-blockers, hydroxy-acids, amino esters, imides, hydantoins, and oxazolidinones, numerous nonsteroidal anti-inflammatory compounds, lactic acids, and other pharmaceuticals and agrochemicals. Both analytical and preparative scale applications are reported *(71)*. As was reported in the 2002 Analytical Chemistry review on Chiral Separations, "The number of publications in the field of chiral separations in high performance liquid chromatography has remained relatively steady, with the exception being a 50% increase in publications using macrocyclic gly-copeptide-based chiral stationary phases" *(92)*.

2. Materials

2.1. Structures and Properties of Macrocyclic Glycopeptide Antibiotics

All macrocyclic glycopeptide chiral selectors are naturally occurring com-pounds. Their molecular masses are between 1000 and 2100. The original macro-cyclic antibiotics were discovered by scientists at the Eli Lilly Company in the 1950s *(4)*. Vancomycin, ristocetin A, and teicoplanin are all active against Gram-positive bacteria *(7)*. They are produced as fermentation products of *Streptomy-ces orientalis*, *Nocardia lurida*, and *Actinoplanes teichomyceticus*, respectively *(56,109–112)*. Macrocyclic glycopeptide antibiotics are soluble in water and acidic aqueous solutions, but are less soluble at neutral pH *(113)*. They are mod-erately soluble in polar aprotic solvents (e.g., dimethyl sulfoxide [DMSO], di-methylformamide [DMF]) but insoluble in most other organic solvents *(113)*.

The three-dimensional molecular structures of three of the commercialized macrocyclic glycopeptides that are used for CSPs have been reported *(10)*. **Figure 1** shows both the molecular "space-filling" model (**Fig. 1A**) and the corresponding "stick" model (**Fig. 1B**) for all four macrocycles. They all pos-sess a characteristic "basket-shape" aglycon, which consists of a peptide core of complex amino acids and linked phenolic moieties. They are positioned to show a profile view of the "C-shaped" aglycon "basket." The colored regions in **Fig. 1A** are hydrophilic groups as specified in the figure legend. The black area denotes the more hydrophobic regions, including amido linkages, aroma-tic rings, and apolar connecting carbons. The aglycon basket of all these mole-cules consists of either three or four fused macrocyclic rings. They differ in their size and shape, as well as the geometric arrangement of their numerous stereo-genic centers and functional groups, which are responsible for their enantio-selective properties. The macrocycles contain both ether and peptide linkages. They make the aglycon semi-rigid, but with some flexibility. Two important characteristics of the aglycon basket are (i) the openness of the C-shaped agly-con basket, and (ii) its degree of helical twist. Both of these features are illu-strated in **Fig. 2**. When the glycopeptides are observed in profile (as seen in **Figs. 1** and **2**), "openness" is related to the distance between opposite ends of the aglycon. Clearly, vancomycin has the most open aglycon, with their short-est end-to-end distance being approx 9.3 Å *(10)*. Teicoplanin and TAG appear to have the most closed aglycon (almost cyclic) with the end-to-end distances varying from approx 4.5 to 5.5 Å *(10)*. Ristocetin A is intermediate, with end-to-end distances varying from approx 5.2 to 8.8 Å *(10)*. These variations in the

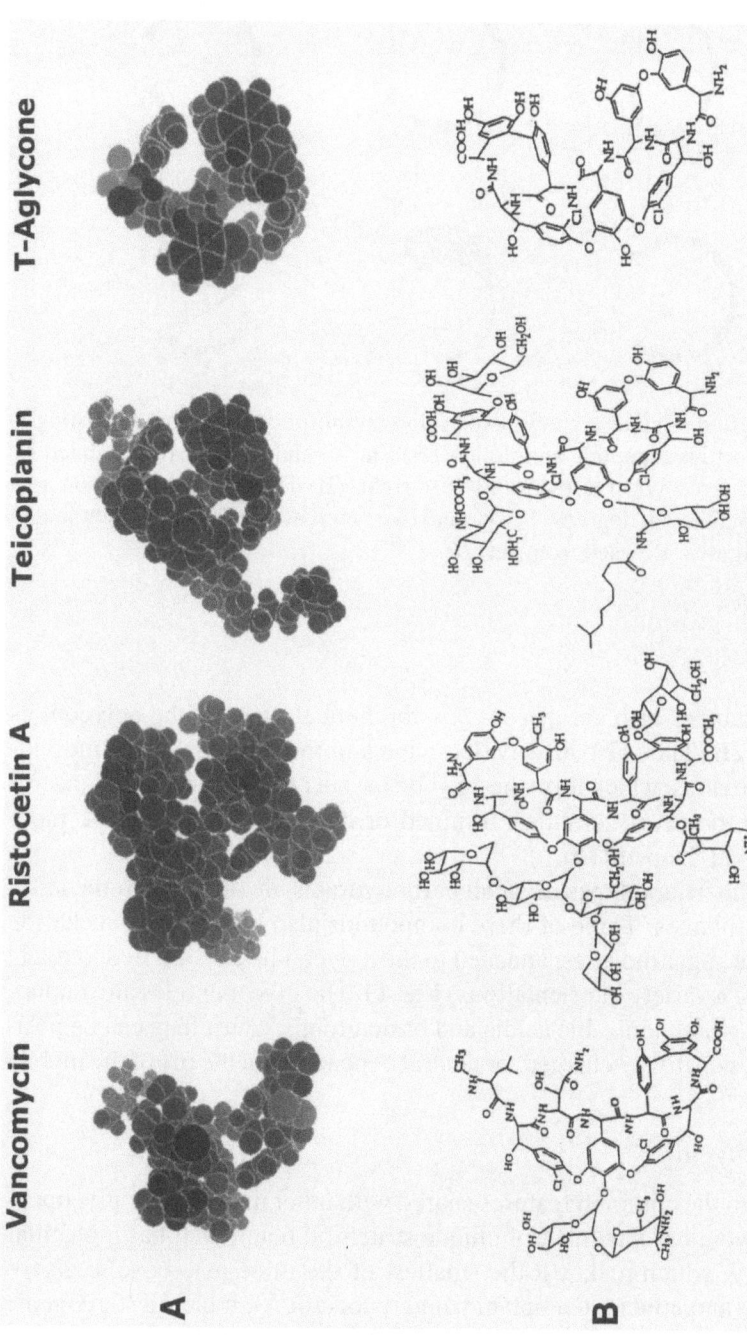

Fig. 1. Structures of the macrocyclic antibiotics vancomycin, teicoplanin, ristocetin A, and teicoplanin aglycon showing a profile view of the aglycon basket using (**A**) space-filling molecular models produced through energy minimization and (**B**) stick figures. The colored atoms in part A denote the hydrophilic moieties, while the black portion designates the more hydrophobic regions. Red represents carboxylate groups, green are ammonium groups, and blue are hydroxyls. Black regions include the aromatic rings, connecting carbons, and amido linkages. Revised from **ref. 10**.

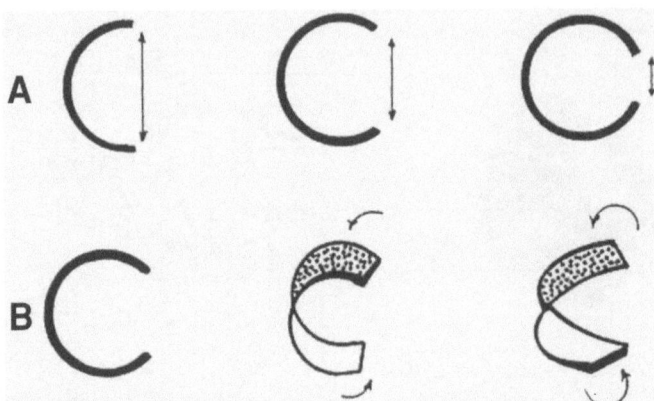

Fig. 2. Simplified schematic shows two important morphological characteristics of the aglycon part of glycopeptide antibiotics. (**A**) End-to-end distance (represented by the length of the arrow) decreases from left to right. (**B**) The C-shaped aglycon also can be twisted to different degrees. The helical twist increases from left to right in this series of three figures. Revised from **ref. 10**.

end-to-end distances also are affected by the helical twist of the aglycon, as shown in **Fig. 2B**. They also affects the shape and morphology of the individual macrocyclic rings, which form the aglycon basket *(10)*. An unstrained macrocyclic ring can be nearly circular. Strained or deformed rings may be more elongated or oval-shaped *(10)*.

Various numbers and types of small carbohydrates are attached to the aglycon via ether linkages. Three of these compounds also have aminosaccharide moieties. These sugar moieties, attached to the aglycon basket, are free to rotate and can assume a variety of orientations (**Fig. 1**). The glycopeptides are amphoteric, containing both ionizable acidic and basic groups. Thus, they can be positively charged, negatively charged, or neutral depending on the pH of the mobile phase.

2.1.1. Vancomycin

In addition to the common features shared with other macrocyclic glycopeptides, vancomycin has a number of unique structural features. It has molecular weight of 1449, which makes it the smallest of the three macrocyclic glycopeptides (i.e., vancomycin, teicoplanin, and ristocetin A). It has 18 stereogenic centers, as well as a pendant, freely rotating, disaccharide moiety consisting of D-glucose and vancosamine, and an *N*-methyl amino acid side chain around three fused macrocyclic rings bridged by five aromatic rings linked by ether and peptide bonds. Vancomycin has nine hydroxyl groups around the "basket-shaped" aglycon and on the attached disaccharide moiety, two amine groups (one primary and one secondary), and one carboxylic acid group. All these polar and ionizable groups are proximate to the ring structure and are able to offer strong hydrogen bonding and electrostatic interactions, respectively, with solute molecules. The pKas for vancomycin are 2.9, 7.2, 8.6, 9.6, 10.4, and 11.7 *(61,114)*. Its isoelectric point (pI) value (i.e., where vancomycin has a net zero charge) is 7.2 *(10)*. Below the pI value (which is the normal HPLC operational pH range of 3.5–7.5), the ionizable groups should be positively charged. The aglycon of vancomycin also contains two chloro-substituted aromatic rings, together with seven amido groups and three phenolic groups. The aromatic rings and peptide linkages provide some rigidity to the aglycon and provide the opportunity for π-π interactions and dipolar interactions, respectively, with small molecules.

2.1.2. Teicoplanin

Teicoplanin has many unique characteristics, which make it complementary to vancomycin. One of the unusual structural characteristics of teicoplanin is that it has a hydrophobic acyl side chain (hydrophobic tail) attached to a 2-amino-2-deoxy-β-D-glucopyranosyl moiety (**Fig. 1** and **Table 1**). Consequently, teicoplanin is surface active and aggregates to form micelles *(10,115)*. None of other glycopeptides has shown this type of behavior (**Table 1**). Teicoplanin's critical micelle concentration in unbuffered aqueous solutions is approx 0.18 m*M* *(10)*. As a consequence of this molecular structure, teicoplanin shows a slightly lower solubility in water than the other macrocyclic glycopeptides *(10)*.

Another unique character of teicoplanin is that it has a pI of approx 3.5. From **Table 1** and **Fig. 3**, we can see that the mobility vs pH curve is nearly flat between pH 3.0 and pH 7.0, which means that under normal operational conditions (approx pH 3.5–7.5), teicoplanin shows a slightly anionic character, while vancomycin and ristocetin A both show cationic behavior. This difference makes teicoplanin somewhat unique among the macrocyclic glycopeptide antibiotics.

Teicoplanin has a single primary amine group and a single carboxylic acid group. The respective pKa values are around 9.2 and 2.5 *(9)*. Furthermore, teicoplanin is unique in that it has two amino saccharides, both of which are *N*-acylated (**Fig. 1**), while both vancomycin and ristocetin A have an additional free amine group in an attached aminosaccharide moiety. Obviously, this additional amine affects the charge and overall ionization behavior of vancomycin and ristocetin A. There is a carboxylic acid moiety on the aglycon of both vancomycin and teicoplanin, while the equivalent group on ristocetin A is esterified (**Fig. 1**).

Table 1
Comparison of the Physicochemical Properties
of Vancomycin, Ristocetin A, and Teicoplanin,
and Teicoplanin Aglycon-Macrocyclic Antibiotics

Characteristic	Vancomycin	Ristocetin A	Teicoplanin	Teicoplanin aglycon
Molecular weight	1449	2066[a]	1877[b]	1197
No. of stereogenic centers	18	37	23	8
Produced from fermentation product of	*Streptomyces orientalis*	*Nocardia lurida*	*Actinoplanes teicomyceticus*	Synthesis from teicoplanin
No. of macrocycles	3	4	4	4
No. of monomer sugar moieties	2	6	3	0
Hydrophobic tail	0	0	1	0
No. of OH group[c]	9(3)	21(4)	14(4)	7(6)
No. of amine groups	2	2	1	1
No. of carboxylic acids	1	0	1	1
No. of amido groups	7	6	8	6
No. of aromatic groups[d]	5(2)	7(0)	7(2)	7(2)
Methyl esters	0	1	0	0
pI[e]	7.2	7.5	4.2, 6.5	NA
Relative stability	1 to 2 wk	3 to 4 wk	2 to 3 wk	NA
Aggregational behavior[f]	No	No	Yes	No

[a]Ristocetin A is a mixture of two structurally similar compounds differing by the number of carbons in one of its sugar moieties. The mixture used in this study is >90% ristocetin A.

[b]Teicoplanin is a mixture of five closely related compounds differing by the number of carbons, i.e., (C_{10}–C_{11}), and substituted groups attached to the fatty acid side chain terminating off the amino sugar (*see* **Fig. 1**).

[c]The number in parentheses correspond to phenolic moieties.

[d] The number in parentheses correspond to the number of chlorinated substituents attached to the aglycon baskets.

[e] The pI/values were determined using dilute solutions of the respective macrocyclic antibiotics in 0.1 *M* phosphate buffer. Methanol was used as the electro-osmotic flow (EOF) marker. These values were taken from **refs. *17–19*** and **33**.

[f] This is the aggregational behavior under the buffered conditions reported in (**10**). This table is revised from **ref. 10**.

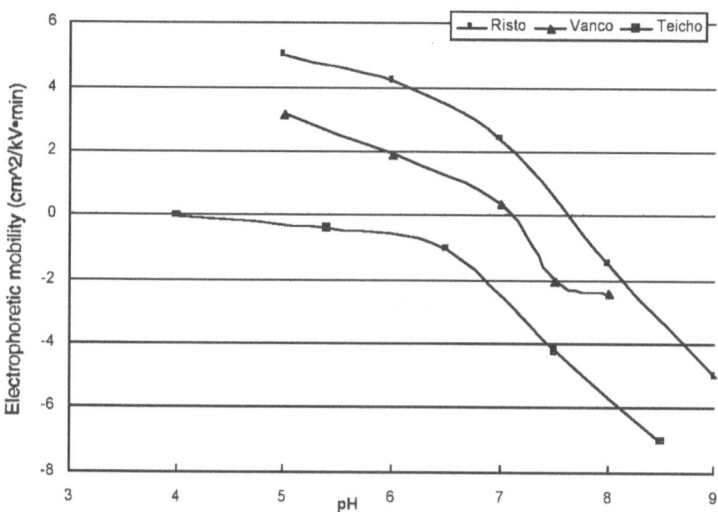

Fig. 3. Plot showing the effect of solution pH on the electrophoretic mobility of ristocetin A (●), vancomycin (▲), and teicoplanin (■) macrocyclic antibiotics using 0.1 *M* phosphate buffer. The capillary for ristocetin A and vancomycin studies was 32.5 cm × 50 µm i.d. (25 cm to the detector window). The voltage was +5 kV. The electrophoretic mobility of teicoplanin was obtained using a 44 cm × 50 µm i.d. capillary (36.5 cm to the detector) and a run voltage of +10 kV. Either acetone or methanol was used as the electro-osmotic flow marker. Revised from **ref. *10***.

In addition to its ionizable groups, teicoplanin has 10 primary and secondary hydroxyl groups and four phenolic groups, which provide additional hydrogen bonding sites for chiral analytes. The less polar character of the interior of the aglycon basket and the ten carbon side chain will provide hydrophobic interaction sites. Teicoplanin has three attached monosaccharides moieties, two of which are D-glucosamine, and one of which is D-mannose.

There are five closely related teicoplanin glycopeptides that have been identified, namely, teicoplanin A_2-1, A_2-2, A_2-3, A_2-4, and A_2-5 *(116)*. The five forms of teicoplanin have different fatty acid chains attached to the amine group of glucosamine, i.e., A_2-1 with (z)-4-decanoic acid, A_2-2 with 8-methynonoic acid, A_2-3 with n-decanoic acid, A_2-4 with 8-methyldecanoic acid, and A_2-5 with 9-methyldecanoic acid, respectively. The most common teicoplanin glycopeptide is A_2-2, which has a molecular weight of 1879.7 *(116)*.

2.1.3. Teicoplanin Aglycon

The Chirobiotic TAG column is made by covalently bonding the aglycon part of teicoplanin to silica gel via linkage chains. The sugar moieties are removed by the following method *(23)*. First, teicoplanin is dissolved in 30:1/DMSO:80% H_2SO_4 (v/v) at concentration around 0.1 *M*. Then the mixture is heated to 65°C for 1.5 h. The phenoxy linkage of the nonylglucosamine is then hydrolyzed. To remove the mannose unit, addition of 80% H_2SO_4 is needed (at the same concentration as the starting solution), and the mixture is kept at 65°C for 3 h. To remove the last *N*-acetyl-β-D-glucosamine unit, the temperature is raised to 80°C for 24 h *(23)*. The aglycon of teicoplanin becomes water insoluble after the removal of the sugar units from teicoplanin. It consists of the four fused macrocyclic rings, which form a semi-rigid basket. It still contains seven aromatic rings of which two are chloro-substituted, and five of which are ionizable phenolic moieties (**Fig. 1**). Because it is just the aglycon portion of teicoplanin, the molecular weight drops to 1197. It has only 8 stereogenic centers, in the meantime, the number of hydroxyl groups also decreases to eight. However, the chiral selectivity is not necessarily decreased by the loss of the sugar moieties. The single primary amine, carboxylic acid group, and the phenolic moieties control the overall charge of the aglycon. These structural features and their effect on enantioselectivity will be discussed later in this chapter.

2.1.4. Ristocetin A

Ristocetin A is the third macrocyclic glycopeptide antibiotic used and commercialized as a CSP for HPLC. It is also the largest member of this class, with a MW of 2066. It has the greatest number of stereogenic centers (i.e., 37). It contains seven aromatic rings, six amide linkages, twenty-one hydroxyl groups, two primary amine groups, and one methyl ester. It has four, rather than three, fused macrocyclic rings (one twelve-membered, one fourteen-membered, and two sixteen-membered rings) and a greater number and different types of attached sugar moieties. Ristocetin A has a pendant tetrasaccharide and two monosaccharide moieties, composed of D-arabinose, D-mannose, D-glucose, and D-rhamnose. As mentioned before, the carboxylic acid group on ristocetin A is esterified (**Fig. 1**). The pI value for ristocetin A is 7.5, which means that under normal operational conditions (approx pH 3.5–7.5), it will show cationic behavior. The only other ionizable groups on ristocetin A are the phenolic moieties. Thus at operational pH value, these are generally protonated and probably serve mainly as hydrogen bonding sites.

2.2. Commercially Available
Glycopeptide Antibiotic Chiral Stationary Phases

2.2.1 Preparation of Glycopeptide Antibiotic CSPs

As mentioned before, the designations for commercialized macrocyclic glycopeptide CSPs for HPLC are Chirobiotic V, Chirobiotic R, Chirobiotic T, and

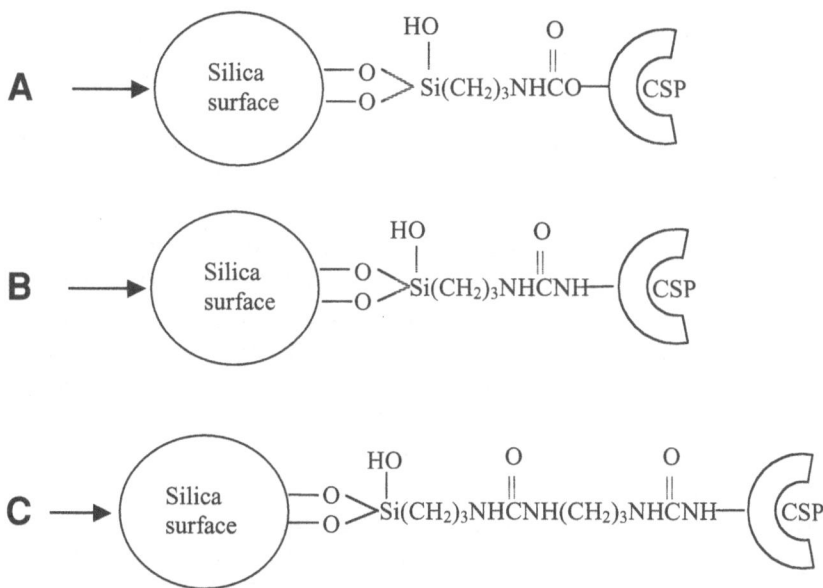

Fig. 4. Schematic diagram of the attachment of glycopeptide to silica.

Chirobiotic TAG based on the first letter of the corresponding glycopeptide chiral selectors attached to a 5-μm spherical silica gel. The bonding reagents or linkage chains may be organosilanes that are terminated by carboxylic groups, amine groups, epoxy groups, and isocyanate groups. Examples of such reagents are 2-(carbomethoxy)-ethyltrichlorosilane, 3-aminopropyldimethylethoxy-silane, and 3-(glycidoxypropyl)-trimethoxysilane, etc. Therefore, the linkage that covalently attaches the macrocycle to the tethers attached to silica may be an ether, thioether, amine, amide, carbamate, or urea *(61,114)*. **Figures 4A** and **B** show examples of the attachment of glycopeptide to silica employing 3-iso-cyanatopropyltriethoxylsilane.

In **Fig. 4**, the CSP stands for the macrocyclic glycopeptide CSPs. Parts (A) and (B) represent the CSPs react with 3-isocyanatopropyltriethoxylsilane through a hydroxyl group or an amino group forming a carbamate or a urea linkage, respectively. Part (C) also shows the chiral selector linked to γ-amino-propylsilanized silica gel via an alkyl diisocyanate moiety *(61,114)*. On average, there are four linkage chains to every chiral selector *(2)*.

Overall, these macrocyclic glycopeptide CSPs are much more stable than traditional protein CSPs and have much higher loading capacities. In compari-son to the cellulose and amylose phases, the glycopeptides CSPs can tolerate a much wider range of solvents and, consequently, have greater versatility, stabil-ity, and longevity *(71)*.

2.2.2. Chiral Selector Coverage

In HPLC, chiral selectors can be used as mobile phase additives or as part of the CSPs. In the case of as CSPs, the column loading (or the coverage of the chiral selectors) can affect retention, selectivity, efficiency, and enantioresolution. A systematic study of the effect of the surface coverage of these chiral selectors on column efficiency and enantiomeric selectivity was done *(7,22)*. Surface coverage can be controlled by altering the initial reaction ratio (of selector to silica gel), the selector concentration, the reaction time, reaction temperature, and the other relevant reaction conditions. Separations were compared based on their capacity factors (k'), selectivity values (α), efficiencies (N), and resolution (Rs) for different coverage columns (high, medium and low) in all three mobile phase modes. It is reasonable that the retention is longer on the higher coverage CSP, which provides a greater number of chiral selector adsorption sites. However, the increase in selectivity factor (α) is relatively smaller. For most of the compounds, both shorter retention times and comparable efficiencies and selectivities were achieved on medium coverage CSP for ristocetin A CSP *(22)*. As can be seen from **Fig. 5**, it is the more retained (i.e., the second eluting) enantiomer of each analyte that is most affected by changes in the loading of the chiral selectors on the CSP.

2.2.3. Chirobiotic Column Care and Stability

In order to get the best results and extend the column lifetime, the following column care steps should be followed:

1. Both new column and used columns, which may display decreased resolution, should be conditioned first with 50/50: CH_3CN/50 m*M* NH_4OAc for 20 column vol (sometimes, a buffer at an opposite range of the pH is good if the column has been used at one specific pH for a while), followed by 10 column vol of HPLC-grade water. Finally, wash with 10 column vol of organic solvent, i.e., acetonitrile or ethanol. This procedure will eliminate removable contaminants and adsorbed components that could block important binding sites on the CSP.
2. A test compound for a specific mobile phase should be selected to evaluate the performance of a specific Chirobiotic column (this is also referred to as the selectivity control test). **Table 2** lists suggested test conditions for different Chirobiotic columns. The test compounds and the corresponding test conditions are chosen in order to achieve barely baseline separations, so that decreased column performance (i.e., resolution and retention times) can be easily detected.
3. Because different mobile phase modes can be employed when using Chirobiotic columns, care should always be exercised in switching from one mobile phase to another. Normally, Chirobiotic columns are best stored in isopropanol due to its weaker polarity, solubility properties, and good compatibility with any kind of mobile phase solvent. When switching from the polar organic mode, methanol

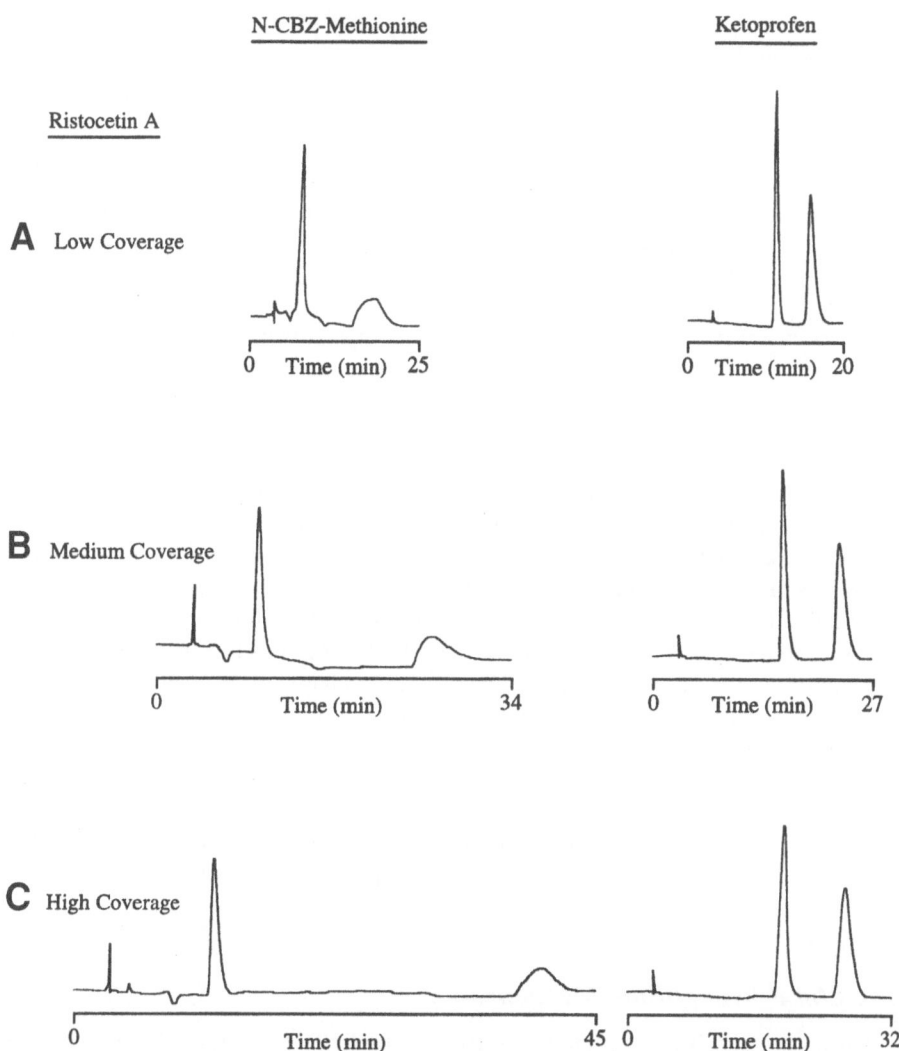

Fig. 5. Reversed-phase separation of N-CBZ methionine and ketoprofen using high (1 g ristocetin A/3.00 g silica gel), medium (0.75 g ristocetin A/3.00 g silica gel), and low coverage (0.5 g ristocetin A/3.00 g silica gel) of the ristocetin A chiral selector. MP: MeOH/0.1%TEAA buffer, pH 7.0 (20/80, v/v). Detection: UV 254 nm. Flow rate: 1.0 mL/min room temperature. Revised from **ref. 22**.

should be first employed to remove acids and bases, especially when trifluoro-acetic acid was used. The column should be washed while it is connected to the system in order to purge all system lines and the detector cell. In switching from the normal phase mode, the column should be washed with ethanol first to get rid

Table 2
Recommended Selectivity
Test Compounds and Conditions for Chirobiotic Columns

CSPs	Test compound	Mobile phase	Retention times (min)
Chirobiotic V	Nefopam	100/0.1%: MeOH/NH$_4$TFA	13.4, 14.4
Chirobiotic T	Phenylalanine	30/70: EtOH/H$_2$O	5.2, 6.1
Chirobiotic TAG	Ornithine	20/80: MeOH/0.1 M NaH$_2$PO$_4$	7.2, 8.3
Chirobiotic R	Ketoprofen	100/0.01%: MeOH/NH$_4$OAc	6.9, 7.4

    of hexane or other organic nonpolar solvents. When switching from the reversed-phase mode, wash the column with water first and then follow with neat ethanol.

4. It is necessary to be aware of the pH stability range of silica gel-based Chirobiotic columns, which is usually from 3.5 to 7.5. Strong acids (pH < 2.0) or bases (pH > 8.0) will cause column damage. Running with a pure water mobile phase should be avoided or at least reduced. A short presaturation column of 40 μm silica gel prior to the injector is useful for the Chirobiotic columns to extend their lifetime when using aqueous or reversed-phase mobile phases. The function of this column is to saturate any water-containing mobile phase with silicic acid. It has an added benefit of filtering the mobile phase.

5. Never store columns, even for a short period of time, in buffer. This may cause the columns to clog or become damaged if the buffered solvent evaporates and precipitation occurs. Wash the column with water after using buffer solvents. Then flush the column with ethanol, methanol, isopropanol, or acetonitrile.

3. Methods

3.1. Chiral Recognition Mechanism

    Chiral recognition refers to the ability of the CSP to interact differently with two enantiomers. In most cases, this also leads their separation. Chromatographic separation depends on the ability of the CSP to preferentially interact with one of the analyte enantiomers when forming somewhat different, relatively short-lived, transient diastereomeric complexes. Separation is achieved based on the difference in retention times.

    The interaction of two enantiomers with a CSP, can be expressed as the difference in free energy $-\Delta_{1,2} \Delta G°$ (i.e., the free energy difference for the transfer of the enantiomers between the mobile phase and the stationary phase). It can be calculated from the separation factor, α, according to the following equations:

$$-\Delta_{1,2}\,\Delta G^{\circ} = \Delta_2 G^{\circ} - \Delta_1 G^{\circ}$$
$$-\Delta_{1,2}\,\Delta G^{\circ} = RT\ \ln k_2\,/\,k_1 = RT\ \ln \alpha$$

Here, k_2 and k_1 are the retention factors of the two enantiomers. Thus, chiral discrimination can be expressed in thermodynamic terms. It usually turns out that only small energy differences are needed for the chromatographic resolution of enantiomers.

Understanding the chiral discrimination mechanism is useful when developing an optimum separation method for HPLC using macrocyclic glycopeptide CSPs. Although there is some debate about its simplistic nature, most chiral recognition models are still based on the three-point interaction rule, where a minimum of three simultaneous interactions between at least one enantiomer and the chiral stationary phase are required. However, the complex structures of macrocyclic glycopeptide CSPs can make it difficult to understand the exact interactions that lead to chiral recognition. Almost all possible intermolecular interactions leading to chiral recognition are available with macrocyclic glycopeptide CSPs. This is one thing that contributes to their wide applicability as chiral selectors. Another factor that increases the applicability of these CSPs, but also makes their mechanism of action harder to define, is that there can be different interactions in different chromatographic modes.

3.1.1. Reversed-Phase Mode

In the reversed-phase mode, as with achiral HPLC stationary phases, the glycopeptide CSP acts as a relatively nonpolar media compared to the polar hydro-organic mobile phase. Hydrophobic interactions between glycopeptide CSPs and chiral analytes were proposed by Armstrong et al. *(2)* for the reversed-phase mode and confirmed by other researchers *(82)*. Thus, hydrophobic interactions between the nonpolar part of an analyte and the interior of the glycopeptide basket, is a dominant factor in retention and may also be one of the important factors contributing to chiral recognition. The possible hydrophobic interaction sites are shown as the black areas in **Fig. 1**. Additional interactions that can lead to chiral recognition include electrostatic interactions, hydrogen bonding, dipole-dipole interactions, and steric repulsive interactions. Which interactions are most important for a specific compound depends on the nature of that analyte (i.e., its size, geometry, number and type of functional groups, flexibility, etc.), as well as the nature of the mobile phase used.

Electrostatic interactions can affect both retention and enantioselectivity. **Figures 6** and **7** show the pH effects on the retention factor, k, for dansyl-amino acids *(82)*. An inflection point was observed for all the k vs pH plots, as well as the selectivity factor vs pH plot. This point is very close to the pKa value of the solute amino group, which is 4.5. Below this point, the amine group of the

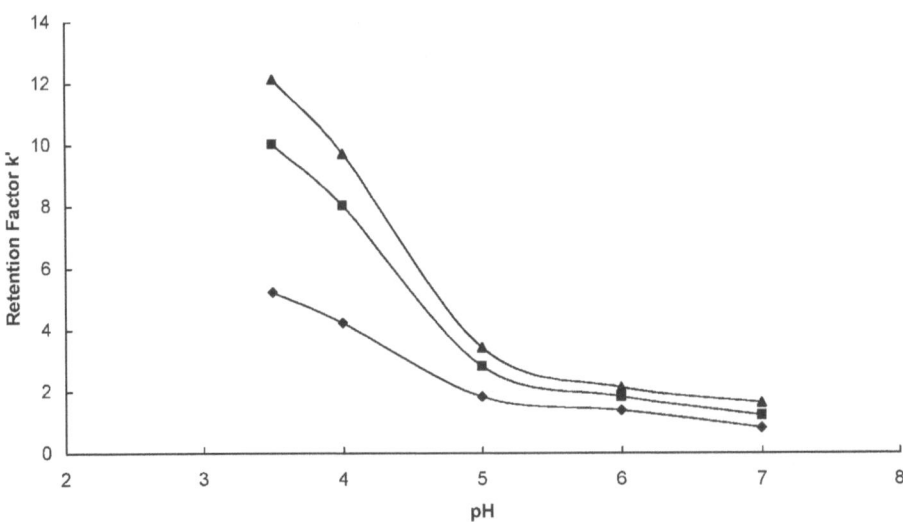

Fig. 6. The pH effect on retention factor (k') for various dansyl amino acids (▲: dansyl leucine; ■: dansyl valine; ♦: dansyl serine) at T = 20°C using Chirobiotic T as CSP, mobile phase: 0.01 *M* citrate buffer/methanol (90/10, v/v). Revised from **ref. 82**.

analyte is protonated and retention is significantly increased. This could be attributed to the ionic interaction between protonated amine group of the analyte and the carboxylate group of the CSP. However, the selectivity factor, α, decreased somewhat (**Fig. 7**). This reveals that the electrostatic interaction between the carboxylate group on the CSPs and the amine group on the analyte is effective only for increasing retention, but not for enhancing chiral recognition of dansyl-amino acid enantiomers. After a comprehensive study, Peyrin et al. *(82)* verified both thermodynamically and experimentally that the interaction between the amine moiety of teicoplanin CSP and the carboxylate group of dansyl-amino acids plays a crucial role in chiral recognition mechanism, which was consistent with previous results *(7,9,11–13,23,42,66,74)*. Nair et al. reached a similar conclusion in 1996 using vancomycin as the chiral selector in a CE study *(13)*. This was done by forming a copper complex that blocked the secondary amine group on the aglycon portion of vancomycin *(13)*.

Additional evidence was also provided by the huge difference in selectivity when separating enantiomers of the dipeptide Ala-Gly and Gly-Ala *(9)*. The selectivity factor (α) was significantly increased from 1.15 for DL-Ala-Gly to 10 for Gly-DL-Ala, while the only difference between these two peptides is the proximity of the terminal carboxylate group to the stereogenic center. In Gly-DL-Ala, the carboxylate group is directly connected to the stereogenic center,

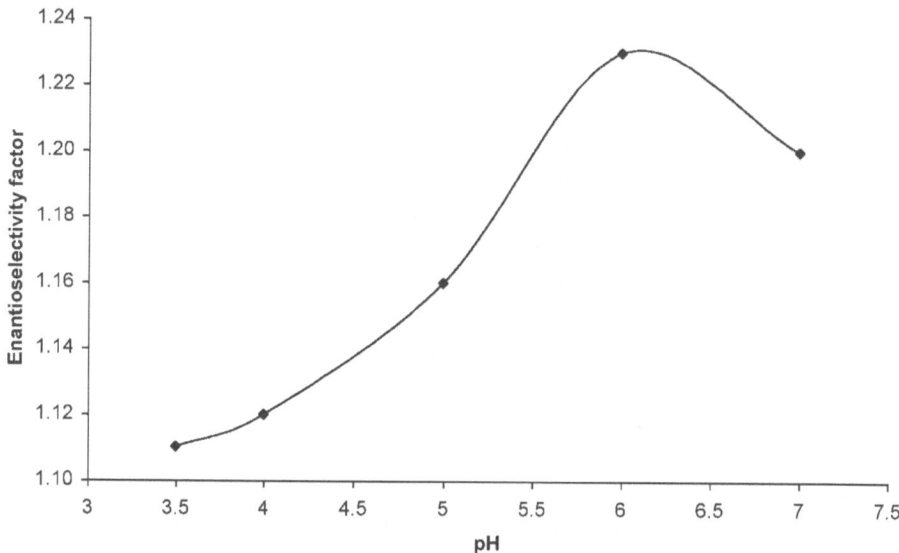

Fig. 7. The pH effect on selectivity factor (α) for dansyl serine enantiomers. All other conditions same as **Fig. 6**. Revised from **ref. 82**.

while that of the DL-Ala-Gly is five atoms away from the stereogenic center. This result clearly showed the essential role of the teicoplanin amine group as well as the analyte carboxylate group (and its proximity to the chiral center) in the chiral recognition mechanism *(9)*. In addition, the fact that chiral compounds with acidic groups (i.e., carboxylate, sulfinate, phosphate, etc.) are most easily resolved, further supports this as one of the important chiral interaction sites for acidic or anionic compounds.

3.1.2. Normal Phase Mode

In the normal phase mode (where the mobile phase is nonpolar), the CSP behaves as a polar stationary phase. The strongly polar functional groups and aromatic rings of the CSP provide the interactions needed for both retention and chiral recognition. Therefore, hydrogen bonding, π-π interactions, dipole stacking, steric repulsion, and, in some cases, electrostatic interactions are the dominant interactions that occur between the CSP and the analytes in the normal phase mode. Note that the absence of water, by definition, precludes the possibility of hydrophobic interactions contributing to either retention or selectivity. However, in the presence of nonpolar solvents (as in the normal phase mode), the enhanced π-π and dipolar interactions often make up for the lost hydrophobic interactions that were important in the reversed-phase mode.

3.1.3. New Polar Organic Mode

In the new polar organic mode, no nonpolar solvents (like hexane) are used. Generally, the main component in the polar organic mobile phase is an alcohol (e.g., methanol, ethanol, or isopropanol). The dominant interactions between the analyte and CSP usually involve hydrogen bonding, electrostatic, dipolar, and steric interactions (or some combination thereof) *(34)*. Chiral analytes suitable for this kind of mobile phase have some special requirements (*see* **Note 1**).

It should be noted that the dominance of different types of interactions in the reversed-phase mode vs the normal phase mode means that the chiral recognition mechanism can be very different in these modes. Consequently, very different types of chiral molecules often can be separated in one mode vs the other.

3.1.4. Role of Carbohydrate Moieties

The role of the pendant carbohydrate moieties in chiral recognition was studied by comparing separations performed on the Chirobiotic TAG and T columns *(23)*. A significant selectivity increase was observed for separation of amphoteric molecules like the α, β, γ, and cyclic amino acids, as well as some other kinds of acids. This indicated the importance of the aglycon portion in chiral recognition. In this case, the carbohydrate moieties of the native teicoplanin molecule may intervene in the chiral recognition process in at least three possible ways: (i) steric hindrance, in which the sugar units occupy room on the aglycon, which limits the access of other molecules to binding sites; (ii) blocking a possible interaction sites on the aglycon, where two sugars are linked to the aglycon through phenol hydroxyl groups, and the third sugar is linked through an alcohol moiety (**Fig. 1**); and (iii) offering competing interaction sites, since the carbohydrates are themselves chiral and have hydroxyl, ether, and amido functional groups available for interaction *(23)*.

Conversely, the resolution of some analytes, such as amino alcohols, which were excellent on Chirobiotic T, decreased on the Chirobiotic TAG column. This indicates the importance of the sugar units for enhancing chiral recognition for specific types of analytes.

As with the TAG CSP, removal of the sugar units from native vancomycin improved the selectivity of some neutral sulfoxides molecules *(43)*. However, it completely destroyed chiral recognition in a series of racemic esters *(117)*. This means that the carbohydrate moieties enhanced the enantioselective separation of some molecules, but decreases it for others.

Steric effects were also found to be very important for chiral recognition in all mobile phase modes. Bulky groups, such as phenyl groups or methyl groups, attached on or next to the stereogenic center can effect chiral separation when using macrocyclic glycopeptide CSPs *(14,43)*.

3.2. Method Development

Before using a glycopeptide CSP, one should first examine the analyte structure. Macrocyclic glycopeptide CSPs are able to separate many kinds of molecules. However, molecules with the following characteristics are very likely to achieve baseline separation on these CSPs, if they have: (i) ionizable groups to provide electrostatic interactions; (ii) hydrogen bonding groups; (iii) hydrophobic groups; (iv) bulky substituents next to or close to the chiral center; (v) one relatively polar functional group attach to or close to (α or β) the stereogenic center; and (vi) aromatic groups capable of π-π interactions.

3.2.1. Optimizing Chiral Separations

Given the variety of the functionality that exists within the macrocyclic glycopeptides, Chirobiotic columns can work well in reversed-phase, normal phase, and polar organic modes. Most of the time, higher efficiencies are observed in the polar organic mode and normal phase mode. The ability to operate in different modes can be advantageous, since different compounds separate best in different modes. If the polar organic mode is applicable (*see* **Note 1**), it is usually the best choice to start with. The solubility of the analytes in different solvents also affects the choice of the mobile phase modes, please *see* **Note 2**.

3.2.1.1. REVERSED-PHASE MODE

Chirobiotic columns separate many compounds in the reversed-phase mode. Optimization of reversed-phase separations are done in much the same way as optimizing achiral separations on C18 columns. It is done by controlling the type and the percentage of organic modifier and other additives, the type and pH of the buffer, ionic strength, flow rate, and the temperature. These factors will be discussed below.

3.2.1.1.1. Organic Modifier. The effect of organic modifier concentration on the retention of 5-methyl-5-phenylhydantoin in the reversed-phase mode is shown in **Fig. 8** *(2)*. A typical U-shaped curve was obtained when plotting retention and resolution vs composition of organic modifier/aqueous solution (as was previously noted with cyclodextrin-based CSPs) *(2)*. Longer retention and better resolution were usually observed in both the high and low concentration regions of organic modifier. This retention behavior clearly indicated that the importance of hydrophobic interactions between the analyte and the macrocycle at higher concentrations of aqueous buffer *(2)*. When using higher concentration of organic modifier, hydrophobic interactions no longer contribute to retention. However, other interactions become increasingly dominant (e.g., hydrogen bonding, dipolar interactions, etc.).

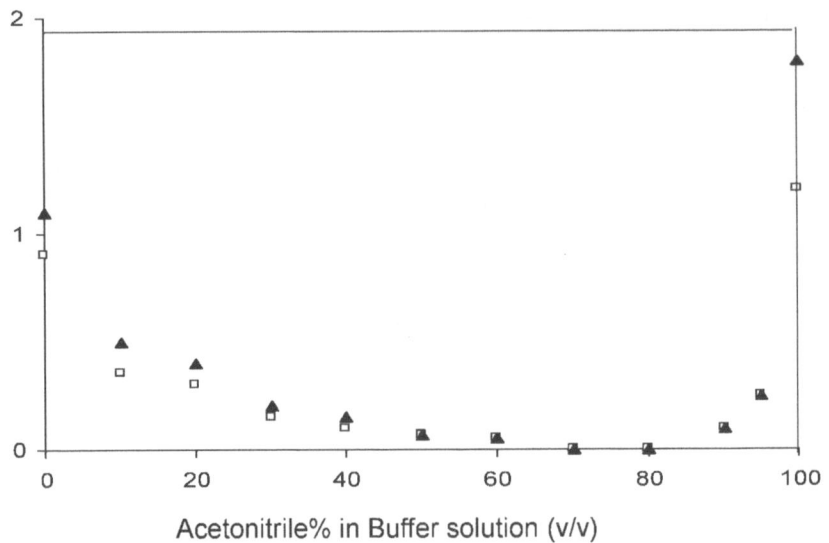

Fig. 8. Reversed-phase retention of the first eluted (□) and second eluted (▲) enanti-omers of 5-methyl-5-phenylhydantoin as a function of mobile phase composition using Chirobiotic V as CSP. Revised from **ref. 2**.

The type of organic modifier can affect both selectivity and efficiency. If no separation is observed with the first solvent choice, switching to a different organic modifier is highly recommended. **Figure 9** shows effects of different types of organic modifiers in the RP mode using vancomycin as the CSP selec-tor. A variety of organic modifiers can be used to affect selectivity including methanol, ethanol, and isopropanol, acetonitrile (ACN), tetrahydrofuran (THF), etc. As shown in **Fig. 9**, different macrocyclic glycopeptide CSPs have a dif-ferent optimum type of organic modifier. For detail about choice of types of organic modifiers, *see* **Note 3**. A typical starting composition ratio is 10/90: ACN/Buffer, pH 3.5–7.0, while the composition of 20/80 is recommended when using alcohol as modifier.

Figure 10 shows that the retention behavior of amino acids in the reversed-phase mode is different from that of most other analytes. Amino acids often have smaller retention factors with water-rich mobile phases. This phenom-enon may be due to the higher solubility of amino acids in water than in almost all other solvents. In addition, it was shown that the electrostatic interaction between the amino acid and the CSP is so strong that organic modifier concen-tration does not affect the retention and selectivity factor as much as it does other analytes *(7)*. The resolution factor (Rs) does change with the mobile phase

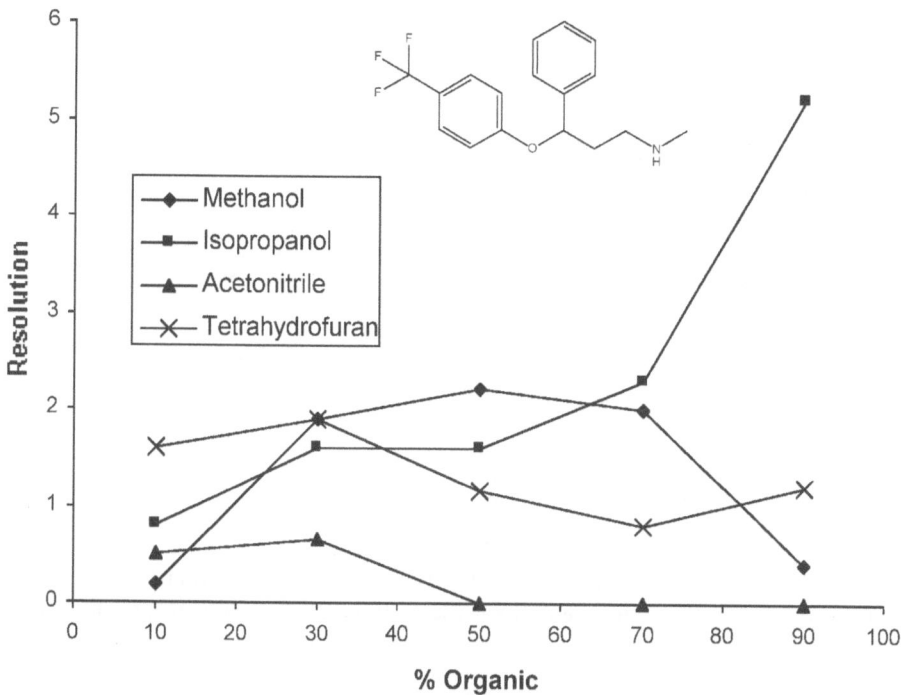

Fig. 9. Effect of different organic modifier on resolution using Chirobiotic V as CSP.
Courtesy of Scott Sharpe, Eli Lilly & Co.

composition. This is because the Rs value depends on both the selectivity and
the efficiency.

3.2.1.1.2. Buffers. Buffers are widely used in the mobile phase for Chirobiotic columns to control the ionization of both the analyte and the CSP. In addition, the buffer has ionic strength effects, and affects the selectivity for certain
compounds via secondary interactions between the buffer and CSP and/or buffer and analyte. The use of buffer can increase the efficiency of the separation significantly *(118)*. To obtain efficient and reproducible separations in the reversed-
phase mode, some buffers are essential for separations even for neutral compounds. For most commonly used buffers, *see* **Note 4**.

Macrocyclic glycopeptides have ionizable groups as proteins do, therefore,
their charge and perhaps their conformation can vary with the pH of the mobile
phase. As mentioned previously in **Subheading 2.1.**, pH will have different
effects with different Chirobiotic columns due to variations in their ionizable
functional groups and their pI values.

Within the operating pH range (3.5–7.5), the strength of the short-lived complex formed by the CSP and the analyte can depend on the charge of the analyte.

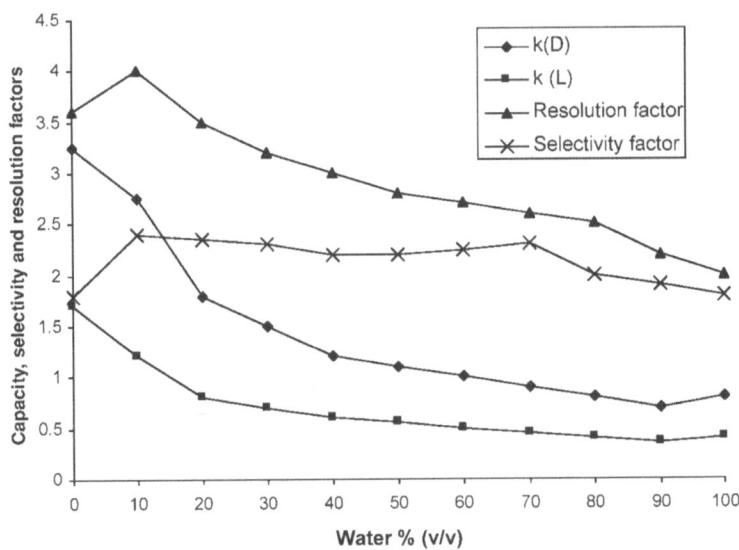

Fig. 10. Effect of organic modifier (EtOH)/water ratio on resolution, selectivity, and retention factor k for the enantiomeric separation of methionine using Chirobiotic T as CSP. Revised from **ref. 7.**

The charge of the analyte will in turn be affected by the pH of the mobile phase. Vancomycin and ristocetin A are both in cationic form, while teicoplanin and most amino acid analytes exist in the zwitterionic form (i.e., with an anionic –COO⁻ group and a cationic –NH₃⁺ group). Changes in the pH of the mobile phase can change the ionization of both the analyte and the CSP. Therefore, pH can affect the interaction mechanism even if the analyte is a neutral molecule. As a general rule, the starting pH should close to the pI value of glycopeptide antibiotics. Alternatively, test runs can be made at pH 4.0 and 7.0. After finding which pH extreme produces the optimum separation, adjust the pH to 0.5 pH unit above or below the pKa of the analyte.

3.2.1.2. NORMAL PHASE MODE

Macrocyclic glycopeptide CSPs effectively separate a variety of compounds in the normal phase mode. There is no solvent-induced denaturation or any irreversible change in these CSPs *(2,43)*. One advantage of using Chirobiotic columns in the normal phase mode is the higher efficiency of most separations. Complementary selectivity to the reversed-phase mode also is found in the normal phase mode (i.e., even if no separation was observed in the reversed-phase mode, it may be obtained by switching to the normal phase mode, or vice versa). In normal phase HPLC, hydrogen bonding, π-π interactions, dipole-dipole

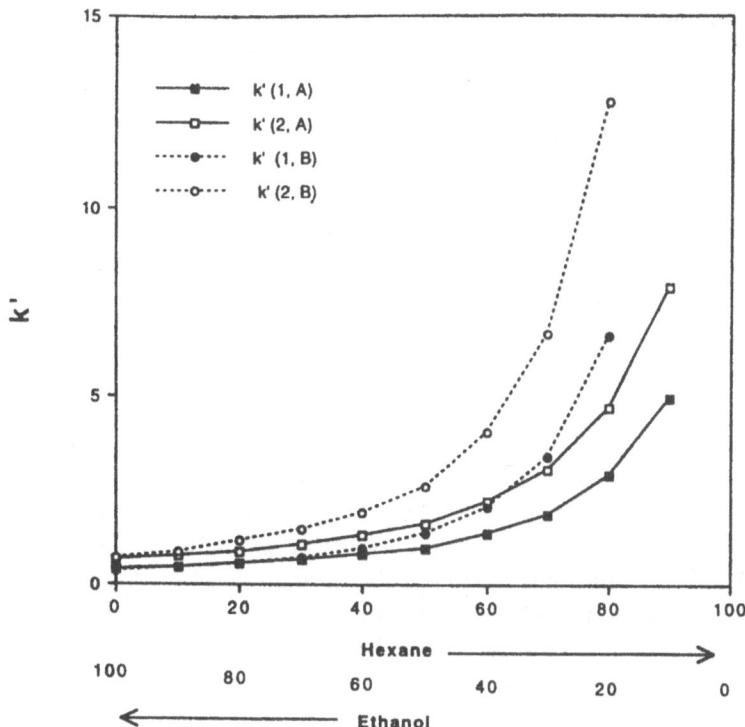

Fig. 11. Effect of organic modifier (EtOH)/hexane ratio on the retention of the first (1) and second (2) enantiomers of γ-phenyl-γ-butyrolactone (solid line, A) and 4-phenyl-2-methoxy-6-oxo-1,4,5,6-tetrahydropyridine-3-carbonitrile (dotted line, B) using teicoplanin as CSP. Revised from **ref. 7**.

stacking, and steric interactions are more important. Obviously, hydrophobic interactions are not relevant.

In normal phase HPLC, retention is controlled by adjusting the percentage of a polar organic modifier such as ethanol or propanol. **Figure 11** shows the effect of added organic modifier on the retention of two pairs of enantiomers in the normal phase mode. For all enantiomers, retention is decreased as the percentage of organic modifier is increased. However, the exact effect of the mobile phase composition varies somewhat from compound to compound. Also, it was observed that resolution can be dramatically changed by using different modifiers. Usually this results from a change in the efficiency of a separation. For example, the separation efficiencies with hexane/ethanol mobile phase mixtures are usually higher than that with hexane/isopropanol mixture on macrocyclic glycopeptide CSPs.

Different combinations of polar and nonpolar solvents can affect the selectivity of a separation as well. The most commonly used nonpolar solvents for Chirobiotic CSPs are hexane or heptane. Methyl *tert*-butyl ether (MtBE) was also found to be a useful nonpolar solvent in some cases, when combined with alcohol or acetonitrile modifiers *(43)*. Recently, this unusual solvent combination was used with Chirobiotic V and VAG columns to separate racemic sulfoxide compounds *(43)*. A significant increase in the number of separations was achieved when switching the mobile phase of hexane/ethanol (EtOH) to MtBE/EtOH, while the opposite effect was observed on Chirobiotic T and Chirobiotic TAG column.

The most common organic modifiers are alcohols (i.e., ethanol and isopropanol). Most of the time, the ethanol is the better organic modifier in terms of efficiency and resolution. However, it is worthwhile to mention that methanol is slightly soluble in hexane (about 1% v/v). Thus, in normal phase separations, which require very little polar organic modifier, this may be an option. It was also reported that halogenated solvents as well as DMF and dioxane are sometimes useful on Chirobiotic CSPs in the normal phase mode *(71)*.

Once a separation is achieved, adjusting other parameters to gain better resolution is necessary. These parameters include percentage and type of modifier, as well as small amounts of other additives, including some acids or bases.

3.2.1.3. New Polar Organic Mode

The new polar organic mode is a modification of the polar organic mode used with cyclodextrin-bonded phases *(119,120)*. Subsequently, this approach was found to be highly effective with macrocyclic glycopeptide CSPs *(22,34,121)*.

This mode is more closely related to the normal phase mode than to the reversed-phase mode. Generally, the main component in the polar organic mobile phase is an alcohol (e.g., methanol, ethanol, or isopropanol) with a very small amount of acid/base added to effect retention and selectivity. This mobile phase can be considered as an "extreme" case of the normal phase mode, or an "extreme" case of the reversed-phase mode. Methanol and ethanol are usually the best polar organic solvents when using the ristocetin A, teicoplanin, and teicoplanin aglycon columns. If not enough retention is obtained, acetonitrile is added to the alcoholic mobile phase at various ratios to gain an appropriate retention time.

There are usually two different factors that need to be adjusted to optimize a separation using polar organic mobile phases. The first one is the absolute amount of acid and base added, which is essential for optimization of retention, and the second one is the relative ratio of acid to base, which controls selectivity (*see* **Note 5**). **Figure 12** shows the effect of added acid and base on resolution in the polar organic mode.

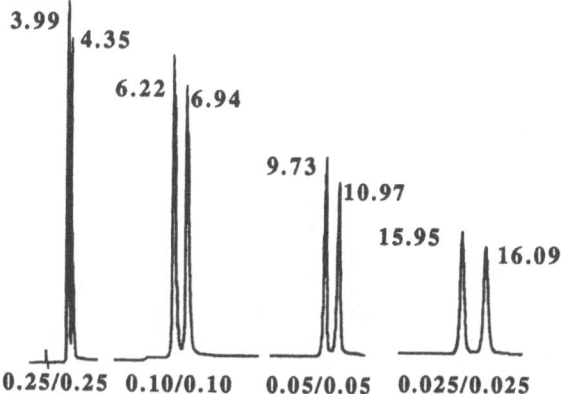

Fig. 12. The separation of enantiomers of propranolol employing different HOAC/ TEAA concentrations on Chirobiotic T. Room temperature and at a flow rate of 2 mL/ min. Revised from **ref.** *14*.

One advantage of the polar organic mobile phase is that it can dissolve many analyte salts (i.e., amine hydrochlorides and sodium carboxylates etc.) that can not be dissolved with traditional normal phase solvents *(119)*. Prior to the development of the polar organic mode, these more polar analytes and salts could only be separated in the reversed-phase mode. Another advantage is that polar organic mobile phases are compatible with reversed-phase solvents when these two formats are used in achiral-chiral column switching procedures *(45,122–124)*.

The polar organic mode should be considered first when doing methods development for analytes that meet the previously mentioned structural criteria (*see* **Note 1**). This mobile phase system offers several advantages, such as simplicity, versatility, and high efficiency. In addition, it uses methanol, which is a relatively inexpensive, less toxic solvent.

3.2.2. Miscellaneous Approaches for Improving Selectivity and Resolution

3.2.2.1. TEMPERATURE EFFECTS

Temperature can effect the retention and resolution of chiral analytes on Chirobiotic columns *(14,20,44,92,125,126)*. The reason for this can be attributed to a change in the binding constant of a solute to the CSP with temperature. Studies on the effect of temperature on separations using Chirobiotic V, Chirobiotic T, and Chirobiotic R columns in both normal phase and reversed-phase modes have been reported *(2,20,44,125)*. A systematic study of temperature effects on the resolution of 4-benzyl-2-oxazolidinone on vancomycin was done by Scott and Beesley *(75)*. It was observed that the temperature effect on

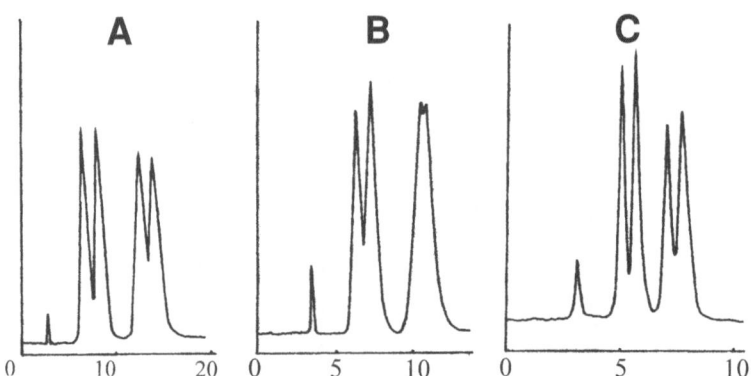

Fig. 13. Temperature effect on separation of β-MePhe enantiomers. Column: Chirobiotic T; MP: H_2O/MeOH (10/90, v/v), detection 202 nm. Flow rate 1 mL/min. (**A**) 1°C, (**B**) 20°C (**C**) 50°C; peaks: 1) erytho-L isomer, 2) erytho-D isomer; 3) threo-L isomer; 4) threo-D isomer. Revised from **ref. *14*.**

the retention factor (k), separation factor (α), peak-to-peak separation distance, and efficiency (i.e., peak width) are all affected by temperature. Temperature and solvent composition are independent in terms of their effects toward the above factors. Usually an increased resolution can be achieved by lowering the separation temperature.

Figure 13 shows the effect of temperature for enantioselectivity of β-methyl-phenylalanine separated on a Chirobiotic T column *(14)*. This figure shows that poor resolution was observed at ambient temperature, but improved resolution was observed at both higher and lower temperatures *(14)*. There are two different temperature-related effects on enantiomeric resolution *(20)*. One is the thermodynamic effect, which is responsible for the observed decrease in the selectivity factor (α) when temperature is increased. This is due to the partition coefficients and, therefore, the Gibbs free energy change (ΔG°) of transfer of the analyte between the mobile phase and the stationary phase with temperature *(125)*. The other kinetic effect produces an increase in efficiency with an increase in temperature. This results from the decrease in viscosity and the increase in the analytes diffusion coefficients. In the case of **Fig. 13**, the latter kinetic effect dominates at higher temperature, and the thermodynamic effect is dominant at lower temperature. There are two different mass transfer effects here. One is mobile phase mass transfer, in which a reduced mobile phase viscosity occurs as temperature increases. However, an increase of temperature also increases the diffusion coefficient of the solute in both the mobile phase and the stationary phase, and it decreases the viscosity of the stationary phase (increasing stationary phase mass transfer). Therefore, a temperature increase usually provides a trade-off in terms of resolution. The increased efficiency is good for resolution, while the lessening of the peak-to-peak separation is bad for resolution.

It has been observed that changing the temperature has a greater effect on the retention of solutes in normal phase chromatography than reversed-phase *(20)*. Higher temperatures (>40°C) can racemize some chiral compounds. Lower temperatures improve the separation for most compounds *(44)*. However, the temperature effect must be determined on a case-by-case basis. The normal starting temperature is ambient temperature. Variations in temperature can be controlled by using a thermostated column temperature control device.

3.2.2.2. FLOW RATES

The effect of flow rate on resolution is less pronounced with Chirobiotic columns than other factors. However, a general phenomenon observed with chiral stationary phases that have inclusion pockets is that there is an increase in resolution with a decrease in flow rate *(71)*.

This effect was also observed for cyclodextrin-based columns particularly in the reversed-phase mode. Decreasing the flow rate increases the retention time, but in return, better resolution is obtained. Usually, the flow rate does not affect the enantioselectivity factor (α), but does affect the separation efficiency. This is reflected by the inverse relationship between resolution (Rs) and flow rate. Flow rates <0.5 mL/min are not very common, since it will not produce further significant increases in resolution.

In the polar organic mode, smaller effects have been observed. While in normal phase mode, no observable effects were shown on selectivity by increasing the flow rate up to three times faster than the original condition *(71)*. However, the resolution is affected slightly due to the change in efficiency. Consequently, higher flow rates are often used for normal phase separations, as long as the back pressure is not excessive. Normal phase flow rates ran as high as 5 mL/min have been used in our laboratory. High efficiency normal phase separations are especially appealing to industry. For reversed-phase LC, flow rates around 0.5–1.5 mL/min are recommended.

3.2.3. Supercritical Fluid Separations

It is worthwhile to mention that SFC, is a high speed chiral separation approach that can substitute for normal phase HPLC separations. It can be advantageous for industrial and preparative-scale chiral separations *(107,127,128)*. SFC separations offer rapid column equilibration times, simple eluent compositions (i.e., CO_2 and an organic modifier, such as methanol and/or organic acids and bases), and less waste solvent generation. Due to the very low viscosity of the SFC eluent, faster flow rates (>4 mL/min) are frequently utilized. Normally, chiral separations using SFC can be achieved in <15 min (*107*). One drawback of this technique is that the observed peak shapes are not as smooth and symmetrical as those obtained in HPLC normal phase separations (*107*). This effect may result from the variation in the density and viscosity of the super/subcritical fluid due to the SFC operating conditions (i.e., temperature and pressure, etc.). The solubility of the analyte in the eluent may be inhomogeneous over the length of the column, which in turn, will affect the signal response.

3.2.4. Complementary Separations

As mentioned previously, complementary selectivity is a very useful feature of the Chirobiotic series of columns. **Figure 14** shows the principle of complementary separations using the Chirobiotic V and Chirobiotic T columns for the separation of warfarin and *N*-carbobenzyloxy (CBZ)-Norvaline. It can be seen that a partial separation was significantly improved when switching from one to another related column in this family using the same mobile phase conditions. This phenomenon also exists between all other Chirobiotic columns.

3.3. Method Development With Individual Chirobiotic Columns

3.3.1. Vancomycin CSP (Chirobiotic V)

Vancomycin shows considerable enantioselectivity for neutral molecules, amides, acids, esters, and cyclic amines. A wide variety of secondary and tertiary amines have been separated on the Chirobiotic V column in the polar organic mode. As discussed before, if a molecule meets the structural criteria in **Note 1**, the polar organic mode is the mobile phase of choice, followed by the normal phase mode and the reversed-phase mode, depending on solubility issues (*see* **Note 2**), etc. **Table 3** and the protocol flow chart in **Fig. 15** provide a generic approach to a chiral separation using a Chirobiotic V column. Optimization is necessary following the starting mobile phase based on the criteria outlined in the previous sections.

3.3.2. Teicoplanin CSP (Chirobiotic T)

The Chirobiotic T column seems to be particularly adept at resolving the following general classes of compounds: native amino acids (most ordinary α and β amino acids are separated on Chirobiotic T) (*7,9*), *N*-blocked amino acids (including fluorenylmethyl chloroformate [FMOC], CBZ, *tert*-butoxy-carbonyl [*t*-Boc] etc.), α-hydroxycarboxylic acids, acidic compounds (including carboxylic acids, phenols, etc.), small peptides (*9*), cyclic amides, sulfoxides (*43*), neural aromatic analytes (*129*), and other neutral cyclic amines containing aromatic moieties. Separations normally obtained on a chiral crown ether or ligand exchange type CSP are also possible on the Chirobiotic T column. In

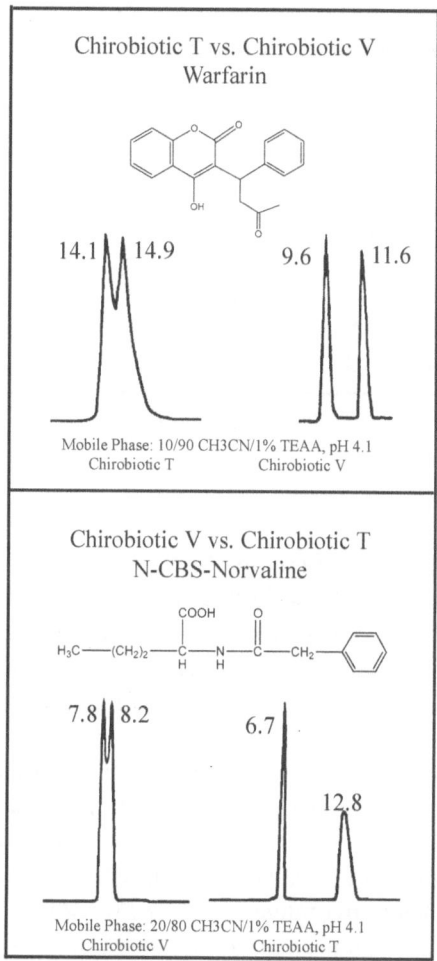

Fig. 14. An example of complementary separation using Chirobiotic T vs Chirobiotic V. Separation conditions as shown on the figure. Revised from **ref.** *71.*

general, there is no better chiral stationary phase for the separation of native amino acid enantiomers than that based on teicoplanin *(61,114)*. In addition, β-blockers (amino alcohols) have been resolved. It also shows the "complementary stereoselectivity" to Chirobiotic V, Chirobiotic R, and Chirobiotic TAG columns.

In methods development, the polar organic mode is the first utilized mode if the analyte meets the structural criteria (*see* **Note 1**). In the reversed-phase mode, the Chirobiotic T appears to achieve the best separations when an alcohol type

Table 3
Starting Mobile Phase Compositions for Chirobiotic V (71)

Mode[a]	Mobile phase composition	Solvent/additive ratios	Types of chiral analytes
PO	MeOH/HOAC/TEA	100/0.1/0.1	Amino alcohols and (cyclic) amines.
RP	THF/20mM NH_4NO_3, pH 5.5	10/90	Amines, imides, acids, profens.
NP	Hex/EtOH	80/20	Hydantoins, barbiturates, imides, and oxazolidinones.

[a]PO, polar organic; RP, reversed-phase; NP, normal phase.

mobile phase modifier is used (*see* **Note 3**). The order of priority for organic modifiers is methanol (MeOH) > EtOH > THF > 2-propanol (IPA). The exceptions are amino acids where EtOH modifier produced higher selectivities. The pH can be adjusted to optimize the reversed-phase mode separation. Lower pHs, to 3.8, produces a significant increase in retention for analytes with free carboxyl groups. In all cases, both selectivity and resolution varies with pH. Nonionizable analytes typically show less variation or a decrease in retention with the decrease in pH. The pH resulting in optimum selectivity (α) for nonionizable analytes, i.e., pH 7.0, rarely corresponds to the pH of optimum resolution, i.e., pH 4.0. Chirobiotic T is very sensitive to acidic conditions, therefore, the safest and most stable pH range is 3.8 to 7.0 (*71*).

Chirobiotic T can be used in any normal phase mode, however, the preference of polar organic modifier for Chirobiotic T is EtOH. There are a few cases where IPA works better. **Table 4** gives a summary of starting mobile phases for the Chirobiotic T column in all three mobile phase modes. **Figure 16** shows the method development protocol for Chirobiotic T.

3.3.3. Teicoplanin Aglycon CSP (Chirobiotic TAG)

The Chirobiotic TAG is a variation of the Chirobiotic T and has shown excellent complementary selectivity to the Chirobiotic T. Much better resolution was observed for separation of amphoteric molecules, including many α, β, γ, and cyclic amino acids, including carnitine, as well as other kind acids (*23*). It also showed remarkable selectivity for neutral molecules such as oxazolidinones, hydantoins, diazepines, coumarine derivatives, and chiral sulfur-containing compounds (*7,36,43,129*).

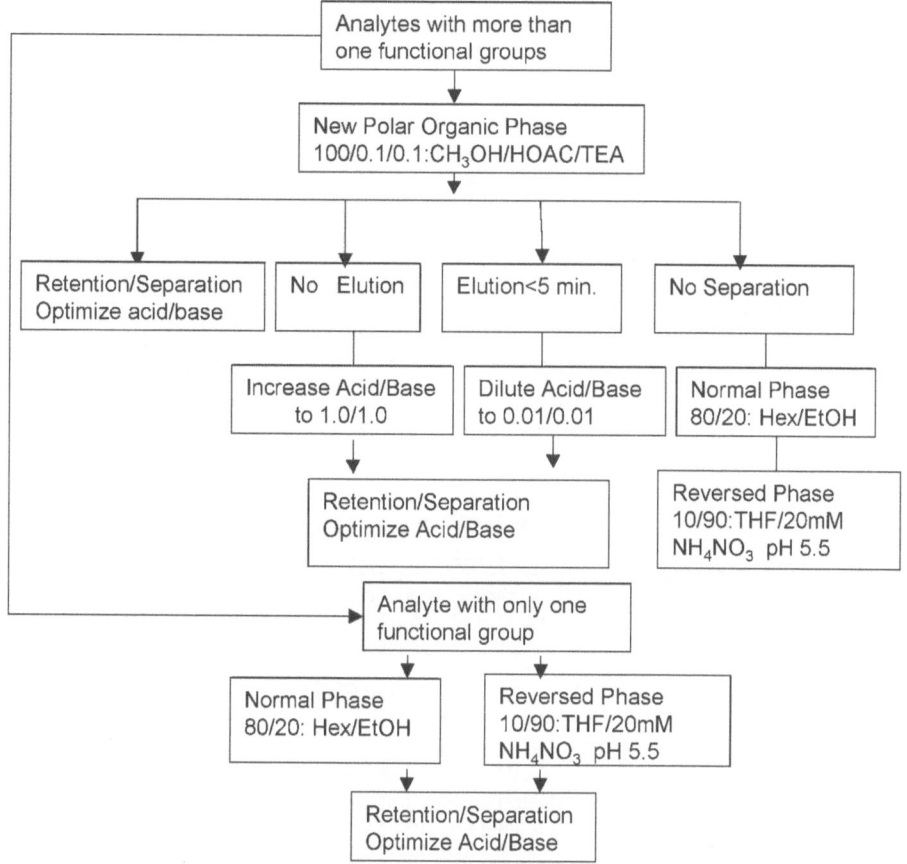

Fig. 15. Method Development Protocol for Chirobiotic V.

Table 4
Recommended Starting Mobile Phase Compositions for Chirobiotic T

Mode[a]	Mobile phase composition	Solvent/additive ratios	Types of chiral analytes
PO	MeOH/HOAC/TEA	100/0.1/0.1	Amino alcohols and N-blocked amino acids.
RP	MeOH/0.1% TEAA, pH 4.1	20/80	α-Hydroxy acids, oxazolidinones.
RP for amino acids		EtOH/H$_2$O	50/50 underivatized and N-blocked amino acids and peptides.
NP	Hex/EtOH	80/20	Hydantoins and imides.

[a]PO, polar organic; RP, reversed-phase; NP, normal phase.

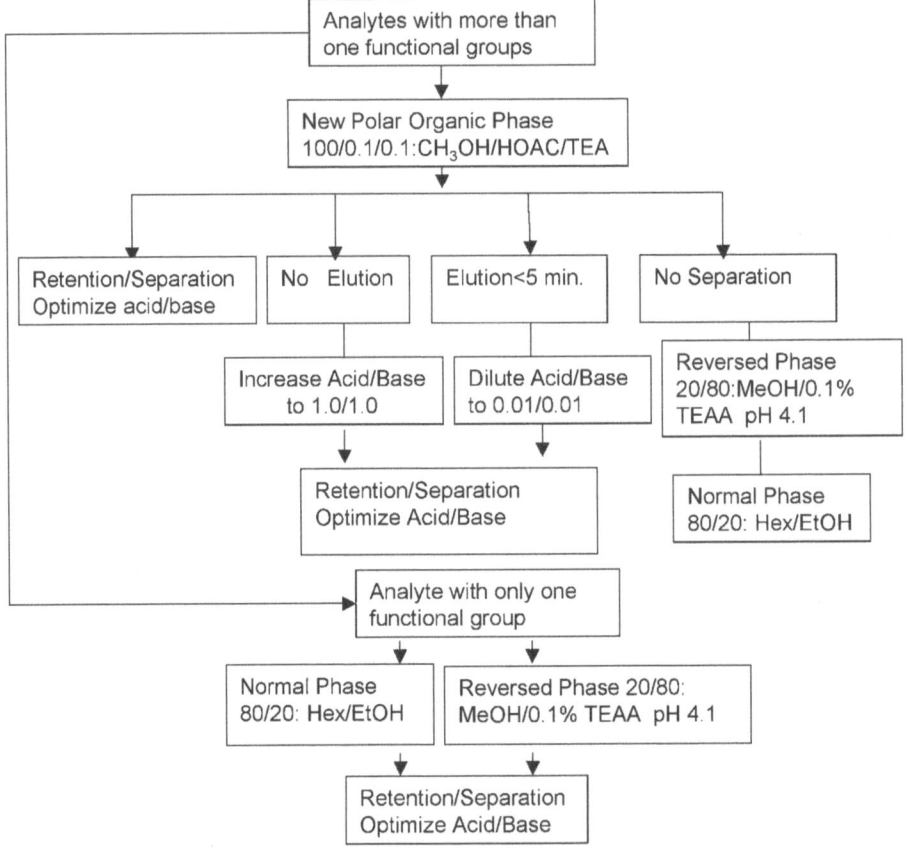

Fig. 16. Method Development Protocol for Chirobiotic T.

The mobile phase selection criteria are similar to that for Chirobiotic T. All of the points made under the Chirobiotic T are applicable to Chirobiotic TAG. The flow chart for method development protocol-Chirobiotic TAG is shown in **Fig. 17. Figure 17** shows that the single solvent (i.e., methanol, ethanol, and acetonitrile) was used for Chirobiotic TAG as the mobile phase, and it showed excellent resolution on some neutral molecules.

3.3.4. Ristocetin A (Chirobiotic R)

Chirobiotic R also is complementary to Chirobiotic V and Chirobiotic T with high selectivity for anionic chiral molecules and many amino acids *(17)*. The mobile phase selection and separation optimization criteria are also similar to those for Chirobiotic T. It also appears to favor alcohol type mobile phase

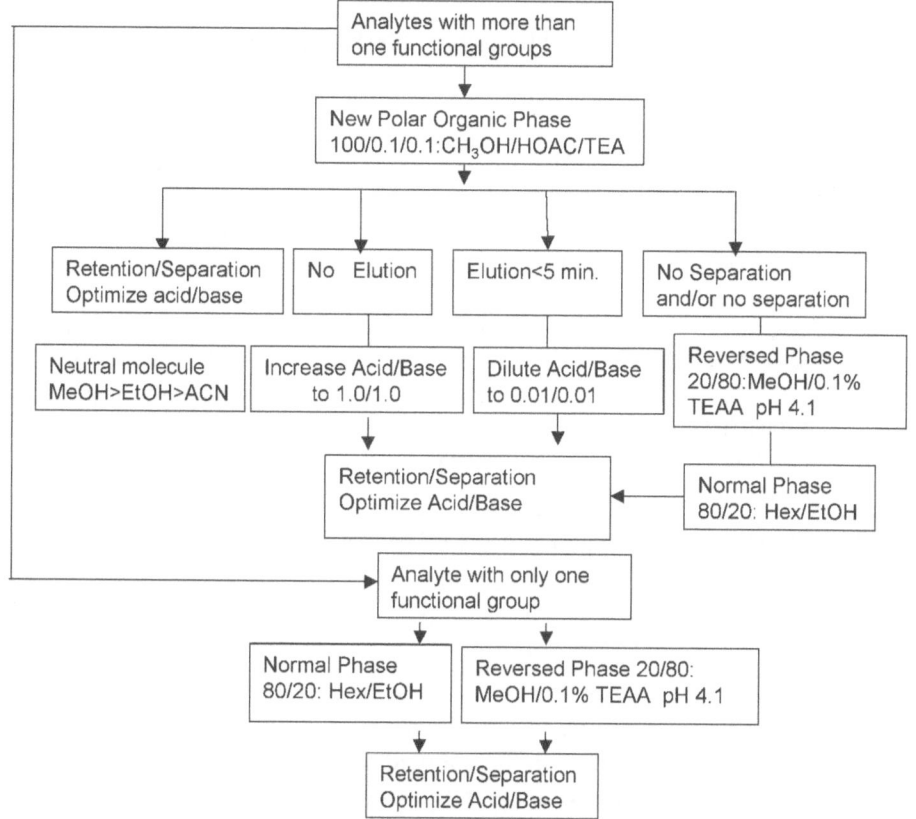

Fig. 17. Method Development Protocol for Chirobiotic TAG.

by a large margin. **Table 5** and flow chart in **Fig. 18** will be very useful to start a separation using Chirobiotic R as a CSP.

3.4. Column Coupling for Rapid Screening Chiral Selectivity

"Column coupling" is a very useful application of the "complementary feature" of Chirobiotic columns. Since each of the Chirobiotic columns (i.e., Chirobiotic V, T, and R) has its unique selectivity in all three mobile phase modes towards different chiral racemates, it should be a very efficient way to put them together and make a combined column to do a fast screening for selectivity. This coupling column was made by ASTEC by combining three 10-cm Chirobiotic columns in the direction of increasing polarity regardless of the mobile phase type, i.e., Chirobiotic R→V→T. This technique is very useful. It

Table 5
Recommended Starting Mobile Phases for Chirobiotic R

Mode[a]	Mobile phase composition	Solvent/additive ratios	Types of chiral analytes
PO	MeOH/HOAC/TEA	100/0.1/0.1	α-Hydroxy acids, profens, N-blocked amino acids.
RP	MeOH/0.1% TEAA, pH 6.8	20/80	α-Hydroxyl/halogenated acids, substituted aliphatic acids, profens, N-blocked amino acids, hydantoins, peptides.
RP	EtOH/H$_2$O	50/50	Amino acids.
NP	Hex/EtOH	40/60	Imides, hydantoins, N-blocked amino acids.

[a]PO, polar organic; RP, reversed-phase; NP, normal phase.

allows for the evaluation of this entire class of chiral selectors with a single coupled column for the ability to separate a molecule. Even if a partial separation or a shoulder is obtained on the coupled column, a baseline separation is guaranteed with one of the columns in this class.

However, the enantiomeric elution order for an analyte on different columns may reverse *(9,14,36,43)*, which may cause cancellation of the overall separation. Therefore, consideration of the elution order on different Chirobiotic columns before screening is necessary. But it is very unlikely that the separation will be completely diminished by elution order reversal on different columns. Eventually, some resolution should be able to be observed, even like a partial split or a shoulder peak. It has been observed that if a partial resolution of 0.6 or greater is obtained in the column coupling screening, a resolution of >1.5 baseline separation can be optimized on a 25-cm column for the selected stationary phases *(62)*. This fast screening method is very straightforward and requires minimum amount of time in searching for an optimum CSP.

Practically, in order to determine whether a chiral separation is achievable on Glycopeptide antibiotics CSP, three runs using this coupling column in the following order will be able to tell.

1. R + V + T in the new polar organic mode: MeOH/acetic acid (HOAC)/triethyl-amine (TEA) = 100/0.02/0.01.

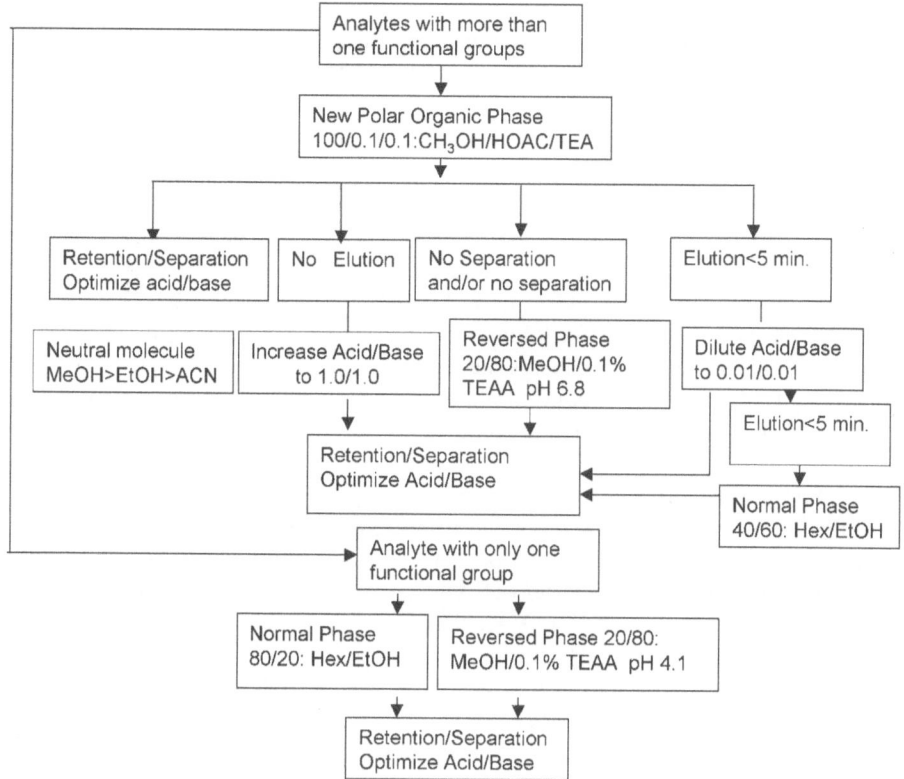

Fig. 18. Method Development Protocol for Chirobiotic R.

2. R + V + T in reversed-phase mode: MeOH/triethylammonium acetate (TEAA) (0.1%, pH 6.0) = 25/75.
3. V + T in normal phase mode: EtOH/Hexane = 60/40.

Once a partial or complete separation is done on the coupling column, the optimization steps in specific mobile phase modes, as shown in **Table 6**, should be followed to get a better resolution on a single CSP *(71)*.

3.5. Applications Using Glycopeptide Antibiotic CSPs

The use of glycopeptide antibiotic CSPs has resulted in the successful separation of most types of neutral, acidic, and basic compounds. **Table 7** summarizes the separations achieved on the Chirobiotic V, T, R, and TAG columns. They are classified as amino acids and peptides, *N*-blocked amino acid, alcohols, acids, sulfoxides, neutral molecules, etc. The name of the compound separated,

Table 6
Recommended Optimization Steps Following the Coupling Column Screening

Optimize in PO mode[a]	Optimize in RP mode[a]	Optimize in NP mode[a]
1. Choose a single analytical column (25 cm R, T, or V).	1. Choose a single analytical column (25 cm R, T, or V).	1. Choose a single analytical column (25 cm R, T, or V).
2. Choose proper acid/base (HOAC, TEA, TFA, NH₄OH, or salts, etc.).	2. Choose proper organic modifier (THF for V, MeOH for R and T).	2. Choose proper polar solvent (EtOH, IPA).
3. Adjust acid/base ratio (4/1 to 1/4) or salt concentration 0.01% to 1%. Higher concentration of salts results in lower retention.	3. Change the concentration of organic modifier. Higher concentration results lower retention.	3. Change the concentration of polar solvent. Higher concentration results lower retention.
4. Change flow rate, lower flow rate often results in higher resolution.	4. Choose proper buffer (TEAA, NH₄OAC, NH₄NO₃, Na citrate).	4. Add small amounts of acid + base as modifiers.
5. Decrease temperature can increase resolution.	5. Change the concentration of aqueous buffer range: 0.05% to 1%.	5. Change temperature (T); selectivity and elution order may change with T. Lower T, increase Rs, higher T may lead to co-elution and finally reversal of elution order. Maximum operation temperature is 65°C.
	6. Change pH of aqueous buffer.	
	7. Change flow rate.	
	8. Decrease temperature.	

[a]PO, polar organic; RP, reversed-phase; NP, normal phase.

the column type, mobile phase conditions, separation parameters, and reference are all listed in **Table 7**. This is not intended to be a complete database for these columns, but rather an attempt to give example of the variety of compounds that have been separated using Chirobiotic columns. This may provide a useful starting point, as well as provide pertinent references for scientists interested in LC enantiomeric separation.

Some general trends concerning types of chiral compounds separated on these columns can be noted from the information listed in **Table 7**.

Table 7
Chiral Separations Achieved on Chirobiotic Columns

Chiral compound	Chirobiotic column	Mobile phase	References	Note
Amino acids, derivatives and peptides				
Native amino acid (including 20 naturally occurring and amino acids not found in proteins)	T, TAG, R	RP	9,17,23,29,35, 74,78,88,130,131	More than 70 compounds
Dopa (DL-3-(3,4-Dihydroxyphenyl) alanine)	T, TAG, R	RP	9,17,23,29	Antiparkisonian
Folinic acid (5-formyl-5,6,7,8-tetrahydrofolate)	T, TAG	RP	23,29	Antianemic
Carnitine and Acetyl carnitine	T, TAG	RP, PO	23,29,35	Fat fighter
N-Blocked amino acids				
Dansyl (5-dimethylamino-1-naphthalenesulfonyl) amino acids	T, R, V	RP, PO, NP	2,7,17,19, 22,36,82,132	
N-2,4-Dinitrophenyl-amino acids	R, T	RP, PO, NP	7,17	
N-2,4-Dinitrophenyl-α-amino-carboxylic acid	R, T	RP, PO	7,17	
N-3,5-Dinitro-2-pyridyl-aminoacids	T, R	PO, NP, RP	7,17,19,22	
N-3,5-Dinitrobenzoyl-amino acids	V, T, R	RP, PO, NP	2,7,17,19,22	
N-Acetyl-amino acids	R, T	RP, PO	7,17,29,35,133	
N-Acetyl-n-fluro-phenylalanine	R, T	RP, PO	7,17,29	
N-Fluoro-amino acids	R	RP	17	
N-Methyl-amino acids	R	RP	17	
N-Benzoyl-amino acids	V, T, R	RP, PO, NP	2,7,17,19,22	
N-Blocked tryptophan analogues	R	RP	125	7 Compounds
N-Carbamyl-amino acids	R, T, V	RP, PO	2,7,17	
N-CBZ (carbobenzyloxy) amino acids	V, T, R	RP, PO, NP	2,7,17,19,22	
N-FMOC (9-Fluorenylmethyl chloroformate) amino acids	T, R	RP, NP	17,19	

(continued)

Table 7 (Continued)

Chiral compound	Chirobiotic column	Mobile phase	References	Note
N-Formyl-amino acids	R, T	RP, PO	7,17	
N-Phthaloyl-amino acids	R	PO, NP	17	
N-Phthaloyl-glycyl-amino acids	R, T	RP, PO, NP	7,17	
N-t-BOC (tert-butoxycarbonyl) amino acids	T, V, R	RP	2,7,17,74,118	
Other amino acids				
Synthetic amino acid analogs containing 1,2,3,4-tetrahydroisoquinoline, tetraline or 1,2,3,4-tetrahydro-2-carboline skeletons	R	RP, PO	31	28 Compounds
Unusual amino acid analogs (tyrosine, phenylalanine, tetrahydroisoquinoline, aminotetralin, tryptophan analogs, etc.)	R	PO, RP	134	25 Compounds
Unusual ring- and α-methyl-substituted phenylalanine analogs	T	RP	38	6 Compounds
Secondary amino acids possessing 1,2,3,4-tetrahydro-isoquinoline and related analogs	T	RP	26	9 Compounds
Unusual secondary aromatic amino acids	T	RP	18	4 Compounds
Unusual β-methyl-substituted amino acids	T	RP	25	13 Compounds
β-Amino acids	T	RP	135	5 Compounds
Unusual cyclic β-substituted α-amino acids	T	RP	80,92,126	5 Compounds
Unusual amino acids (phenylalanine, tyrosine analogues and analogs containing 1,2,3,4-tetrahydroisoquinoline, tetraline, 1,2,3,4-tetrahydro-2-carboline, cyclopentane, cyclohexane, bicycle[2.2.1]heptane or heptene skeletons)	T	RP	14	31 Compounds
β-Substituted—β-alanines	T, R	RP, PO	33	8 Compounds

β-Substituted—tryptophan analogues	T	RP	27	5 Compounds
β-Methyl-amino acids (tyrosine, phenylalanine, tryptophan, and 1,2,3,4-tetrahydroisoquinoline-3-carboxylic acid)	T	RP	14,20	4 Compounds
2-Pyrrolidone-5-carboxylic acid	T, TAG	RP	23,29	
3-(n-Naphthyl) alanine	V, R	NP, RP	2,17	
4-Bromo-phenylalanine	R	RP	17	
4-Chloro-phenylalanine	R	RP	17	
5- or 7-Benzyloxy-tryptophan	R	RP, PO	17	
Baclofen	T	RP	7	
Carnitine and O-acylcarnitine derivatives	T	RP	40	13 Compounds
Carnitine and O-acylcarnitine derivatives	T	RP	136	13 Compounds
Penicillamine	R	RP	17	
Selenomethionine	T	RP	89	HPLC-ICP-MS
Dopa and 3-O-methyl-dopa	T	RP	49	Antiparkisonian
Theanine	R	RP	17	
Both N and Carboxy-Protected amino acids (PAAs)				
Fmoc-Ser-OH, Fmoc-Asp-OH, Fmoc-Arg-(Pmc)-OH, Fmoc-Asp-(OtBu)-OH, Fmoc-Glu-(OtBu)-OH, etc.	T	RP	46	21 Compounds
Di- and tripeptides	R, T	RP	7,9,17	60 Compounds
Amino alcohols,				
(β-blockers, β-adrenoreceptor blocking drugs)				
R- and S-atenolol (elution order determined)	T			
Alprenolol	T, V, R	PO	45	Anticardiovascular
Arotinolol	T, V	PO	2,137,138	Antihypertensive
Atenolol	T, V, TAG	PO	23,29	Antihypertensive
Labeltalol	V	PO	28	Antiarhytmic
Metoprolol	T, R	PO	7,17	Achieved in CEC
				Antihypertensive

(continued)

Table 7 (Continued)

Chiral compound	Chirobiotic column	Mobile phase	References	Note
Oxprenolol	T, V	PO	2,7	Antihypertensive
Pindolol	T, TAG, V	PO	23,29,71,87	Antihypertensive
Practolol	T	PO	139	Antihypertensive
Propranolol	T, V, R	PO	2,137	Antihypertensive
Propranolol, metoprolol	V, T	PO	71,87	Antihypertensive
Sotalol	V	PO	28	Achieved in CEC
Calcium channel blockers (modulators)				Antihypertensive
4-Aryldihydropyrimidine derivatives (DHPMs)	V, T	RP	47	27 Compounds
Nicardipine	V, T	PO	87	
Other cardiovascular drugs				
Albuterol (salbutamol)	V, T	PO	71,72	β₂ Adrenoreceptor agonist
Alkylamino derivatives of aryloxypropanols	V, T	PO	34	62 Compounds anticardiovascular disorder
Clenbuterol	T	PO	139	β₂ Adrenoreceptor agonist
Denopamine				Cardiotonic
Flosequinan				Vasodilator
Formoterol				β₂ Adrenoreceptor agonist
Pinacidil				Antihypertensive
Simendan				Intotropic drug
Terfenadine	T	RP, PO		Antihistaminic
Valsartan,				Angiotensin II
Verapamil	T	RP, PO	29	Antianginal

	T, TAG, R, V	RP	2,7,17,23,29	Anticoagulant
Warfarin	T, TAG, R, V	RP	2,7,17,23,29	Anticoagulant
Nonsteroidal anti-inflammatory				
Benoxaprofen	V	RP	71	
Carprofen	T	RP	7	
Fenoprofen	V	RP	71	
Fenoprofen methyl ester	V	RP	71	
Flurbiprofen	T, V	RP	7,19,71	
Ibuprofen	T, V, R	RP	7,19,22	
Indoprofen	T, R, V	RP	2,7,17,19	
Ketoprofen	T, V, R	RP	7,17,19,22,71	
Surfrofen	T	RP	7,19	
Chiral acids				
1,1-Binaphthyl-2,2'-diyl hydrogen phosphate	V	RP	2	
2-(2,4-Dichlorophenoxy)propionic acid	T	RP	7,140	
2-(2-Chlorophenoxy)propionic acid	T	RP, NP	7	
2-(3-Chlorophenoxy)propionic acid	T	RP	7	
2-(4-Chloro-2-methylphenoxy)propionic acid	T	RP	140	
2-(4-Chlorophenoxy)propionic acid	T	NP, RP	7,19	
2-(4-Hydroxyphenoxy)propionic acid	T	RP	7	
2-(4-Nitrophenyl)propionic acid	T	RP	7	
2-Imidazolidone-4-carboxylic acid	T	PO	7,19	
2-Phenoxypropionic acid	T	RP	7	
2-Phenylpropionic acid	T, R	RP, PO	19	
2-Phenylpropionic acid	T	RP	7	
3-(4-Hydroxyphenyl) lactic acid	T	RP	7	
3-(Benzyloxycarbonyl)-4-oxazolidine carboxylic acid	T	RP	21	
3-Amino-3-phenylpropionic acid	R	PO	19	
3-Hydroxy-4-methoxymandelic acid	T, R	RP, PO	7,17	

(continued)

Table 7 (Continued)

Chiral compound	Chirobiotic column	Mobile phase	References	Note
3-Indolelacedtic acid	T, R	RP, PO	7,17	Urinary Antiseptic
4-Hydroxy mandelic acid	T, TAG, R	RP, PO	7,17,22,23,29,35	
6-Methoxy-1,2,3,4-tetrahydro-9H-pyrido[3,4-b] indole-1-carboxylic acid	R	RP	19	
Alkoxy-substituted esters of phenylcarbamic acid	V, T	PO	32,42	30 Compounds local anesthetics
Aryloxyphenoxypropanoic acid	T	RP	48	Antitumor agent
Atrolactic acid	R, T	PO, RP	7,22	
Benzocyclobutenecarboxylic acid	T	RP	7	
Cis-2-amino cyclohexane carboxylic acid	T, TAG	RP, PO	23,29	
Iopanic acid	T	RP	7	
Mandelic acid	T, R	RP, PO	7,17,21,22,35	
Neproxen	R	RP	19	
p-Chloromandelic acid	T	RP	7	
Ritalinic acid (RA)	V, T	PO	87	
trans-4-Cartinine-carboxylic acid	T	RP	7	
Tropic acid	T	RP, PO	7,29	
α-Amino-2-thiopheneacetic acid	R	RP	19	
α-Methoxyphenylacetic acid	T	RP	7	
β-Phenyllactic acid	R, T	PO, RP	7,22	
Mecoprop, mecoprop-methyl	T, TAG	RP, PO	23,29	Herbicide
Chiral Amines				
N-Benzyl-α-methyl-benzylamine	V	PO	21	
N-Benzyl-1-(1-naphthyl)-ethylamine hydrochloride	V	PO	21	
Phenyl propanolamine (norephedrine)	T	RP, PO	29	

Idazoxan	V	RP	2	
α-(1-Aminoethyl)-hydroxybenzyl Alcohol	V	RP	2	
Bupivacaine	V, T	RP	2,7	
N-(3,5-Dinitrobenzoyl)-α-methylbenzylamine	V	RP	2	
β-Hydroxyphenethylamine	R	RP, NP	17,19	
Penicillamine	R	RP	19	
Nα-Benzoylarginine-β-naphthylamine	T	PO	7	
Plant growth regulators and related indole compounds				
3-(3-Indolyl)-butyric acid, abscisic acid and structural related compounds including a variety of substituted tryptophan, etc.	T, R	RP	52	18 Compounds
Organometallic complexes				
Tris-diimine ruthenium(II) complexes	T, TAG	RP	141	9 Compounds
Ferrocenylalkyl polyfluoroalkyl benzimidazoles	V, T	RP, NP	41	3 Compounds
Other heterocyclic compounds				
Polyfluoroalkyl benzimidazoles	V, T	RP, NP	41	
4-Benzyl-2-oxazolidinone	T, V	RP, NP	16,19,75	
4-, or 5-Substituted racemic pyridones (Substituted 2-methoxy-6-oxo-1,4,5,6-tetrahydro-pyridine-3-carbonitriles)	V, T	RP	68	9 Compounds
dl-threo-Methylphenidate	V	RP	86	Antihyperactive
Oxazepam	V, T	PO	87	
Temazepam	V	RP	2	
Cyclic imidic compouds (barbiturates, piperidine-2,6-diones, and mephenytoin) including mephobarbital and thalidomide, hexobarbital, etc.)	V	RP, NP	142	11 Compounds
Citalopram and its 2 N-demethylated metabolites demethylcitalopram and didemethylcitalopram	V	RP	138	

(continued)

Table 7 (Continued)

Chiral compound	Chirobiotic column	Mobile phase	References	Note
RS-4-Phenyl-2-oxazolidinone	V, R, T	RP, NP, PO	36	
RS-4-Benzyl-2-oxazolidinone	V, R, T	RP, NP, PO	2,7,19,36	
RS-4-Benzyl-3-propionyl- oxazolidinone	V, R, T	RP, NP, PO	36	
RS-5,5,Dimethyl-4-phenyl-2-oxazolidionone	V, R, T	RP, NP, PO	36	
RS-3-Benzyloxy carbonyl-4-oxazolidine carboxylic acid	V, R, T	RP, NP, PO	36	
4S,5R(+)-cis-4,5-diphenyl-2-oxazolidinone	V, R, T	RP, NP, PO	36	
Coumachlor	T, TAG, V	RP, PO	2,7,19,23,35	Rodenticide
Coumafuryl	V,T	RP, NP, PO	2,7	
5-Methyl-5-phenyl hydantoin	T, TAG, V, R	RP, PO, NP	2,7,17,19,22, 23,29,35	Anticonvulsant
Styrene oxide	T, TAG	RP, PO	23	
Thioridazine	T, TAG	RP, PO	23	Antipsychotic
Tetrakis[1-[(4-tert-butyl-phenyl)sulfonyl]-pyrro-lidine-carboxylate] dirhodium(II)	V	PO	21	Synthesis intermediate
2-Methyl-4-phenyl indanone	T	RP	35	
Phenylphthalide	T	RP	35	
γ-(2-Naphthyl)-butyrolactone	T	RP	35	
γ-Phenyl-γ-butyrolactone	T	NP	7	
Althiazide	T, V	RP, NP	2,29	Diuretic
3-Methyl-5-cano-6-methoxy-3,4-dihydro-2-pyridone	V	RP	2	
Thioridazine	V	RP	2	
5-(4-Hydroxyphenyl)-5-phenylhydandoin	V, R, T	RP, NP	2,7,17,22	
5-(3-Hydroxyphenyl)-5-phenylhydandoin	R	NP	17	
3-Benzoylphthalide	V	RP	2	

Compound			
3-Phenylphthalide	T	RP	7
2,2,2-Trifluoro-1-(9-anthryl)ethanol	V	RP	2
Mephobarbital	V	NP	2
Hexobarbital	V	RP	2
3a,4,5,6-Tetrahydrosuccinimide[3,4-b]acenaphthen-10-one	V, T, R	RP, NP	2,7,17,19,22
1-Benzoyl-2-tert-butyl-3-methyl-4-imideazolidinone	V, T	RP, NP	2,7
3-[2-(2-Bromoacetamido)acetamido] PROXYL	V	RP	2
Ethyl-2-pyrrolidone-5-carboxylate	V	RP, NP	2,19
α-Carbethoxy-γ-phenyl-γ-butyrolactone	V	RP, NP	2,19
CGA-40919	V	RP	2
Ftorafur	V	RP, NP	2,19
5-(4-Methylphenyl)-5-phenylhydantoin	V, T	RP, NP	2,7,19
1,1'-Bi-2-naphthol	V	RP	2
γ-Phenyl-γ-butyrolactone	V, R	RP, PO	2,17
Ethyl-2-pyrrolidone-5-carboxylate	V	RP	2
Chlorthalidone	R	NP	22
5-Phenyl-2-(2-propynylamino)2-oxazolin-4-one	R, T	NP	7,22
Althiazide	T, R	RP, NP	7,17
4-Benzyl-2-methoxy-6-oxo-1,4,5,6-tetrahydro-phyridine-3-carbonitrile	T	RP	7
4-Phenyl-2-methoxy-6-oxo-1,4,5,6-tetrahydro-phyridine-3-carbonitrile	T	NP	7
4-Cyclohexyl-2-methoxy-6-oxo-1,4,5,6-tetrahydro-phyridine-3-carbonitrile	T	RP, NP	7
4-Methyl-2-methoxy-6-oxo-1,4,5,6-tetrahydro-phyridine-3-carbonitrile	T	NP	7
Devrinol	T	RP	7

(continued)

Table 7 (Continued)

Chiral compound	Chirobiotic column	Mobile phase	Reference	Note
Tetrahydropapaveroline hydrochloride	T	PO	7	
Mephentoin	T	NP	7	
TAPA	R	NP	17	
Tetrantoin	R	NP, PO	17	
Chlorthalidone	R	NP	17	
1-Benzocyclobutenecarbonitrile	R	NP	17	
1-Acetoxy-8-hydroxy-1,4,4a,9a-tetra-hydro-anthraquinone	R	NP	17	
Fluoxetine, Terbutaline	V, T	PO	87	
Bromacil	T, TAG, V	RP, PO	2,7,19,23,29	Herbicide
Mephenytoin	V, T	RP, NP	2,19	
Devrinol	V	RP	2	
Norverapamil	V	RP	2	
Verapamil	V	RP	2,19	
Semisynthetic ergot alkaloids				
Nicergoline (α-adrenergic blocking agent), lisuride (serotonin antagonist), terguride (mixed D_2 agonist/antagonist of the pituitary) meluol.	V, T	RP	76	
Other neutral compounds				
Aminoglutethimide	V, R	RP, NP	2,17	
Bendroflumethiazide	V, T	RP	2,7	
Benzoin methyl ester	V	RP	2	
Captopril diastereoisomers	T	RP	44	Antihypertensive
Furo-Coumarine derivatives	T, R, V, TAG, VAG	RP, PO, NP	129	27 Compounds

Indapamide	V	RP, NP	2	
Mandelamide	V, T	RP, NP	2,7	
Methsuximide	V, T	RP	2,7	
N-(1-Phenylethyl)maleimide	V	RP	21	
N-(α-Methylbenzyl)phthlic acid monoamide	T	RP	7	
N,N'-Bis(α-methylbenzyl)sulfamide	V	RP	2	
N,N¹-Bis(α-metylbenzyl) sulfamide	T	NP	7	
N-Benzoylalanine methyl ester	V	RP	2	
Phensuximide	V, R, T	RP, NP	2,7,17,22	
Proglumide	V, T	RP	2,7,19	
Promethazine	V, R	RP, PO	50	Antidepressive and antiallergic
Pyridoglutethimide	V	RP	2	
Salbutamol and its 4-O-Sulfate Metabolites	T	PO	85	
Sulfoxides, sulfinate esters, and tosylated sulfilimines	T, R, V, TAG, VAG	RP, PO, NP	43	42 Compounds
Tropicamide	T	RP	7	
α,α-Dimethyl-β-methyl succinimide	T	NP	7	
α-Methyl-α-phenyl succinimide	T, TAG, V, R	RP, PO, NP	2,7,17,22,23,29	Antiurolithic
α-Methyl-α-propyl succinimide	R, T	NP	7,17,22	

RP, reversed-phase; NP, normal phase; PO, polar organic.

1. First of all, almost all amino acids and *N*-blocked amino acids are easily separated on these columns. The Chirobiotic TAG provided the best selectivity for native amino acids, followed by the Chirobiotic T column (which usually produces higher efficiency separations). *N*-blocked amino acids usually are best separated by the Chirobiotic T and Chirobiotic V columns. Chirobiotic V column works best for esters compared to other Chirobiotic columns.
2. Neutral aromatic molecules are best separated in the normal phase mode and sometimes in the polar organic mode (if they have at least two polar functional groups). The separation efficiency in these modes is usually good. Reversed-phase separations are also possible.
3. Acidic or anionic molecules, including many nonsteroidal anti-inflammatory compounds, are best separated in the polar organic mode or reversed-phase mode. The Chirobiotic V column separates the most amine-containing compounds (particularly, 2°, 3°, and cyclic amines) as does the Chirobiotic T column. None of the Chirobiotic columns separate primary amine compounds well unless they also have other polar functional groups (e.g., amino acids, amino alcohols, etc.).
4. When compounds are separated by both the Chirobiotic T and TAG columns, usually the TAG column has greater selectivity, and the T column shows greater efficiency.

Given the wide applicability of these columns, it is clear that use will expand in the future.

4. Notes

1. Chiral analytes suitable for the new polar organic mobile phase mode should have at least two polar functional groups. These functional groups include alcohols, halogens (F, Cl, Br, I), nitrogen in any form (primary, secondary, and tertiary amines), carbonyl, carboxyl, oxidized forms of sulfur and phosphorus, for example. At least one of the analyte's polar functional groups must be on or near the stereogenic center. The other polar group can be located anywhere in the molecule. It is also beneficial if the analyte has some steric bulk or aromatic rings close to the stereogenic center.
2. To choose the best mobile phase mode for a separation, the solubility of the sample in the mobile phase is a key issue, particularly for preparative scale separations. If the analyte is soluble only in organic solvents, either the normal phase or polar organic mode can be used (depending on the number and locations of functional groups on the chiral molecule, *see* **Note 1**). When the solute is water soluble only, the reversed-phase mode is required. There are some solutes that can be separated in both reversed-phase and polar organic modes. This gives the analyst a choice of conditions. Likewise, some other analytes can be separated in both polar organic and normal phase modes. Usually one mode is superior to the other in terms of separation speed, resolution, sample loadability, and compatibility to prior sample work-up procedures.

3. In the reversed-phase mode, different macrocyclic glycopeptide CSPs prefer different type of organic modifiers. For example, THF and ACN work best on Chirobiotic V and VAG, while Chirobiotic T, TAG, and R produce better selectivity and efficiencies with alcohol-type modifiers. The elutropic strength for ACN and THF is about twice that of alcohols on Chirobiotic columns.

4. Recommended buffers in the reversed-phase mode in the order of their usefulness are as follows: TEAA, ammonium acetate, ammonium nitrate, and sodium citrate. The percentage of the buffer salts can be varied from 0.01% to 1% depending on the retention factor of specific analytes. Normally, the higher the buffer concentration, the shorter is the observed retention. Buffer solutions with 0.1% TEAA are most frequently used. They are prepared by titrating a 0.1% solution of triethylamine with glacial acetic acid, to the appropriate pH.

5. The absolute amount of added acid and base in the polar organic mode is essential for optimization of the retention time. If the analyte elutes too fast, the concentration of the acid/base pair is reduced, or acetonitrile can be added to the mobile phase. On the other hand, if the analyte is strongly retained, the acid/base concentration is increased. The range of concentrations for the acid/base pair are between 1% and 0.001%. If an acid/base concentration >1% is needed, this indicates that the analyte is too polar and that a reversed-phase separation may be preferred. Concentration below 0.001% indicates a normal phase system may be preferred.

The ratio of acid to base controls the degree to which the ionizable solutes are protonated or deprotonated *(119)*. It is a key factor that affects the selectivity. By adjusting the ratio of acid to base and the overall percentage of both acid and base, retention and resolution both can be affected. The typical starting ratio is 1:1 (mol/mol), and then a 1:2 or 2:1 ratio are used to find the most improved resolution. The ratio of acid to base can be as high as 5:1. Acids and bases that can be used with Chirobiotic columns include triethylamine, ammonia, acetic acid, trifluoroacetic acid (TFA), etc. TFA is usually used in 50% of the amount of acetic acid due to its greater acidity. Ammonium acetate, ammonium trifluoroacetate, and ammonium formate are very popular mobile phase additives in both HPLC and LC-MS.

References

1. Armstrong, D. W. and Zhang, B. (2001) Chiral stationary phases for HPLC. *Anal. Chem.* **73,** 557A–561A.
2. Armstrong, D. W., Tang, Y., Chen, S., Zhou, Y., Bagwill, C., and Chen, J.-R. (1994) Macrocyclic antibiotics as a new class of chiral selectors for liquid chromatography. *Anal. Chem.* **66,** 1473–1484.
3. Armstrong, D. W. and Zhou, Y. (1994) Use of a macrocyclic antibiotic as the chiral selector for enantiomeric separations by TLC. *J. Liq. Chromatogr.* **17,** 1695–1707.
4. Armstrong, D. W., Rundlett, K. L., and Chen, J.-R. (1994) Evaluation of the macrocyclic antibiotic vancomycin as a chiral selector for capillary electrophoresis. *Chirality* **6,** 496–505.

5. Armstrong, D. W., Gasper, M. P., and Rundlett, K. L. (1995) Highly enantio-selective capillary electrophoretic separations with dilute solutions of the macro-cyclic antibiotic ristocetin A. *J. Chromatogr. A* **689,** 285–304.
6. Chen, S., Liu, Y., Armstrong, D. W., Borrell, J. I., Martinez-Teipel, B., and Matallana, J. L. (1995) Enantioresolution of substituted 2-methoxy-6-oxo-1,4,5,6-tetrahydropyridine-3-carbonitriles on macrocyclic antibiotic and cyclodextrin sta-tionary phases. *J. Liq. Chromatogr.* **18,** 1495–1507.
7. Armstrong, D. W., Liu, Y., and Ekborgott, K. H. (1995) A covalently bonded teicoplanin chiral stationary phase for HPLC enantioseparations. *Chirality* **7,** 474–197.
8. Rundlett, K. L., Gasper, M. P., Zhou, E. Y., and Armstrong, D. W. (1996) Capil-lary electrophoretic enantiomeric separations using the glycopeptide antibiotic, teicoplanin. *Chirality* **8,** 88–107.
9. Berthod, A., Liu, Y., Bagwill, C., and Armstrong, D. W. (1996) Facile LC enan-tioresolution of native amino acids and peptides using a teicoplanin chiral station-ary phase. *J. Chromatogr. A* **731,** 123–137.
10. Gasper, M. P., Berthod, A., Nair, U. B., and Armstrong, D. W. (1996) Comparison and modeling study of vancomycin, ristocetin A, and teicoplanin for CE enantio-separations. *Anal. Chem.* **68,** 2501–2514.
11. Tesarová, E. and Armstrong, D. W. (1998) Enantioselective separations, in *Advanced Chromatographic and Electromigration Methods in Biosciences, Vol. 60* (Deyl, Z., ed.), Elsevier, Amsterdam, pp. 197–256.
12. Berthod, A., Nair, U. B., Bagwill, C., and Armstrong, D. W. (1996) Derivatized vancomycin stationary phases for LC chiral separations. *Talanta* **43,** 1767–1782.
13. Nair, U. B., Chang, S. S. C., Armstrong, D. W., Rawjee, Y. Y., Eggleston, D. S., and McArdle, J. V. (1996) Elucidation of vancomycin's enantioselective binding site using its copper complex. *Chirality* **8,** 590–595.
14. Peter, A., Torok, G., and Armstrong, D. W. (1998) High-performance liquid chro-matographic separation of enantiomers of unusual amino acids on a teicoplanin chiral stationary phase. *J. Chromatogr. A* **793,** 283–296.
15. Armstrong, D. W. and Nair, U. B. (1997) Capillary electrophoretic enantiosep-arations using macrocyclic antibiotics as chiral selectors. *Electrophoresis* **18,** 2331–2342.
16. Armstrong, D. W., Lee, J. T., and Chang, L. W. (1998) Enantiomeric impurities in chiral catalysts, auxiliaries and synthons used in enantioselective synthesis. *Tetrahedron Asymmetry* **9,** 2043–2064.
17. Ekborg-Ott, K., Liu, Y., and Armstrong, D. W. (1998) Highly enantioselective HPLC separations using the covalently bonded macrocyclic antibiotic, ristocetin A, chiral stationary phase. *Chirality* **10,** 434–483.
18. Peter, A., Torok, G., Toth, G., et al. (1998) Enantiomeric separation of unusual secondary aromatic amino acids. *Chromatographia* **48,** 53–58.
19. Ekborg-Ott, K. H., Kullman, J. P., Wang, X., Gahm, K., He, L., and Armstrong, D. W. (1998) Evaluation of the macrocyclic antibiotic avoparcin as a new chiral selector for HPLC. *Chirality* **10,** 627–660.

20. Peter, A., Torok, G., Armstrong, D. W., Toth, G., and Tourwe, D. (1998) Effect of temperature on retention of enantiomers of beta-methyl amino acids on a teicoplanin chiral stationary phase. *J. Chromatogr. A* **828,** 177–190.

21. Armstrong, D. W., He, L., Yu, T., Lee, J. T., and Liu, Y.-S. (1999) Enantiomeric impurities in chiral catalysts, auxiliaries, synthons and resolving agents. Part 2. *Tetrahedron Asymmetry* **10,** 37–60.

22. Ekborg-Ott, K. H., Wang, X., and Armstrong, D. W. (1999) Effect of selector coverage and mobile phase composition on enantiomeric separations with ristocetin A chiral stationary phases. *Microchem. J.* **62,** 26–49.

23. Berthod, A., Chen, X., Kullman, J. P., et al. (2000) Role of the carbohydrate moieties in chiral recognition on teicoplanin-based LC stationary phases. *Anal. Chem.* **72,** 1767–1780.

24. Peter, A., Olajos, E., Casimir, R., et al. (2000) High-performance liquid chromatographic separation of the enantiomers of unusual alpha-amino acid analogues. *J. Chromatogr. A* **871,** 105–113.

25. Peter, A., Torok, G., Toth, G., et al. (1998) Chiral separation of unusual beta-methyl amino acids. *Proc. Eur. Pept. Symp. 25th.* **25,** 300–301.

26. Torok, G., Peter, A., Toth, G., et al. (1998) Chiral separation of secondary amino acids possessing 1,2,3,4-tetrahydroisoquinoline and related structures. *Proc. Eur. Pept. Symp.* 25th. **25,** 302–303.

27. Torok, G., Peter, A., Vekes, E., et al. (2000) Enantiomeric high-performance liquid chromatographic separation of beta-substituted tryptophan analogues. *Chromatographia* **51,** S165–S174.

28. Karlsson, C., Karlsson, L., Armstrong, D. W., and Owens, P. K. (2000) Evaluation of a vancomycin chiral stationary phase in capillary electrochromatography using polar organic and reversed-phase modes. *Anal. Chem.* **72,** 4394–4401.

29. Berthod, A., Yu, T., Kullman, J. P., et al. (2000) Evaluation of the macrocyclic glycopeptide A-40,926 as a high-performance liquid chromatographic chiral selector and comparison with teicoplanin chiral stationary phase. *J. Chromatogr. A* **897,** 113–129.

30. Karlsson, C., Karlsson, L., Armstrong, D. W., and Owens, P. K. (2000) Enantioselective reversed-phase and non-aqueous capillary electrochromatography using a teicoplanin chiral stationary phase. *J. Chromatogr. A* **897,** 349–363.

31. Peter, A., Torok, G., Armstrong, D. W., Toth, G., and Tourwe, D. (2000) High-performance liquid chromatographic separation of enantiomers of synthetic amino acids on a ristocetin A chiral stationary phase. *J. Chromatogr. A* **904,** 1–15.

32. Lehotay, J., Hrobonova, K., Cizmarik, J., Reneova, M., and Armstrong, D. W. (2001) Modification of the chiral bonding properties of teicoplanin chiral stationary phase by organic additives. HPLC separation of enantiomers of alkoxysubstituted esters of phenylcarbamic acid. *J. Liq. Chromatogr. Relat. Technol.* **24,** 609–624.

33. Peter, A., Lazar, L., Fulop, F., and Armstrong, D. W. (2001) High-performance liquid chromatographic enantioseparation of beta-amino acids. *J. Chromatogr. A* **926,** 229–238.

34. Hrobonova, K., Lehotay, J., Cizmarikova, R., and Armstrong, D. W. (2001) Study of the mechanism of enantioseparation. I. Chiral analysis of alkylamino derivatives of aryloxypropanols by HPLC using macrocyclic antibiotics as chiral selectors. *J. Liq. Chromatogr. Relat. Technol.* **24,** 2225–2237.

35. Berthod, A., Valleix, A., Tizon, V., Leonce, E., Caussignac, C., and Armstrong, D. W. (2001) Retention and selectivity of teicoplanin stationary phases after copper complexation and isotopic exchange. *Anal. Chem.* **73,** 5499–5508.

36. Xiao, T. L., Zhang, B., Lee, J. T., Hui, F., and Armstrong, D. W. (2001) Reversal of enantiomeric elution order on macrocyclic glycopeptide chiral stationary phases. *J. Liq. Chromatogr. Relat. Technol.* **24,** 2673–2684.

37. Anan'eva, I. A., Shapovalova, E. N., Shpigun, O. A., and Armstrong, D. W. (2001) Separation of amino acid enantiomers and enantiomers of their derivatives on macrocyclic antibiotic teicoplanin. *Vestn. Mosk. Univ., Ser. 2: Khim.* **42,** 278–280.

38. Peter, A., Olajos, E., Casimir, R., et al. (2000) High-performance liquid chromatographic separation of the enantiomers of unusual alpha-amino acid analogues. *J. Chromatogr. A* **871,** 105–113.

39. Torok, G., Peter, A., Armstrong, D. W., Tourwe, D., Toth, G., and Sapi, J. (2001) Direct chiral separation of unnatural amino acids by high-performance liquid chromatography on a ristocetin A-bonded stationary phase. *Chirality* **13,** 648–656.

40. D'Acquarica, I., Gasparrini, F., Misiti, D., et al. (1999) Direct chromatographic resolution of carnitine and O-acylcarnitine enantiomers on a teicoplanin-bonded chiral stationary phase. *J. Chromatogr. A* **857,** 145–155.

41. Snegur, L. V., Boev, V. I., Nekrasov, Y. S., et al. (1999) Synthesis and structure of biologically active ferrocenylalkyl polyfluoro benzimidazoles. *J. Organometallic Chem.* **580,** 26–35.

42. Iungelova, J., Lehotay, J., Hrobonova, K., Cizmarik, J., and Armstrong, D. W. (2002) Study of local anaesthetics. CLVIII. Chromatographic separation of some derivatives of substituted phenylcarbamic acid on a vancomycin-based stationary phase. *J. Liq. Chromatogr. Relat. Technol.* **25,** 299–312.

43. Berthod, A., Xiao, T. L., Liu, Y., Jenks, W. S., and Armstrong, D. W. (2002) Separation of chiral sulfoxides by liquid chromatography using macrocyclic glycopeptide chiral stationary phases. *J. Chromatogr. A* **955,** 53–69.

44. Owens, P. K., Svensson, L. A., and Vessman, J. (2001) Direct separation of captopril diastereoisomers including their rotational isomers by RP-LC using a teicoplanin column. *J. Pharm. Biomed. Anal.* **25,** 453–464.

45. Lamprecht, G., Kraushofer, T., Stoschitzky, K., and Lindner, W. (2000) Enantioselective analysis of (R)-and (S)-atenolol in urine samples by a high-performance liquid chromatography column-switching setup. *J. Chromatogr. B* **740,** 219–226.

46. Esquivel, J. B., Sanchez, C., and Fazio, M. J. (1998) Chiral HPLC separation of protected amino acids. *J. Liq. Chromatogr. Rel. Technol.* **21,** 777–791.

47. Kleidernigg, O. P. and Kappe, C. O. (1997) Separation of enantiomers of 4-aryl-dihydropyrimidines by direct enantioselective HPLC. A critical comparison of chiral stationary phases. *Tetrahedron Asymmetry* **8,** 2057–2067.

48. He, J., Cheung, A. P., Struble, E., Wang, E., and Liu, P. (2000) Enantiomeric separation of an aryloxyphenoxypropanoic acid by CE and LC. *J. Pharm. Biomed. Anal.* **22,** 583–595.
49. Wu, G. and Furlanut, M. (1999) Hydrogen bonding as a possible interaction for the chiral separation of DL-dopa and DL-3-O-methyl-dopa in a chiral teicoplanin column. *Farmaco* **54,** 188–190.
50. Bosakova, Z., Klouckova, I., and Tesarova, E. (2002) Study of the stability of promethazine enantiomers by liquid chromatography using a vancomycin-bonded chiral stationary phase. *J. Chromatogr. B Anal. Technol. Biomed. Life Sci.* **770,** 63–69.
51. Schneiderheinze, J. M., Armstrong, D. W., and Berthod, A. (1999) Plant and soil enantioselective biodegradation of racemic phenoxyalkanoic herbicides. *Chirality* **11,** 330–337.
52. Hui, F., Ekborg-Ott, K. H., and Armstrong, D. W. (2001) High-performance liquid chromatographic and capillary electrophoretic enantioseparation of plant growth regulators and related indole compounds using macrocyclic antibiotics as chiral selectors. *J. Chromatogr. A* **906,** 91–103.
53. Ekborg-Ott, K. H. and Armstrong, D. W. (1997) Stereochemical analyses of food components, in *Chiral Separations: Application and Technology, Ch. 9* American Chemical Society, Washington, DC, pp. 201–270.
54. Nieto, M. and Perkins, H. R. (1971) The specificity of combination between ristocetins and peptides related to bacterial cell wall mucopeptide precursors. *Biochem. J.* **124,** 845–852.
55. Nieto, M. and Perkins, H. R. (1971) Modifications of the acyl-D-alanyl-D-alanine terminus affecting complex-formation with vancomycin. *Biochem. J.* **123,** 789–803.
56. Nieto, M. and Perkins, H. R. (1971) Physicochemical properties of vancomycin and iodovancomycin and their complexes with diacetyl-L-lysyl-D-alanyl-D-alanine. *Biochem. J.* **123,** 773–787.
57. Armstrong, D. W., Rundlett, K., and Reid, G. L., III. (1994) Use of a macrocyclic antibiotic, rifamycin B, and indirect detection for the resolution of racemic amino alcohols by CE. *Anal. Chem.* **66,** 1690–1695.
58. Ward, T. J., Dann, C. I., and Blaylock, A. (1995) Enantiomeric resolution using the macrocyclic antibiotics rifamycin B and rifamycin SV as chiral selectors for capillary electrophoresis. *J. Chromatogr. A* **715,** 337–344.
59. Ward, T. J. (1994) For capillary electrophoresis. *Anal. Chem.* **66,** 632A–640A.
60. Armstrong, D. (1995) Macrocyclic antibiotics as separation agents. *PCT Int. Appl.* 71.
61. Aboul-Enein, H. Y. and Ali, I. (2000) Macrocyclic antibiotics as effective chiral selectors for enantiomeric resolution by liquid chromatography and capillary electrophoresis. *Chromatographia* **52,** 679–691.
62. Wang, A. X., Lee, J. T., and Beesley, T. E. (2000) Coupling chiral stationary phases as a fast screening approach for HPLC method development. *LC GC* **18,** 626–639.

63. Armstrong, D. W. (1999) Complicating chromatography to make it work more effectively. *Book of Abstracts. 217th ACS National Meeting,* Anaheim, CA, March 21–25.
64. Armstrong, D. W. (1998) The evolution of chiral stationary phases for liquid chromatography. *J. Chin. Chem. Soc. (Taipei)* **45,** 581–590.
65. Tesarova, E. and Armstrong, D. W. (1998) Enantioselective separations. *J. Chromatogr. Libr.* **60,** 197–256.
66. Armstrong, D. W. (1996) Enantioselective interactions and separations with macrocyclic antibiotics. *Book of Abstracts. 211th ACS National Meeting,* New Orleans, LA, March 24–28.
67. Anan'eva, I. A., Shapovalova, E. N., Shpigun, O. A., and Armstrong, D. W. (2003) Separation of β-blockers on chiral stationary phases based on the teicoplanin macrocyclic antibiotic. *Zhurnal Analiticheskoi Khimii* **58,** 663–664.
68. Chen, S., Armstrong, D. W., Borrell, J. I., Martinez-Taipel, B., and Matallana, J. L. (1995) Enantioresolution of substituted 2-methoxy-6-oxo-1,4,5,6-tetrahydropyridine-3-carbonitriles on macrocyclic antibiotic and cyclodextrin stationary phases. *J. Liq. Chromatogr.* **18,** 1495–1507.
69. Bojarski, J., Zakl. Chem. Org., Collegium Medicum, Uniw. Jagiellonski and Krakow, P. (1999) Antibiotics as electrophoretic and chromatographic chiral selectors. *Wiad. Chem.* **53,** 235–247.
70. Kleidernigg, O. P. and Kappe, C. O. (1997) Separation of enantiomers of 4-aryldihydropyrimidines by direct enantioselective HPLC. A critical comparison of chiral stationary phases. *Tetrahedron Asymmetry* **8,** 2057–2067.
71. Advanced Separation Technologies Inc. (2002) *A Guide to Using Macrocyclic Glycopeptide bonded phases for Chiral LC Separations.* 4th ed. Advanced Separation Technologies, Whippany, NJ, pp. 1–65.
72. Fried, K. M., Koch, P., and Wainer, I. W. (1998) Determination of the enantiomers of albuterol in human and canine plasma by enantioselective high-performance liquid chromatography on a teicoplanin-based chiral stationary phase. *Chirality* **10,** 484–491.
73. Aboul-Enein, H. Y. and Serignese, V. (1998) Enantiomeric separation of several cyclic imides on a macrocyclic antibiotic (vancomycin) chiral stationary phase under normal and reversed phase conditions. *Chirality* **10,** 358–361.
74. Tesarova, E., Bosakova, A., and Pacakova, V. (1999) Comparison of enantioselective separation of N-tert-butyloxycarbonyl amino acids and their non-blocked analogues on teicoplanin-based chiral stationary phase. *J. Chromatogr. A* **838,** 121–129.
75. Scott, R. P. W. and Beesley, T. E. (1999) Optimum operating conditions for chiral separations in liquid chromatography. *Analyst* **124,** 713–719.
76. Tesarova, E., Zaruba, K., and Flieger, M. (1999) Enantioseparation of semisynthetic ergo alkaloids on vancomycin and teicoplanin stationary phases. *J. Chromatogr. A* **844,** 137–147.
77. Aboul-Enein, H. Y. and Serignese, V. (1999) Quantitative determination of clenbuterol enantiomers in human plasma by high-performance liquid chromatogra-

phy using the macrocyclic antibiotic chiral stationary phase teicoplanin. *Chromatographia* **13,** 520–524.

78. Risley, D. S. and Strege, M. A. (2000) Chiral separations of polar compounds by hydrophilic interaction chromatography with evaporative light scattering detection. *Anal. Chem.* **72,** 1736–1739.

79. Yu, Y.-P. and Wu, W.-H. (2001) Simultaneous analysis of enantiomeric composition of amino acids and N-acetyl-amino acids by enantioselective chromatography. *Chirality* **13,** 231–235.

80. Schlauch, M., Kos, O., and Frahm, A. W. (2002) Comparison of three chiral stationary phases with respect to their enantio- and diastereoselectivity for cyclic beta-substituted alpha-amino acids. *J. Pharm. Biomed. Anal.* **27,** 409–419.

81. Ward, T. J. and Farris, A. B., III. (2001) 53. Chiral separations using the macrocyclic antibiotics: a review. *J. Chromatogr. A* **906,** 73–89.

82. Peyrin, E., Ravelet, C., Nicolle, E., et al. (2001) Dansyl amino acid enantiomer separation on a teicoplanin chiral stationary phase: effect of eluent pH. *J. Chromatogr. A* **923,** 37–43.

83. Ward, T. J. (1996) Macrocyclic antibiotics—the newest class of chiral selectors. *LC GC* **14, 886,** 890–894.

84. Bojarski, J. and Aboul-Enein, H. Y. (1999) Recent chromatographic and electrophoretic enantiosepararrtions of cardiovascular drugs. *Biomed. Chromatogr.* **13,** 197–208.

85. Joyce, K. B., Jones, A. E., Scott, R. J., Biddlecombe, R. A., and Pleasance, S. (1998) Determination of the enantiomers of salbutamol and its 4-O-sulphate metabolites in biological matrices by chiral liguid chromatography tandem mass spectrometry. *Rapid Commun. Mass Spectrom.* **12,** 1899–1910.

86. Ramos, L., Bakhtiar, R., Majumdar, T., Hayes, M., and Tse, F. (1999) Liquid chromatographic/atmospheric pressure chemical ionization tandem mass spectrometry enantiomeric separation of dl-threo-methylphenidate, (ritalinâ) using a macrocyclic antibiotic as the chiral selector. *Rapid Commun. Mass Spectrom.* **13,** 2054–2062.

87. Bakhtiar, R. and Tse, F. L. S. (2000) High-throughput chiral liquid chromatography/tandem mass spectrometry. *Rapid Commun. Mass Spectrom.* **14,** 1128–1135.

88. Petritis, K., Valleix, A., Elfakir, C., and Dreux, M. (2001) Simutaneous analysis of underivatized chiral amino acids by liquid chromatography-ionspray tandem mass spectrometry using a teicoplanin chiral stationary phase. *J. Chromatogr. A* **913,** 331–340.

89. Mendez, S. P., Gonzalez, E. B., and Medel, A. S. (2000) Chiral speciation and determination of selenomethionine enantiomers in selenized yeast by HPLC-ICP-MS using a teicoplanin-based chiral stationary phase. *J. Anal. At. Spectrom.* **15,** 1109–1114.

90. Armstrong, D. W. and Rundlett, K. L. (1995) CE resolution of neutral and anionic racemates with glycopeptide antibiotics and micelles. *J. Liq. Chromatogr.* **18,** 3659–3674.

91. Carter-Finch, A. S. and Smith, N. W. (1999) Enantiomeric separations by capillary electrochromatography using a macrocyclic antibiotic chiral stationary phases. *J. Chromatogr. A* **848,** 375–385.
92. Ward, T. J. (2000) Chiral separations. *Anal. Chem.* **72,** 4521–4528.
93. Desiderio, C. and Fanali, S. (1998) Chiral analysis by capillary electrophoresis using antibiotics as chiral selector. *J. Chromatogr. A* **807,** 37–56.
94. Wan, H. and Blomberg, L. G. (1997) Enantiomeric separation of small chiral peptides by capillary electrophoresis. *J. Chromatogr. A* **792,** 393–400.
95. Carotti, A. and Gioia, F. D. (1999) Teicoplanin-based enantiomeric separations in CZE using partial filling technique. *J. High Resol. Chromatogr.* **22,** 315–321.
96. Ward, T. J. (1994) Chiral media for capillary electrophoresis. *Anal. Chem.* **66,** 632A–640A.
97. Ward, T. J., Dann, C. I., and Brown, A. P. (1996) Separation of enantiomers using vancomycin in a countercurrent process by suppression of electroosmosis. *Chirality* **8,** 77–83.
98. Vespalec, R., Corstjens, H., Billiet, H. A. H., Frank, J., and Luyben, K. C. A. M. (1995) Enantiomeric separation of sulfur- and selenium-containing amino acids by capillary electrophoresis using vancomycin as a chiral selector. *Anal. Chem.* **67,** 3223–3228.
99. Sharp, V. S., Risley, D. S., Mccarthy, S., Huff, B. E., and Strege, M. A. (1997) Evaluation of a new macrocyclic antibiotic as a chiral selector for use in capillary electrophoresis. *J. Liq. Chromatogr.* **20,** 887–898.
100. Strege, M. A., Huff, B. E., and Risely, D. S. (1996) Evaluation of macrocyclic antibiotic A82846B as a chiral selector for capillary electrophoresis separations. *LC GC* **14,** 144–150.
101. Ward, T. J., Farris, A. B., III, and Woodling, K. (2001) Synergistic chiral separations using the glycopeptides ristocetin A and vancomycin. *J. Biochem. Biophys. Methods* **48,** 163–174.
102. Wikstrom, H., Svensson, L. A., Torstensson, A., and Owens, P. K. (2000) Immobilisation and evaluation of a vancomycin chiral stationary phase for capillary electrochromatography. *J. Chromatogr. A* **869,** 395–409.
103. Carlsson, E., Wikström, H., and Owens, P. K. (2001) Validation of a chiral capillary electrochromatographic method for metoprolol on a teicoplanin stationary phase. *Chromatographia* **53,** 419–424.
104. Medvedovicia, A., Sandraa, P., Toribiob, L., and David, F. (1997) Chiral packed column subcritical fluid chromatography on polysaccharide and macrocyclic antibiotic chiral stationary phases. *J. Chromatogr. A* **785,** 159–171.
105. Svensson, L. A. and Owens, P. K. (2000) Enantioselective supercritical fluid chromatography using ristocetin A chiral stationary phases. *Analyst* **125,** 1037–1039.
106. Toribio, L., David, F., and Sandra, P. (1999) Enantiomeric separation of some cyclic ketones and dioxalene derivatives by chiral SFC. *Quimica Analitica (Barcelona)* **18,** 269–273.

107. Liu, Y., Berthod, A., Mitchell, C. R., Xiao, T. L., Zhang, B., and Armstrong, D. W. (2002) Super/subcritical fluid chromatography chiral separations with macrocyclic glycopeptide stationary phases. *J. Chromatogr. A* **978,** 185–204.

108. Sun, F. Q. and Olesik, S. V. (1999) Chiral separations performed by enhanced fluidity liquid chromatography on a macrocyclic antibioitc chiral stationary phase. *Anal. Chem.* **71,** 2139–2145.

109. Jin, Z. H., Wang, M. R., and Cen, P. L. (2002) Production of teicoplanin by valine analogue-resistant mutant strains of *actinoplanes teichomyceticus*. *Appl. Microbiol. Biotechnol.* **58,** 63–66.

110. Harris, C. M., Kopecka, H., and Harris, T. M. (1983) Vancomycin: structure and transformation to CDP-I. *J. Am. Chem. Soc.* **105,** 6915–6922.

111. Higgins, H. M., Harrison, W. H., Wild, G. M., Bungay, H. R., McCormick, M. H., and Eli Lilly & Co., I., IN. (1957–1958) Vancomycin, a new antibiotic. VI. Purification and properties of vancomycin. *Antibiot. Annu.* 906.

112. Parenti, F., Beretta, G., Berti, M., and Arioli, V. (1978) Teichomycins, new antibiotics from *actinoplanes teichomyceticus* Nov. Sp. I. Description of the producer strain, fermentation studies and biological properties. *J. Antibiot.* **1978,** 276–283.

113. Philip, J. E., Schenck, J. R., Hargie, M. P., Abbott Labs., and Chicago, N. (1957) Ristocetins A and B, two new antibiotics. Isolation and properties. *Antibiot. Annu.* **1956–1957,** 699–705.

114. Beesley, T. E. and Scott, R. P. W. (1998) *Chiral Chromatography.* John Wiley & Sons, West Sussex, England.

115. Tesarova, E., Tuzar, Z., Nesmerak, K., Bosakova, Z., and Gas, B. (2001) Study on the aggregation of teicoplanin. *Talanta* **54,** 643–653.

116. O'Neil, M. J., Smith, A., and Heckelman, P., eds. (2001) *The Merck Index. 13th Edition,* Merck, Rahway, NJ, pp. 9194–9195.

117. Xiao, T. L. and Armstrong, D. W. (2003) Chiral separation using glycopeptide antibiotics chiral stationary phases, unpublished.

118. Tesarova, E., Bosakova, Z., and Zuskova, I. (2000) Enantioseparation of selected N-tert.-butyloxycarbonyl amino acids in high-performance liquid chromatography and capillary electrophoresis with a teicoplanin chiral selector. *J. Chromatogr. A* **879,** 147–156.

119. Chang, S. C., Reid, G. L., Chen, S., Chang, C. D., and Armstrong, D. W. (1993) Evaluation of a new polar-organic high-performance liquid chromatographic mobile phase for cyclodextrin-bonded chiral stationary phases. *Trends Anal. Chem.* **12,** 144–153.

120. Armstrong, D. W., Chen, S., Chang, C., and Chang, S. (1992) A new approach for the direct resolution of racemic beta adrenergic blocking agents by HPLC. *J. Liq. Chromatogr. Relat. Technol.* **15,** 545–556.

121. Mislanova, C., Stefancova, A., Oravcova, J., Horecky, J., Trnovec, T., and Lindner, W. (2000) Direct high-performance liquid chromatographic determination of (R)- and (S)-propranolol in rat microdialysate using on-line column switching procedures. *J. Chromatogr. B* **739,** 151–161.

122. Armstrong, D. W., Kullman, J. P., Chen, X., and Rowe, M. (2001) Composition and chirality of amino acids in aerosol/dust from laboratory and residential enclosures. *Chirality* **13,** 153–158.
123. Ekborg-Ott, K. H. and Armstrong, D. W. (1996) Evaluation of the concentration and enantiomeric purity of selected free amino acids in fermented malt beverages (beers). *Chirality* **8,** 49–57.
124. Pawlowska, M. and Armstrong, D. W. (1994) Evaluation of enantiomeric purity of selected amino acids in honey. *Chirality* **6,** 270–276.
125. Peter, A., Vekes, E., and Armstrong, D. W. (2002) Effects of temperature on retention of chiral compounds on a ristocetin A chiral stationary phase. *J. Chromatogr. A* **958,** 89–107.
126. Schlauch, M. and Frahm, A. W. (2000) Enantiomeric and diastereomeric high-performance liquid chromatographic separation of cyclic beta-substituted alpha-amino acids on a teicoplanin chiral stationary phase. *J. Chromatogr. A* **868,** 197–207.
127. Schurig, V. and Fluck, M. (2000) Enantiomer separation by complexation SFC on immobilized chirasil-nickel and chirasil-zinc. *J. Biochem. Biophys. Methods* **43,** 223–240.
128. Welch, C. J. (2002) Presentation at Iowa State University on Rapid Chiral separation methods. Merck, Rahway, NJ, Feb. 15.
129. Xiao, T. L., Rozhkov, R. V., Larock, R. C., and Armstrong, D. W. (2003) Enantiomeric separation of substituted dihydrofurocoumarin compounds by HPLC using macrocyclic glycopeptide chiral stationary phases. *Anal. Bioanalyt. Chem.,* in press.
130. Jandera, P., Backovska, V., and Felinger, A. (2001) Analysis of the band profiles of the anantiomers of phenylglycine in liquid chromatography on bonded teicoplanin columns using the stochastic theory of chromatography. *J. Chromatogr. A* **919,** 66–77.
131. Jandera, P., Skavrada, M., Klemmova, K., Backovska, V., and Guiochon, G. (2001) Effect of the mobile phase on the retention behaviour of optical isomers of carboxylic acids and amino acids in liquid chromatography on bonded Teicoplanin columns. *J. Chromatogr. A* **917,** 123–133.
132. Courderot, C. M., Perrin, F. X., Guillaume, Y. C., et al. (2002) Chiral discrimination of dansyl-amino-acid enantiomerson teicoplanin phase: sucrose-perchlorate anion dependence. *Anal. Chim. Acta* **457,** 149–155.
133. Yu, Y.-P. and Wu, S.-H. (2001) Simultaneous analysis of enantiomeric composition of amino acids and N-acetyl-amino acids by enantioselective chromatography. *Chirality* **13,** 231–235.
134. Torok, G., Peter, A., Armstrong, D. W., Tourwe, D., Toth, G., and Sapi, J. (2001) Direct chiral separation of unnatural amino acids by high-performance liquid chromatography on a ristocetin A-bonded stationary phase. *Chirality* **13,** 648–656.
135. D'Acquarica, I., Gasparrini, F., Misiti, D., et al. (2000) Application of a new chiral stationary phase containing the glycopeptide antibiotic A-40,926 in the direct chromatographic resolution of beta-amino acids. *Tetrahedron Asymmetry* **11,** 2375–2385.

136. D'Acquarica, I. (2000) New synthetic strategies for the preparation of novel chiral stationary phases for high-performance liquid chromatography containing natrual pool selectors. *J. Pharm. Biomed. Anal.* **23,** 3–13.

137. Wang, A. X., Lee, J. T., and Beesley, T. E. (2000) Coupling chiral stationary phases as a fast screening approach for HPLC method development. *LC GC* **18,** 626–639.

138. Kosel, M., Eap, C. B., Amey, M., and Baumann, P. (1998) Analysis of the enantiomers of citalopram and its demethylated metabolites using chiral liquid chromatography. *J. Chromatogr. B: Biomed. Sci. Appl.* **719,** 234–238.

139. Aboul-Enein, H. Y. and Serignese, V. (1999) Quantitative determination of clenbuterol enantiomers in human plasma by high-performance liquid chromatography using the macrocyclic antibiotic chiral stationary phase teicoplanin. *Chromatographia* **13,** 520–524.

140. Schneiderheinze, J. M., Armstrong, D. W., and Berthod, A. (1999) Plant and soil enantioselective biodegradation of racemic phenoxyalkanoic herbicides. *Chirality* **11,** 330–337.

141. Gasparrini, F., D'Acquarica, I., Vos, J. G., O'Connor, C. M., and Villani, C. (2000) Efficient enantiorecognition of ruthenium(II) complexes by silica-bound teicoplanin. *Tetrahedron Asymmetry* **11,** 3535–3541.

142. Aboul-Enein, H. Y. and Serignese, V. (1998) Enantiomeric separation of several cyclic imides on a macrocyclic antibiotic (vancomycin) chiral stationary phase under normal and reversed phase conditions. *Chirality* **10,** 358–361.

5

Chiral Separation by HPLC
Using Polysaccharide-Based Chiral Stationary Phases

Chiyo Yamamoto and Yoshio Okamoto

1. Introduction

Chiral separation by high-performance liquid chromatography (HPLC) using a chiral stationary phase (CSP) is one of the most efficient methods for separating enantiomers, not only on an analytical scale, but also on a preparative scale, and in the past two decades, many CSPs have been developed. Polysaccharides such as cellulose, amylose, and chitin (**Fig. 1**) are the most abundant optically active polymers on the earth and can be readily modified to carbamates and esters through the reaction with isocyanates and acid chlorides, respectively. The CSPs based on polysaccharide derivatives are some of the most popular ones and can separate a wide range of chiral compounds (*1–4*).

Table 1 lists the polysaccharide derivatives that have a high chiral resolving power (*5–14*). The commercial names are also listed for some of the derivatives. Among these CSPs, the tris(3,5-dimethylphenylcarbamate)s (Chiralcel® OD and Chiralpak® AD) of cellulose and amylose, cellulose tris(4-methylbenzoate) (Chiralcel® OJ), and amylose tris[S-1-phenylethylcarbamate] (Chiralpak® AS) most frequently appear in the literature. By using hexane-alcohol eluent systems on these CSPs, 80–90% of the chiral compounds may be resolved (*3,4*). Polar eluents containing a buffer can also be used (*15,16*).

2. Materials

2.1. Preparation of Phenylcarbamates of Polysaccharides

1. Phenylisocyanate, 4-methylphenylisocyanate, 4-chlorophenylisocyanate, 3,5-dimethylphenylisocyanate, S-1-phenylethylisocyanate, cyclohexylisocyanate (*see* **Note 1**).

From: *Methods in Molecular Biology, Vol. 243: Chiral Separations: Methods and Protocols*
Edited by: G. Gübitz and M. G. Schmid © Humana Press Inc., Totowa, NJ

Table 1
Structures of Polysaccharide Derivatives with a High Chiral Resolving Power

Polysaccharide derivatives	R =	Commercial names	References
Cellulose esters	−CH$_3$	Chiralcel CTA-I (microcrystalline) / Chiralcel OA (coated-type)	5, 6
	(phenyl)	Chiralcel OB	5
	(phenyl)−CH$_3$	Chiralcel OJ	7
	−CH=CH−(phenyl)	Chiralcel OK	6
Cellulose carbamates	(phenyl)	Chiralcel OC	8
	(3-methylphenyl) CH$_3$	Chiralcel OD	8
	(phenyl)−Cl	Chiralcel OF	8
	(phenyl)−CH$_3$	Chiralcel OG	8
	(3,4-dichlorophenyl) Cl Cl	——	8, 9
	(cyclohexyl)	——	10
Amylose carbamates	(phenyl)	——	11
	(3,5-dimethylphenyl) CH$_3$ CH$_3$	Chiralpak AD	11
	(phenyl)−Cl CH$_3$	——	12
	CH$_3$−CH−(phenyl)	Chiralpak AS	13
	(cyclohexyl)	——	10
Chitin carbamates	(phenyl)	——	14
	(3,5-dimethylphenyl) CH$_3$ CH$_3$	——	14
	(3,4-dichlorophenyl) Cl Cl	——	14

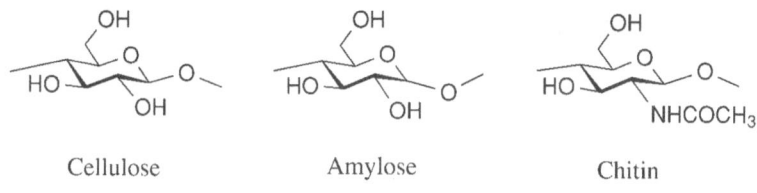

Cellulose   Amylose   Chitin

Fig. 1. Structures of polysaccharides.

2. Cellulose: degree of polymerization is about 200.
3. Amylose: degree of polymerization is about 100–300.
4. Chitin: purified powder from crab shells.
5. Pyridine: anhydrous.
6. *N,N*-Dimethylacetamide: anhydrous.
7. LiCl.
8. Methanol.

2.2. Preparation of Benzoates of Polysaccharides

1. Benzoyl chloride, 4-methylbenzoyl chloride.
2. Cellulose: degree of polymerization is about 200.
3. Pyridine: anhydrous.
4. Methanol.

2.3. Preparation of Packing Materials

2.3.1. Derivatization of Silica Gel With 3-Aminopropyltriethoxysilane

1. Silica gel: macroporous spherical silica gel with a mean particle size of 7 μm and a mean pore diameter of 100 nm.
2. 3-Aminopropyltriethoxysilane.
3. Benzene: anhydrous.
4. Pyridine: anhydrous.
5. Washing solutions: methanol, acetone, hexane.

2.3.2. Coating of Polysaccharide Derivative on Silica Gel

1. Solution of polysaccharide derivative: polysaccharide derivative (0.75 g) dissolved in tetrahydrofuran (10 mL).
2. 3-Aminopropylsilanized silica gel.

2.4. Packing in an HPLC Column

1. Hexane/2-propanol (9:1, v/v).
2. Liquid paraffin.
3. Benzene.

2.5. Separation of Enantiomers by HPLC

1. Hexane.
2. 2-Propanol, ethanol.
3. Trifluoroacetic acid, formic acid.
4. Diethylamine, isopropylamine.
5. Polar eluents (buffer): water/methanol, water/acetonitrile, borate buffer, phosphate buffer, perchlorate buffer, etc.
6. 1,3,5-Tri(*tert*-butyl)benzene: a nonretained compound for estimating the dead time under normal phase.
7. Solution of racemate: concentration is about 1–5 mg/mL.

3. Methods

3.1. Preparation of Phenylcarbamates of Polysaccharides

3.1.1. Cellulose Tris(3,5-Dimethylphenylcarbamate)

1. Dry cellulose (1.0 g) under vacuum at 80°C for 2 h.
2. Add dry pyridine (20 mL) and 3,5-dimethylphenylisocyanate (1.3 equivalents of the hydroxy groups of cellulose) to the dried cellulose and stir the mixture under nitrogen at 80°C for 24 h (*see* **Note 2**).
3. Pour the reaction mixture into methanol (400 mL) and separate the cellulose phenylcarbamate as the methanol-insoluble part by filtration or centrifugation and wash with methanol until the pyridine is removed (*see* **Note 3**).
4. Dry the obtained cellulose phenylcarbamate under vacuum at 60°C for 2 h (*see* **Note 4**).

3.1.2. Amylose Tris(3,5-Dimethylphenylcarbamate)

1. Dry amylose (1.0 g) under vacuum at 80°C for 2 h.
2. Add dry pyridine (20 mL) and 3,5-dimethylphenylisocyanate (1.3 equivalents of the hydroxy groups of amylose) to the dried amylose and stir the mixture under nitrogen at 80°C for 24 h (*see* **Note 2**).
3. Pour the reaction mixture into methanol (400 mL) and separate the amylose phenylcarbamate as the methanol-insoluble part by filtration or centrifugation and wash with methanol until the pyridine is removed (*see* **Note 3**).
4. Dry the obtained amylose phenylcarbamate under vacuum at 60°C for 2 h (*see* **Note 4**).

3.1.3. Chitin Bis(3,5-Dimethylphenylcarbamate)

1. Dry chitin (1.0 g) under vacuum at 80°C for 2 h.
2. To swell chitin before addition of the isocyanates, add dry *N,N*-dimethylacetamide (15 mL) and lithium chloride (1.5 g) to the dried chitin and stir the solution under nitrogen at 80°C for 24 h (*see* **Note 5**).

3. Add dry pyridine (5 mL) and 3,5-dimethylphenylisocyanate (1.3 equivalents of the hydroxy groups of chitin) to the chitin solution and stir the mixture at 80°C for 24 h (*see* **Note 2**).
4. Pour the reaction mixture into methanol (400 mL) and separate the chitin phenylcarbamate as the methanol-insoluble part by filtration or centrifugation and wash with methanol until the pyridine is removed.
5. Dry the obtained chitin phenylcarbamate under vacuum at 60°C for 2 h (*see* **Note 4**).

3.2. Preparation of Benzoates of Polysaccharides

1. Dry cellulose (1.0 g) under vacuum at 80°C for 2 h.
2. Add dry pyridine (20 mL) and benzoyl chloride (1.3 equivalents of the hydroxy groups of cellulose) to the dried cellulose, and stir the mixture under nitrogen at 80°C for 24 h (*see* **Note 2**).
3. Pour the reaction mixture into methanol (400 mL), and separate the cellulose benzoate as the methanol-insoluble part by filtration or centrifugation and wash with methanol until the pyridine is removed (*see* **Note 3**).
4. Dry the obtained cellulose benzoate under vacuum at 60°C for 2 h.

3.3. Preparation of Packing Materials

3.3.1. Derivatization of Silica Gel
With 3-Aminopropyltriethoxysilane (see **Note 6**)

1. Dry silica gel (particle size, 7 μm; pore size; 100 nm) (10 g) under vacuum at 180°C for 2 h.
2. Treat the silica gel with a large excess of 3-aminopropyltriethoxysilane (2 mL) in dry benzene (60 mL) in the presence of a catalytic amount of dry pyridine (0.6 mL) at 80°C for 12 h.
3. Pour the mixture into methanol (600 mL) and leave it for 2 h.
4. Separate the silanized silica gel by filtration with a glass filter and wash with methanol, acetone, and finally hexane.
5. Dry the silanized silica gel under vacuum at 60°C for 12 h.

3.3.2. Coating of Polysaccharide
Derivatives on 3-Aminopropyl Silica Gel

1. Dissolve cellulose tris(3,5-dimethylphenylcarbamate) (0.75 g) in tetrahydrofuran (10 mL) (*see* **Note 7**).
2. Place the silanized silica gel (3.0 g) in a round-bottom flask. Divide the polysaccharide derivative solution into three or four portions. Add several drops of the solution of one portion to the silica gel and shake the flask to uniformly coat the polysaccharide derivative on the silica surface. Repeat this coating process for the remaining solution of the one portion.

3. After the addition of the one portion, dry the silica gel under vacuum at room temperature.
4. Repeat this process for the remaining portions.

3.4. Packing in an HPLC Column

1. Disperse the polysaccharide derivative-coated silica gel in the mixture of hexane and 2-propanol (9:1, v/v).
2. To collect the same size silica particles, remove the supernatant and the lowest layer of the silica gel by decantation containing the small and large silica particles, respectively (*see* **Note 8**).
3. Disperse the silica gel in hexane/2-propanol/liquid paraffin (25:2:3, v/v/v) and pack this slurry in a stainless-steel column (25 × 0.46 cm inner diameter [id]) at 400 kg/cm^2 using hexane/2-propanol (9:1). After a few minutes, reduce the pressure to 100 kg/cm^2 and wash away the liquid paraffin (for approx 20 min).
4. After setting up the HPLC system, estimate the plate number of the column using the benzene peak.

3.5. Separation of Enantiomers (see Note 9)

3.5.1. Neutral Compounds

1. Select a suitable column for efficient resolution of enantiomers. Examine the commercially available columns of Chiralcel OD, Chiralpak AD, Chiralcel OJ, and Chiralpak AS in this order (*see* **Note 10**).
2. Select a suitable eluent. For the normal phase separation of neutral compounds, a mixture of hexane and 2-propanol or ethanol is often the most suitable eluent (*see* **Note 11**). For the separation under a reversed-phase condition, water/alcohol or water/acetonitrile is an effective eluent. The eluent should be degassed before the use.
3. Dissolve racemates in the solvent to be used as the eluent for chromatography (1–5 mg/mL) (*see* **Note 12**).
4. Estimate the dead time (t_0) using 1,3,5-tri-*tert*-butylbenzene as a nonretained compound under normal phase conditions (*see* **Note 13**).
5. Inject a racemic sample (1–10 µL) with a microsyringe. Based on the retention times of each enantiomer (t_1 and t_2) and t_0, the capacity factors k_1' and k_2' can be estimated as $(t_1 - t_0)/t_0$ and $(t_2 - t_0)/t_0$, respectively, and separation factor (α) as k_2'/k_1'.

3.5.2. Acidic Compounds

1. To resolve acidic compounds under normal phase conditions, the addition of a small amount of a strong acid, such as trifluoroacetic acid or formic acid (approx 0.5%) to an eluent may result in better separation (*see* **Note 14**).

3.5.3. Basic Compounds

1. To resolve basic compounds under normal phase conditions, the addition of a small amount of amine such as diethylamine or isopropylamine (approx 0.1%) to an eluent leads to a decrease in tailing of the chromatograms (*see* **Note 15**).

4. Notes

1. Aniline derivatives can be converted to the corresponding isocyanates through the reaction with triphosgene. Benzoic acids can also be converted to isocyanates by the Curtius rearrangement of acyl azide.
2. During the course of the reaction, the infrared (IR) spectrum of the reaction mixture should be measured to check whether isocyanate or benzoyl chloride remains. If a peak around 2200–2300/cm (NCO) or 1850–1900/cm (COCl) cannot be observed, additional isocyanates or benzoyl chloride should be added, respectively.
3. If a polysaccharide derivative does not precipitate in methanol, due to its high solubility, pour the reaction mixture into a methanol/water mixture.
4. The obtained polysaccharide derivative can be characterized by elemental analysis, IR and nuclear magnetic resonance (NMR) spectroscopies. If there exist urea derivatives in the product, it must be purified by reprecipitation or Soxhlet extraction using methanol.
5. It is difficult to derivatize chitin due to its poor solubility. Therefore, chitin is first dissolved in *N,N*-dimethylacetamide/lithium chloride and then allowed to react with isocyanates in the presence of pyridine.
6. Stirring of the silica gel with a magnetic stirrer should be avoided, because it will be easily broken.
7. The coating solvent depends on the polysaccharide derivatives. Amylose tris(3,5-dimethylphenylcarbamate) dissolved in tetrahydrofuran/*N,N*-dimethylformamide (5:1, v/v), cellulose tribenzoate in methylene chloride, and chitin bis(3,5-dimethyl-phenycarbamate) in dimethylsulfoxide are coated in a similar manner. The chiral recognition ability of the polysaccharide derivatives often depends on the coating conditions, particularly the solvent used for the dissolution of the polysaccharide derivatives *(17)*.
8. If the size distribution of the silica particles is broad, the column with a high plate number is difficult to obtain.
9. Polarimetric and circular dichroism (CD) detectors, which can respond only to chiral compounds, are conveniently used in addition to a UV or a refractive index (RI) detector.
10. Chiralcel OD and Chiralpak AD show excellent resolving power for a variety of racemates, and 80% of the racemates may be resolved using these columns *(1–4)*.
11. In some cases, the alteration of an alcohol from 2-propanol to ethanol brings about a better separation and shorter elution time. The structures of alcohols as an additive influence the enantioselectivity *(18)*. For chiral aromatic hydrocarbons with no polar functional groups, iso-octane may be effective as an eluent *(19)*. The CSPs are prepared by coating the polysaccharide derivatives on macroporous silica gel,

and therefore, the solvents such as tetrahydrofuran, chloroform, and acetone, which dissolve or swell the polysaccharides, cannot be added to the mobile phase. In addition, Chiralpak AD cannot be used with hexane/ethanol mixtures ranging from 50:50 to 85:15 (v/v).

12. Dissolution of racemates in the same solvent as the eluent is recommended, but other solvents can be added if the racemates do not precipitate upon mixing with the eluent. (Only a small amount of chloroform and dichloromethane can also be added to the solvent for dissolving racemates, but it may lead to a deterioration of the chiral column.)

13. Under a reversed-phase condition, acetone, methanol, ethanol, or acetonitrile is used as a nonretained compound.

14. Under a reversed-phase condition, it is essential to use an acidic mobile phase to suppress the dissociation of the analyte. A pH 2.0 aqueous solution or buffer containing an organic modifier (alcohol or acetonitrile), such as $HClO_4$ aqueous (pH 2.0)/CH_3CN (60:40) and 0.5 M $NaClO_4$–$HClO_4$ aqueous (pH 2.0)/CH_3CN (60: 40), is effective *(16)*. The use of a perchlorate solution with a high concentration should be avoided, because it may explode when it is heated and evaporated with an organic solvent.

15. Under a reversed-phase condition, it is important to use a suitable buffer with the proper pH. A basic mobile phase is not recommended for a column, because the silica supports are unstable at pH > 7.0. Under neutral and acidic mobile phase conditions, basic analytes are positively charged, and therefore, it is effective to add a considerable amount of anions, for instance, PF_6^-, BF_4^-, and ClO_4^-, in the mobile phase to form an ion pair, because the positively charged analyte cannot efficiently interact with the CSP *(16)*. At first, a 0.5 M $NaClO_4$ aqueous /CH_3CN (60:40) is recommended.

References

1. Okamoto, Y. and Kaida, Y. (1994) Resolution by high-performance liquid chromatography using polysaccharide carbamates and benzoates as chiral stationary phases. *J. Chromatogr. A* **666**, 403–419.

2. Yashima, E. and Okamoto, Y. (1995) Chiral discrimination on polysaccharides derivatives. *Bull. Chem. Soc. Jpn.* **68**, 3289–3307.

3. Okamoto, Y. and Yashima, E. (1998) Polysaccharide derivatives for chromatographic separation of enantiomers. *Angew. Chem. Int. Ed.* **37**, 1020–1043.

4. Yashima, E., Yamamoto, C., and Okamoto, Y. (1998) Polysaccharide-based chiral LC columns. *Synlett* 344–360.

5. Okamoto, Y., Kawashima, M., Yamamoto, K., and Hatada, K. (1984) Useful chiral packing materials for high-performance liquid chromatographic resolution. Cellulose triacetate and tribenzoate coated on macroporous silica gel. *Chem. Lett.* 739–742.

6. Ichida, A., Shibata, T., Okamoto, I., Yuki, Y., Namikoshi, H., and Toda, Y. (1984) Resolution of enantiomers by HPLC on cellulose derivatives. *Chromatographia* **19**, 280–284.

7. Okamoto, Y., Aburatani, R., and Hatada, K. (1987) Cellulose tribenzoate derivatives as chiral stationary phases for high-performance liquid chromatography. *J. Chromatogr.* **389,** 95–102.

8. Okamoto, Y., Kawashima, M., and Hatada, K. (1986) Controlled chiral recognition of cellulose triphenylcarbamate derivatives supported on silica gel. *J. Chromatogr.* **363,** 173–186.

9. Chankvetadze, B., Yamamoto, C., and Okamoto, Y. (2000) Extremely high enantiomer recognition in HPLC separation of racemic 2-(benzylsulfinyl)benzamide using cellulose tris(3,5-dichlorophenylcarbamate) as a chiral stationary phase. *Chem. Lett.* 67–77.

10. Kubota, T., Yamamoto, C., and Okamoto, Y. (2000) Tris(cyclohexylcarbamate)s of cellulose and amylose as potential chiral stationary phases for high-performance liquid chromatography and thin-layer chromatography. *J. Am. Chem. Soc.* **122,** 4056–4059.

11. Okamoto, Y., Aburatani, R., Fukumoto, T., and Hatada, K. (1987) Useful chiral stationary phases for HPLC. Amylose tris(3,5-dimethylphenylcarbamate) and tris(3,5-dichlorophenylcarbamate) supported on silica gel. *Chem. Lett.* 1857–1860.

12. Yashima, E., Kasashima, E., and Okamoto, Y. (1997) Enantioseparation on 4-halogen-substituted phenylcarbamates of amylose as chiral stationary phases for high-performance liquid chromatography. *Chirality* **9,** 63–68.

13. Okamoto, Y., Kaida, Y., Hayashida, H., and Hatada, K. (1990) Tris(1-phenylethylcarbamate)s of cellulose and amylose as useful chiral stationary phases for chromatographic optical resolution. *Chem. Lett.* 909–912.

14. Yamamoto, C., Hayashi, T., Okamoto, Y., and Kobayashi, S. (2000) Enantioseparation by using chitin phenylcarbamates as chiral stationary phases for high-performance liquid chromatography. *Chem. Lett.* 12–13.

15. Ishikawa, A. and Shibata, T. (1993) Cellulosic chiral stationary phase under reversed-phase condition. *J. Liq. Chromatogr.* **16,** 859–878.

16. Tachibana, K. and Ohnishi, A. (2001) Reversed-phase liquid chromatographic separation of enantiomers on polysaccharide type chiral stationary phases. *J. Chromatogr. A* **906,** 127–154.

17. Shibata, T., Okamoto, I., and Ishii, K. (1986) Chromatographic optical resolution on polysaccharides and their derivatives. *J. Liq. Chromatogr.* **9,** 313–340.

18. Dingene, J. (1994) Polysaccharide phases in enantioseparations, in *A Practical Approach to Chiral Separations by Liquid Chromatography* (Subramanian, G., ed.), VCH, New York, pp. 115–181.

19. Maeda, K., Okamoto, Y., Toledano, O., Becker, D., Biali, S. E., and Rappoport, Z. (1994) Multiple buttressing interactions: enantiomerization barrier of tetrakis(pentamethylphenyl)ethane. *J. Org. Chem.* **59,** 5473–5475.

6

Applications of Polysaccharide-Based Chiral Stationary Phases for Resolution of Different Compound Classes

Hassan Y. Aboul-Enein and Imran Ali

1. Introduction

The impact of chirality on drug development and use has been well documented *(1–6)*. Therefore, the chiral resolution is essential in pharmaceutical, agriculture, and food industries. In view of this, the U.S. Food and Drug Administration has issued certain guidelines for the marketing of racemic compounds *(7)*. In last two decades, high-performance liquid chromatography (HPLC) has become one of the most applied modalities in the chiral resolution of different racemates *(8,9)*. The most important aspect in chiral resolution by HPLC is the development of several chiral stationary phases (CSPs). The important CSPs include native or derivatized amino acids, derivatized polysaccharides (cellulose or amylose), cyclodextrin and its derivatives, protein phases, chiral crown ethers, macrocyclic antibiotics, and other chiral compounds *(10)*. Among these, polysaccharide-based CSPs are very important, as they have achieved a great reputation in the field of chiral resolution. The importance of polysaccharide-based CSPs include their ease of use, reproducible results, and wide range of applications *(8,9,11–13)*.

Among the various polysaccharides, polymers such as cellulose, amylose, chitosan, xylan, curdlan, dextran, and inulin, cellulose and amylose have been used for the preparation of commercial CSPs *(11)*. Cellulose and amylose themselves could not be used as commercial CSPs because of their poor resolution capacity and problem in handling *(9)*. Therefore, these polymers have been derivatized as their tricarbamates or triesters *(11,14–16)*. The polymeric chains of D-(+) glucose units contain β-1,4 linkage in cellulose and α-1,4 linkage in amylose, respectively. These chains lie side by side in a linear fashion in cellulose and in helical fashion in amylose. Polysaccharide-based CSPs are available

From: *Methods in Molecular Biology, Vol. 243: Chiral Separations: Methods and Protocols*
Edited by: G. Gübitz and M. G. Schmid © Humana Press Inc., Totowa, NJ

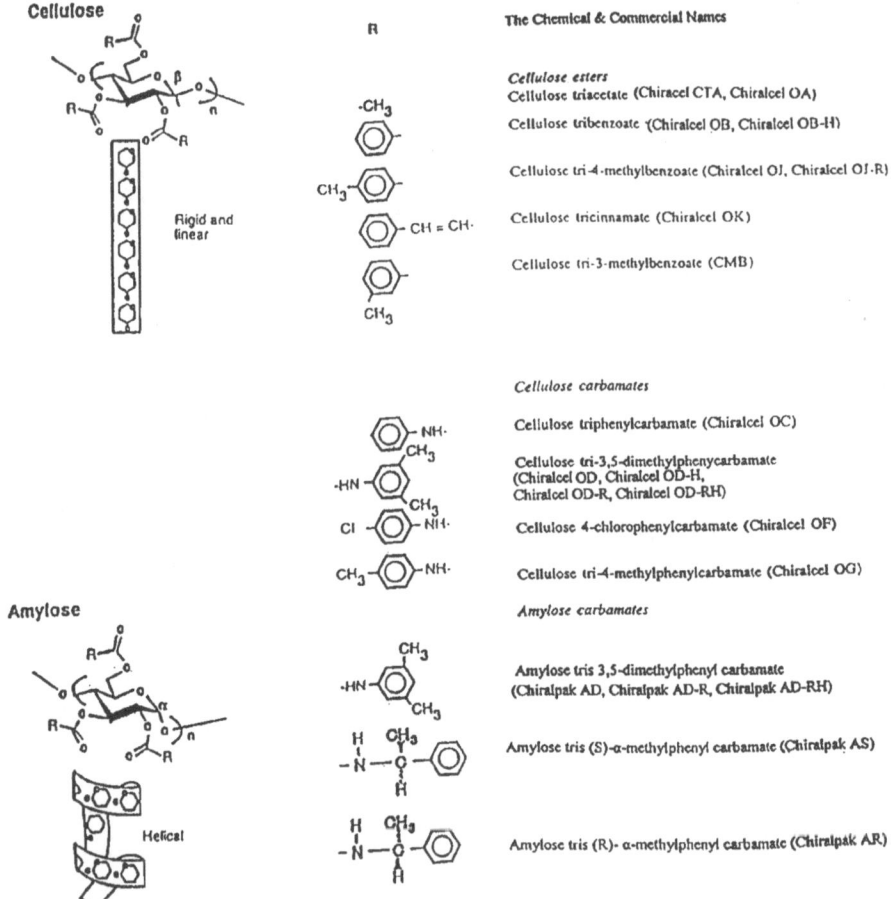

Fig. 1. The chemical structures and the chemical and trade names of most commercially available polysaccharide-based CSPs.

in normal and reversed-phase modes. About 20 derivatives of cellulose and amylose are commercially available and shown in **Fig. 1** with their chemical and trade name. The trade name of the cellulose and amylose derivatives are Chiralcel and Chiralpak, respectively. To denote the reversed-phase nature of the CSPs, R is added in the last of Chiralcel and Chiralpak trade names. The other specifications such as commercial names, column length, and particle size of these CSPs are provided in **Table 1**.

Table 1
The Commercially Available Polysaccharides CSPs

Trade name	Chemical name	Applications
Cellulose CSPs		
Chiralcel OB	Cellulose tris benzoate	Small aliphatic and aromatic compounds
Chiralcel OB-H[a]	Cellulose tris benzoate	Small aliphatic and aromatic compounds
Chiralcel OJ	Cellulose tris 4-methyl benzoate	Aryl methyl esters, aryl methoxy esters
Chiralcel OJ-R[b]	Cellulose tris 4-methyl benzoate	Aryl methyl esters, aryl methoxy esters
Chiralcel CMB	Cellulose tris 3-methylbenzoate	Aryl esters and arylalkoxy esters
Chiralcel OC	Cellulose tris phenylcarbamate	Cyclopentenones
Chiralcel OD	Cellulose tris 3,5-dimethylphenylcarbamate	Alkaloids, tropines, amines, β-blockers
Chiralcel OD-H[a]	Cellulose tris 3,5-dimethylphenylcarbamate	Alkaloids, tropines, amines, β-blockers
Chiralcel OD-R[b]	Cellulose tris 3,5-dimethylphenylcarbamate	Alkaloids, tropines, amines, β-blockers
Chiralcel OD-RH[c]	Cellulose tris 3,5-dimethylphenylcarbamate	Alkaloids, tropines, amines, β-blockers
Chiralcel OF	Cellulose tris 4-chlorophenylcarbamate	β-Lactams, dihydroxypryidines, alkaloids
Chiralcel OG	Cellulose tri 4-methylphenylcarbamate	β-Lactams, alkaloids
Chiralcel OA	Cellulose triacetate on silica gel	Small aliphatic compounds
Chiralcel CTA	Cellulose triacetate microcrystaline	Amides, biaryl compounds
Chiralcel OK	Cellulose tris cinnamate	Aromatic compounds
Amylose CSPs		
Chiralpak AD	Amylose tris 3,5-dimethylphenylcarbamate	Alkaloids, tropines, amines, β-blockers
Chiralpak AD-R[b]	Amylose tris 3,5-dimethylphenylcarbamate	Alkaloids, tropines, amines, β-blockers
Chiralpak AD-RH[a]	Amylose tris 3,5-dimethylphenylcarbamate	Alkaloids, tropines, amines, β-blockers
Chiralpak AR	Amylose tris (R)-1-phenylethylcarbamate	Alkaloids, tropines, amines
Chiralpak AS	Amylose tris (S)-1-methylphenylcarbamate	Alkaloids, tropines, amines

Column Specifications:
[a]25 × 0.46 cm, particle size 5 µm.
[b]15 × 0.46 cm, particle size 10 µm.
[c]15 × 0.46 cm, particle size 5 µm.
Others: 25 × 0.46 cm, particle size 10 µm.
Supplier: Daicel Chemical Industries, Tokyo, Japan.

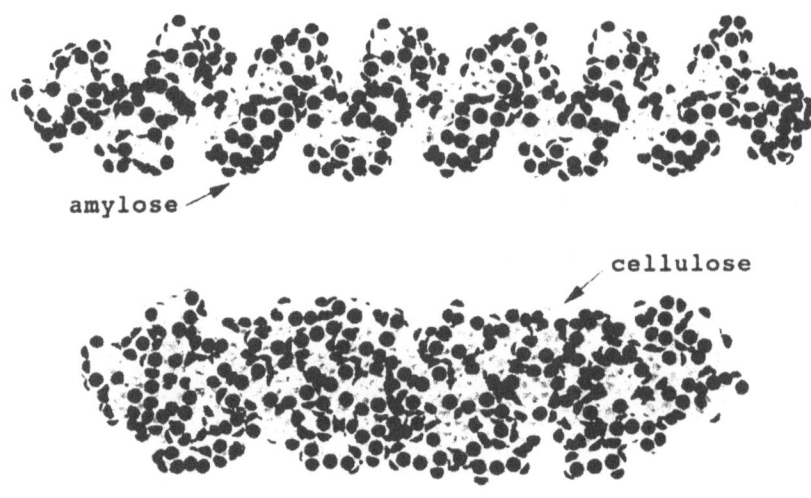

Fig. 2. Three-dimensional structures of amylose and cellulose.

1.1. Mechanism of Chiral Resolution

For optimizing the chromatographic conditions, it is essential to have the knowledge of chiral resolution mechanism on polysaccharide CSPs. The chiral recognition mechanism at a molecular level on the polysaccharide-based CSPs is still unclear although it has been reported that the chiral resolution by these CSPs is achieved through the different hydrogen, π-π, and dipole-induced dipole interactions between the CSP and the enantiomers *(17,18)*. Amylose and cellulose are the semisynthetic polymers that contain the polymeric chains of derivatized D-(+) glucose residues in α-1,4 linkage in amylose and β-1,4 linkage in cellulose, and the chains of these units lie side by side in a linear fashion in case of cellulose and in a helical fashion in case of amylose. It has been observed that amylose has greater resolution capacity than cellulose, which could be attributed to the fact that the amylose CSPs are more helical in nature and possess well-defined grooves. This makes it different from the corresponding cellulose analogues, which appeared to be more linear and rigid in nature *(19)*, and hence provides the greater chiral environment to the analytes. The three-dimensional structures of cellulose and amylose are shown in **Fig. 2**. The structure of amylose clearly indicates the well-defined cavities in a regular fashion. A look on the structures of the reported CSPs (**Fig. 1**) clearly shows the presence of chiral grooves/cavities on these CSPs. The electronegative atoms, such as oxygen, nitrogen, and halogens in the CSPs, are responsible for the formation of hydrogen bonding with the enantiomers. The π-π interactions occur between the phenyl rings of the CSPs and the enantiomers (in case of aromatic racemates). During the chiral resolution, the enantiomers fit stereogenically in different fashions into the chiral grooves of the CSP, which is stabilized by various types of bondings (discussed above) of different magnitudes, and thus the resolution of enantiomers occurs. In addition to these bondings, steric effect also governs the chiral resolution on polysaccharide CSPs *(17,20)*. Recently, it has been observed that coordination bonding is also contributing in the chiral resolution of sulfur-containing enantiomers *(21)*. Besides, some other weak bondings, like Van der Waal forces and ionic bondings, may also contribute in the chiral resolution. A search of literature *(11–16,22)* indicates that most aromatic racemates have been resolved on these CSPs, and it may be due to the significant contribution of π-π interactions in chiral resolution on these CSPs. However, some publications have also been appeared describing the chiral resolution of nonaromatic racemates.

In view of all these, attempts have been made to describe the chiral resolution by polysaccharide CSPs. Only the important aspects of chiral resolution by polysaccharide CSPs are discussed briefly in this chapter. However, to make the chapter more useful for the reader, the experimental part has been described in detail.

2. Materials

2.1. Instruments

A complete HPLC system is required that consists of mobile phase reservoir, pump (two pumps in case of gradient elution is required), injector, chiral column (polysaccharide CSP), detector, and recorder or a work station with an appropriate software program. In addition to this, mobile phase filtration and degassing assemblies are also required. Hamilton syringe is used for loading the sample onto the manual injector. In some HPLC systems, an auto-sampler is supplied with the help of which the required volume of the sample can be loaded.

2.2. Chemicals and Reagents

1. All solvents and reagents should be HPLC grade.
2. Water as the mobile phase constituent may be purchased from chemical suppliers (HPLC grade) or may be prepared in the laboratory using Milli-Q (Millipore, Bedford, MA, USA) water purification unit.
3. For columns and mobile phases see **Tables 1–3**.

3. Methods

The intention of this article is to provide the experimental methodology in detail.

Table 2
The Correlation of Separation
Conditions of Neutral, Acidic, and Basic Compounds

	Systems	
Compounds	Normal phase	Reversed-phase[a]
Neutral	MP = IPA/Hexane pH has no effect on the resolution	MP = water/ACN, pH has no effect on the resolution
Acidic	MP = IPA/hexane/TFA pH near 2.0	MP = pH 2.0 perchlorate, acid/ACN
Basic	MP = IPA/hexane/DEA, IPA/hexane/TFA with pH near 2.0, ion-pair separation	MP = pH <7.0 buffer/ACN. Typical buffer is 0.5 M NaClO$_4$ with pH ranging 4.0–4.5, ion-pair separation

[a]Columns are normally not run under basic conditions.
MP, mobile phase; IPA, isopropanol; ACN, acetonitrile; TFA, trifluoroacetic acid; and DEA, diethylamine.

Table 3
The Most Commonly Used Mobile Phases With Polysaccharide CSPs

Solvents	Compos. (v/v)	Racemates	CSPs
Cellulose CSPs			
Hexane-2-PrOH	95:5	Aromatase inhibitors	Chiralcel OD
Hexane-2-PrOH-DEA	50:50:0.4, 20:80:0.4 and 15:85:0.4	β-Blockers	Chiralcel OD
Hexane-EtOH-DEA	80:20:0.2	Naftopidil	Chiralcel OD
Cyclohexane-MeOH-EtOH	2:95:5	Aminoglutethimide	Chiralcel OD
Hexane-EtOH-DEA	90:10:0.4	β-Blockers	Chiralcel OD
Hexane-1-PrOH	95:5, 97:3	Aromatic amides	Chiralcel OB
Hexane-1-PrOH-MeCN	96:3:1	Aromatic amides	Chiralcel OB
Hexane-2-PrOH	90:10	Aromatic alcohols	Chiralcel OB
Hexane-2PrOH-DEA	425:74:1	Antifungal agents	Chiralcel OD, OJ, OB, OK, OC and OF
Perchlorate solution-MeCN	75:25 and others	β-Blockers	Chiracel OD-R
Hexane-2-PrOH-DEA	15:85:0.4 and others	β-Blockers	Chiralcel OD and OD-R
MeCN-water	50:50	o,p-DDT and o,p-DDD	Chiralcel OD-R and Chiralcel OJ-R
MeCN-2-PrOH	50:50	o,p-DDT and o,p-DDD	Chiralcel OD-R and Chiralcel OJ-R
MeCN-18 mM NH$_4$NO$_3$ (pH 7.0)	25:75, 55:45	Pipiridine derivs.	Chiralcel OD-R

Table 3 (Continued)

Solvents	Compos. (v/v)	Racemates	CSPs
Water-MeCN-TEA	80:20:0.08	Clenbuterol, cimaterol and mabuterol	Chiralcel OD-R and Chiralcel OJ-R
Water-MeCN	45:55	Aromatase inhibitors	Chiralcel OD-R and OJ-R
Hexane-EtOH-MeOH-TFA	480:9.75:9.75:0.5	MPH	Chiralcel OD and OB
2-PrOH-MeCN	90:10, 50:50	Aromatase inhibitors	Chiralpak OD-R and OJ-R
MeCN-water	50:50, 80:20, and 95:5	Aromatase inhibitors	Chiralpak OD-R and OJ-R
EtOH	Pure	Mandelic amide	Chiralcel CTA
EtOH	Pure or 10%	Biaryl compounds	Chiarlcel CTA
Hexane-EtOH	90:10	Biaryl compounds.	Chiarlcel OF and OG
EtOH or MeOH	Pure	Polycyclic aromatics	Chiralcel CA and OB
Hexane-EtOH	90:10	Sulfer compounds	Chiralcel OB and OC
Hexane-2-PrOH	90:10	Phosphorous compounds	Chiarlcel OB, OK and OC
Hexane-EtOH	90:10	Nitrogen and cyano compounds	Chiarlcel OA, OB, OC and OK
Hexane-EtOH	95:5	Amines	Chiarlcel CA,OA, OC, OB and OK
Hexane-2-PrOH	95:5–10	Carboxylic acids and derives.	Chiarlcel OC, OB, OF and OG
EtOH	Pure	Alphatic alcohols	Chiarlcel CA, OB, OK and OK
Hexane-EtOH	95:5, 90:10	Alphatic alcohols	Chiarlcel CA, OB, OK and OK
Amylose CSPs			
Hexane-EtOH	95:5, 90:10 and 88:12	β-Blockers	Chiralpak AD
Hexane-2-PrOH	90:10	β-Blockers	Chiralpak AD
Hexane-EtOH-MeOH-TFA	480:9.75:9.75:0.5	MPH	Chiralpak AD
Hexane-2-PrOH-DEA	400:99:1	Antifungal agents	Chiralpak AD, AS and AR
MeCN-water	50:50	o,p-DDT and o,p-DDD	Chiralpak AD-R
MeCN-2-PrOH	50:50	o,p-DDT and o,p-DDD	Chiralpak AD-R
Water-MeCN-TEA	80:20:0.08	Clenbuterole cimaterol and mabuterol	Chiralpak AD-R
Water-MeCN	60:40	Flurbiprofen	Chiralpak AD-R
MeCN-water-TEA	50:50:0.03	Tetralone deves.	Chiralpak AD-R
MeCN-water-AcOH	60:40:0.03	Tetralone deves.	Chiralpak AD-R
2-PrOH-MeCN	90:10, 50:50	Aromatase inhibitors	Chiralpak AD-R
MeCN-water	50:50, 80:20, and 95:5	Aromatase inhibitors	Chiralpak AD-R
EtOH, 1-PrOH and 2-PrOH separately	Pure	Nebivolol	Chiralpak AD and AD-R

AcOH, acetic acid; DEA, diethylamine; MeCN, acetonitrile; MeOH, methanol; MPH, methylphenidate; EtOH, ethanol; 1-PrOH, 1-propanol; 2-PrOH, 2-propanol; TEA, triethylamine; TFA, trifluoroacetic acid.

1. Prior to start of the experiment, set up HPLC system in the order of solvent reservoir, pump, injector, chiral column (CSP), detector, and recorder.
2. Select the racemic compound and determine its solubility and its appropriate method of detection, e.g., UV, fluorescent, etc.
3. Prepare the standard solution (1 mg/mL) of the racemic compound in appropriate solvent and filter it through 0.22 μm filter paper.

3.1. Selection of Detector

1. In most of the chiral resolution by HPLC, a UV detector has been used frequently because most of the racemic drugs and pharmaceuticals are UV-sensitive.
2. Some other detectors such as conductivity, fluorescent, circular dichroism, etc. have also been used. The use of a detector depends on the properties of the racemic compound to be resolved.
3. After the selection of detector, fix the wavelength (λ max) on the detector and allow it undisturbed for about 30 min for its saturation.
4. The identification of the enantiomers can be determined by using the optical detector if the individual enantiomers of the compound under study is not available.

3.2. Selection of Mobile Phase

1. The selection of mobile phase is the key aspect in the chiral resolution by HPLC. The selection of mobile phase depends on the properties of the racemic compound. The basic knowledge of chemistry is also helpful for the selection of mobile phase.
2. The mobile phase should be selected in which the racemic compound is soluble.
3. The correlation of separation conditions of neutral, acidic, and basic compounds is presented in **Table 2**. However, the most commonly used mobile phases with polysaccharides CSPs (normal and reversed-phases) are summarized in **Table 3**. The protocol for the selection of mobile phases is also presented in **Fig. 3**.
4. After the selection and preparation of mobile phase, filter it through 0.22 mm filter paper, degas for about 5–10 min, and transfer into the solvent/mobile phase reservoir of HPLC system.

3.3. Chromatographic Conditions

1. Allow the HPLC machine to run for about 60 min or until the base line in recorder is constant at a flow rate of 1.0 mL/min.
2. After getting the constant base line in the recorder, inject an appropriate vol of the solution of the compound under study, namely 20 μL of the sample in the injector (in case of manual injector).
3. If the injector is automatic, then fix the required parameters, such as vol of injection, running time, sample number, etc., in the auto-injector and then press the start switch of the auto-injector.
4. Wait until the chromatograms appeared in the recorder and observe the chiral resolution.
5. To identify the (+)- and (–)-enantiomers, optically active pure enantiomers should be run under the identical conditions of HPLC or use the optical detector.

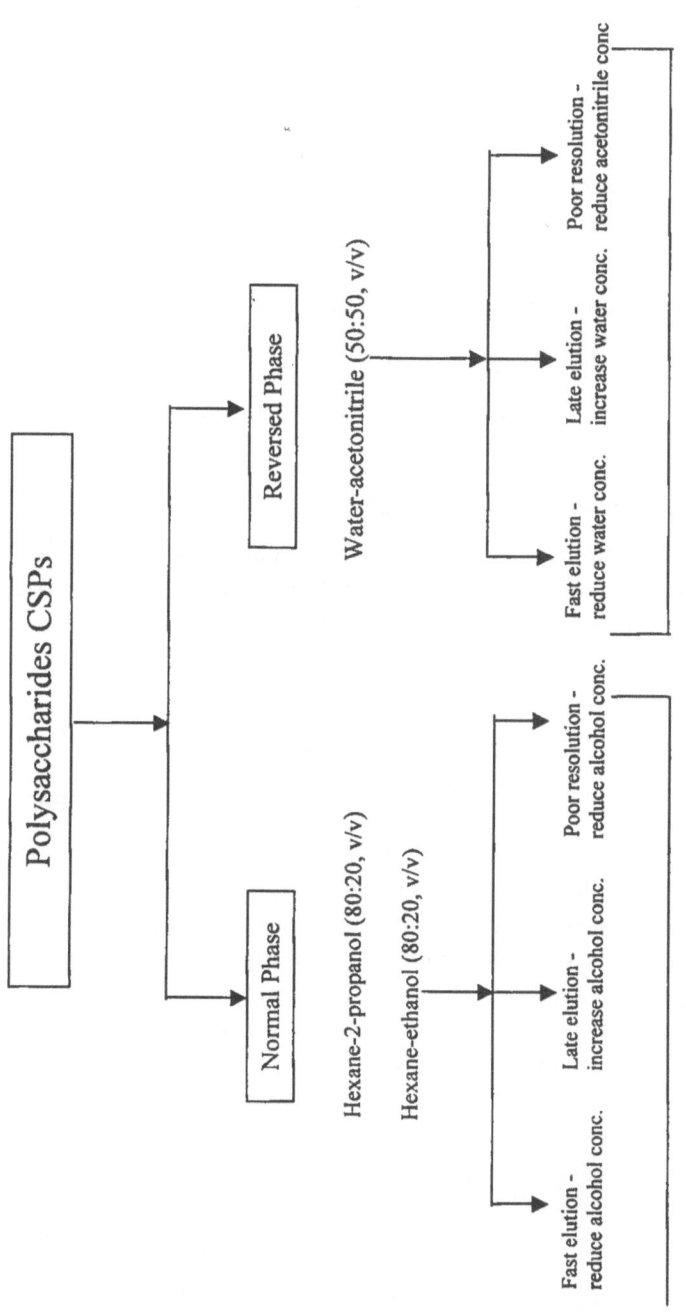

Fig. 3. The protocol for the development of mobile phase for polysaccharide CSPs.

3.4. Optimization of HPLC Conditions

To achieve the maximum chiral resolution, the selection of the suitable CSP and mobile phase is very important. **Tables 1–3** give some aids for selecting the right conditions (*see* **Notes 1, 3,** and **8**). Also, the instrumental parameters should be carefully adapted (*see* **Notes 2–7**).

3.5. Identification of the Enantiomers

The resolved enantiomers are identified by comparing their retention times with those of the optically pure individual enantiomers under the identical conditions of HPLC. If the optically pure active enantiomers are not available, then the identification of resolved enantiomers should be carried out by optical detector. The optical detector record the peaks in (+) and (−) scale, which corresponds to dextro (+)- and levo rotatory (−)-enantiomers, but does not assign the configuration.

3.6. Calculations of Chromatographic Factors

After achieving the complete chiral resolution, the calculation of the chromatographic factors is very essential. The most important factors to be calculated are retention factor (k), separation factor (α), and resolution factor (Rs) for the resolved enantiomers. The values of these parameters can be calculated by the following standard equations *(22)*.

$$k = (t_r - t_0)/t_0 \qquad \text{[Eq. 1]}$$
$$\alpha = k_1/k_2 \qquad \text{[Eq. 2]}$$
$$Rs = 2\,\Delta t/(w_1 + w_2) \qquad \text{[Eq. 3]}$$

Where, t_r and t_0 are the retention times of the enantiomers peak and dead time (solvent front) of the chiral column in minutes, Δt is the difference of the retention times of the two peaks of the resolved enantiomers, w_1 and w_2 correspond to base width of the two peaks of the two enantiomers. If the values of α and Rs are one or greater separately, the resolution is supposed to be the complete. If the values of these parameters are lower than one, then the chiral resolution is supposed to be partial or incomplete.

The quantitative determination of enantiomeric resolution is carried out by the following equation:

$$C_{samp} = (C_{std} \times A_{std})/A_{samp} \qquad \text{[Eq. 4]}$$

Where, C_{samp} and C_{std} are the concentrations of enantiomers in mg/mL in sample and standard solutions, respectively, while A_{std} and A_{samp} correspond to the areas of the peaks of enantiomers of the standard and the sample solutions, respectively. As an example, the actual chromatograms of the resolved

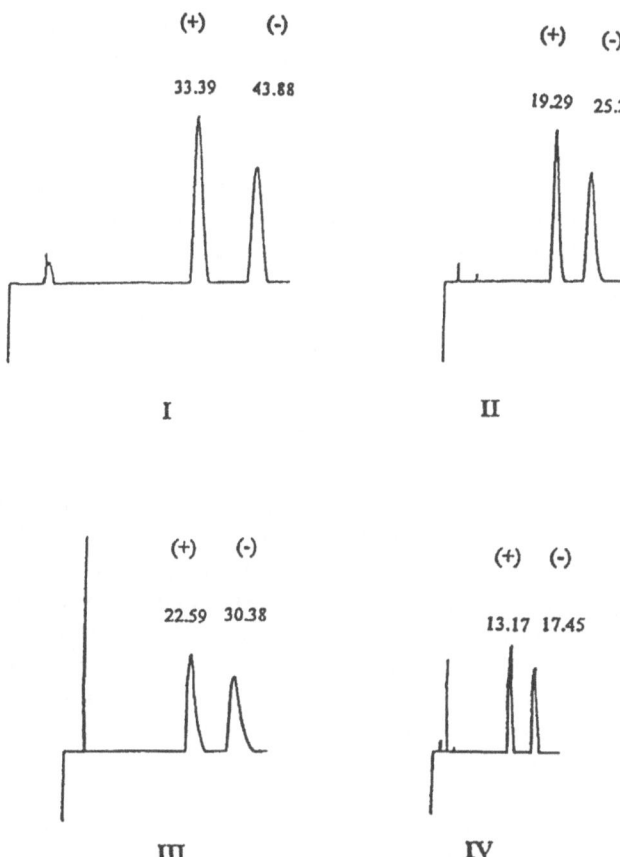

Fig. 4. The chromatograms of the resolved enantiomers of (±)-nebivolol on Chiralpak AD (I and II) column using I: ethanol; II: 1-propanol; and Chiralpak AD-RH column (III and IV) using III: 1-propanol as the mobile phases with 0.5 mL/min as flow rate.

enantiomers of nebivolol on Chiralpak AD and Chiralpak AD-R CSPs are shown in **Fig. 4** *(23)*.

4. Notes

1. If no chiral resolution occurred, then change the mobile phase or stationary phase (CSP).
2. In case of detection problem, vary the wavelength and increase the concentration of sample.
3. If the chiral resolution is partial then: (i) modify the mobile phase (see **Fig. 3**); (ii) decrease the flow rate; (iii) decrease the sample concentration; or (iv) all of the above.

4. If the chromatograms are appearing out of the scale, then increase the attenuation of the recorder or decrease the concentration of the sample.
5. In case of complete resolution, but the peaks are appearing very close to each other, the chart speed should be increased.
6. If the elution is late then: (i) increase flow rate, but it must not be increased greater than 2.0 mL/min (note that backpressure should not exceed the recommended limit); (ii) increase the polarity of the mobile phase; or (iii) both.
7. If the peaks are broad then: (i) increase flow rate but not more than 2.0 mL/min; (ii) increase the polarity of the mobile phase; or (iii) both.

Today, there are several commercial CSPs available for chiral resolution. However, polysaccharide-based CSPs are supposed to be one of most used, because of their greater efficiencies and wide range of applications. Although many of the racemic compounds can be resolved on these CSPs, these CSPs are not in their full development stage, as they are not capable of resolving all racemic compounds. A disadvantage of the polysaccharide-based CSPs remains, however, the limited choice of solvents. The polysaccharide-based CSPs that are most often used, and usually achieve 85% successful resolution, are Chiralcel OD, Chiralcel OD-R, Chiralcel OJ, Chiralcel OJ-R, Chiralpak AD, and Chiralpak AD-R CSPs, while the use of the other polysaccharide-based CSPs is limited. Recently, Okamoto et al. *(24)* have reported the lower chiral recognition capacities of these phases perhaps due to the loss of a regular high-ordered structure of these phases during the immobilization process on silica gel. They prepared a cellulose tris 3,5-dichlorophenylcarbamate derivative and coated it on silica gel and reported its better chiral resolution capacities due to its intact structure. Research is still carried out to improve the chiral resolution capacities of these phases preparing new derivatives of cellulose and amylose. We expect that these CSPs will gain more and more interest in the field of chiral resolution in the near future.

Acknowledgments

The authors (I.A. and H.Y.A-E.) would like to thank the King Faisal Specialist Hospital and Research Center administration for their support to the Pharmaceutical Analysis Laboratory research programme.

References

1. Günther, K. (1991) Enantiomers separations, in *Hand Book of TLC* (Sherma, J. and Fried, B., eds.), Marcel Dekker, New York, pp. 541–591.
2. Stevenson, D. and Wilson, I. D. (1989) *Chiral Separations*. Plennum Press, New York.
3. Waldeck, B. (1993) Biological significance of the enantiomeric purity of drugs. *Chirality* **5,** 350–355.
4. Millership, J. S. and Fitzpatrick, A. (1993) Commonly used chiral drugs: a survey. *Chirality* **5,** 573–576.

5. Ariens, E. J. (1984) Stereochemistry, a basis for sophisticated nonsense in pharmacokinetics and clinical pharmacology. *Eur. J. Clin. Pharmacol.* **26,** 663–668.

6. Knabe, J., Buich, H. P., and Kirsch, G. A. (1987) Barbituric acid derivatives. 36. Racemates and enantiomers of 1-methyl-5-pentyl-5-phenylbarbituric acid synthesis and central nervous action. *Arch. Pharm.* **320,** 323–328.

7. FDA (1992) Policy statements for the development of new stereoisomeric drugs. FDA, Rockville, MD.

8. Aboul-Enein, H. Y. (2001) High performance liquid chromatographic enantioseparation of drugs containing multiple chiral centres on polysaccharide type chiral stationary phases. *J. Chromatogr. A.* **906,** 185–193.

9. Okamoto, Y. and Yashima, E. (1997) Chiral recognition by optically active polymers, in *Molecular Design of Polymeric Materials* (Hatada, K., Kitayama, T., and Vogl, O., eds.), Marcel Dekker, New York, pp. 731–746.

10. Aboul-Enein, H. Y. and Wainer, I. W. (eds.). (1997) *The Impact of Stereochemistry on Drugs Development and Use.* John Wiley & Sons, New York.

11. Okamoto, Y. and Yashima, E. (1997) Chiral recognition mechanism of polysaccharides chiral stationary phases, in *The Impact of Stereochemistry on Drugs Development and Use* (Aboul-Enein, H. Y. and Wainer, I. W., eds.), John Wiley & Sons, New York, pp. 345–375.

12. Shibata, T., Mori, K., and Okamoto, Y. (1989) Polysaccharides phases, in *Chiral Separations by HPLC* (Krstulovic, A. M., ed.), Ellis Horwood, New York, pp. 336–398.

13. Dingenen, J. (1994) Polysaccharide phases in enantioseparations, in *A Practical Approach to Chiral Separations by Liquid Chromatography* (Subramanian, G., ed.), VCH Verlagsgesellschaft mbH, Weinheim, Germany, pp. 115–179.

14. Allenmark, S. (ed.). (1991) *Chromatographic Enantioseparation: Methods and Applications. 2nd ed.,* Ellis Horwood, New York.

15. Beesley, T. E. and Scott, R. P. W. (eds.). (1998) *Chiral Chromatography.* John Wiley & Sons, New York.

16. Okamoto, Y., Kaida, Y., Aburatani, R., and Hatada, K. (1991) Chromatographic optical resolution on polysaccharide carbamate phases, in *Chiral Separations by Liquid Chromatography* (Ahuja, S., ed.), American Chemical Society, Washington, DC.

17. Wainer, I. W. and Alembic, M. C. (1986) Resolution of enantiomeric amides on a cellulose based chiral stationary phase—steric and electronic effects. *J. Chromatogr.* **358,** 85–93.

18. Yamamoto, C., Yashima, E., and Okamoto, Y. (1999) Computational studies on chiral discrimination mechanism of phenylcarbamate derivatives of cellulose. *Bull. Chem. Soc. Jpn.* **72,** 1815–1825.

19. Ronden, N. G., Nyquist, R. A., Gillie, J. K., Nicholson, L. W., and Goralski, C. T. (1993) Theoretical elucidation of recognition mechanisms between aminoalcohol enantiomers and an amylose based chiral stationary phase [abstract]. *4th Int. Symposium., Montreal, Canada* p. 162, Abstract no. 90.

20. Aboul-Enein, H. Y., Ali, I., Simons, C., and Gubitz, G. (2000) HPLC enantio-
    meric resolution of novel aromatase inhibitors on cellulose and amylose based
    reversed and chiral stationary phases under reversed phase mode. *Chirality* **12,**
    727–733.
21. Aboul-Enein, H. Y. and Ali, I. (2001) A comparison of chiral resolution of econ-
    azole, miconazole and sulconazole by HPLC using normal phase amylose CSPs.
    Fresenius J. Anal. Chem. **370,** 951–955.
22. Aboul-Enein, H. Y. and Ali, I. (2003) *Chiral Separation by Liquid Chromatogra-
    phy and Related Technologies.* Marcel Dekker, New York, pp. 69–75.
23. Aboul-Enein, H. Y. and Ali, I. (2001) Studies on the effect of alcohols on the
    chiral discrimination mechanisms of amylose stationary phase on the enantiosep-
    aration of nebivolol by HPLC. *J. Biochem. Biophys. Methods* **48,** 175–188.
24. Chankvetadze, B., Yamamoto, C., and Okamoto, Y. (2000) Enantioseparation
    using tris(3,5-dichlorophenylcarbamate) during high performance liquid chroma-
    tography with analytical and capillary columns: potential for screening of chiral
    compounds. *Combi. Chem. High Throu. Screen.* **3,** 497–508.

7

Chiral Separation by HPLC
With Pirkle-Type Chiral Stationary Phases

Myung Ho Hyun and Yoon Jae Cho

1. Introduction

Liquid chromatographic resolution of enantiomers on chiral stationary phases (CSPs) has been known as the most accurate and convenient means of determining the enantiomeric composition of chiral compounds including chiral drugs. As a result of significant effort devoted to the development of effective CSPs during the past decades, various CSPs are now available *(1–4)*.

Pirkle-type CSPs have been known to separate the two enantiomers of racemic compounds through a minimum of three simultaneous interactions between the CSP and the racemic solute with at least one interaction being stereochemically dependent *(5–7)*. The interactions between the CSP and the solute can be either attractive or repulsive in nature, including any of the molecular interactions, such as π-π donor-acceptor interaction, hydrogen bonding, dipole stacking, and steric repulsion. Especially, π-π donor-acceptor interaction is essential for the chiral recognition on Pirkle-type CSPs *(5–7)*. For effective π-π donor-acceptor interaction with racemic solutes, Pirkle-type CSPs have been usually designed to contain π-acidic and/or π-basic aromatic groups. Similarly, racemic solutes that are resolvable on Pirkle-type CSPs should contain π-acidic and/or π-basic aromatic groups or should be derivatized with derivatizing agents containing π-acidic or π-basic aromatic groups.

For example, CSP 1 (leucine), containing a π-acidic aromatic group (**Fig. 1**), has been utilized mostly in the resolution of π-basic racemates *(8)*. On the contrary, CSP 2 (napthylleucine), containing a π-basic aromatic group (**Fig. 1**), has been utilized in the resolution of π-acidic racemates *(8)*. CSP 3 (whelk-O 1), containing both π-acidic (3,5-dinitrophenyl) and π-basic (1-naphthyl) aromatic groups (**Fig. 1**), has been utilized in the resolution of various racemic compounds containing π-acidic or π-basic aromatic groups *(9,10)*.

From: *Methods in Molecular Biology, Vol. 243: Chiral Separations: Methods and Protocols*
Edited by: G. Gübitz and M. G. Schmid © Humana Press Inc., Totowa, NJ

The purpose of this chapter is to provide the practice, as an example, for the chiral separation of α-amino acids and α-arylpropionic acids (anti-inflammatory drugs) by high-performance liquid chromatography (HPLC) with commercially available Pirkle-type CSPs such as CSPs 1, 2, and 3 (**Fig. 1**). For the enantioseparation of α-amino acids on Pirkle-type CSPs, achiral derivatization with π-basic or π-acidic derivatizing agents is essential to obtain the enantioselectivity and to improve the detectability and chromatographic resolution behaviors. α-Arylpropionic acids can be resolved on Pirkle-type CSPs without derivatization (9). However, achiral derivatization of naproxen with π-basic or π-acidic derivatizing agents also improves the enantioselectivity and chromatographic resolution behaviors. In this chapter, the processes for the π-basic or π-acidic derivatization of leucine and naproxen, and their chiral separations on CSPs 1, 2, and 3 are described (*see* **Fig. 2** for the structures of the π-basic and π-acidic derivatives of leucine and naproxen).

2. Materials

2.1. Derivatization of α-Amino Acids

2.1.1. π-Basic Derivatization of α-Amino Acids
(Preparation of 3,5-Dimethylanilide of N-Boc-Leucine) **(Fig. 2, 4)**

1. Racemic and *R*- or *S*-leucine.
2. Triethylamine.
3. Mixed solvent of dioxane and water (1:1, v/v).
4. di-*tert*-Butyldicarbonate.
5. Ethyl acetate.
6. Na$_2$SO$_4$ anhydrous.
7. 2-Ethoxy-1-ethoxycarbonyl-1,2-dihydroquinoline (EEDQ).
8. Methylene chloride.
9. 3,5-Dimethylaniline.
10. 1 *N* HCl solution; 1 N NaOH solution.
11. Silica gel for column chromatography.
12. Mixed solvent of ethyl acetate and hexane (1:3, v/v).

2.1.2. π-Acidic Derivatization of α-Amino Acids
(Preparation of N-[3,5-Dinitrobenzoyl]Leucine n-Propylamide) **(Fig. 2, 5)**

1. Racemic and *R*- or *S*-leucine.
2. Triethylamine.
3. Mixed solvent of dioxane and water (1:1, v/v).
4. di-*tert*-Butyldicarbonate.
5. Ethyl acetate.
6. Na$_2$SO$_4$ anhydrous.

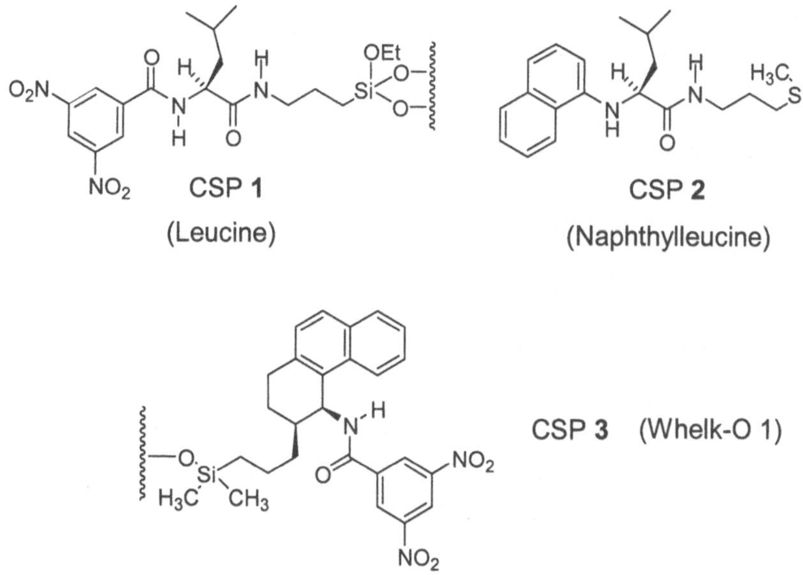

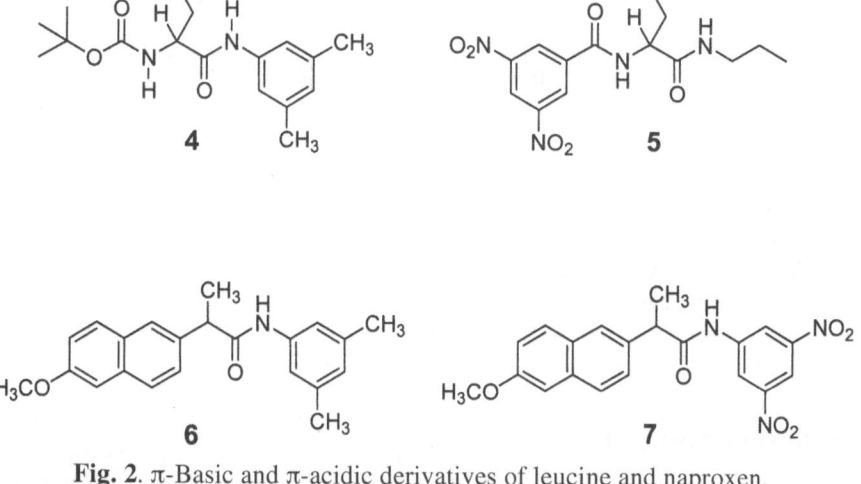

Fig. 1. Structures of CSP 1, CSP 2, and CSP 3.

Fig. 2. π-Basic and π-acidic derivatives of leucine and naproxen.

7. EEDQ.
8. Methylene chloride.
9. *n*-Propylamine.
10. 1 *N* HCl solution, 1 N NaOH solution.
11. Na$_2$SO$_4$ anhydrous.
12. Silica gel for column chromatography.
13. Mixed solvent of ethyl acetate and hexane (1:3, v/v).
14. Trifluoroacetic acid.
15. 3,5-Dinitrobenzoyl chloride.

2.2. Derivatization of Naproxen

2.2.1. π-Basic Derivatization of Naproxen
(Preparation of 3,5-Dimethylanilide of Naproxen) **(Fig. 2, 6)**

1. Racemic naproxen and (*S*)-naproxen
2. Thionyl chloride distilled freshly.
3. Benzene.
4. Methylene chloride.
5. 3,5-Dimethylaniline.
6. Propylene oxide.
7. Saturated NaHCO$_3$ solution.
8. 1 *N* HCl solution.
9. Na$_2$SO$_4$ anhydrous.
10. Silica gel for column chromatography.
11. Mixed solvent of ethyl acetate and hexane (1:3, v/v).

2.2.2. π-Acidic Derivatization of Naproxen
(Preparation of 3,5-Dinitroanilide of Naproxen) **(Fig. 2, 7)**

1. Racemic naproxen and (*S*)-naproxen.
2. Thionyl chloride distilled freshly.
3. Benzene.
4. Methylene chloride.
5. 3,5-Dinitroaniline.
6. Propylene oxide.
7. Saturated NaHCO$_3$ solution.
8. 1 *N* HCl solution.
9. Na$_2$SO$_4$ anhydrous.
10. Silica gel for column chromatography.
11. Mixed solvent of ethyl acetate and hexane (1:3, v/v).

2.3. Chromatography

1. 3,5-Dimethylanilide of *N-t*-BOC-leucine (**Fig. 2, 4**) (prepared in **Subheading 3.1.1.**).

2. *N*-(3,5-Dinitrobenzoyl)leucine n-propylamide (**Fig. 2, 5**) (prepared in **Subheading 3.1.2.**).
3. 3,5-Dimethylanilide of naproxen (**Fig. 2, 6**) (prepared in **Subheading 3.2.1.**).
4. 3,5-Dinitroanilide of naproxen (**Fig. 2, 7**) (prepared in **Subheading 3.2.2.**).
5. 1,3,5-tri-*tert*-Butylbenzene.
6. 20% Isopropyl alcohol in hexane.
7. Chiral columns packed with CSP 1, CSP 2, or CSP 3 (*see* **Fig. 1**) (commercially available from Regis Technologies, Morton Grove, IL, USA).

3. Methods (see *Note 1*)

3.1. Derivatization of α-Amino Acids (see *Note 2*)

3.1.1. π-Basic Derivatization of α-Amino Acids
(Preparation of 3,5-Dimethylanilide of N-t-BOC-Leuine) (*Fig. 2, 4*)

1. Dissolve 1.05 g 8 mmol leucine (racemic or optically active) and 2.2 mL 16 mmol triethylamine in 40 mL of 1:1 mixed solvent of dioxane and water in a 100-mL round bottom flask. Stir the whole mixture until the solution becomes homogeneous with a magnetic stirrer.
2. Add 2.0 mL 8.8 mmol di-*tert*-butyldicarbonate to the solution and then stir the whole solution for 6 h at room temperature. Extract the solution with 50 mL of ethylacetate. Dry the organic layer over anhydrous Na_2SO_4 and then remove the solvent by using a rotary evaporator to afford *N-t*-BOC-leucine (*see* **Note 3**).
3. Dissolve 150 mg 0.65 mmol *N-t*-BOC-leucine and 0.178 g 0.72 mmol EEDQ in 15 mL of methylene chloride in a 100-mL round bottom flask. Stir the solution for 10 min and then add 80 mg 0.66 mmol 3,5-dimethylaniline. Stir the whole mixture for 12 h.
4. Wash the whole reaction mixture with 20 mL 1 *N* HCl solution and 20 mL 1 *N* NaOH solution and then dry the organic solution over anhydrous Na_2SO_4. Remove the solvent by using a rotary evaporator and then purify the residue by silica gel chromatography (ethyl acetate:hexane, 1:3, v/v) to afford 3,5-dimethylanilide of *N-t*-BOC-leucine (*see* **Note 4**).

3.1.2. π-Acidic Derivatization of α-Amino Acids
(Preparation of N-[3,5-Dinitrobenzoyl]Leucine n-Propylamide)
(*Fig. 2, 5*) (see *Note 5*)

1. Dissolve 150 mg 0.65 mmol *N-t*-BOC-leucine prepared in **Subheading 3.1.1.** and 0.178 g 0.72 mmol EEDQ in 15 mL of methylene chloride in a 100-mL round bottom flask. Stir the solution for 10 min and then add 0.059 mL 0.72 mmol *n*-propylamine. Stir the whole mixture for 12 h.
2. Wash the whole reaction mixture with 20 mL 1 *N* HCl solution and 20 mL 1 *N* NaOH solution and then dry the organic solution over anhydrous Na_2SO_4. Remove the organic solvent by using a rotary evaporator and purify the residue by silica gel chromatography (ethyl acetate:hexane, 1:3, v/v) to afford *N-t*-BOC-leucine propylamide.

3. Dissolve 109 mg 0.4 mmol N-t-BOC-leucine propylamide in 10 ml of methylene chloride and then add 0.31 mL 4 mmol trifluoroacetic acid. Stir the solution for 6 h.
4. Wash the reaction mixture with 15 mL of 1 N NaOH solution and then dry the organic solution over anhydrous Na_2SO_4. Remove the organic solvent by using a rotary evaporator to afford leucine n-propylamide.
5. Dissolve 60 mg 0.35 mmol leucine n-propylamide and 0.054 mL 0.39 mmol triethylamine in 15 mL of methylene chloride and then cool the whole solution in ice-water bath. Add 81 mg 0.35 mmol 3,5-dinitrobenzoylchloride dissolved in 15 mL of methylene chloride slowly and stir the reaction mixture for 30 min at room temperature.
6. Wash the whole reaction mixture with 15 mL of 1 N HCl solution and then with 15 mL of 1 N NaOH solution. Dry the organic solution over anhydrous Na_2SO_4 and then remove the organic solvent by using a rotary evaporator. Purify the residue by silica gel chromatography (ethyl acetate:hexane, 1:1, v/v) to afford N-(3,5-dinitrobenzoyl)leucine n-propylamide (*see* **Note 4**).

3.2. Derivatization of Naproxen (see Note 6)

3.2.1. π-Basic Derivatization of Naproxen
(Preparation of 3,5-Dimethylanilide of Naproxen) (Fig. 2, 6)

1. Dissolve 0.2 g 0.87 mmol naproxen and freshly distilled thionyl chloride in 20 mL of benzene. Heat the mixture to reflux for 2 h under an argon atmosphere. Cool the reaction mixture to room temperature and then dry the solvent by using a rotary evaporator.
2. Dissolve the residue in 20 mL of dry methylene chloride. Slowly add a mixture of 105 mg 0.87 mmol 3,5-dimethylaniline and 0.07 mL (1.0 mmol) propylene oxide diluted in 10 mL of dry methylene chloride to the stirred solution (*see* **Note 7**). Stir the reaction mixture for 12 h at room temperature under an argon atmosphere and wash it successively with 60 mL of saturated $NaHCO_3$ solution, 50 mL of 1 N HCl solution, and brine. Dry the organic solution over anhydrous Na_2SO_4 and then evaporate the solution to dryness.
3. Purify the residue by silica gel column chromatography (ethyl acetate:hexane:methylene chloride, 1:3:1, v/v/v) to afford 3,5-dinitroanilide of naproxen (*see* **Note 4**).

3.2.2. π-Acidic Derivatization of Naproxen
(Preparation of 3,5-Dinitroanilide of Naproxen) (Fig. 2, 7)

1. Dissolve 0.2 g 0.87 mmol naproxen and freshly distilled thionyl chloride in 20 mL of benzene. Heat the mixture to reflux for 2 h under an argon atmosphere. Cool the reaction mixture to room temperature and then dry the solvent by using a rotary evaporator.
2. Dissolve the residue in 20 mL of dry methylene chloride. Slowly add a mixture of 160 mg 0.87 mmol 3,5-dinitroaniline and 0.07 mL (1.0 mmol) propylene oxide

diluted in 10 mL of dry methylene chloride to the stirred solution (*see* **Note 7**). Stir the reaction mixture for 12 h at room temperature under an argon atmosphere and wash it successively with 60 mL of saturated NaHCO$_3$ solution, 50 mL of 1 *N* HCl solution, and brine. Dry the organic solution over anhydrous Na$_2$SO$_4$ and then evaporate to dryness.

3. Purify the residue by silica gel column chromatography (ethyl acetate:hexane: methylene chloride, 1:3:1, v/v/v) to afford 3,5-dinitroanilide of naproxen (*see* **Note 4**).

3.3. Chromatography

1. Prepare eight sample solutions by dissolving 10 mg of racemic and optically active 3,5-dimethylanilide of *N-t*-BOC-leucine, *N*-(3,5-dinitrobenzoyl)leucine *n*-propyl-amide, 3,5-dimethylanilide of naproxen, and 3,5-dinitroanilide of naproxen in 10 mL of methylene chloride (*see* **Note 8**).
2. Prepare a solution of an unretained solute by dissolving 10 mg of 1,3,5-tri-*tert*-butylbenzene in 10 mL of methylene chloride (*see* **Note 8**).
3. Let the mobile phase of 20% isopropyl alcohol in hexane flow through the HPLC system installed with the chiral column of CSP 1, 2, or 3 (**Fig. 1**) with a flow rate of 2 mL/min (*see* **Note 9**).
4. Inject 3 µL of the solution of 1,3,5-tri-*tert*-butylbenzene and determine the t_0 value (retention time of the void vol marker) from the peak.
5. Inject 3 µL of one racemic sample solution and record the chromatogram (*see* **Note 10**). When resolving π-basic derivatives of analytes (**Fig. 2, 4** and **6**), use CSP 1 or 3 (**Fig. 1**). Similarily, when resolving π-acidic derivatives of analytes (**Fig. 2, 5** and **7**), use CSP 2 or 3 (**Fig. 1**) (*see* **Note 11**).
6. Inject 3 µL of the corresponding optically active sample solution and record the chromatogram (*see* **Note 10**).
7. After recording the chromatograms, stop the HPLC system or inject the next sample (*see* **Note 12**).
8. Determine the elution order by comparing the two chromatograms for the resolution of racemic and optically active samples.
9. Calculate the chromatographic parameters such as retention factors (k_1 and k_2) and separation factors (α) or, if necessary, resolution factors (R$_S$) (*see* **Note 13**).

4. Notes

1. All reactions must be done under a well-ventilated fume hood.
2. Both racemic and optically active leucine [*S*- or *R*-leucine] should be derivatized in order to determine the elution order of the two enantiomers. The method can be applied equally to the derivatization of racemic and optically active leucine.
3. Since *N-t*-BOC-leucine is commercially available (Sigma, St. Louis, MO, USA), it can be used instead of the preparation.
4. A short pass column packed with silica gel is enough for the chromatographic sample.

5. An alternative procedure for the preparation of *N*-(3,5-dinitrobenzoyl)leucine *n*-propylamide is also possible. The alternative procedure includes the direct treatment of leucine with 3,5-dinitrobenzoylchloride in the presence of propylene oxide and then treatment with *n*-propylamine in the presence of EEDQ. However, a certain degree of racemization of optically pure leucine has been experienced in our laboratory during the alternative derivatization process.

6. Both racemic naproxen and *S*-naproxen should be derivatized in order to determine the elution order of the two enantiomers. The method can be applied equally to the derivatization of racemic naproxen and optically pure *S*-naproxen. Naproxen contains a π-basic aromatic functional group such as 6-methoxynaphthyl. Consequently, it should be noted that naproxen is resolved without derivatization on Pirkle-type CSPs, such as CSP 3 (**Fig. 1**), with the use of a somewhat modified mobile phase *(9)*. In addition, the simple n-propylamide derivative of naproxen can also be resolved on π-acidic CSPs.

7. The use of other bases instead of propylene oxide to quench HCl evolved during the reaction results in the partial racemization of optically pure *S*-naproxen.

8. The concentration of sample solution can be lowered to the detection limit of the detector (usually UV detector) used.

9. By increasing the content of isopropyl alcohol in hexane, the retention times of the two enantiomers can be diminished and vice versa. Instead of isopropyl alcohol, other polar solvents such as chloroform, methylene chloride, ethanol, tetrahydrofuran (THF), or ethyl acetate can be used. However, a good mobile phase for starting method development with Pirkle-type CSPs is 10% isopropyl alcohol in hexane. The usual flow rate is 2 mL/min. However, the retention times increase by decreasing the flow rate. If possible, for the reproducibilities, chiral separations must be done at constant temperature, usually at 20°C. By decreasing the separation temperature, the retention and the separation of the two enantiomers usually improves.

10. The vol of the injection sample solution can be varied according to the sample concentration. When the concentration of the sample solution is very low, the maximum vol can be injected (e.g., 20 µL with a 20-µL sample loop).

11. To resolve π-basic analytes, other π-acidic CSPs can be used. Similarly, to resolve π-acidic analytes, other π-basic CSPs can be used.

12. After use, the chiral column can be in stored in the mobile phase (in this case, 20% isopropyl alcohol in hexane). However, when the columns are used under reverse-phase conditions containing water, acidic modifiers, inorganic modifiers, or buffers (Pirkle-type CSPs tolerates a wide range of solvents used for either normal or reversed-phases), the columns should be washed with water and then flushed with a compatible organic solvent, such as ethanol. When the column is removed from the HPLC system, it should be installed with plugs to be kept wet during storage.

13. Modern HPLC systems are equipped with a computer data processing program, and chromatographic parameters are calculated automatically. However, when

calculating chromatographic parameters manually, the chromatographic parameters can be calculated according to Equations 1–4.

$$\text{Retention factor for the first eluted enantiomer, } k_1 = (t_1 - t_0)/t_0 \qquad \text{[Eq. 1]}$$
$$\text{Retention factor for the second eluted enantiomer, } k_2 = (t_2 - t_0)/t_0 \qquad \text{[Eq. 2]}$$
$$\text{Separation factor, } \alpha = k_2/k_1 \qquad \text{[Eq. 3]}$$
$$\text{Resolution factor, } R_S = 2(t_2 - t_1)/(w_1 + w_2) \qquad \text{[Eq. 4]}$$

In the above equations, t_1 and t_2 are the respective retention times of the first and the second eluted enantiomers, and t_0 is the retention time of the unretained compound measured by injecting 1,3,5-tri-*tert*-butylbenzene (void vol marker). In addition, w_1 and w_2 are the respective peak widths at the base of the first and the second eluted enantiomers, measured in the same scale as t_1 and t_2.

References

1. Subramanian, G. (ed.). (1994) *A Practical Approach to Chiral Separations by Liquid Chromatography.* VCH, Weiheim.
2. Beesley, T. E. and Scott, R. P. W. (1998) *Chiral Chromatography.* John Wiley & Sons, Chichester.
3. Ahuja, S. (2000) *Chiral Separations by Chromatography.* Oxford University Press, American Chemical Society, Washington, D.C.
4. Gasparrini, F., Mistini, D., and Villani, C. (2001) High-performance liquid chromatography chiral stationary phases based on low-molecular-mass selectors. *J. Chromatogr. A* **906,** 35–50.
5. Finn, J. M. (1988) Rational design of pirkle-type chiral stationary phases, in *Chromatographic Chiral Separations* (Zief, M. and Crane L. J., eds.), Marcel Dekker, New York, pp. 53–90.
6. Macaudiere, P., Lienne, M., Tambute, A., and Caude, M. (1989) Pirkle-type and related chiral stationary phases for enantiomeric resolution, in *Chiral Separations by HPLC: Applications to Pharmaceutical Compounds* (Krstulovic, A. M., ed.), Ellis Horwood, Chichester, pp. 399–445.
7. Pirkle, W. H. and Pochapsky, T. C. (1989) Considerations of chiral recognition relevant to the liquid chromatographic separation of enantiomers. *Chem. Rev.* **89,** 347–362.
8. Perrin, S. R. and Pirkle, W. H. (1991) Commercially available brush-type chiral selectors for the direct resolution of enantiomers, in *Chiral Separations by Liquid Chromatography* (Ahuja, S., ed.), ACS Symposium Series 471, American Chemical Society, Washington, D.C., pp. 43–66.
9. Pirkle, W. H., Welch, C. J., and Lamm, B. (1992) Design, synthesis, and evaluation of an improved enantioselective naproxen selector. *J. Org. Chem.* **57,** 3854–3860.
10. Pirkle, W. H. and Welch, C. J. (1994) Use of simultaneous face to face π-π interactions to facilitate chiral recognition. *Tetrahedron: Asymm.* **5,** 777–780.

8

Chiral Separation by HPLC
Using the Ligand-Exchange Principle

Vadim A. Davankov

1. Introduction

Ligand exchange was the first enantioselective (chiral) liquid chromatography technique *(1)* that, in the late 1960s *(2)*, allowed a complete separation of enantiomers of a racemic analyte.

Basic principle of ligand exchange *(3)* is the involvement of a complexing metal ion into interaction between the analyte enantiomers to be resolved and the chiral selector, namely, through the formation of diastereomeric ternary complexes selector/metal ion/analyte. It is essential that the complexes be kinetically labile, i.e., they must form and dissociate at a high rate; otherwise the chromatographic column efficiency would be compromised. Complexes of Cu(II), Zn(II), Ni(II), and few other ions meet this condition while coordinating amino, carboxy, hydroxy, amido, thio, and few other electron donating functional groups. Herewith, the lone electron pairs of the hetero atoms (N, O, S) of the functional groups, belonging to the analyte and selector, occupy definite positions in the coordination sphere of the central metal ion, to result in the formation of the ternary complex. During the chromatography process, the coordinated ligands are reversibly replaced by other ligands, such as molecules of water, ammonia, or other components of the eluent. Quick exchange of ligands in the metal ion coordination sphere dictates the name of the technique—ligand-exchange chromatography (LEC).

Recognition and discrimination of two enantiomers of an analyte by a chiral selector requires the interaction between the selector and analyte molecules to proceed simultaneously on at least three positions *(4)*. Therefore, analytes having three or two electron-donating functional groups, that are suitably positioned in the molecule to simultaneously enter the coordination sphere of the complexing metal ion, thus functioning as tridentate or bidentate chelating ligands, are generally better resolved by the LEC technique than monodentate ligands.

From: *Methods in Molecular Biology, Vol. 243: Chiral Separations: Methods and Protocols*
Edited by: G. Gübitz and M. G. Schmid © Humana Press Inc., Totowa, NJ

Accordingly, the best classes of organic compounds for the analytical or preparative scale resolution by LEC are α- and β-amino acids, hydroxy acids, amino alcohols, diamines, dicarboxylic acids, amino amides, etc. In the case of bifunctional analytes, the third interaction position that is required for the chiral recognition is provided by steric or dipol-type interaction with the selector. Therefore, bulky groups in the analyte molecule situated close to the center of asymmetry usually enhance the enantioselectivity of separation, i.e., the separation factor α. The latter is defined as the ratio of capacity factors of the column for two enantiomers (R) and (S), $\alpha = k_R/k_S$. Analytes possessing only one electron-donating functional group in the molecule resolve in the LEC technique less frequently and with smaller enantioselectivity α values. The same is true for polyfunctional compounds that fail to form chelates with one metal ion because of the long distance between their functional groups.

In accordance with the above chelation requirements, most efficient chiral selectors in ligand exchanging chromatographic systems are represented by bi- or trifunctional chiral compounds selected from the group of α-amino acids, amino amides, and amino alcohols. Depending on the location of the chiral selector in the chromatographic system, three types of approaches in LEC can be distinguished: use of chiral stationary phases (CSP), chiral-coated phases (CCP), and chiral mobile phases (CMP). In the first class of procedures, the chiral selector is covalently bound to the column packing, and the formation of ternary complexes takes place within the stationary phase. The analyte enantiomer that forms a stronger ternary complex with the selector is longer retained in the chromatographic column. The second technique uses conventional achiral high-performance liquid chromatography (HPLC) columns, which are then dynamically coated with a suitable chiral complexing selector to convert the packing into a CCP. Contrary to CSPs, which are stable in all eluents, the CCPs should be only used under elution conditions that do not cause a noticeable desorption of the chiral coating. In the CCP mode, formation of ternary complexes takes place on the surface of the packing. Additional interactions of the analyte with that surface start to play a significant role in the recognition process. Finally, third technique, CMP, uses chiral selectors that are soluble in the mobile phase. Diastereomeric ternary complexes now form in the liquid phase, but they have to be noticeably retained by the stationary phase in order to be resolved in the chromatographic mode. The analyte enantiomer, which forms a more stable complex with the selector in the mobile phase, elutes first in the CMP process. Differentiation between the CCP and CMP modes and prediction of the enantiomer elution sequence becomes ambiguous in systems where the chiral selector partitions between the stationary and mobile phases in comparable amounts.

Since the analytes and the selectors used in chiral LEC incorporate strongly polar functional groups, they are usually better dissolved in water, alcohols, or other strongly polar solvents. Thus, LEC uses aqueous or aqueous-organic eluents. The eluents are doped with small concentrations of the complexing metal (usually copper) salt, in order to compensate for the metal removal from the chromatographic system by the complexing analytes and the eluent. On the other hand, complexation with the metal ion significantly improves the detection possibility for the majority of analytes, which strongly absorb at 254 nm, when in the form of a copper(II) complex.

Of many polymeric and silica-bonded CSPs for LEC described in the literature, several are commercially available. These are: Chiral-Si 100 L-ProCu, Chiral-Si 100 L-ValCu, Chiral-Si 100 L-HyProCu (Serva, Heidelberg, Germany), Nucleosil Chiral-1 (Macherey-Nagel, Dueren, Germany), Chiralpak WH and WM (Daicel, Osaka, Japan), TSK gel Enantio L1 (Toso, Kyoto, Japan), MCl gel CRS l0W (Japan), Chirosolve L-Proline, Chirosolve L-Valine (JPS Chimie, Neuchâtel, Switzerland). Any scientific evaluation of results obtained by using these phases is, however, complicated, since the exact structure of immobilized chiral ligands on many commercially available CSPs is not specified by the manufacturers.

Among many CCPs described, those prepared by coating N-decyl-L-hydroxyproline onto a C18 column are commercially available from Regis Technologies (Morton Grove, IL, USA) as Davankov columns. They display a high resolving power toward numerous amino acids and their derivatives, racemic glycyl-di- and tripeptides, 3-amino-ε-caprolactam, as well as norephedrine and its analogous amino alcohols. Reversed phases coated with N,S-dioctyl-(D)-penicillamine and (R,R)-tartaric acid mono-(R)-1-(α-naphthyl)ethylamide are also available under the name Sumichiral OA-5000 and 6000 (Sumica Chemical Analysis Service, Kansai, Japan).

In the CMP mode, especially popular are amino acid-type chiral selectors (proline, hydroxyproline, phenylalanine) as well as their amides and other derivatives.

Due to the simplicity of the procedure, extremely high enantioselectivity of complexation, LEC remains one of most reliable and inexpensive chiral HPLC techniques for resolving and analyzing enantiomeric composition of several important classes of biologically relevant compounds, such as amino acids and hydroxy acids.

In the following sections, three examples of analytical-scale resolutions using LEC, one for each of the CSP, CCP, and CMP modes, are given in details. In the first case, dealing with the resolution of racemic α-hydroxy acids (*5*), a commercial silica-bonded chiral phase is used that is prepared by reacting L-hydroxyproline with glycidoxypropyl-activated silica. The second example presents resolution of racemic amino acids on a reversed-phase column dynamically coated with a hydrophobic derivative of L-hydroxyproline (*6*). Noteworthy, a similar technique of chiral coating is used in preparing chiral thin layer plates that operate in accordance with the LEC principle. The last example is for the resolution of racemic antiparkinsonian drugs, of the amino acid-type, too. An achiral reversed-phase column is used here in combination with an aqueous eluent doped with a copper complex of L-phenylalanine *(7)*.

Additional information on chiral LEC can be found in several review papers *(2,3,8)*.

2. Materials

2.1. Resolution of Racemic Hydroxy Acids on a CSP Containing Bonded S-Hydroxyproline

1. HPLC column Chiral-Si 100 L-HyProCu.
2. Racemic α-hydroxy acids (D,L-2-hydroxyglutaric acid, D,L-malic acid).
3. Copper(II) sulfate, $Cu(SO_4) \cdot 5H_2O$.
4. Potassium dihydrogen phosphate, KH_2PO_4.

2.2. Resolution of Racemic Amino Acids on a Reversed-Phase Column With a Chiral Coating of N-Decyl-S-Hydroxyproline

2.2.1. Preparation of N-Decyl-S-Hydroxyproline

1. (*S*)-Hydroxyproline.
2. *n*-Decylbromide (*see* **Note 1**).
3. 1.0 *M* Sodium hydroxide solution (*see* **Note 2**).
4. 2 *M* Hydrochloric acid (*see* **Note 2**).
5. Diethyl ether (*see* **Note 1**).
6. Ethanol (*see* **Note 1**).

2.2.2. Dynamic Coating of a Reversed-Phase Column With Chiral Ligand

1. Conventional HPLC reversed-phase C18 column, 250 × 4.5 mm-inner diameter (ID).
2. *N*-Decyl-(*S*)-hydroxyproline.
3. Copper(II) acetate, $Cu(AcO)_2 \cdot H_2O$.
4. Methanol (*see* **Note 1**).
5. 0.45-μm Membrane filter.

2.2.3. Chromatography on the Dynamically Coated Column

1. Racemic α-amino acids (D,L-alanine, D,L-arginine, D,L-valine).
2. Eluent: a 0.1 m*M* copper acetate solution in a methanol/water mixture (15:85, v/v), pH 5.0.

2.3. Resolution of Racemic Amino Acids on a Reversed-Phase Column With a CMP Containing S-Phenylalanine

1. A reversed-phase HPLC column LiChrosphere C_{18}, 5 μm, 125 × 4.0 mm I.D.
2. L-Phenylalanine.

3. Copper sulfate, $Cu(SO_4) \cdot 5H_2O$.
4. 0.45-µm Membrane filter.

3. Methods

3.1. Resolution of Racemic Hydroxy Acids on a CSP Containing Bonded S-Hydroxyproline

1. Make the eluent: a 0.1 mM copper sulfate solution in 0.01 mM KH_2PO_4, filtrate it through a 0.45-µm membrane filter, and degas before use in an ultrasonic bath.
2. Make sample solutions of racemic hydroxy acids (D,L-2-hydroxyglutaric acid, D,L-malic acid) in the eluent at a concentration of approx 0.5 mg/mL, as well as a mixture of equal amounts of the two solutions, and filtrate them through a 0.45-µm membrane filter.
3. Equilibrate the column with the eluent.
4. Carry out chromatography at a flow rate of the eluent of 1.0 mL/min and detecting the effluent photometrically at 223 nm. Resolve the racemates of each hydroxy acid individually and then the mixture of two pairs of enantiomers. A typical chromatogram is presented in **Fig. 1**.

3.2. Resolution of Racemic Amino Acids on a Reversed-Phase Column With a Chiral Coating of N-Decyl-S-Hydroxyproline

3.2.1. Preparation of N-Decyl-S-Hydroxyproline

Dissolve S-hydroxyproline (0.1 mol) in a mixture of 100 mL water, 100 mL 1.0 M sodium hydroxide solution and 100 mL dioxane. Heat the mixture to 40°–50°C. Under constant and intensive stirring, add, in alternating small portions, a solution of 0.1 mol n-decylbromide in 50 mL dioxane and 100 mL 1.0 M sodium hydroxide solution. Heat and stir the mixture until a clear solution is obtained (about 2 h). Cool the reaction mixture to room temperature. If the mixture is opaque, extract it with a portion of diethyl ether, and filtrate the aqueous layer through a cellulose filter. Neutralize the solution with approx 50 mL 2 M HCl to pH about 7.0. Filtrate the white precipitate. Redissolve it in minimum amount of methanol and precipitate with a 10-fold vol of water. Crystallization from hot aqueous methanol (3:1, v/v) results in white plates with a melting point 161.5°–162°C.

3.2.2. Dynamic Coating of a Reversed-Phase Column With Chiral Ligand

1. Wash the reversed-phase column with methanol at a flow rate of 1.5 mL/min for 10 min.
2. Make the chiral coating solution: a solution of 150 mg N-decyl-S-hydroxyproline in 5 mL of a methanol/water mixture (80:20, v/v) and filtrate it through a 0.45-µm membrane filter.

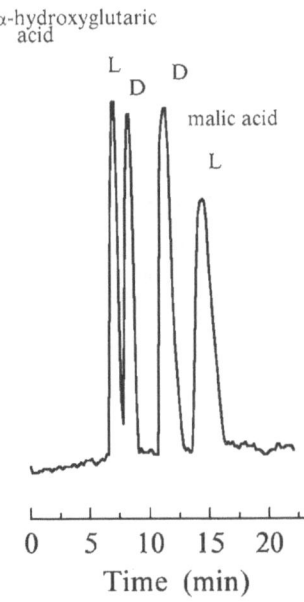

α-hydroxyglutaric acid

L D D

malic acid

L

Time (min)

Fig. 1. Chromatogram of a mixture of DL-2-hydroxyglutaric acid and DL-malic acid (5 μg of each). Redrawn from **ref. 5** with permission. Conditions: column, Chiral-Si 100 L-HyProCu, 5 μm, 250 × 4.6 mm I.D.; eluent, 0.1 mM copper sulfate in 0.01 mM KH$_2$PO$_4$; flow rate, 2 mL/min; temperature, 50°C; detection, 223 nm.

3. Make the copper loading solution: a solution of 200 mg copper acetate in 6 mL of a methanol/water mixture (15:85, v/v) and filtrate it through a 0.45-μm membrane filter.
4. Make the eluent: a 0.1 mM copper acetate solution in a methanol/water mixture (15:85, v/v), pH 5.0, and filtrate it through a 0.45-μm membrane filter.
5. Pump the chiral coating solution through the reversed-phase column at a flow rate of 0.3 mL/min until N-decyl-L-hydroxyproline appears in the effluent (*see* **Notes 3** and **4**).
6. Pump the copper loading solution through the column at a flow rate of 0.5 mL/min until copper ions appear in the effluent (*see* **Notes 5** and **6**).
7. Equilibrate the column with the eluent at a flow rate of 1.5 mL/min for 30 min. To avoid generation of bubbles in the detector cell, degas the eluent before use in an ultrasonic bath.

3.2.3. Chromatography on the Dynamically Coated Column

1. Make sample solutions of racemic amino acids (D,L-alanine, D,L-arginine, D,L-valine) in the eluent at a concentration of approx 0.5 mg/mL, as well as a mixture of equal amounts of the three solutions.

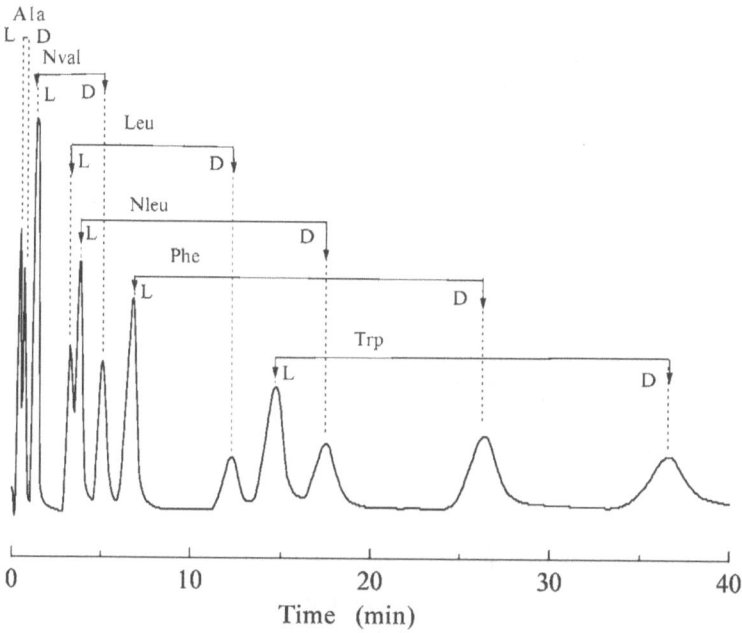

Fig. 2. Chromatogram of a mixture of 6 racemic amino acids. Redrawn from **ref. 6** with permission. Conditions: column, LiChrosorb RP 18, 5 µm, 100 × 4.2 mm I.D., dynamically coated with *N*-hexadecyl-L-hydroxyproline; eluent, 0.1 m*M* copper acetate in methanol/water, 15:85 (v/v), pH 5.0; flow rate, 2 mL/min; temperature, 20°C; detection, 254 nm; enantiomer elution sequence, L before D.

2. Carry out chromatography at a flow rate of the eluent of 1.0 mL/min and detecting the effluent photometrically at 254 nm. Resolve the racemates of each amino acid individually and then the whole mixture of three pairs of enantiomers. A typical chromatogram is presented in **Fig. 2** (*see* **Note 7**).

3.3. Resolution of Racemic Amino Acids
on a Reversed-Phase Column With a CMP Containing S-Phenylalanine

1. Make the eluent: a 3.0 m*M* copper sulfate, 6.0 m*M* L-phenylalanine in double-distilled deionized water and filtrate it through a 0.45-µm membrane filter. To avoid generation of bubbles in the detector cell, degas the eluent before use in an ultrasonic bath.
2. Make sample solutions by dissolving 10 mg of racemic dihydroxyphenylalanine (Dopa) and Carbidopa in 10 mL of the eluent and filtrate them through a 0.45-µm membrane filter.

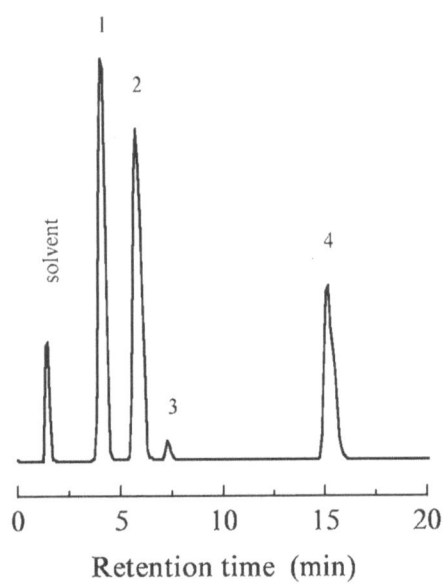

Retention time (min)

Fig. 3. Chromatogram of a typical mixture containing (1) D-Dopa (12 µg), (2) L-Dopa (10 µg), (3) D-Carbidopa (0.1 µg), and (4) L-Carbidopa (0.9 µg). Redrawn from **ref. (7)** with permission. Conditions: column, LiChrospher C18, 5 µm, 125 × 4.0 mm I.D.; eluent, 3.0 m*M* aqueous copper sulfate, 6.0 m*M* L-phenylalanine; flow rate, 1 mL/min; temperature, 27°C; detection, 280 nm. Dopa, β-3,4-dihydroxyphenylalanine; Carbidopa, α-hydrazino-3,4-dihydroxy-α-methyl-benzenepropanoic acid.

3. Equilibrate the column with the eluent.
4. Carry out chromatography at a flow rate of the eluent of 1.0 mL/min and detecting the effluent photometrically at 280 nm (*see* **Note 8**). A typical chromatogram is presented in **Fig. 3**.

4. Notes

1. Many organic solvents and reagents, including methanol, *n*-decylbromide, may be harmful if swallowed, inhaled, or absorbed through skin. Besides, they are inflammable. Diethyl ether is extremely dangerous from this point of view. All organic liquids should be manipulated with extreme caution.
2. Alkaline and acidic aqueous solutions are especially dangerous for eyes and should be manipulated with caution.
3. It is easy to observe the completion of the coating procedure by letting the column effluent drop into a beaker filled with water. *N*-Decyl-*S*-hydroxyproline is insoluble in water and forms white precipitate right after the breakthrough. By measuring the vol of the chiral coating solution passed through the column to the breakthrough moment, it is possible to estimate the amount of the chiral selector dynamically

coated on the reversed-phase packing material. One can reduce the amount of the chiral selector adsorbed in the column by reducing the water percent in the coating solution or by reducing concentration of the selector, or vice versa, by increasing the selector concentration or water proportion in the coating solution, one can increase the density of chiral coating (take into account that water reduces solubility of the selector in aqueous methanol).

4. One can remove the chiral coating and regenerate the reversed-phase column by washing the latter with methanol or ethanol.

5. It is easy to observe the completion of the copper loading procedure by letting the column effluent drop into a beaker filled with aqueous ammonia solution. Copper ions form dark blue tetraammine complexes with ammoniac molecules.

6. One can remove the copper ions from the coated column by rinsing the column with an acidic aqueous solution.

7. Since hydrophobic interactions with the reversed-phase packing significantly contribute to the retention, the retention factors of hydrophobic amino acids can be conveniently reduced by increasing the proportion of organic modifier in the mobile phase. The latter, however, should not exceed the threshold of approx 20%; otherwise the chiral coating could be stripped from the column.

8. Contrary to systems with CSP and CCP, the CMP transports the resolved enantiomers into the detector in the form of ternary copper complexes with the chiral selector (L-phenylalanine). These complexes are diastereomeric and can exhibit different UV absorption properties. Therefore, quantitation of the enantiomers requires separate calibration for each of the two enantiomers.

References

1. Davankov, V. A. (1997) Analytical chiral separation methods. *Pure Appl. Chem.* **69,** 1469–1474.
2. Davankov, V. A. (2000) 30 years of chiral ligand exchange. *Enantiomer* **5,** 209–223.
3. Davankov, V. A., Navratil, J. D., and Walton, H. F. (1988) *Ligand Exchange Chromatography.* CRC Press, Boca Raton.
4. Davankov, V. A. (1997) The nature of chiral recognition: is it a 3-point interaction? *Chirality* **9,** 99–102.
5. Gübitz, G. and Mihellyes, S. (1994) Resolution of 2-hydroxycarboxylic acid enantiomers by ligand exchange chromatography on chemically bonded chiral phases. *J. High Resolut. Chromatogr.* **17,** 733–734.
6. Davankov, V. A., Bochkov, A. S., Kurganov, A. A., Roumeliotis, P., and Unger, K. K. (1980) Separation of unmodified α-amino acid enantiomers by reversed phase HPLC. *Chromatographia* **13,** 677–685.
7. Husain, S., Sekar, R., and Rao, R. (1994) Enantiomeric separation and determination of antiparkinsonian drugs by reversed-phase ligand-exchange high-performance liquid chromatogrphy. *J. Chromatogr. A* **687,** 351–355.
8. Davankov, V. A. (2003) Enantioselective ligand exchange in modern separation techniques. *J. Chromatogr. A* **1000,** 891–915.

9

Chiral Separations by HPLC
Using Molecularly Imprinted Polymers

Peter Spégel, Lars I. Andersson, and Staffan Nilsson

1. Introduction

In the past decade the molecular imprinting technology (MIT) has developed to become a promising approach for the synthesis of artificial receptors (1,2). The technique is based on a template-assisted polymerization, and the resultant molecularly imprinted polymer (MIP) is able to rebind the template molecule with a high selectivity. Two main approaches to prepare MIPs can be recognized, i.e., the covalent imprinting (3) and the noncovalent imprinting (4). In the covalent MIP the interactions are based on weak covalent bonding, which offers strong interactions, however, the kinetics of this recognition mechanism are slow. Today, the noncovalent approach is by far the most utilized for analytical applications, and therefore, covalent imprinting will not be covered in this chapter.

In separation science, the MIP has successfully been used as an affinity phase in capillary electrochromatography (CEC) (5) (see also Chapter 25, Chiral Separations by Capillary Electrochromatography using Molecularly Imprinted Polymers), high-performance liquid chromatography (HPLC) (6), thin-layer chromatography (TLC) (7), and supercritical fluid chromatography (SFC) (8). Of these separation techniques, the MIP-HPLC is the oldest and most widely explored. The use of the MIP as a stationary phase in HPLC offers a means of obtaining a predetermined selectivity, with the imprinted analyte being the most retained when compared to closely resembling molecules.

1.1. MIP Synthesis

The preparation of an MIP is straightforward and can be divided into three main steps (Fig. 1), i.e., complexation, polymerization, and extraction. The preparation

From: Methods in Molecular Biology, Vol. 243: Chiral Separations: Methods and Protocols
Edited by: G. Gübitz and M. G. Schmid © Humana Press Inc., Totowa, NJ

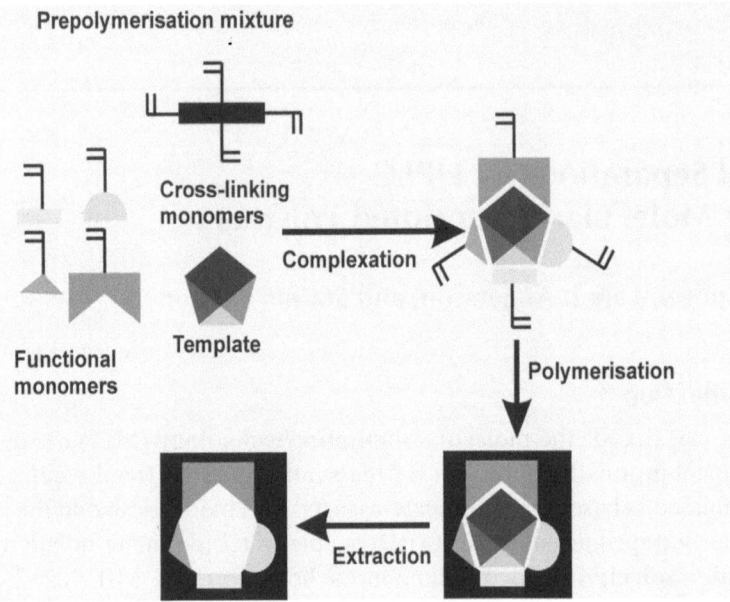

Fig. 1. The three steps in MIP synthesis. A prepolymerization mixture is prepared, and complexes between the functional monomers and the template are formed. These complexes are polymerized together with crosslinking monomers in order to obtain a rigid polymer network. The template molecule is extracted from the resultant polymer, and a cavity is revealed. This cavity is complementary to the template molecule in shape, size, and chemical functionality.

procedure is simple in terms of laboratory equipment demands. However, in order to succeed with the preparation of an MIP, it is important to have close control of all steps in the preparation procedure. It is also important to decide what type of MIP format is desired and to optimize the MIP preparation steps in this aspect. MIPs have been prepared in the form of superporous monoliths *(9)* (*see* Chapter 25), surfaces for open tubular chromatography *(10)* (*see* Chapter 25), and particles of different shapes and sizes *(11)*. For ordinary HPLC columns, the use of bulk polymerized, ground, sieved, and slurry packed MIPs (**Fig. 2**) have been dominating, although other techniques have recently been developed *(6)*. The use of the bulk polymerization approach suffers from the irregular shapes of the resulting particles yielding poor column packing and, thus, poor flow performance in chromatography. Recently, different approaches have been developed for synthesis of spherical particles, including suspension polymerization *(12)*, two-step swelling and polymerization *(13)*, and precipitation polymerization *(14)*. These very promising techniques will be described briefly. The

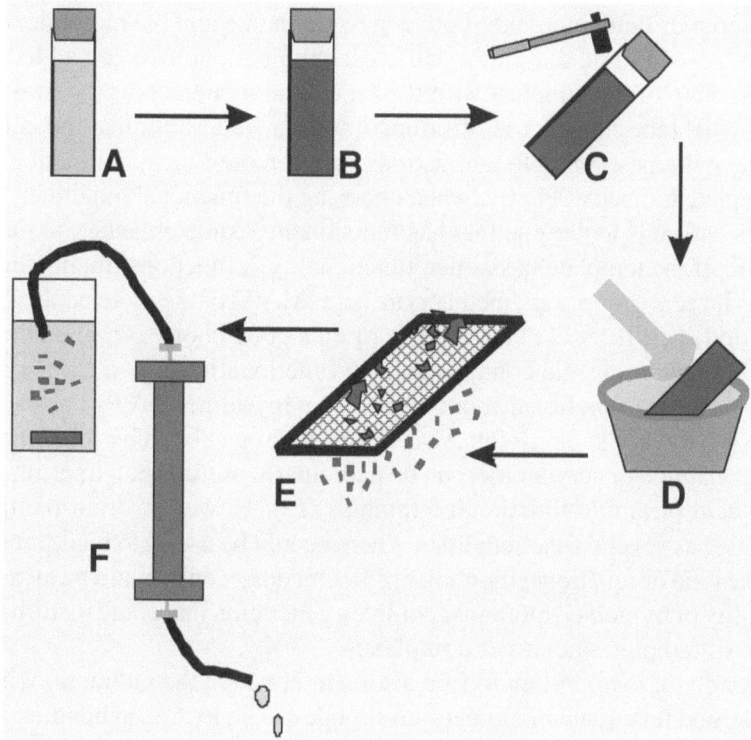

Fig. 2. Schematics of MIP preparation for use in HPLC. (**A**) The prepolymerization mixture is introduced into a borosilicate test tube and degassed. (**B**) Polymerization is initiated, and a bulk polymer is obtained. (**C**) The tube is smashed, and the MIP is recovered. (**D**) The MIP is ground using mechanical mortar. (**E**) The desired particle size distribution is collected by sieving and repeated sedimentation. (**F**) The resultant particles are subsequently slurry packed into a stainless steel HPLC-column, washed on-line to get rid of the template molecule and conditioned in mobile phase.

main point of this article will, however, be the bulk polymerization approach, as it is the simpler system and allows for a straightforward explanation of the basic features of MIP preparation. The main steps in MIP synthesis are covered below, and these steps do apply for most MIP formats.

1.1.1. Complexation

The first step is the preparation of a prepolymerization mixture containing all components needed for the MIP synthesis (**Fig. 1**). The template molecule, i.e., the molecule against which an MIP is desired, is mixed with functional monomers in an appropriate solvent. The functional monomers should possess such characteristics that complexes between the template and the monomers will be formed. Also, a radical initiator and crosslinking monomers are added. As the self-assembly of the template with the functional monomers is the most crucial step in MIP fabrication, it is of utmost importance to optimize the conditions to achieve the best possible interactions between the functional monomers and the template molecule. Firstly, when choosing the functional monomer, the first obvious choice is to use one that has functionality complementary to that of the template. If the template has amine functionality, a functional monomer having carboxylic acid group, e.g., methacrylic acid (MAA) or the more acidic 2-(trifluoromethyl)-acrylic acid (TFMAA), is often a good choice. On the other hand, if the template molecule contains an acid functionality, e.g., a carboxylic acid group, the basic functional monomers 2-vinylpyridine (2-VPy) or 4-vinylpyridine (4-VPy) could be useful. Also, strong hydrogen bonding monomers, e.g., methacrylamide or acrylamide, can be particularly useful when imprinting carboxylic acid or amide functional templates *(15)*. However, often the template molecule has several functionalities. Then it could be useful to incorporate more than one type of functional monomer *(16)*. Of course, one should be aware of the possibility of monomer-monomer complex generation that could disturb the generation of template-monomer complexes.

Secondly, it is important to find a solvent in which the interactions between template and functional monomers are strongest, e.g., hydrogen bonding is strong in nonpolar solvents such as toluene and dichloromethane. Thus, protic and highly polar solvents, which disturb hydrogen bonding interactions and other electrostatic interactions, should be avoided if the template-monomer interactions are not of a greater magnitude. Also, the solvent should give rise to the formation of a porous structure in the resultant polymer, exposing a large number of imprinted sites to the analytes and facilitating mass-transfer between the bulk and the imprint *(6)*.

1.1.2. Polymerization

Polymerization is initiated using either heat or UV-light mediated decomposition of the radical initiator, which is typically an azo-initiator. The functional monomers, in complex with the template, will be locked in place by reaction with the crosslinking monomers resulting in a highly crosslinked and rigid polymer. The most frequently used crosslinking monomers are ethylene glycol dimethacrylate (EDMA), containing two vinyl groups, and trimethylolpropane trimethacrylate (TRIM), containing three vinyl groups *(4)*. The crosslinking ratio employed is rather high, often 50%, when using TRIM, or 75%, when using EDMA, of the total monomer concentration.

It has repeatedly been shown that MIPs prepared at low temperature (<0°C) show higher selectivities towards the template, when compared to MIPs prepared using heat initiated polymerization *(17)*. Also, the polymer morphology is affected by the temperature of polymerization, e.g., MIPs prepared utilizing heat-initiated polymerization tend to have larger pore volumes *(17)*.

1.1.3. Extraction

When the template is removed, it leaves behind cavities complementary to the template in shape, size, and chemical functionality (**Fig. 1**). If the MIP has been prepared as a bulk polymer, it has to be ground, sieved, and sedimentated to achieve the appropriate particle size distribution prior to use in HPLC (**Fig. 2**). The resultant particles are slurry packed into an HPLC column and thoroughly washed on-line, most frequently with methanol containing 5–20% acetic acid, to remove the template. If the MIP is to be used in trace analysis, there could be a problem with template leakage as it is difficult to wash out all the template used *(18)*. In order to avoid these problems, a structural analog of the analyte can be used as template *(19)*. Template leaking is not a major problem in MIP-HPLC, however it is a larger problem in solid phase extraction (SPE) and pseudo-immunoassays, where this problem needs to be addressed *(20)*.

1.2. Separations Using MIP

Retention of an analyte is due both to selective interactions with the imprints and nonspecific interactions with the polymer surface. For chiral compounds, nonspecific interactions affect both enantiomers equally, and any difference in retention between the enantiomers is due solely to analyte-imprint interactions. Hence, in a number of studies, MIPs have been prepared against pure enantiomer for the purpose of investigating the imprinting process in detail as the enantiomer separation ability of the resultant MIP.

Eluent optimization aims at increasing the selective imprint-derived retention giving high enantiomer separation and reducing nonspecific retention interactions. Two main approaches in choosing the eluent to be used in MIP-HPLC can be recognized. One opinion is that the best eluent is the one that most closely resembles the polymerization solvent used in the preparation of the MIP. This approach is based on the fact that the same interactions will be present between the analyte and the MIP as was present between the template and the functional monomers. Also, the polymer will be in the same level of swelling as when it was formed. Swelling and shrinking may affect the imprints and change their recognition properties. However, in practice, retention is a more important parameter, and swelling of the polymer is rarely considered in optimizing MIP-HPLC separations. The interactions between the analyte and the MIP most often are very strong in such a mobile phase, especially if the MIP synthesis is carefully optimized, and thus the elution times tend to be very long. The other opinion is to regard the MIP as a mixed-mode separation phase and optimize the chromatographic separation using three components, i.e., a nonpolar and a polar solvent together with a component competing with the MIP. A good starting point is to use pure acetonitrile. If the analyte does not elute or is too strongly retained, a component competing with the MIP can be added to the mobile phase. A good choice, when using, e.g., MAA as functional monomer, is to add acetic acid or methanol to compete with the imprints. If, on the other hand, the analyte elutes in the void, a nonpolar component, e.g., chloroform or toluene, can be added. When using systems in which ionic interactions can be present, e.g., when imprinting an amine using acidic functional monomers, it might be useful to optimize the pH of the eluent as well. In practice, it seems that optimization using acetic acid is the best choice for these systems also. An alternative approach, applicable to MIPs prepared with templates containing acidic or basic groups, e.g., amino acids, imprinted using basic or acidic functional monomers, e.g., MAA or 2-VPy, is to optimize the mobile phase for recognition based on ion exchange *(21)*. In this approach, buffer is added to the acetonitrile phase, and optimization of the pH of the eluent is a key factor. However, when increasing the aqueous buffer content, retention due to nonspecific hydrophobic effects will appear.

One of the drawbacks of the use of MIPs in chromatographic separations is the tailing of the peaks, especially of the peak corresponding to the imprinted analyte (**Fig. 3**). Several plausible reasons for this have been identified during the past decade. One of them originates from variances in template-monomer self-assembly structures, and another is owing to the amorphous structure of the polymer, which affects the accessibility of the imprints. The result of these effects is a continuous distribution of binding site energies resulting in tailing peaks *(22)*. Gradient elution has been shown to improve the chromatography (**Fig. 4**) *(23)*.

1.3. New Approaches in MIP-HPLC

The bulk polymerization approach for synthesis of MIPs for HPLC suffers from several disadvantages. One is the low yield of MIP, as large amounts of the material are rejected during the crushing, grinding, and sieving process. Most often the amount of MIP in the appropriate particle size for packing a HPLC column is in the range of 40–60% of the original MIP amount achieved after polymerization *(20)*. Also, the obtained particles are irregular in size and shape, which results in poor flow properties in the resulting MIP-HPLC column. It would, thus, be beneficial to synthesize spherical particles with a narrow particle size distribution. Two techniques that have successfully been used to synthesize spherical MIP particles are discussed here in order to introduce the reader to new approaches used to improve MIP-HPLC separations.

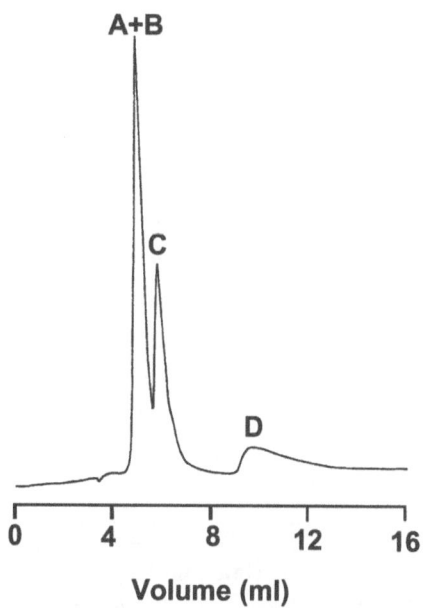

Fig. 3. Typical chromatogram obtained when using MIPs in HPLC. The polymer was imprinted against Cbz-L-glutamic acid (peak D). The peaks corresponds to (**A**) Cbz-L-aspartic acid, (**B**) Cbz-D-aspartic acid, (**C**) Cbz-D-glutamic acid, and (**D**) Cbz-L-glutamic acid. One can identify the characteristic tailing of the peak corresponding to the imprinted analyte. Reprinted with permission from **ref. 25**.

1.3.1. Suspension Polymerization

One promising technique developed to produce spherical MIP particles is suspension polymerization using a liquid perfluorocarbon as the dispersing phase. By using this technique, highly polar dispersing phases can be avoided. Difficulties with polar dispersing phases include saturation of the polymerization solution with the polar phase, which is present in large amounts, as well as partitioning of the template, functional monomers, radical initiators, and crosslinking monomers into the dispersing phase. The perfluorocarbon solution, on the other hand, can be considered immiscible with most organic compounds and, thus, forms an inert dispersing phase. Indeed, this technique results in highly efficient MIP-HPLC phases. A detailed description of this MIP synthesis technique is available in **ref. 12**.

1.3.2. Two-Step Swelling and Polymerization

This technique utilizes polymeric seed particles as shape templates to produce spherical MIP particles. The seed particles are first swelled using a highly

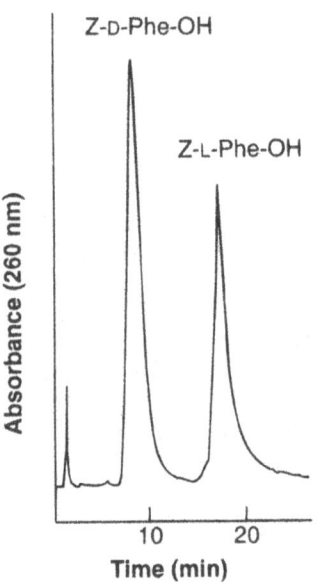

Fig. 4. Improvements of MIP chromatography can be achieved by using gradient elution. Here, racemic Z-Phe-OH is separated on a column containing a Z-L-Phe-OH-imprinted polymer using a gradient with an increasing amount of acetic acid. Reprinted with permission from **ref. 22**.

water-insoluble solvent. Second, the particles are further swelled using the prepolymerization mixture, containing functional monomers, polymerization solvent, crosslinking monomers, radical initiator, and template molecule, followed by polymerization. As this technique is performed in water dispersion, it is important to be aware of the potential partitioning of the prepolymerization components into the aqueous dispersing phase. Surface modification of these particles has been achieved by adding hydrophilic monomers during polymerization. This hydrophilic external layer of polymer could be formed without losing the imprint effect. Detailed descriptions of the two-step swelling and polymerization technique for MIP preparation is given in **refs. 13,24**.

2. Materials

2.1. Preparation of MIP-HPLC-Column

1. Dry acetonitrile (HPLC grade) (polymerization solvent), MAA and 2-VPy (functional monomers), EDMA (crosslinking monomer), Boc-L-tryptophan (template molecule), 2,2'-azobis(2,4-dimethylvaleronitrile) (ABDV) (radical initiator), and a 50-mL borosilicate glass ampoule to be used as polymerization container. Be aware of that all monomers are toxic and potential skin sensitizers.

2. Water bath thermostated at 45°C for polymerization initiation.
3. Mechanical mortar, sieve with 25 μm cut off and acetone.
4. Air driven fluid pump, chloroform/acetonitrile (17:3, v/v), acetonitrile, a 30-mL packing bomb, and an empty HPLC column.
5. MIP washing solution consisting of methanol/acetic acid (9:1, v/v).

2.2. HPLC Analysis

1. Isocratic HPLC pump, variable wavelength UV detector, and a 20-μL injection loop. Mobile phase consisting of acetonitrile with 1% (v/v) acetic acid. Samples consisting of the racemate, *R* and *S* enantiomer of the imprint molecule at 10 g/L in mobile phase.

3. Methods

3.1. MIP-HPLC Column Preparation

1. The prepolymerization mixture is prepared by mixing 6.55 mM of each functional monomers (MAA and 2-VPy), 65.5 mM cross-linking monomer (EDMA), 1.64 mM template molecule (Boc-L-tryptophan), and 150 mg radical initiator (ABDV) in 20 mL of dry acetonitrile in a 50-mL borosilicate glass test tube. The mixture is degassed by sonication under vacuum and purging with nitrogen for 5 min. Pre-polymerization solutions should be prepared fresh as required (*see* **Notes 1–5**).
2. The borosilicate glass test tube is carefully sealed and put in a waterbath at 45°C for 15 h to initiate the polymerization (*see* **Notes 6** and **7**).
3. To collect the MIP, the borosilicate glass test tube is smashed, and the MIP is transferred to and ground in a mechanical mortar. The ground MIP particles are wet sieved in water using a 25-μm sieve. Particles to large to pass the sieve are re-transferred to the mortar, and this procedure is repeated until all particles are able to pass the sieve. Fines are removed by repeated sedimentation in a large vol of acetone (approx 5 g of MIP/L of acetone). The obtained particles are dried under vacuum over silica gel and stored dry until HPLC column packing or other use (*see* **Notes 8–11** and **12**).
4. To pack the column, the MIP particles are slurried in chloroform/acetonitrile (17:3, v/v) and filled into a 30-mL packing bomb. For a 20 × 4.6 mm inner diameter (i.d.) HPLC column, 0.9–1 g MIP particles are needed. The packing bomb is connected to the HPLC column and an air-driven fluid pump. The pressure is allowed to rise from 0 bar to 300 bar in 2–4 s and kept constant at 300 bar until the packing is complete and at least 200 mL of packing solvent has passed at constant pressure. Pure acetonitrile is used as packing solvent (*see* **Notes 13** and **14**).
5. The HPLC column is washed extensively using methanol/acetic acid (9:1, v/v) at 1 mL/min until a stable baseline is obtained when monitored at 280 nm (*see* **Note 15**).
6. The MIP-HPLC column can be stored at room temperature and shielded from light until use. The column must be kept wet in order not to shrink the MIP and thus destroy the packing (*see* **Note 12**).

3.2. HPLC Analysis

1. Condition the column with acetonitrile containing 1% acetic acid (v/v) at 1 mL/min until a stable baseline is obtained at 280 nm. Inject the sample and perform the separation at a flow rate of 1 mL/min (*see* **Notes 16–18**).

4. Notes

1. The polymerization solvent can be varied. The use of more nonpolar solvents such as dichloromethane, chloroform, and toluene can improve the imprinting process. However, when using chlorinated solvents, the specific surface area tends to be smaller.
2. The functional monomer 2-VPy can be exchanged for 4-VPy. This monomer has shown improved recognition when compared to 2-VPy in some systems.
3. The crosslinking monomer can be exchanged for TRIM. This monomer has three vinyl groups and can thus be used at lower concentrations. A good starting point for a system containing TRIM is 39.3 mM TRIM, 19.65 mM 2-VPy, 19.65 mM MAA, 4.92 mM Boc-L-tryptophan, and 150 mg ABDV in 20 mL of dry acetonitrile.
4. The monomers obtained from commercial sources often contain polymerization inhibitors. To remove these the monomers can either be distilled or passed through a basic alumina column. 2-VPy, 4-VPy, and MAA can be vacuum distilled at low temperature. EDMA should not be distilled, as it will cause the formation of small oligomer and polymer segments. Purified monomers should be used within 3 or 4 d. It is, however, possible to polymerize with these inhibitors present.
5. The radical initiator ABDV can be substituted for 2,2'-azobisisobutyronitrile (AIBN). AIBN is more stable and easier to handle than ABDV, however, it requires a higher polymerization temperature (65°C) than does ABDV (45°C).
6. It is important to degas the prepolymerization solution carefully as dissolved oxygen will inhibit polymerization.
7. Both ABDV and AIBN can be used for UV-initiated polymerization ($\lambda = 366$ nm) at low temperatures, e.g., −28°–0°C. Owing to its better stability, AIBN is probably the radical initiator to prefer. A lower polymerization temperature has proven beneficial in terms of imprint quality. Some template molecules, such as Boc-L-tryptophan, might decompose upon UV-irradiation and form colored products that inhibit the polymerization. For such templates, heat-initiated polymerization should be used.
8. After completion of polymerization, prior to grinding, it is often beneficial, for safety reasons, to leave the MIP in a methanol solution overnight to extract unreacted monomers and loosely bound templates. This is especially recommended if toxic templates, monomers, or polymerization solvents have been used in the MIP synthesis.
9. A standard mortar and pestle might be used if a mechanical mortar is not available. However, this is extremely time-consuming.
10. Finer sieves may be used. If sieves with a cut-off <15 µm are used, it is recommended to use an ultrasonic sieving machine.

11. Acetonitrile and ethanol can also be used for sedimentation of the MIP particles.
12. MIPs can be stored in the dry state for several years at ambient temperature. However, it should be protected from direct exposure to sunlight, as the MIP might turn brittle upon light exposure.
13. A solution consisting of chloroform/acetone (17:3, v/v) can also be used to slurry the MIP particles.
14. If the pressure is allowed to immediately reach 300 bar, there is a risk of obtaining uneven packing of MIP particles with denser packing closest to the outlet frit. This might have disadvantageous effects on the chromatography.
15. The amount of acetic acid in the washing solution can be increased up to 50%. The more acetic acid in the washing solution, the stronger it is as template eluent.
16. The mobile phase can be varied in order to alter the chromatography. If the k' values are undesirably high, a more practical elution time can be obtained by increasing the amount of acetic acid in the mobile phase. If the analyte elutes in the void, the amount of acetic acid can be decreased or a more nonpolar solvent, e.g., chloroform, can be added to the mobile phase.
17. Gradient elution can be used in order to improve the chromatography. A gradient employing an increasing amount of acetic acid in chloroform or acetonitrile can be useful.
18. The MIP often shows some cross-selectivity, i.e., it can recognize structural analogs of the imprinted molecule. This extends the applications of the produced MIP to also include enantiomer separations of closely resembling molecules. The selectivity of the MIP against a structurally related analog is, for obvious reasons, owing to a less perfect fit in the imprint, less than for the imprinted analyte.

References

1. Sellergren, B. (ed.) (2001) *Molecularly Imprinted Polymers: Man Made Mimics of Antibodies and Their Applications in Analytical Chemistry.* Elsevier, New York.
2. Spégel, P., Schweitz, L., and Nilsson, S. (2002) Molecularly imprinted polymers. *J. Anal. Bioanal. Chem.* **372,** 37–38.
3. Wulff, G. and Biffis, A. (2001) Molecular imprinting with covalent or stoichiometric non-covalent interactions, in *Molecularly Imprinted Polymers: Man Made Mimics of Antibodies and Their Applications in Analytical Chemistry.* (Sellergren, B., ed.), Elsevier, New York, pp. 71–111.
4. Sellergren, B. (2001) The non-covalent approach to molecular imprinting, in *Molecularly Imprinted Polymers: Man Made Mimics of Antibodies and Their Applications in Analytical Chemistry.* (Sellergren, B., ed.), Elsevier, New York, pp. 113–184.
5. Schweitz, L., Spégel, P., and Nilsson, S. (2001) Approaches to molecular imprinting based selectivity in capillary electrochromatography. *Electrophoresis* **22,** 4053–4063.
6. Sellergren, B. (2001) Imprinted chiral stationary phases in high-performance liquid chromatography. *J. Chromatogr. A* **906,** 227–252.

7. Kriz, D., Berggren Kriz, C., Andersson, L. I., and Mosbach, K. (1994) Thin-layer chromatography based on molecular imprinting technique. *Anal. Chem.* **66,** 2636–2639.

8. Ellwanger, A., Owens, P. K., Karlsson, L., et al. (2000) Application of molecularly imprinted polymers in supercritical fluid chromatography. *J. Chromtogr. A* **897,** 317–327.

9. Schweitz, L., Andersson, L. I., and Nilsson, S. (1997) Capillary electrochromatography with predetermined selectivity obtained through molecular imprinting. *Anal. Chem.* **69,** 1179–1183.

10. Tan, J. Z. and Remcho, V. T. (1998) Molecular imprint polymers as highly selective stationary phases for open tubular liquid chromatography and capillary electrochromatography. *Electrophoresis* **19,** 2055–2060.

11. Mayes, A. G. (2001) Polymerisation techniques for the formation of imprinted beads, in *Molecularly Imprinted Polymers: Man Made Mimics of Antibodies and Their Applications in Analytical Chemistry.* (Sellergren, B., ed.), Elsevier, New York, pp. 305–324.

12. Mayes, A. G. and Mosbach, K. (1996) Molecularly imprinted polymer beads: suspension polymerisation using a liquid perfluorocarbon as the dispersing phase. *Anal. Chem.* **68,** 3769–3774.

13. Haginaka, J., Takehira, H., Hosoya, K., and Tanaka, N. (1999) Uniform-sized molecularly imprinted polymer for (S)-naproxen slectively modified with hydrophilic external layer. *J. Chromatogr. A* **849,** 331–339.

14. Schweitz, L., Spégel, P., and Nilsson, S. (2000) Molecularly imprinted microparticles for capillary electrochromatographic enantiomer separation of propranolol. *Analyst* **125,** 1899–1901.

15. Yu, C. and Mosbach, K. (1997) Molecular imprinting utilizing an amide functional group for hydrogen bondingn leading to highly efficient polymers. *J. Org. Chem.* **62,** 4057–4064.

16. Ramström, O., Andersson, L. I., and Mosbach, K. (1993) Recognition sites incorporating both pyridinyl and carboxy functionalities prepared by molecular imprinting. *J. Org. Chem.* **58,** 7562–7564.

17. O'Shannessy, D., Ekberg, B., and Mosbach, K. (1989) Molecular imprinting of amino acid derivatives at low temperature (0°C) using photolytic homolysis of azobisnitriles. *Anal. Biochem.* **177,** 144–149.

18. Ellwanger, A., Berggren, C., Bayoudh, S., et al. (2001) Evaluation of methods aimed at complete removal of template from molecularly imprinted polymers. *Analyst* **126,** 784–792.

19. Andersson, L. I., Paprica, A., and Arvidsson, T. (1997) A highly selective solid-phase extraction sorbent for preconcentration of sameridine made by molecular imprinting. *Chromatographia* **46,** 57–62.

20. Andersson, L. I. (2000) Molecular imprinting for drug bioanalysis: a review on the application of imprinted polymers to solid-phase extraction and binding assay. *J. Chromatogr. B* **739,** 163–173.

21. Sellergren, B. and Shea, K. J. (1993) Chiral ion-exchange chromatography: correlation between solute retention and a theoretical ion-exchange model using imprinted polymers. *J. Chromatogr. A* **654,** 17–28.
22. Sellergren, B. and Shea, K. J. (1995) Origin of peak asymmetry and the effect of temperature on solute retention in enantiomer separations on imprinted chiral stationary phases. *J. Chromatogr. A* **690,** 29–39.
23. Kempe, M. (1996) Antibody-mimicking polymers as chiral stationary phases in HPLC. *Anal. Chem.* **68,** 1948–1953.
24. Masci, G., Aulenta, F., and Crescenzi, V. (2002) Uniform-sized clenbuterol molecularly imprinted polymers prepared with methacrylic acid or acrylamide as an interacting monomer. *J. Appl. Polym. Sci.* **83,** 2660–2668.
25. Andersson, L. I. and Mosbach, K. (1990) Enantiomeric resolution on molecularly imprinted polymers prepared with only non-covalent and non-ionic interactions. *J. Chromatogr.* **516,** 313–322.

10

Indirect Enantioseparation by HPLC Using Chiral Benzofurazan-Bearing Reagents

Toshimasa Toyo'oka

1. Introduction

The chiral resolutions by high-performance liquid chromatography (HPLC) are divided into the following three: one is direct resolution using a chiral stationary phase (CSP); another relies on diastereomer formation with a suitable chiral derivatization reagent; a third technique is a minor method utilizing chiral mobile phase additives.

Many chiral compounds have been determined with the direct resolution that employ CSP columns containing immobilized chiral selectors. The separation mechanism is owing to the stability difference of the diastereomeric complexes formed between CSP and each enantiomer in the flow system. Since the method requires no pretreatment, such as derivatization, possible racemization during the separation seems to be negligible. Consequently, the direct resolution using a CSP column may be preferable for trace analysis of antipode enantiomer in main component. However, the separation and the detection sensitivity are sometimes insufficient in real sample analysis. The separation is highly influenced on the interaction between CSP and enantiomer. Thus, the choice of the best column for separation of each racemate is fairly difficult. The elution order of a pair of enantiomers is also dependent upon the CSP column used and can not be changed easily. Furthermore, the sensitivity is usually not adequate for trace analysis in biological specimens.

The indirect resolution involving a derivatization step with a chiral tagging reagent is an efficient technique for separation of many racemates. The separation is based upon diastereomer formation by the reaction with a chiral derivatization reagent. In spite of many problems associated with the indirect method, i.e., optical purity of the reagent, stability of the reagent, possibility of racemization during tagging reaction, and supply of the reagent, good sensitivity and

From: *Methods in Molecular Biology, Vol. 243: Chiral Separations: Methods and Protocols*
Edited by: G. Gübitz and M. G. Schmid © Humana Press Inc., Totowa, NJ

selectivity of indirect method coupled with an efficient detection system are attractive means for the determination of chiral molecules in real sample analysis. Therefore, this indirect derivatization method is suitable for trace analysis of enantiomers in biological samples, such as blood and urine, because highly sensitive detection is performed with the option of coupling with suitable reagents that have high molar absorptivity (ε) or high fluorescence (FL) quantum yield (ϕ).

Tagging of analytes with reagents that afford the structures absorbing ultraviolet (UV) or visible (VIS) region is the most popular means of derivatization. Because almost all laboratories possess a UV-VIS detector, and the analysts are experiencing the manipulation, there are various chiral derivatization reagents for HPLC that provide UV-VIS absorption. It is predominant that the derivatives derived from the reagents have strong absorption in long wavelength region. There are many endogenous substances in samples, which absorb in relatively short wavelengths. Since interference of impurities absorbing the detection wavelength is considered in real samples, especially in complex matricies such as biological specimens, the reagents absorbed in VIS band are preferable in terms of selectivity. Although a number of UV labels have been applied to the tagging of various functional groups, the sensitivity of the derivatives is not good enough in some real samples.

To solve this disadvantage, various types of FL labels have been developed, and a great number of papers concerning FL tagging have been published. As fluorometry is both sensitive and selective, many FL tagging reagents developed are applied to the analysis of real samples. The FL properties of the resulting derivatives tend to be greatly affected by temperature, viscosity of the solvent, pH of the medium, and the contamination of halide ions such as Cl^- and Br^-. It should be also noted that undesirable FL materials contaminated in tested samples, especially in biological specimens, interferred with the determination. The pretreatment of real samples is another important topic in trace analysis. In the analysis of real samples such as biological, environmental, and food, most significant and major part of the procedure involves how effectively to obtain trace analytes of interest from complicated matrix of proteins, fats, and minerals. Sample pretreatment, i.e., clean-up and concentration of analytes, is inevitable for HPLC measurement with derivatization. The FL label is the most effective for determinations in biological specimens, in terms of sensitivity and/or selectivity.

Different types of FL labeling reagents have been developed for the enantioseparation of biological importances. The selection of the reagent dominates accuracy, precision, and repeatability of the quantitative analysis. Several important points worthy of consideration for the choice of chiral derivatization reagent are: (i) The optical purity of the reagent should be as great as the chemical purity. Since the opposite enantiomer contaminating the reagent also produces a corresponding diastereomer, it is obvious that erroneous results will be obtained with the use of impure reagents. (ii) The degree of racemization during the labeling reaction and storage of the reagent itself is an another important issue for quantitative determinations. Furthermore, the chemical stability of the resulting diastereomers also influences the results. Good stability (at least 1 d) is required for many analyses, because autoanalysis overnight is usually planed. (iii) The reactivity of the reagent for each enantiomer and the FL properties (wavelength and intensity) of the resulting derivatives are essentially the same. When the reaction rates are extremely different in both enantiomers, the reaction condition should be optimized carefully. In the case of difference of FL properties, each calibration curve should be drawn for the determination of a pair of enantiomers. (iv) The reagent possesses specificity for the target functional group and quantitatively labels the analyte under mild conditions. (v) The resulting diastereomers exhibit an adequate detector response for sample analysis. (vi) Another important point is the solubility of the reagent, whether it is freely soluble in water or miscible in aqueous solvents, such as methanol and acetonitrile, because many bioactive chiral molecules are in aqueous solutions. (vii) It is another importance with practical mean that both enantiomers of the reagent are commercially available or easily obtained by simple synthesis. Because the elution order can be controlled with the selection of the reagent enantiomer. This is necessary when the determination of a trace enantiomer is required in the presence of a large amount of antipode enantiomer. These items listed here are of general importance for all chiral tagging reagents, not only for FL but also UV-VIS.

Although the FL detection in HPLC provides excellent sensitivity and selectivity, the sensitivity is often insufficient for trace determination in real samples. Laser-induced fluorescence (LIF) detection is adopted in such case. Various laser sources such as Ar-ion, He-Cd ion, and semiconductor laser are commercially available. The minimum detectable concentrations with LIF are typically one to five orders of magnitude lower than those of FL and UV detection. However, the excitation wavelengths for possible use are limited with each laser source selected, which is an important disadvantage of LIF detection.

Particular fluorophores emit light in chemical reactions without the need of optical excitation with lamps such as xenon arc. As the flicker noise based upon the lamp is negligible, extremely high sensitivity is theoretically obtained from chemiluminescence (CL) derived from chemical reaction. Indeed, trace analysis at attomole levels has been achieved with this technique in the reaction of CL reagents, such as luminol, and the combination of oxalate esters and hydrogen peroxide (H_2O_2). The CL method is also possible to apply the sensitive detection of distereomers derived from FL chiral tagging reagents.

The labeling with chiral derivatization reagent is carried out with the reactions of reactive functional groups in substrates, e.g., amines (primary and secondary), carboxyl, carbonyl, hydroxyls (alcohol and phenol), and thiol. Many organic reactions are adopted for the labeling of various functional groups in substrate.

A number of optically active reagents have been developed for HPLC analysis of chiral molecules having various functional groups (1–3). Among the functional groups, the tagging reactions of primary and secondary amines have been extensively investigated, because chiral primary and secondary amines are easy to label with carboxyls, chloroformates, isocyanates, and isothiocyanates to yield corresponding amides, carbamates, ureas, and thioureas, respectively. Racemic carboxylic acids are usually labeled with chiral primary amines in the presence of activation reagents, e.g., diethyl phosphorocyanidate (DEPC) and 1-ethyl-3-(3-dimethylaminopropyl)carbodiimide (EDC). The reaction proceeds under mild conditions at room temperature. Ester formation with chiral alcohol is also adopted for derivatization of carboxyls. Since the reaction conditions are generally drastic, possible racemization should be monitored during the reaction. The labeling of alcoholic OH is the most difficult, because the reaction competes with water in the medium. Hence, the reaction with acid chloride-type reagents is mainly performed in anhydrous solvents such as chloroform and benzene. A variety of organic reactions is possible for the carbonyl compounds, aldehydes, and ketones. However, few labels are reported for the resolution of carbonyl enantiomers. Those are amine- and hydrazine-type reagents to produce corresponding nitriles and hydrazones.

A pair of enantiomers is labeled with a chiral derivatization reagents to generate a couple of diastereomers, correspondingly. The separation is performed with the differences from physicochemical properties (e.g., stereochemistry and stability) with achiral stationary phase. The elution order and degree of separation of the diastereomers derived from each reagent are not easily predicted with conventional achiral stationary phase columns, such as ODS. As the separation is influenced with the distance between the two asymetric centers of substrate and the reagent, the distance should be minimized to get good separation. The conformational rigidity around the chiral centers is another important factor for the separation. It is recommended that a resolving reagent that freely rotates near the asymetric center in the substrate is interfered with the formation of the diastereomer. However, there is no obvious rule concerning the separation of both diastereomers.

The aim of this chapter is to show the methodologies for indirect resolution of chiral compounds based upon diastereomer formation. Since there is not enough pages to describe all of the procedures utilizing various chiral tagging

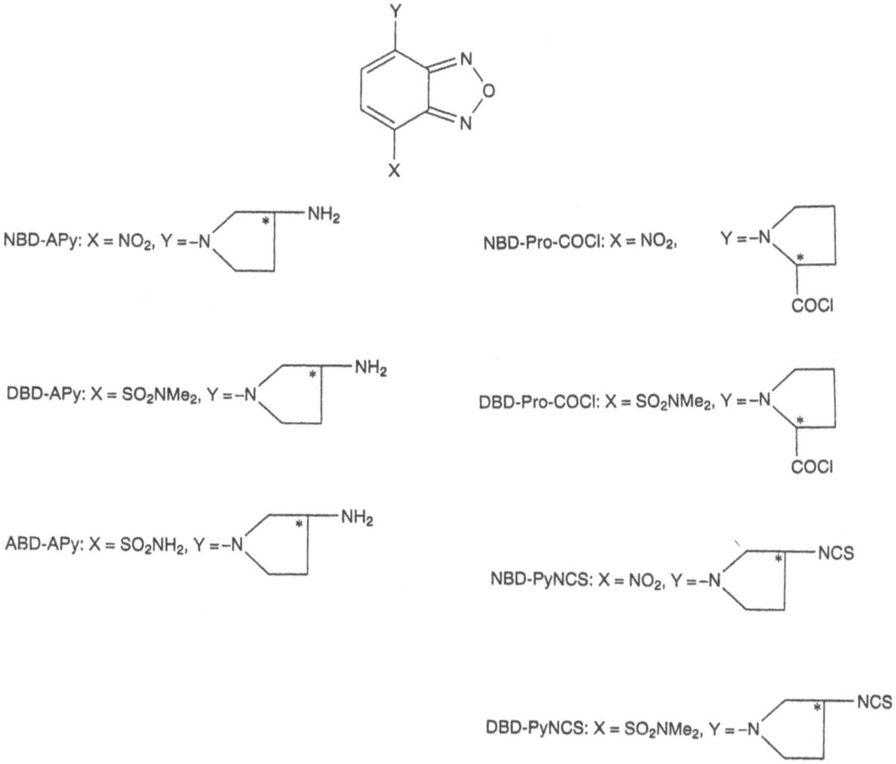

Fig. 1. Structures of fluorescent chiral derivatization reagents having benzofurazan moiety.

reagents reported up until now, the derivatization, separation, and detection methods with the chiral benzofurazan-bearing reagents, i.e., 4-(3-aminopyrrolidin-1-yl)-7-(*N,N*-dimethylaminosulfonyl)-2,1,3-benzoxadiazole (DBD-APy), 4-nitro-7-(*N,N*-dimethylaminosulfonyl)-2,1,3-benzoxadiazole (NBD-APy), 4-(3-aminopyrrolidin-1-yl)-7-(aminosulfonyl)-2,1,3-benzoxadiazole (ABD-APy), 4-(3-isothiocyanatopyrrolidin-1-yl)-7-(*N,N*-dimethylaminosulfonyl)-2,1,3-benzoxadiazole (DBD-PyNCS), 4-(3-isothiocyanatopyrrolidin-1-yl)-7-nitro-2,1,3-benzoxadiazole (NBD-PyNCS), 4-(2-chloroformylpyrrolidin-1-yl)-7-(*N,N*-dimethylaminosulfonyl)-2,1,3-benzoxadiazole (DBD-Pro-COCl) and 4-(2-chloroformylpyrrolidin-1-yl)-7-nitro-2,1,3-benzoxadiazole (NBD-Pro-COCl), are described as the representatives (**Fig. 1**). When the reactive functional groups in the reagent and the analyte are the same, the conditions described in this chapter are essentially adopted for the other reagents.

2. Materials

2.1. Derivatization and Resolution of Carboxylic Acid Enantiomers (4–7)

2.1.1. Derivatization of Carboxylic Acids

1. Chiral reagent solution: 10 mM chiral reagent [R- or S-enantiomer of DBD-APy (*4*), NBD-APy (*4*), or ABD-APy (*4*)] in acetonitrile. The reagents are now on a market (Tokyo Kasei Ltd, Tokyo, Japan).
2. Chiral carboxylic acid solution: 2 μM anti-inflammatory drugs (naproxen, ibuprofen, and loxoprofen, etc.) in acetonitrile.
3. Activation solution: mixed solution of 10 mM 2,2'-dipyridyl disulfide (DPDS) and 10 mM triphenylphosphine (TPP) in acetonitrile. The activation reagents for carboxylic acids are prepared just prior to use.

2.1.2. Separation and Detection of Diastereomers

1. HPLC system: LC-10A series (Shimadzu, Kyoto, Japan).
2. FL detector: FL-10A$_{XL}$ (Shimadzu).
3. LIF detector: LF-8010 (Tosoh, Tokyo, Japan).
4. CL detector: CL-2 (Shodex, Tokyo, Japan).
5. Reversed-phase column: Inertsil ODS-2 (150 × 4.6 mm, inner diameter [i.d.], 5 μm) or TSK-gel ODS-80TM (250 × 2.0 mm, i.d., 5 μm).
6. Column temperature: 40°C.
7. Detection: 580 nm (excitation at 470 nm) for the derivatives of DBD-APy, 585 nm (excitation at 470 nm) for the derivatives of ABD-APy, and 540 nm (excitation at 470 nm) for the derivatives of NBD-APy.
8. Mobile phase for FL and LIF detection: water/acetonitrile mixture.
9. 0.1 M Imidazole-HNO$_3$ buffer for CL detection: 0.1 M imidazole water solution adjusted to pH 7.0 with 61% HNO$_3$.
10. CL reagents: 0.5 mM bis(2,4,6-trichlorophenyl)oxalate (TCPO) or bis[4-nitro-2-(3,6,9-trioxadecyloxy)oxalate (TDPO) and 15 mM hydrogen peroxide (H$_2$O$_2$) in acetonitrile. The CL reagents are prepared every day.

2.2. Derivatization and Resolution of Amine and Amino Acid Enantiomers (8–15)

2.2.1. Derivatization of Amines and Amino acids

1. Chiral reagent solution: 12 mM chiral reagent [R- or S-enantiomer of DBD-PyNCS (*8*) or NBD-PyNCS (*8*)] in acetonitrile. The reagents are now on a market (Tokyo Kasei).
2. Chiral amine and amino acid solutions: 50 μM amino acids in water/acetonitrile (1:1).
3. Activation solution: 3% triethylamine (TEA) in water/acetonitrile (1:1).

4. Internal standard (IS) solutions: 0.2 mM β-Ala and 6-amino-n-caproic acid in water.

2.2.2. Separation and Detection of Diastereomers

1. HPLC system, FL detector and column temperature: described in **Subheading 2.1.2.**
2. Reversed-phase column: Wakosil-II 3C18RS (150 × 4.6 mm, i.d., 3 μm).
3. Mobile phase: water/acetonitrile containing 0.1% trifluoroacetic acid (TFA).
4. Detection: 550 nm (excitation at 460 nm) for DBD-PyNCS derivatives and 530 nm (excitation at 490 nm) for NBD-PyNCS derivatives.

2.2.3. Pretreatment of Real Samples

1. Yogurt and cream: methanol (1.0 mL) is added to 1 g yogurt (or 0.5 g cream) and mixed vigorously. After centrifugation at 1600g for 10 min, 0.1 mL supernatant is evaporated under reduced pressure. The residue is redissolved in 0.1 mL water and used for derivatization.
2. Milk: methanol (1.8 mL) is added to 0.2 mL milk and mixed vigorously. After centrifugation at 1600g for 10 min, the supernatant is used for derivatization.
3. Yakult™: Yakult (1.0 mL) is centrifuged at 1600g for 10 min, and the supernatant is passed through a 0.2-μm membrane filter HLC DISK 25 (Kanto Chemical, Tokyo, Japan). The filtrate is diluted two times with water and used for derivatization.
4. Ketchup and tomato puree: water (1.0 mL) is added to 0.4 g ketchup (or tomato puree) and mixed vigorously. After centrifugation at 1600g for 10 min, the supernatant is passed through a 0.2-μm membrane filter HLC DISK 25. The filtrate is used for derivatization.
5. Tomato juice, beer, and wine: tomato juice, beer, or wine is passed through a 0.2-μm membrane filter HLC DISK 25. The filtrate is used for derivatization.
6. Urine: methanol (0.27 mL) is added to 30 μL urine, and centrifuged at 1600g for 5 min. The supernatant (0.2 mL) is evaporated under reduced pressure. The residue is redissolved in 20 μL water and used for derivatization.

2.3. Resolution of Chiral Sequential Analysis of Peptides (16–19)

2.3.1. Chiral Sequential Analysis of Peptides

1. Chiral reagent solution: 10 mM R- or S-DBD-PyNCS in acetonitrile.
2. Peptide solutions: 1 mM peptide in water.
3. Derivatization solution: water/acetonitrile/pyridine (2:1:1).
4. Reagent extraction solution: n-heptane/dichloromethane (8:2).
5. Cleavage reagent: TFA.
6. Cyclization and conversion solution: 50% TFA.
7. Extraction solution of thiohydantoin-amino acid: ethyl acetate (AcOEt).

2.3.2. Separation and Detection of Thiohydantoin-Amino Acids

1. HPLC system, detection, column, column temperature, and mobile phase: described in **Subheading 2.2.2.**

2.4. Derivatization and Resolution of Alcohol Enantiomers (20–22)

2.4.1. Derivatization of Alcohols

1. Chiral reagent solution: 10 mM chiral reagent [R- or S-enantiomer of DBD-Pro-COCl (20) or NBD-Pro-COCl (22)] in anhydrous benzene. The reagents are now on a market (Tokyo Kasei).
2. Chiral alcohol solution: 10 μM alcohols in benzene.
3. Activation solution: 2% pyridine in benzene.
4. Reaction stop solution: 1% methylamine in benzene.

2.4.2. Separation and Detection of Diastereomers

1. HPLC system, FL detector, LIF detector, and column temperature: described in **Subheading 2.1.2.**
2. Normal-phase column: Inertsil SIL (150 × 4.6 mm, i.d., 5 μm).
3. Mobile-phase: n-hexane-AcOEt for normal-phase chromatography.
4. Reversed-phase column: Inertsil ODS-2 (150 × 4.6 mm, i.d., 5 μm).
5. Mobile phase: water/acetonitrile containing 0.1% TFA for reversed-phase chromatography.
6. Detection: 560 nm (excitation at 450 nm) for DBD-Pro-COCl derivatives and 540 nm (excitation at 470 nm) for NBD-Pro-COCl derivatives.

3. Methods (see Note 1)

3.1. Derivatization and Resolution of Carboxylic Acid Enantiomers (4–7)

3.1.1. Derivatization of Carboxylic Acids

1. Mix 0.25 mL chiral carboxylic acid (e.g., DL-naproxen) solution and 0.15 mL activation solution (*see* **Note 2**).
2. Add 0.1 mL chiral reagent solution and mix thoroughly (*see* **Note 3**).
3. Allow to stand for a few hours (e.g., 2 h) at room temperature in the dark (*see* **Note 4**).

3.1.2. Separation of Diastereomers and Detection With FL (4,7) or LIF (5,7)

1. After the derivatization, inject an aliquot (e.g., 10 μL) of reaction solution to reversed-phase and normal-phase columns, and separate with water/acetonitrile and n-hexane/AcOEt (*see* **Note 5**).
2. Monitor the FL intensity at 580 nm (excitation at 470 nm) for DBD-APy derivative. The FL maximal wavelengths of ABD-APy derivative and NBD-APy derivative are 600 nm (ex. 470 nm) and 560 nm (ex. 470 nm), respectively (*see* **Note 6**).
3. Monitor the LIF intensity of NBD-APy derivatives with argon-ion laser (*see* **Note 7**).

3.1.3. Separation of Diastereomers and Detection with CL (6)

1. After the derivatization, inject an aliquot (e.g., 10 μL) of reaction solution containing DBD-APy derivatives to HPLC column and separate with mobile phase [e.g., 0.1 M imidazole-HNO$_3$ buffer (pH 7.0)/acetonitrile (1:1)] (*see* **Note 8**).
2. Flow 1.0 mL/min of the mobile-phase and 1.5 mL/min of CL reagents solution. The CL solution was vigorously mixed to the mobile-phase outlet of column (*see* **Note 9**).
3. Monitor CL intensity of DBD-APy derivatives (*see* **Note 10**).

3.2. Derivatization and Resolution
of Amine and Amino Acid Enantiomers (8–15)

3.2.1. Derivatization of Amines and Amino Acids

1. Mix 40 μL sample pretreatment solution containing 10 μL IS solution, and 50 μL activation solution (*see* **Note 11**).
2. Add 50 μL chiral reagent solution, and mix thoroughly (*see* **Note 12**).
3. Allow to stand for 20 min at 55°C in the dark (*see* **Note 13**).

3.2.2. Separation of Diastereomers and Detection With FL

1. After the derivatization, inject an aliquot (e.g., 5 μL) of reaction solution to HPLC column and separate with mobile phase (e.g., water/acetonitrile containing 0.1% TFA) (*see* **Note 14**).
2. Gradient elutions are adopted for simultaneous separation of multicomponents, such as a mixture of DL-amino acids (*see* **Note 15**).
3. Monitor the FL intensity at 550 nm (excitation at 460 nm) for DBD-PyNCS derivatives and at 530 nm (excitation at 490 nm) for NBD-PyNCS derivatives (*see* **Note 16**).

3.3. Resolution of Chiral Sequential Analysis of Peptides (16–19)

3.3.1. Derivatization for Chiral Sequential Analysis of Peptide

1. Mix 50 μL peptide solution and 100 μL chiral reagent solution (*see* **Note 17**).
2. Dry the solution in a stream N$_2$ gas.
3. Redissolve the residues in 200 μL derivatization solution.
4. Mix thoroughly and heat at 65°C for 60 min in the dark (*see* **Note 18**).
5. Extract out nonreacted chiral reagent and hydrophobic substances with reagent extraction solution (*see* **Note 19**).
6. Dry the aqueous solution containing labeled peptide in a stream N$_2$ gas.
7. Add 100 μL TFA to the residues (*see* **Note 20**).
8. Heat at 55°C for 1 min (*see* **Note 21**).
9. Dry the solution in a stream N$_2$ gas.
10. Add 100 μL water and 200 μL AcOEt to the residues.
11. Mix vigorously and centrifuge at 1600g for 2 min.

12. Repeat two times the same extraction procedure.
13. Dry the aqueous phase, and use the residues for the analysis of next cycle (*see* **Note 22**).
14. Dry the organic phase in a stream N_2 gas.
15. Add 100 μL of 50% TFA to the residues.
16. Heat at 65°C for 30 min (*see* **Note 23**).
17. Dry the solution in a stream N_2 gas.
18. Add 100 μL water and 200 μL AcOEt.
19. Mix vigorously and centrifuge at 1600g for 2 min.
20. Repeat two times the same extraction procedure (*see* **Note 24**).
21. Dry the organic phase in a stream N_2 gas.
22. Add 100 μL acetonitrile to the residues and dilute a suitable concentration with acetonitrile.

3.3.2. Separation of Thiohydantoin-Amino Acids and Detection With FL

1. Inject an aliquot (e.g., 10 μL) of diluted acetonitrile solution to HPLC column and separate with mobile phase (e.g., water/acetonitrile containing 0.1% TFA) (*see* **Note 25**).
2. Monitor the fluorescence intensity at 550 nm (excitation at 460 nm) (*see* **Note 26**).

3.4. Derivatization and Resolution of Alcohol Enantiomers (20–22)

3.4.1. Derivatization of Alcohols

1. Mix 0.5 mL chiral alcohol (e.g., 2-hexanol) solution containing 2% pyridine and 0.5 mL chiral reagent solution (*see* **Note 27**).
2. Capped tightly and heat at 80°C for 1 to 2 h in the dark (*see* **Note 28**).
3. Add 1% methylamine solution to the reaction solution for stop the reaction (*see* **Note 29**).
4. Inject an aliquot of the diluted solution to columns.

3.4.2. Separation of Diastereomers and Detection With FL or LIF

1. After dilution, inject an aliquot (e.g., 10 μL) of reaction solution to normal phase or reversed-phase column (*see* **Note 30**).
2. Monitor the fluorescence intensity at 580 nm (excitation at 470 nm).
3. Monitor the LIF intensity of NBD-Pro-COCl derivatives with argon-ion laser (*see* **Note 31**).

4. Notes

1. Since the optical purity of chiral derivatization reagents is generally less than 99.5%, strict assay of trace quantity of enantiomer in large amount of antipode is relatively difficult in an indirect method. Thus, the methods described in the text

are recommended for the analysis, such as metabolic study in biological specimens, because the percent coefficient of variation (% CV) is usually in the acceptable range of error.

2. The other reagents, such as DEPC and EDC, are also applicable for the activation of carboxylic acids. The choice is dependent upon the target carboxylic acid.

3. The optical purities of DBD-APy, ABD-APy, and NBD-APy (*R*- and *S*-enantiomers) are 99.8, 99.8, and 99.5%, respectively. According to Ames test using *Salmonella typhimurium* strains TA100 and TA98, no mutagenicities of the reagents are observed with TA98 with and without activation with S9 mix. *RS*-NBD-APys show slight mutagenic activity with TA100. Judging from the results of positive controls, furylfuramide (AF-2) and 2-aminoanthracene (2-AA), the mutagenicities of the reagents including NBD-APy are of very low activity.

4. Chiral carboxylic acid is converted to the corresponding amide diastereomer. The reactivities of *R*- and *S*-DBD-APys to each enantiomer of carboxylic acid (naproxen) are comparable for both enantiomers of DBD-APy. However, slight difference of the FL intensities is observed: the derivative obtained from D-naproxen and *S*-DBD-APy is the highest. Although the reaction time to obtain quantitative yield is different from each carboxylic acid, 1-h reaction at room temperature is usually enough. The labeling at higher temperature sometimes decreases the reaction yield.

5. The *R* values by reversed-phase and normal phase chromatography are 1.62–6.96 and 2.58–7.60, respectively. Although the complete resolutions of the enantiomers of carboxylic acids are achieved by the normal phase chromatography, the technique may not be suitable for biological specimens because of sample handling difficulties and the use of harmful organic solvents. Use of a micro- or semimicrocolumn is recommended to increase detection sensitivity because of the condensation of derivatized analyte. The use is a general aspect for highly sensitive detection.

6. The emission maxima shift to the shorter wavelength region together with an increase of organic solvent in the mobile phase. The fluorescence intensities also elevate with the increment.

7. The peak areas of the derivatives depend upon the laser source used. Although higher laser power produces higher peaks, the baseline noises also increase. Thus, the detection sensitivity has to be considered as the signal-to-noise ratio. The detection limits of naproxen derivatives of DBD-APy, NBD-APy, and ABD-APy on reversed-phase chromatography are 2.7, 1.0, and 10 fmol, respectively.

8. The CL reaction depends upon reaction pH. The optimum pH is different with each oxalate. The CL intensity obtained from TCPO-H_2O_2 is strongest at pH 7.0, whereas the highest intensity with TDPO-H_2O_2 is at pH 6.5. As the catalytic effect of imidazole in the CL reaction is stronger than that of phosphate buffer, which is commonly used as an eluent, the imidazole buffer is selected for the separation of the derivatives. The purity of imidazole has a great influence on both the level of background emission and the variation of baseline noise. Therefore, high-quality imidazole must be used for the preparation of the buffer. Of course, the purities of water and the organic modifier (CH_3CN) are important for high sensitive detection.

Since halogen ions, such as Cl⁻ and Br⁻, in the mobile phase quench the CL intensity, the pH is adjusted with HNO_3 instead of HCl and HBr. The imidazole concentration is also one of the important factors to obtain high sensitivity.

9. Thorough mixing of the effluent from column outlet and CL reagents provides stable baseline and reproducible peaks. Thus, a rotating mixing device is recommended instead of usual T-type mixer. Stainless-steel tubes of 40 cm × 0.1 mm, i.d. and 100 cm × 0.1 mm, i.d. are used as a reaction delay and damper coils for the CL reagent solution, respectively. The flow rate of the CL reagents solution depends on the reagent concentrations. The total flow rate of the mobile phase and the reagents solution influences the CL intensity because of the difference of the CL reaction time.

10. The order of the detection sensitivity is DBD-APy > ABD-APy > NBD-APy. The detection limits of naproxen derivatives of DBD-APy, ABD-APy, and NBD-APy are 0.5, 1.9, and 15 fmol, respectively.

11. The chiral reagents react with primary and secondary amino functional groups in the presence of TEA to produce the corresponding fluorescent thioureas. Instead of TEA, 0.5% quinuclidine, 0.5% 1,8-diazabicyclo[5.4.0]undecene (DBU), and 0.05 M borate buffer (pH 10.0) are usable as the activation reagent.

12. The optical purities are more than 99.5% for all four reagent enantiomers. These reagents exhibit excellent stability, not only as the solid, but also in solution. No significant degradation is observed in acetonitrile solution after storage for 2 wk at room temperature and 1 mo at 5°C in a refrigerator.

13. The procedure is applicable to the resolution of a pair of enantiomers of primary and secondary amines, including amino acids, peptides, and some drugs (e.g., β-blockers). β-Blockers with the isopropylamino moiety (i.e., oxoprenolol, propranolol, alprenolol, atenolol, indenolol, and pindolol) can be detected at lower levels (16–320 fmol) than β-blockers with *tert*-butylamino moiety (i.e., bupranolol, bucmolol, carteolol, timolol) (1.25–8.0 pmol). The lower sensitivity may be due to low derivatization yield because of steric hindrance of the reaction site of the β-blockers. Consequently, care should be taken when tagging to such hindered compounds.

   A pair of the enantiomers of thiol also react with the reagents in the presence of pyridine to produce the corresponding dithiocarbamate diastereomers *(23,24)*. Several thiols derivatized are efficiently separated by an ODS column with water/acetonitrile containing 0.1% TFA. The R values are in the range 1.05–3.33 for the diastereomers obtained with R-DBD-PyNCS. The detection limits are 0.4–2.4 pmol. Since the chiral reagents label with not only thiol but also amines under similar reaction conditions, care should be taken in the derivatization of analytes containing both thiol and amino functional groups in the structure, because it is possible to transfer the FL moiety from S to N.

14. The R values of amino acids are in the range 0.55–3.57 for the diastereomers obtained from NBD-PyNCS and 0.68–2.57 for those from DBD-PyNCS. The R values obtained from neutral and/or aromatic amino acids are larger than those of

basic and acidic amino acids. The diastereomers corresponding to the *R* configuration elute faster than those of the *S* configuration with *R*-enantiomer of the reagent. The opposite elution order is obtained with the use of the *S*-enantiomers of the reagents.

15. β-Ala and 6-amino-*n*-caproic acid are used as IS for hydrophilic (i.e., His, Arg, Ser, Thr, Gly, Glu, Asp, Ala, and Pro) and hydrophobic (Tyr, Val, Met, Ile, Leu, Phe, Trp, and Lys) amino acids, respectively. Best elution profile for the separation of hydrophilic amino acids is by an isocratic elution with water/30% methanol in acetonitrile (72:28) containing 0.1% TFA as eluent. That of hydrophobic amino acids is by linear gradients using 25 m*M* sodium acetate (pH 5.2) (eluent A) and acetonitrile (eluent B): i.e., a linear gradient from A-B (80:20) to A-B (75:25) for 25 min, a linear gradient from A-B (75:25) to A-B (72:28) for 35 min, followed by a linear gradient from A-B (72:28) to A-B (60:40) for 10 min, and then isocratically at A-B (60:40) for 10 min.

16. The maximal excitation and emission wavelengths of the diastereomers obtained from NBD-PyNCS are 490 and 530 nm, respectively. The diastereomers are also separated by capillary electrophoresis (CE) and determined with LIF detection. The CE running buffer is 25 m*M* acetate buffer (pH 4.0) containing 10 m*M* Triton® X-100 *(15)*.

17. The Edman degradation method for sequential analysis of peptide with the reagent is divided into following four steps, as same as those with phenylisothiocyanate (PITC); (i) labeling of N-terminal amino group; (ii) cleavage of labeled N-terminal amino acid; (iii) cyclization of liberated amino acid to thiazolinone derivative; and (iv) recyclization to thiohydantoin derivative. Since the conversion of unstable thiazolinone to thiohydantoin is very fast, the reaction from step (ii) to step (iv) simultaneously proceed in short time.

18. The thiocarbamoyl-peptide is produced in the reaction. No racemization occurrs during the reaction.

19. The chiral reagent should be completely removed in this extraction.

20. TFA is convenient as an acid source for cleavage and cyclization because of the high volatility under normal temperature and atmospheric pressure.

21. The N-terminal amide bond of peptide is cleaved, and the thiocarbamoyl derivative of N-terminal amino acid is liberated with the reaction.

22. The aqueous phase contains the residual peptide that is an N-terminal amino acid shorter than former peptide. The yield in next cycle depends upon the recovery ratio of the peptide. The yields at every step (e.g., yield of labeling, efficiency of the extraction of thiohydantoins, and loss of peptide during the extraction) dictate the detection limit and measurable cycle numbers of analyte peptide.

23. The thiocarbamoyl-amino acid is converted to the corresponding thiohydantoin derivative of N-terminal amino acid via the thiazolinone. Long heating time not only decreases the production, but also increases the racemization. The degree of racemization depends on the stereo structure of the peptide analyzed. It is not easy to predict the degree of racemization ratios of amino acids in every peptide.

Although the use of BF_3 (a Lewis acid), instead of TFA, decreases the racemization, overall yield of the reaction is less than that of TFA.

24. The thiohydantoin-amino acid is efficiently extracted with AcOEt. In the case of Edman degradation method using PITC, the thiohydantoin-amino acids derived from hydroxy amino acids, such as Ser and Thr, are easily oxidized during the conversion reaction. However, the derivatives in the procedures using DBD-PyNCS retain the structure of thiohydantoins without oxydation.

25. The simultaneous separation of all pairs of DL-amino acids is difficult by single chromatographic run using gradient elutions. Therefore, optimal isocratic elutions toward each pair are used for the resolution of DL-amino acids.

26. The proposed degradation method using DBD-PyNCS is also adopted to autoanalysis by gas phase sequencer. The separation and the detection (UV 254 nm) conditions of the derivatives are possible to use without any change from those for the method using PITC as the tagging reagent. However, FL and LIF detection are recommended for the resolution of trace amount of DL-amino acid residues in peptide.

27. The chiral reagents are fairly stable as solids. Although the acyl chloride group exhibits excellent reactivity with alcohols, the reagents are also reactive with moisture. Thus, the reagents must be prepared just prior to use. The derivatization in a hydrophobic solvent gives higher yield than in hydrophilic solvent (benzene > tetrahydrofuran [THF] > acetonitrile). The benzene used in the reaction should be thoroughly dried with suitable reagents such as molecular sieves and sodium wire. Although quinuclidine and TEA also accelerate the derivatization reaction, their efficiencies are approximately one-fourth that of pyridine. These reagents are equally reactive with alcohols, amines, phenols, and aromatic amines, whereas carboxylic acids and thiols do not yield the derivative.

28. The derivatization reaction proceeds even in the absence of the pyridine in the medium. However, the reaction rate is accelerated by the addition of pyridine. The derivatives are stable for at least 240 min at 80°C. The reactivities of a pair of chiral reagent enantiomers are essentially the same for both enantiomers of heptan-2-ol, representative as chiral alcohols. The FL intensity (detected as a peak area) of the diastereomer obtained from R-reagent and S-alcohol (or S-reagent and R-alcohol) is slightly higher than that of RR (or SS).

29. The methylamine is added to the reaction medium to scavenge excess amount of the chiral reagent.

30. DBD-Pro-COCl is more suitable than NBD-Pro-COCl by normal phase chromatography. However, the derivatives derived from alcohols and NBD-Pro-COCl can be separated by reversed-phase chromatography. When S-enantiomer is used as the chiral derivatization reagent, the corresponding S-enantiomers of the alcohols elute more rapidly than the R-enantiomers. The R values of the derivatives, obtained from NBD-PyNCS and DBD-PyNCS, by normal phase chromatography are 3.0–4.1 and 3.3–4.5, respectively.

31. The detection limits by the method are in the range of 10–50 fmol. The sensitivity is improved by two-orders of magnitude with LIF detection and attomole level detection is possible.

References

1. Toyo'oka, T. (1999) Derivatization for resolution of chiral compounds, in *Modern Derivatization Methods for Separation Sciences* (Toyo'oka, T., ed.), Wiley & Sons, Chichester, UK, pp. 217–289.
2. Toyo'oka, T. (1996) Recent progress in liquid chromatographic enantioseparation based upon diastereomer formation with fluorescent chiral derivatization reagents. *Biomed. Chromatogr.* **10,** 265–277.
3. Sun, X. X., Sun, L. Z., and Aboul-Enein, H. Y. (2001) Chiral derivatization reagents for drug enentioseparation by high-performance liquid chromatography based upon pre-column derivatization and formation of diastereomers: enantioselectivity and related structure. *Biomed. Chromatogr.* **15,** 116–132.
4. Toyo'oka, T., Ishibashi, M., and Terao, T. (1992) Fluorescence chiral derivatization reagents for carboxylic acid enantiomers in high-performance liquid chromatography. *Analyst* **117,** 727–733.
5. Toyo'oka, T., Ishibashi, M., and Terao, T. (1992) Resolution of carboxylic acid enantiomers by high-performance liquid chromatography with highly sensitive laser-induced fluorescence detection. *J. Chromatogr.* **625,** 357–361.
6. Toyo'oka, T., Ishibashi, M., and Terao, T. (1992) Resolution of carboxylic acid enantiomers by high-performance liquid chromatography with peroxyoxalate chemiluminescence detection. *J. Chromatogr.* **627,** 75–86.
7. Toyo'oka, T., Ishibashi, M., and Terao, T. (1993) Further studies for the resolution of carboxylic acid enantiomers by high-performance liquid chromatography with fluorescence and laser-induced fluorescence detection. *Anal. Chim. Acta* **278,** 71–81.
8. Toyo'oka, T. and Liu, Y.-M. (1995) Development of optically active fluorescent "Edman-type" reagents. *Analyst* **120,** 385–390.
9. Toyo'oka, T. and Liu, Y.-M. (1995) Resolution of amino acid enantiomers by high-performance liquid chromatography with fluorescent chiral Edman reagents. *J. Chromatogr. A* **689,** 23–30.
10. Jin, D., Nagakura, K., Murofushi, S., Miyahara, Y., and Toyo'oka, T. (1998) Total resolution of 17 DL-amino acids labelled with a fluorescent chiral reagent, *R*(-)-4-(3-isothiocyanatopyrrolidin-1-yl)-7-(*N,N*-dimethylaminosulfonyl)-2,1,3-benzoxadiazole, by high-performance liquid chromatography. *J. Chromatogr. A* **822,** 215–224.
11. Jin, D., Miyahara, Y., Oe, T., and Toyo'oka, T. (1999) Determination of D-amino acids labelled with fluorescent chiral reagents, *R*(-) and *S*(+)-4-(3-isothiocyanatopyrrolidin-1-yl)-7-(*N,N*-dimethylaminosulfonyl)-2,1,3-benzoxadiazole, in biological and food samples by liquid chromatography. *Anal. Biochem.* **269,** 124–132.
12. Toyo'oka, T., Toriumi, M., and Ishii, Y. (1997) Enantioseparation of β-blockers labelled with a chiral fluorescent reagent, *R*(-)-DBD-PyNCS, by reversed-phase liquid chromatography. *J. Pharm. Biomed. Anal.* **15,** 1467–1476.
13. Liu, Y.-M. and Toyo'oka, T. (1995) Determination of D- and L-amino acid residues in peptides with fluorescent chiral tagging reagents by high-performance liquid chromatography. *Chromatographia* **40,** 645–651.

14. Liu, Y.-M., Miao, J.-R., and Toyo'oka, T. (1995) Enantiomeric separation of di- and tripeptides with chiral fluorescence labelling reagents by liquid chromatography. *Anal. Chim. Acta* **314,** 169–173.

15. Liu, Y.-M., Schneider, M., Sticha, C. M., Toyo'oka, T., and Sweedler, J. V. (1998) Separation of amino acid and peptide stereoisomers by nonionic micelle-mediated capillary electrophoresis after chiral derivatization. *J. Chromatogr. A* **800,** 345–354.

16. Toyo'oka, T., Suzuki, T., Watanabe, T., and Liu, Y.-M. (1996) Sequential analysis of DL-amino acid in peptide with a novel chiral Edman degradation method. *Anal. Sci.* **12,** 779–782.

17. Suzuki, T., Watanabe, T., and Toyo'oka, T. (1997) Descrimination of DL-amino acid in peptide sequences based on fluorescent chiral derivatization by reversed-phase liquid chromatography. *Anal. Chim. Acta* **352,** 357–363.

18. Toyo'oka, T., Tomoi, N., Oe, T., and Miyahara, T. (1999) Separation of 17 DL-amino acids and chiral sequential analysis of peptides by reversed-phase liquid chromatography after labeling with *R*(-)-4-(3-isothiocyanatopyrrolidin-1-yl)-7-(*N,N*-dimethylaminosulfonyl)-2,1,3-benzoxadiazole. *Anal. Biochem.* **276,** 48–58.

19. Toyo'oka, T., Jin, D., Tomoi, N., Oe, T., and Hiranuma, H. (2001) *R*(-)-4-(3-isothiocyanatopyrrolidin-1-yl)-7-(*N,N*-dimethylaminosulfonyl)-2,1,3-benzoxadiazole, a fluorescent chiral tagging reagent: sensitive resolution of chiral amines and amino acids by reversed-phase liquid chromatography. *Biomed. Chromatogr.* **15,** 56–67.

20. Toyo'oka, T., Ishibashi, M., Terao, T., and Imai, K. (1993) 4-(*N,N*-dimethyl-aminosulfonyl)-7-(2-chloroformylpyrrolidine-1-yl)-2,1,3-benzoxadiazole: novel fluorescent chiral derivatization reagents for the resolution of alcohol enantiomers by high-performance liquid chromatography. *Analyst* **118,** 759–763.

21. Toyo'oka, T., Liu, Y.-M., Hanioka, N., Jinno, H., and Ando, M. (1994) Determination of hydroxyls and amines, labelled with 4-(*N,N*-dimethylaminosulfonyl)-7-(2-chloroformylpyrrolidine-1-yl)-2,1,3-benzoxadiazole, by high-performance liquid chromatography with fluorescence and laser-induced fluorescence detection. *Anal. Chim. Acta* **285,** 343–351.

22. Toyo'oka, T., Liu, Y.-M., Hanioka, N., Jinno, H., Ando, M., and Imai, K. (1994) Resolution of enantiomers of alcohols and amines by high-performance liquid chromatography after derivatization with a novel fluorescent chiral reagent. *J. Chromatogr. A* **675,** 79–88.

23. Jin, D., Takehana, K., and Toyo'oka, T. (1997) Chiral separation of racemic thiols based on diastereomer formation with a fluorescent chiral tagging reagent by reversed-phase liquid chromatography. *Anal. Sci.* **13,** 113–115.

24. Jin, D. and Toyo'oka, T. (1998) Indirect resolution of thiol enantiomers by high-performance liquid chromatography with a fluorescent chiral tagging reagent. *Analyst* **123,** 1271–1277.

11

Separation of Racemic Trans-Stilbene Oxide by Sub-/Supercritical Fluid Chromatography

Leo Hsu, Genevieve Kennedy, Gerald Terfloth

1. Introduction

1.1. Background

Chromatographic methods are commonly used in the pharmaceutical environment for the qualitative and quantitative analysis of raw materials, active pharmaceutical ingredient, drug products, and compounds in biological fluids. Regulatory requirements *(1)* mandate that "the stereoisomeric composition of a drug with a chiral center should be known and the quantitative isomeric composition of the material used in pharmacologic, toxicologic, and clinical studies known. Specifications for the final product should assure identity, strength, quality, and purity from a stereochemical viewpoint."

1.2. Characteristics and Advantages of Super-Critical Fluid Chromatography (see Note 1)

The separation of chiral compounds by sub- and supercritical fluid chromatography has been a field of great progress since the first demonstration of a chiral separation by supercritical fluid chromatography (SFC) by Mourier et al. in 1985 *(2)*. Easier and faster method development, high efficiency, superior and rapid separations of a wide variety of analytes, extended-temperature capability, analytical and preparative-scale equipment improvements, and a selection of detection options have been reported *(3,4)*. Subcritical fluid chromatography (subFC), SFC, and enhanced fluidity chromatography are commonly used terms to describe the use of mobile phases operated near or above the critical temperature and pressure parameters. Chiral SFC has been reported in packed

From: *Methods in Molecular Biology, Vol. 243: Chiral Separations: Methods and Protocols*
Edited by: G. Gübitz and M. G. Schmid © Humana Press Inc., Totowa, NJ

column, open-tubular column, packed capillary, and ion-pair modes *(5–7)*. Operating conditions typically are mild, at temperatures below 40°C, affording long column lifetime (>2 yr) and highly reproducible separations.

Virtually all chiral separations by subFC/SFC published have used carbon dioxide as the primary mobile phase component. The advantages of using carbon dioxide as a mobile phase component have long been recognized. Carbon dioxide, when compared with most commonly used organic solvents, is environmentally friendly and has a viscosity that is about one order of magnitude less than that of water (0.93 centiPoise [cP] at 20°C), allowing for high flow rates and low pressure drops. In addition, diffusion coefficients of dissolved compounds are increased by one order of magnitude, resulting in high efficiency separations due to improved mass transfer. The eluent strength can be varied by controlling the density of the mobile phase through adjustments in pressure and temperature. A wider polarity range becomes available by adding organic modifiers, such as alcohols, and additives, such as acids and bases. Binary or ternary mobile phases are commonly used.

1.3. Method Development

Chiral method development can be automated using commercially available equipment, greatly reducing the time requirement to identify the best chiral stationary phase (CSP)/mobile phase combination. CSPs have been prepared by modifying compounds from the chiral pool, such as amino acids or alkaloids, by the derivatization of polymers, such as peptides, proteins, and carbohydrates, by bonding of macrocycles, or are based on synthetic selectors, such as Pirkle-phases, poly(meth)acrylates, polysiloxanes, polysiloxane copolymers, and imprinted polymers. The selectors typically are coated and/or bonded to a pressure-stable support, such as silica *(5,6)*.

If partial selectivity is observed after the first injection, it is advisable to first adjust the modifier concentration. If the peak shape is not satisfactory, then the addition of 0.1% acetic or trifluoroacetic acid or 0.1% Huenig's base, diethyl or triethyl amine to the modifier can bring an improvement. In case the selectivity cannot be improved by the previous measures, decreasing the operating temperature can result in the desired separation because of the effect on the eluent density and elution strength. If all of these adjustments should fail, a different CSP should be investigated. Due to the low viscosity of carbon dioxide-based mobile phases, multiple columns can be coupled without a significant change in column backpressure. This provides the opportunity to increase chemical selectivity for the analysis of complex samples by coupling an initial achiral column with a chiral column. Also, the successful coupling of multiple chiral columns has been reported *(7,8)*.

2. Materials

2.1. Instrumentation

All measurements were obtained with:

1. Berger analytical SFC system (Berger Instrument, Newark, DE, USA).
2. Agilent diode array detector (Agilent, Palo Alto, CA, USA).
3. JMBS chromatographic software-SFC PRONTO (version 1.5.305.15) was used for system control, data collection, and analysis (JMBS, Newark, DE, USA).
4. The sampling rate was 5 points/s and the samples were injected a minimum of 5 times.

2.2. Column, Reagents, and Solutions

1. Chromatographic separations were carried out on:
   250 × 4.6 mm inner diameter (i.d.) ChiralPak AD column (Chiral Technologies, Exton, PA, USA).
2. SFC grade CO_2 (PRAXAIR, Danbury, CT, USA) and high-performance liquid chromatography (HPLC)-grade methanol (J. T. Baker, Phillipsburg, NJ, USA) were used as mobile phase components.
3. A 20 mg/mL *trans*-stilbene oxide stock solution was prepared by accurately weighing about 500 mg of *trans*-stilbene oxide (98% pure; Aldrich, Milwaukee, WI, USA) into a 20-mL volumetric flask. Ten milliliters of methanol were added to totally dissolve the compound, and the solution was diluted to vol with methanol.
4. The stock solution was stored at 4°C and protected from light.

3. Methods

1. The elution profile was 30% methanol in carbon dioxide under isocratic conditions for 3 min.
2. Separation was performed on a 250 × 4.6 mm i.d. ChiralPack AD column (*see* **Note 2**).
3. Injection vol was 2 µL (*see* **Note 3**).
4. The flow rate was set at 2 mL/min.
5. Outlet pressure was set at 150 bars.
6. Column oven temperature was set at 40°C.
7. UV detection at 254 nm was used because it provided the best detection response for the concentration range under investigation (*see* **Notes 4–6**).
8. Standard dilutions of the stock solution were prepared as needed using volumetric glassware and methanol as diluent (*see* **Note 7**).

4. Notes

1. This chapter describes an efficient, sensitive and specific SFC method for a chiral separation of *trans*-stilbene oxide. This compound was chosen as it is easily available and practitioners should be in the position to reproduce the results reported.

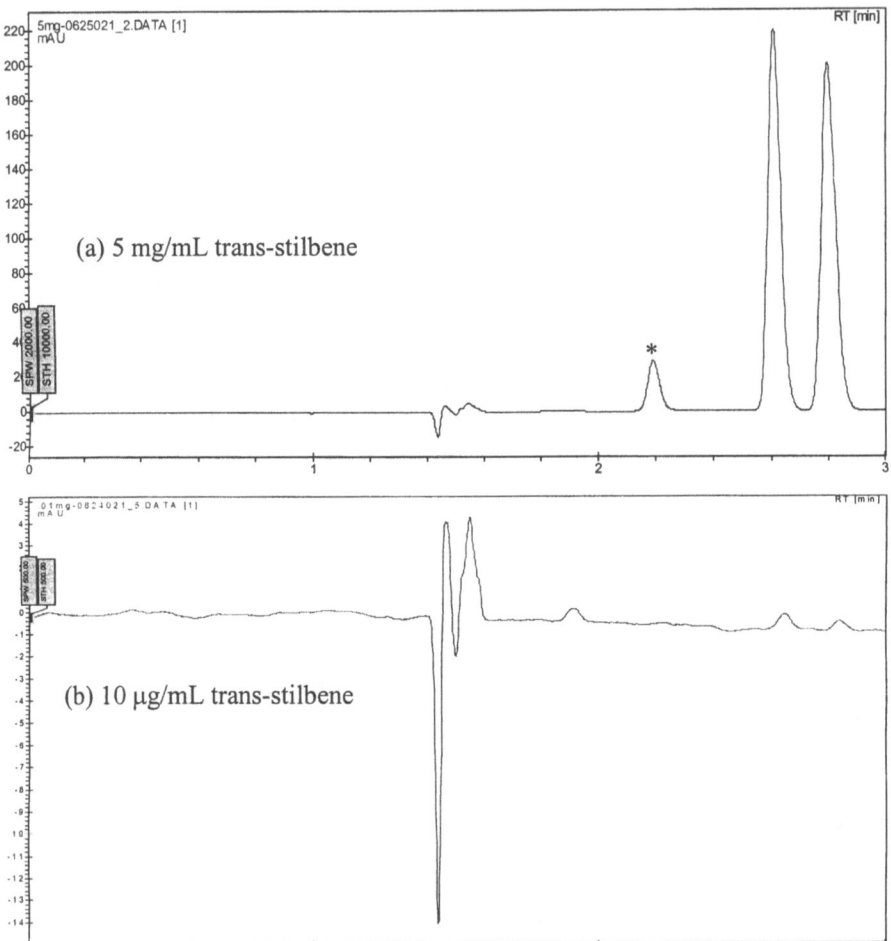

Fig. 1. Chromatogram of 5 mg/mL and 10 µg/mL *trans*-stilbene oxide standard. SFC conditions: injection vol, 2 µL; column temperature, 40°C; Berger Analytical SFC with an Agilent diode array detector; mobile phase, 30% (v/v) methanol/CO_2; run time, 3 min; detection, UV at 254 nm. *Indicates a chemical impurity.

2. After the column was installed and checked for any leaks (frozen condensation at the source of the leak) following the pressurization of the system, the system was equilibrated with the mobile phase for 5 min prior to sample injections.
3. For accurate quantitative analysis, the sample loop needs to be primed prior to any injection to ensure no air bubbles are present in the sample loop.
4. Since enantiomers have identical physical properties, only the results for the first eluted isomer will be used for the following discussions. The limit of detection and the limit of quantification for racemic *trans*-stilbene oxide were found to be 0.3 and 10 µg/mL, respectively.

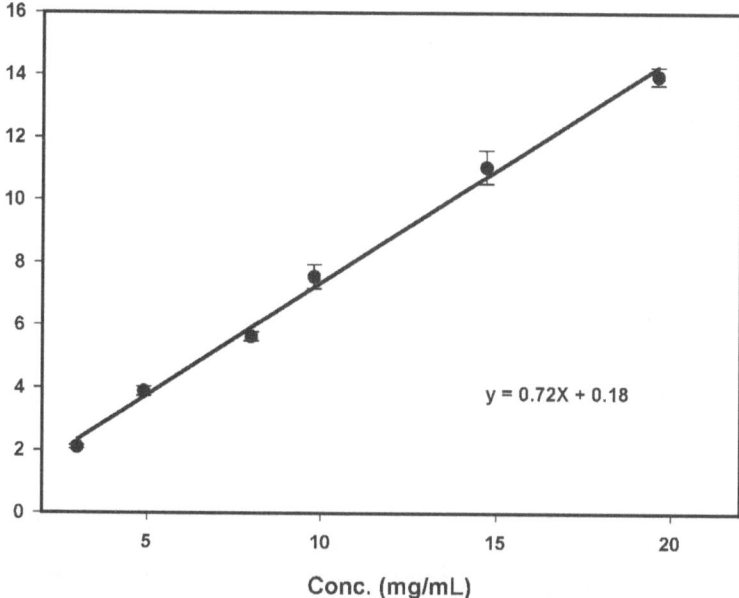

Fig. 2. A 4-d *trans*-stilbene oxide calibration curve from 3 to 20 mg/mL. The fitted equation y = 0.72x + 0.18 with a correlation coefficient of 0.99. The error bar is the 95% confidence interval.

5. The limit of detection (LOD) was determined using Equation 1 derived by Foley and Dorsey *(9)*:

$$LOD = 3s_B/S \qquad \text{[Eq. 1]}$$

Where s_B and S are the standard deviation of the noise and the analytical sensitivity (or calibration factor), respectively. The analytical sensitivity is defined as the slope of the calibration curve (signal output per unit concentration). Based on these calculations, the LOD for *trans*-stilbene oxide at the current separation conditions is 0.3 µg/mL. The limit of quantification (LOQ) is defined as the amount of analyte detected at 10 times the standard deviation of the noise *(10)* and was calculated using Equation 2:

$$LOQ = 10s_B/S \qquad \text{[Eq. 2]}$$

6. Using this calculation, the LOQ for *trans*-stilbene oxide is 10 µg/mL at the current separation conditions. **Figure 1** illustrates the chromatograms of 5.0 mg/mL and 10 µg/mL *trans*-stilbene oxide.
7. An extensive method validation was performed. The calibration curve obtained was linear over the range of 3–20 mg/mL of *trans*-stilbene oxide. Linear regression analysis on the 4-d calibration data provided the equation y = 0.72x + 0.18 and a correlation coefficient greater than 0.99 for trans-stilbene oxide (**Fig. 2**).

Table 1
Accuracy and Precision Data
for *Trans*-Stilbene Oxide[a] in a 4-Day Validation (*see* Note 8)

Parameter	Concentration (mg/mL)				
R.S.D. (%)[b]	3	5	10	15	20
Day 1	1.5	1.6	0.6	3.0	1.1
Day 2	1.6	2.3	0.6	3.0	2.9
Day 3	0.7	2.8	3.7	1.1	0.7
Day 4	6.7	2.1	1.2	5.2	3.3
Error (%)[c]					
Day 1	5.5	3.9	2.5	1.8	1.1
Day 2	9.2	6.0	3.3	1.9	1.4
Day 3	2.9	5.1	5.6	3.9	2.7
Day 4	3.1	5.6	6.4	4.6	3.1
Day-to-day R.S.D.[d]	2.6	2.3	1.5	3.1	2.0
Mean accuracy (%)	94.8	94.9	95.6	97.0	97.9

[a]The first eluted isomer.

[b]%RSD = (Standard deviation [σ] divided by the average) × 100.

[c](Calculated concentration − actual concentration)/actual concentration × 100.

[d]Mean of the daily R.S.D.s.

The accuracy, estimated by the average concentration back calculated from the composite standard calibration curve, was within 6% of the original value at each concentration.

8. **Table 1** summarizes the results obtained from a 4-d validation study in which five replicate standards at six concentrations, 3, 5, 8, 10, 15, and 20 mg/mL, were analyzed each day. The mean accuracy of the assay at these concentrations ranged from 94.8% to 97.9%, whereas the day-to-day precision, indicated by the mean of the daily relative standard deviations (R.S.D.s), varied from 1.5% to 3.1%. The reproducibility of the assay was with within-day precision, indicated by the R.S.Ds of the daily means, ranging 0.6–6.7%

Acknowledgments

The authors would like to thank Dr. M. Kersey for his help in preparing this manuscript.

References

1. FDA (1992) *FDA's Policy Statement for the Development of New Stereoisomer Drugs.* FDA, Rockville, MD.

2. Mourier, P. A., Eliot, E., Caude, M. H., Rosset, R. H., and Tambuté, A. G. (1985) Supercritical and subcritical fluid chromatography on a chiral stationary phase for the resolution of phosphine oxide enantiomers. *Anal. Chem.* **57,** 2819–2823.
3. Berger, T. A. (1995) *Packed Column SFC.* The Royal Society of Chemistry, Cambridge, pp. 22–41.
4. Anton, K. and Berger, C. (1998) *Supercritical Fluid Chromatography with Packed Columns.* Marcel Dekker, New York, pp. 223–249.
5. Terfloth, G. J., Pirkle, W. H., Lynam, K. G., and Nicolas, E. C. (1995) Broadly applicable polysiloxane-based chiral stationary phase for high-performance liquid chromatography and supercritical fluid chromatography. *J. Chromatogr.* **705,** 185–194.
6. Terfloth, G. (2001) Chiral chromatography by subcritical and supercritical fluid chromatography, in *Encyclopedia of Chromatography* (Cazes, J., ed.), Marcel Dekker, New York, pp. 158–160.
7. Pirkle, W. H. and Welch, C. J. (1996) Some thoughts on the coupling of dissimilar chiral columns or the mixing of chiral stationary phases for the separation of enantiomers. *J. Chromatogr. A* **731,** 322–326.
8. Zhang, T. and Francotte, E. (1995) Chromatographic properties of composite chiral stationary phase based on cellulose derivatives. *Chirality* **7,** 425–433.
9. Foley, J. P. and Dorsey, J. G. (1984) Detection of the limit of detection in chromatography. *Chromatographia* **18,** 503.
10. (1980) Guidelines for Data Acquisition and Data Quality Evaluation in Environmental Chemistry. *Anal. Chem.* **52,** 2242–2249.

12

Chiral Separations Using the Macrocyclic Antibiotics in Capillary Electrophoresis

Timothy J. Ward and Colette M. Rabai

1. Introduction

Although capillary electrophoresis (CE) is a relatively new technique as compared to high-performance liquid chromatography (HPLC) or thin-layer chromatography (TLC), it has increased significantly in popularity over the last decade. The increased attention and use of CE for chiral analysis has occurred for several reasons. The narrow bore fused silica capillaries used in CE efficiently dissipates heat, allowing for the use of high voltage that results in rapid and efficient separations. The amount of reagents and materials consumed in CE is minute, resulting in a tremendous reduction in waste disposal. Also, the small amount of waste generated by CE is mostly aqueous buffers and can often be discarded without any danger to the environment. This is a significant advantage over HPLC and TLC, where large volumes of organic solvent waste are generated and must be disposed. For chiral analysis, the resolving agents used in CE are dissolved in the running buffer to affect separation of chiral compounds.

The macrocyclic glycopeptides have been shown to be an effective and powerful resolving agent for chiral anionic solutes *(1–6)*. The macrocyclic glycopeptides are effective chiral selectors in CE for several reasons: (i) they contain ionizable functional groups, which can be either acidic or basic depending on pH; (ii) they have multiple stereogenic centers; (iii) they possess numerous functional groups conducive to stereoselectivity; and (iv) they contain both hydrophobic and hydrophilic groups, making them soluble in water and aqueous buffers and slightly soluble in hydroorganic solvents. Separation of enantiomeric analytes results from the formation of transient noncovalent diastereomeric complexes with the chiral selector *(7)*. Since at least three points of interaction

From: *Methods in Molecular Biology, Vol. 243: Chiral Separations: Methods and Protocols*
Edited by: G. Gübitz and M. G. Schmid © Humana Press Inc., Totowa, NJ

between chiral selector and enantiomer are necessary for separation, one primary and at least two secondary interactions must occur. Multiple interactions can occur between the macrocyclic antibiotic and the analyte as a result of the numerous stereogenic centers and functional groups of the macrocyclic antibiotic. As shown in **Fig. 1A**, vancomycin possesses eighteen stereogenic centers and is composed of three fused macrocyclic rings, which make up the aglycone basket. Attached to the aglycone basket via a single phenolic linkage are two pendant sugar moieties. Vancomycin possesses nine hydroxyl groups, two amine functions, seven amido groups, five aromatic esters, and a carboxylic acid. Ristocetin A consists of four fused macrocyclic rings in its aglycone basket and has six pendant sugar moieties attached **(Fig. 1B)**. Ristocetin A has 38 stereogenic centers, 21 hydroxyl groups, two amine groups, six amido groups, seven aromatic groups, and a methyl ester *(5)*. The primary interactions with analytes are ionic or charge-charge interactions produced by the ionizable functional groups *(4,8)*. Secondary interactions arise from hydrophobic, dipole-dipole, π-π interactions, hydrogen bonding, and steric repulsion *(4,8)*. Although the aromatic rings in the aglycone portion of the compounds are UV absorbing, at the concentrations used (1–5 mM), the background absorption remains relatively low allowing direct detection with good sensitivity. In aqueous solutions at a pH range of 5.0–7.0, vancomycin deteriorates within 2–4 d at room temperature and 6 to 7 d when stored at 4°C *(see* **Note 1**). In aqueous solutions at a pH range of 4.0–7.0, ristocetin A has a relative stability of 1 to 2 wk at room temperature and 3 to 4 wk when stored at 4°C. Both are indefinitely stable in the solid anhydrous form at 0°C *(5,6)*.

As mentioned previously, CE utilizes narrow bore fused silica capillaries, which have silanol groups at the inner surface of the capillary. Therefore, at pH values above approx 2.5, the inside of the capillary is negatively charged due to the deprotonation of the acidic silanol groups. Positive charged ions in the running buffer adsorb to the inner capillary wall. When a potential is applied across the capillary, the positive ions at the capillary surface break away and move towards the cathode dragging with them bulk solution of the electrolyte due to viscous drag. This phenomenon is called electroosmotic flow (EOF). To calculate the electrophoretic mobility of an analyte, the effect of EOF on the solute's mobility must be taken into account. The movement or migration of a charged species under the influence of an applied potential is characterized by its electrophoretic mobility, which generally has units of cm^2/(min) (kV). In the presence of electroosmotic flow, the apparent mobility, μ_a, is the sum of the electrophoretic mobility of the analyte, μ_e, and the electrophoretic mobility of the electroosmotic flow, μ_{eo}:

$$\mu_a = \mu_e + \mu_{eo}$$

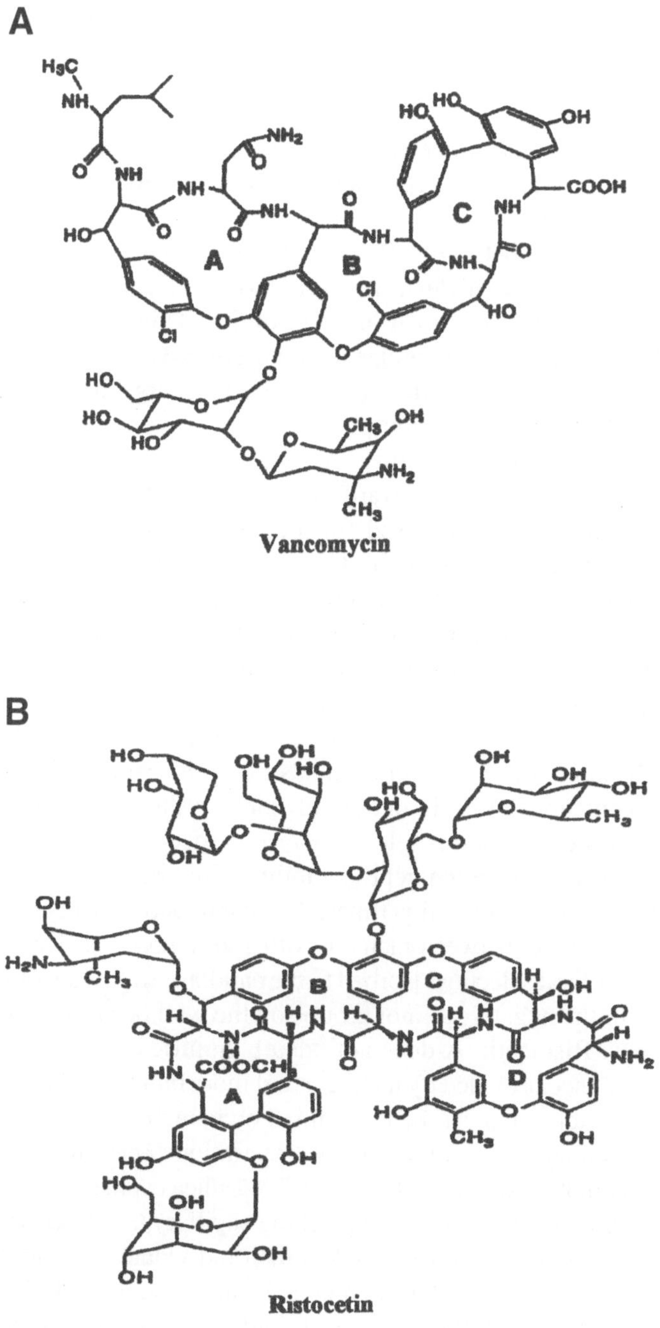

Fig. 1. Structure of (**A**) vancomycin and (**B**) ristocetin A.

The apparent electrophoretic mobility, μ_a, is determined experimentally by the equation:

$$\mu_a = (l)(L)/(V)(t)$$

where l is the length in centimeters to the detection window, L is the total capillary length in centimeters, V is applied potential in volts, and t is the migration time in min (*see* **Note 2**).

At operational pHs between 5.0–7.0, the charge on the glycopeptides is positive resulting in a positive electrophoretic mobility. Vancomycin's isoelectric point, pI, is 7.2, and ristocetin A has a pI of 7.5. Thus, in a buffer whose pH value is at the pI, the glycopeptides have no effective mobility. At pH values below their pI, the glycopeptides have a positive charge and a positive mobility, while at pH values above their pI, they have an overall negative charge and a negative electrophoretic mobility. A compound possessing a positive electrophoretic mobility migrates toward the capillary outlet and has its migration through the capillary superimposed or added to the EOF. A compound possessing a negative electrophoretic mobility migrates towards the inlet, opposite the EOF. Since the EOF is generally much greater than the solute's electrophoretic mobility, the solute still moves toward the outlet, though with a longer migration time than a solute with a positive mobility.

The resulting EOF in fused silica capillaries requires considerations besides its effect on a solute's apparent mobility. The positively charged macrocyclic antibiotics used as chiral selectors can adsorb to the capillary wall. This interaction between capillary wall and chiral selector should be minimized, since wall adsorption of the chiral selector causes a decrease in efficiency, chiral selector-analyte complexation, reproducibility of migration times, detection sensitivity, and increase in migration times. These interactions can be minimized by flushing the capillaries between each run with a strong base, such as 0.1 M NaOH, to displace the chiral selector adsorbed to the capillary wall. Vancomycin adsorbs appreciably to the negative silinol groups on the wall because of its protonated amine groups. Ristocetin A does not adsorb significantly, most likely due to the steric hindrance provided by its larger and more bulky pendant sugar groups.

Wall interactions also can be minimized by using coated capillaries, which have become an attractive alternative to separations performed with uncoated capillaries. Available commercially, these fused silica capillaries have been derivatized with a polymer to coat the inside of the capillary (*see* **Note 3**). The coating suppresses EOF, thus ions move based on their individual electrophoretic mobility and are not affected by EOF. In a coated capillary, the polarity of the electrodes are reversed, with the anode at the inlet and the cathode located at the capillary end. In this technique, the positively charged chiral selector migrates toward the capillary inlet, while the anionic analyte migrates toward the cathode or outlet. This separation method, in which the chiral selector and analyte migrate in opposite directions, is beneficial for several reasons. Since the cationic chiral selector is attracted to the inlet cathode, the UV absorbing chiral selector passes the detection window before the analyte migrates into the detection cell window. The analyte is separated efficiently in the length between the inlet and the window, and the detection of the analyte is more sensitive, because no background absorbance of the macrocyclic antibiotic occurs. This is the basis for the partial filling method, which utilizes a coated capillary in a countercurrent process *(9,10)*.

The glycopeptide macrocyclic antibiotics have been shown to separate a wide variety of compounds with high efficiency. Acidic or anionic analytes are best suited for separation with the glycopeptides vancomycin and ristocetin A. Though vancomycin and ristocetin A behave similarly because of their characteristic structure, they differ with respect to cost and enantioselectivity for a number of analytes. Classes of compounds that have been separated effectively by these chiral selectors include nonsteroidal anti-inflammatory drugs, antineoplastics, *N*-blocked amino acids, lactic acids, herbicides, and rodenticides, to name a few classes.

2. Materials

2.1. Preparation of Capillary

2.2.1. Preparation of Virgin Capillary

1. Fused silica capillary, 50 µm (inner diameter).
2. Acetone.
3. 0.1 M NaOH.

2.2.2. Preparation of Coated Capillary

1. Fused silica capillary coated with polymer, 50 µm (inner diameter).

2.2. Buffer

1. The aqueous buffer solutions are prepared by adjusting the pH of a solution containing the appropriate amount of sodium phosphate monobasic with NaOH. Prepare a 0.1 *M* phosphate buffer by dissolving 0.69 g of sodium dihydrogen phosphate (NaH_2PO_4) in 50 mL of deionized water.
2. Sonicate solution to dissolve.
3. Using a pH meter (*see* **Note 4**), add 0.1 *M* NaOH dropwise until the pH of the buffer (*see* **Note 5**) is 6.00 ± 0.01 (*see* **Note 6**).
4. Filter the buffer solution using the 0.45-µm nylon filter (*see* **Note 7**).

2.3. Chiral Selector in Run Buffer

1. Run buffers containing vancomycin are prepared by weighing the proper amount of the macrocyclic antibiotic into a volumetric flask, adding the phosphate buffer,

and sonicating to dissolve the antibiotic. Most separations can be achieved using 2 mM chiral selector.

2. The molecular weight of vancomycin is 1449 g/mol; therefore add 0.29 g of vancomycin to a 100-mL volumetric flask and dilute with prepared sodium phosphate buffer, pH 6.0 (*see* **Note 8**).
3. Sonicate solution for 1–3 min to dissolve minute particles.
4. Filter the chiral run buffer using the 0.45-µm nylon filter (*see* **Note 7**).

2.4. Analyte

Dissolve 0.10 mg of analyte in a 1.0-mL volumetric flask, diluting with deionized water (*see* **Note 9**).

3. Methods

3.1. Preparation of Capillary

3.1.1. Preparation of Virgin Capillary

1. Cut the capillary according to the recommended length in the instrument manual, typically, 30 cm total length, 25 cm to detection window (*see* **Note 10**).
2. Burn a detection window 1 to 2 cm in length by passing the capillary over the top of a flame about 5 to 6 cm above the burner itself for 3 to 4 s. The capillary polyimide coating should turn from dark brown to black in appearance.
3. Remove the charred polyimide coating by carefully and gently rubbing a Kimwipe, moistened with acetone, over the burned area.
4. To condition the column, flush the capillary with 0.1 M NaOH and leave it overnight (*see* **Note 11**).

3.1.2. Preparation of Coated Capillary

1. Cut the capillary according to the recommended length in the instrument manual, typically, 30 cm total length, 25 cm to detection window (*see* **Note 10**).
2. Burn a detection window 1 to 2 cm in length by passing the capillary over the top of a flame about 5 to 6 cm above the burner itself for 3 to 4 s. The capillary polyimide coating should turn from dark brown to black in appearance.
3. Remove the charred polyimide coating by carefully and gently rubbing a Kimwipe, moistened with acetone, over the burned area.
4. To condition the column, flush the capillary with water followed by the running buffer (*see* **Note 3**).

Capillary electropherograms demonstrating the separation of racemic dansylvaline on an uncoated and coated capillary are shown in **Fig. 2**. Initial starting conditions of 2 mM chiral selector in 0.1 M phosphate buffer at a pH value between 5.0 and 6.0, will usually result in a successful separation. If necessary, separations may be further optimized by adjusting the concentration of chiral selector, since it has the greatest impact on enantioresolution. In general, increasing

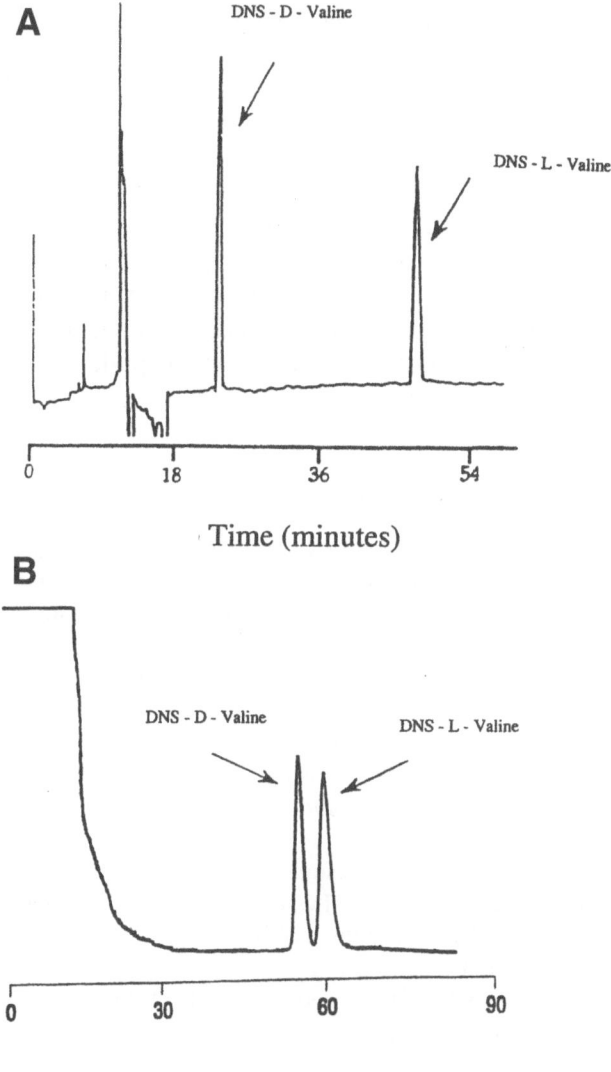

Fig. 2. Capillary electropherograms showing the separation of racemic dansyl-valine on (**A**) an uncoated 50-μm inner diameter fused silica column. Conditions: 50 μm × 32.5 cm (25 cm to cell window) capillary, 0.1 M phosphate buffer, pH 4.9, and 5 mM vancomycin, voltage was +5 kV, UV absorbance at 254 nm; (**B**) a polyacrylamide-coated 50-μm inner diameter fused silica column. Conditions: 50 μm × 60 cm (32 cm to cell window) capillary, 0.1 M phosphate buffer, pH 7.0 and 5 mM vancomycin, voltage was −10 kV, UV absorbance at 254 nm (*see* **Note 12**). Panel **A** reprinted from *Chirality* **6,** 496–509 (1994) by permission of Wiley-Liss, Inc., a subsidiary of John Wiley & Sons, Inc.

chiral selector concentration results in an increase in enantioresolution as well as a small increase in migration time for the analyte. If further optimization is required after increasing the chiral selector concentration, pH can also be adjusted to improve separations. Optimization may be achieved by increasing or decreasing the pH of the running buffer by 0.5 U at time until a satisfactory separation is achieved (*see* **Note 6**). In the event a separation is still unsuccessful after optimizing chiral selector concentration and pH, one should attempt the separation again using the other macrocyclic glycopeptide. It has been shown that the selectivity of vancomycin and ristocetin A are complimentary to one other and a compound not resolved using one glycopeptide has a high probability of being resolved successfully when using the other glycopeptide *(6)*.

4. Notes

1. For the stability of the glycopeptide, store the stock solution in the refrigerator between runs and overnight.
2. While the units generally are $cm^2/(kV)$ (min), the voltage can also be measured in V, and the time can be measured in seconds. This would give the units as $cm^2/(V)$ (s). Regardless of which units are chosen, the measured units from one run to another must be consistent.
3. Do not leave coated capillaries in aqueous solution; the coating is hydrolyzed by water and can be stripped off if stored in water. To prevent degradation, flush the coated capillary with water, followed by methanol or ethanol or air.
4. The pH meter should be calibrated using pH 4.0 and pH 10.0 standards.
5. The sodium phosphate buffer should be approximately pH 4.7, when prepared using sodium dihydrogen phosphate (NaH_2PO_4).
6. If the buffer pH needs further adjustment, use 0.1 M HCl to lower the pH to 6.0 if the pH is too high and 0.1 M NaOH when increasing pH. Buffer pH values above pH 7.5 should be avoided, since the macrocyclic glycopeptides are readilty hydrolyzed in basic solutions.
7. Before filtering the buffer, filter approx 10 mL of deionized water in order to remove any old solutions that may remain if the filter has been used previously to filter the buffer solution. Next, remove the filter and remove excess water by filling the syringe with air and force out any water remaining in the filter. Then pull buffer into the syringe and replace the filter. Allow the first few drops of filtered buffer to go to waste and retain the rest of the buffer in a clean dry container.
8. The molecular weight of ristocetin A is 2066 g/mol; therefore, 0.41 g would be necessary for 100 mL. The number of grams of chiral selector needed can be calculated from the product of the molecular weight of the chiral selector, the desired molarity, and the volume needed, i.e., grams needed = MW × M × V. The salt associated with the chiral selector also must be noted, since this will alter the MW of the desired chiral selector. For example, vancomycin • hydrochloride has a molar mass of 1485.

9. A concentration of 0.1 mg/mL can be detected and resolved in most cases; however, initially, 1.0 mg/mL can be tested in order to determine chemical and instrumental parameters. If the analyte is slightly insoluble in water, add a drop or two of methanol first to affect dissolution and then dilute with deionized water.

10. For illustrative purposes instructions are based on using the Bio-Rad BioFocus 3000 (Bio-Rad, Hercules, CA, USA) and Waters Quanta 4000 (Waters, Milford, MA, USA), which use capillaries 30 cm in length and 25 cm to the detector and capillaries 32.5 cm in length and 25 cm to the detector, respectively. For appropriate dimensions with other instruments, consult the instrument manual.

11. Conditioning the column overnight ensures that the walls of the column are uniform, thus ensuring reproducibility. After conditioning overnight, flush the capillary thoroughly with deionized water, followed by running buffer before beginning a separation.

12. The number of theoretical plates can be determined experimentally by the equation $N = 16 \, (t_r / W)^2$, where N is the number of theoretical plates, t_r is the migration time in min, and W is the width of the peak in min.

References

1. Ward, T. J. (2000) Chiral separations. *Anal. Chem.* **72,** 4521–4528.
2. Gubitz, G. and Schmid, M. G. (1997) Chiral separation principles in capillary electrophoresis. *J. Chromatogr. A.* **792,** 179–225.
3. Verleysen, K. and Sandra, P. (1998) Separation of chiral compounds by capillary electrophoresis. *Electrophoresis* **19,** 2798–2833.
4. Gasper, M. P., Berthod, A., Nair, U. B., and Armstrong, D. W. (1996) Comparison and modeling study of vancomycin, ristocetin A, and teicoplanin. *Anal. Chem.* **68,** 2501–2514.
5. Armstrong, D. W. and Nair, U. B. (1997) Capillary electrophoretic separations using macrocyclic antibiotics as chiral selectors. *Electrophoresis* **18,** 2331–2342.
6. Ward, T. J. and Farris, A. B. (2001) Chiral separations using the macrocyclic antibiotics: a review. *J. Chromatogr. A.* **906,** 73–89.
7. Chankvetadze, B. (1999) Recent trends in enantioseparations using capillary electromigration techniques. *Trends Anal. Chem.* **18,** 485–498.
8. Nair, U. B., Chang, S. S. C., Armstrong, D. W., Rawjee, Y. Y., Egglester, D. S., and McArdle, J. V. (1996) Elucidation of vancomycin's enantioselective binding site using its copper complex. *Chirality* **8,** 590–595.
9. Ward, T. J. and Oswald, T. M. (1997) Enantioselectivity in capillary electrophoresis using the macrocyclic antibiotics. *J. Chromatogr. A.* **792,** 309–325.
10. Ward, T. J., Dann, C., III, and Brown, A. P. (1996) Separation of enantiomers using vancomycin in a countercurrent process by suppression of electroosmosis. *Chirality* **8,** 77–83.

13

Enantioresolutions by Capillary Electrophoresis Using Glycopeptide Antibiotics

Salvatore Fanali

1. Introduction

A wide number of compounds belonging to the pharmaceutical, biochemical, environmental, and agrochemical fields, due to the presence in their chemical structure of one or more stereogenic center, exhibit two or more optical isomers (at least two enantiomers). Very often, the two enantiomers of a certain compound can exhibit very different biological and/or pharmacological properties, e.g., (-)-epinephrine is a sympathomimetic drug currently used for cardiac stimulation and is 10 times more potent than its enantiomer *(1)*. Consequently, the control (qualitative and quantitative) of chiral compounds is a very important topic in the different fields, because health and pollution are, directly or indirectly, strongly involved. Therefore, analytical methods possessing high resolution capability, high efficiency, and low costs are requested for the separation and quantification of enantiomers present in, e.g., biological fluids, drugs, product of synthetic reactions, etc.

Analytical methods so far employed for the enantiomer separation include gas chromatography (GC), high-performance liquid chromatography (HPLC) and recently, capillary electrophoresis (CE). In the field of chiral separation, CE is recognized as a challenging and powerful tool possessing not only the above mentioned properties, but also allowing the use of minute amounts of expensive chiral selectors mainly using the simple and feasible direct method of enantiomeric separation *(2,3)*.

The list of chiral selectors employed in CE includes: copper-amino acid complexes, chiral micelles, antibiotics, crown ethers, proteins, cyclodextrins (CDs) and their derivatives, etc. Among them, CDs and their derivatives are very popular and were widely investigated in capillary electrophoretic techniques employing

From: *Methods in Molecular Biology, Vol. 243: Chiral Separations: Methods and Protocols*
Edited by: G. Gübitz and M. G. Schmid © Humana Press Inc., Totowa, NJ

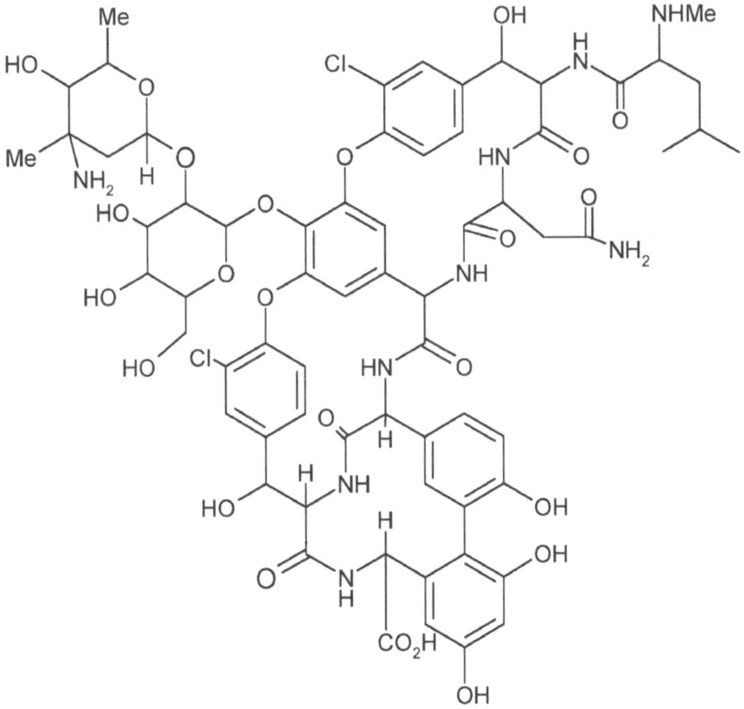

Fig. 1. Chemical structure of vancomycin.

the different modes (free zone electrophoresis [CZE], micellar electrokinetic chromatography [MEKC], isotachophoresis [ITP], capillary gel electrophoresis [CGE], and capillary electrochromatography [CEC].

Macrocyclic antibiotics (MAs), e.g., vancomycin, teicoplanin, ryfamycin, ristocetin, A82846B, MDL 63,246, were used as chiral selectors as buffer additives in CZE for the enantiomeric resolution of a wide number of compounds including chargeable (positive and negative) and uncharged species. Furthermore, some of the above mentioned antibiotics were also used in CEC *(4–6)*.

MAs were firstly introduced by Armstrong as enantioselective agent in HPLC *(7)* and later on widely applied in CE for the separation of enantiomers because their high enantioresolution capability due to their unique chemical structure *(4–6)*. As an example, **Fig. 1** shows the chemical structure of vancomycin containing 18 asymmetric centers and several functional groups such as carboxylic, amino, amido, hydroxyl, aromatic, etc., responsible for the interactions (affinity) with analytes. The three chargeable groups allow the vancomycin molecule to be either charged or uncharged depending on the medium pH. The isoelectric point (pI) of this antibiotic is about 7.2. Due to the presence of three fused macrocyclic rings and two side chains, it seems that a characteristic basket shape is formed, which is responsible for inclusion complexation with enantiomers.

Vancomycin is very soluble in water and/or polar organic solvents, and therefore, it was widely applied in CE resulting an excellent chiral selector for acidic compounds using background electrolytes at acidic pHs (range of pH 3.0–6.0) *(8,9)*.

Although the excellent enantiorecognition capabilities of vancomycin, its use in CE presents some drawbacks, such as strong adsorption on the capillary wall and strong absorption at the common wavelengths used. Therefore, in order to avoid low sensitivity as well as strong peak dispersion, it is necessary to apply some technical strategies. The use of coated capillaries seems to be very useful in order to avoid vancomycin adsorption on the capillary wall; this approach is also minimizing the electroosmotic flow (EOF). Sensitivity improvement can be easily achieved employing two different approaches, namely partial filling-counter current method in CZE and chiral stationary phases containing vancomycin in CEC (*see* **Note 1**). It is noteworthy to mention that vancomycin-modified silica was packed into capillaries and used for the separation of mainly basic enantiomers.

Here, we describe the two different modes of CZE and CEC, which employ vancomycin as the chiral selector for the enantiomeric resolution of acidic or basic compounds.

2. Materials

2.1. Separation of Chiral Compounds by CZE (10)

2.1.1. Preparation of Polyacrylamide-Coated Capillaries

1. Fused silica capillary 50 cm × 50 μm inner diameter (I.D.).
2. Acrylamide.
3. *N,N,N',N'*-tetraethylenediamine (TEMED).
4. Ammonium persulfate.
5. 3-(Trimethoxysilyl)propylmethacrylate 1% (v/v) in 50% water/acetone mixture.

2.1.2. Capillary Zone Electrophoresis

1. Vancomycin.
2. Racemic samples: nonsteroidal anti-inflammatory drugs (ibuprofen, indoprofen, naproxen, ketoprofen, suprofen, carprofen, flurbiprofen, cicloprofen).
3. Background electrolyte: Britton Robinson Buffer (B.R.B): mixture of 50 mM each phosphoric acid, acetic acid, and boric acid titrated with sodium hydroxide 1 *M* pH 4.0–7.0 and diluted with water, 1:1 (v/v) to obtain 75 mM B.R.B.
4. Chiral separation buffer: 75 m*M* B.R.B. containing 2.5 or 5 m*M* vancomycin.

2.2. Separation of Chiral Compounds by CEC (11)

2.2.1. Synthesis of Vancomycin Silica Stationary Phase (see **Fig. 2**).

1. 200 mg LiChrospher diol silica 5 μm particle diameter.
2. 15 mL 60 mM sodium periodate (water/methanol, 4:1, v/v).
3. Mixture 3 mM vancomycin and 10 mM cyanoborohydride, respectively dissolved in 50 mM phosphate buffer, pH 7.04.
4. 10 mM Cyanoborohydride in 50 mM phosphate, pH 3.1.

2.2.2. Preparation of Vancomycin-Packed Capillaries

1. Fused silica capillaries 75 or 100 μm I.D., 365 μm outer diameter (O.D.).
2. 2-μm Silica particles.
3. LiChrospher diol silica 5 μm particle diameter.
4. LiChrospher C60 silica 5 μm particle diameter.

2.2.3. Capillary Electrochromatography

1. Racemic compounds (acebutolol, atenolol, clenbuterol, mefloquine, metoprolol, mianserin, oxprenolol, pindolol, propranolol, terbutaline, tolperisone, venlafaxine).
2. 100 mM Ammonium acetate, pH 4.0–7.0.
3. Methanol.
4. Acetonitrile.

3. Methods

3.1. Capillary Zone Electrophoresis

3.1.1. Preparation of Polyacrylamide-Coated Capillaries

1. The solution reported in **Subheading 2.1.1.** was sucked up into the capillary; after 1 h, the solution was withdrawn and the capillary washed with water.
2. Fill the capillary with 3% (w/v) acrylamide in water containing 0.04% (v/v) TEMED and 0.05% (w/v) ammonium persulfate.
3. After 15–20 min, flush the capillary with water and dry by aspiration.
4. Cut the capillary at the desired length (35 cm) and prepare the window removing the polyimide layer with a razor.

3.1.2. Enantiomeric Separation by CZE

1. Prepare stock standard racemic sample solutions in methanol (10^{-3} M). For injection, the sample solutions were diluted with 7.5 mM B.R.B. (5×10^{-5} M).
2. Flush the capillary with 75 mM B.R.B.
3. Inject at 175 p.s.i. * s the chiral background electrolyte (75 mM B.R.B. containing 2.5 or 5 mM vancomycin) (*see* **Notes 2** and **3**).
4. Inject sample solutions at 10 p.s.i. * s.
5. Apply –18 kV and run the electrophoretic separations achieving the chiral separation of the studied analytes as reported in **Fig. 3** (*see* **Note 4**).

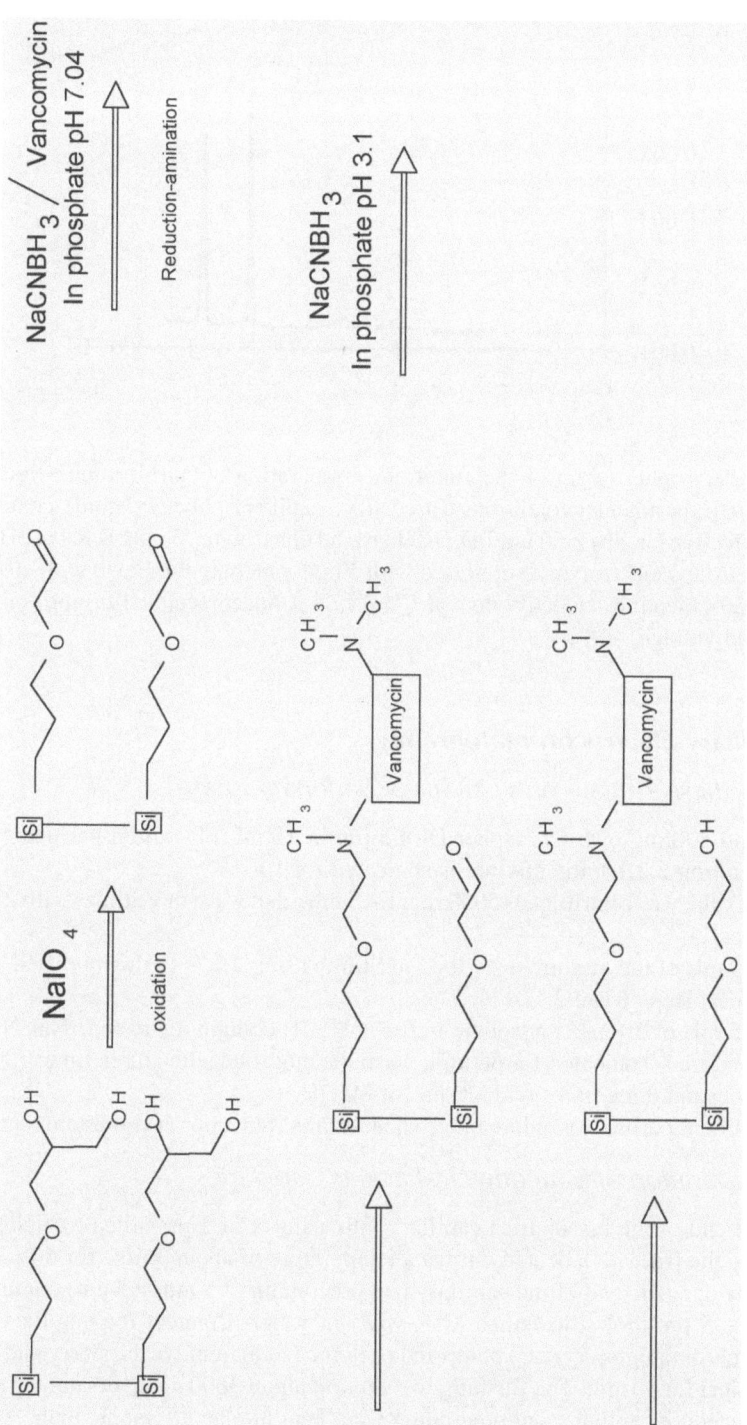

Fig. 2. Scheme of the synthesis of vancomycin silica stationary phase.

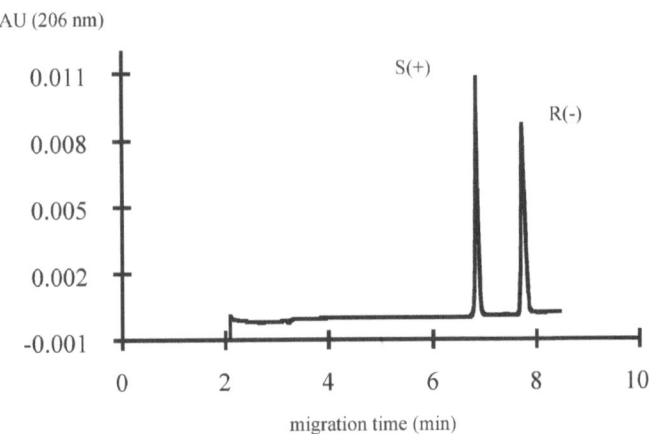

Fig. 3. Electropherogram of the enantiomer separation of flurbiprofen. Modified from **ref. *10***. Experimental conditions: fused silica capillary polyacrylamide-coated 35 cm (31.5 effective length) × 50 μm I.D.; background electrolyte 75 m*M* B.R.B., pH 5.0, chiral background electrolyte a supported with 5 m*M* vancomycin. Flush with electrolyte a for 120 s then inject electrolyte b at 175 p.s.i. * s; inject racemic flurbiprofen, and run. Applied voltage, −18 kV.

3.2. Capillary Electrochromatography

3.2.1. Synthesis of Vancomycin-Silica Stationary Phase

1. Suspend 200 mg of LiChrospher Diol 5 μm in 15 mL of sodium periodate (*see* **Subheading 2.2.1.**); the mixture is sonicated for 1 h.
2. The mixture was centrifuged (5000 rpm for 5 min) and wash three times with 20 mL of water.
3. Add 15 mL of the mixture described in **Subheading 2.2.1.** to the modified silica and repeat steps 1 and 2.
4. Add 15 mL of 50 m*M* phosphate buffer, pH 3.1, containing 10 m*M* NaCNBH$_3$. The mixture is sonicated for 60 min; wash the modified silica three times (20 mL of water) and three times with 20 mL of MeOH.
5. Evaporate the MeOH residue under vacuum in a rotavapor at room temperature.

3.2.2. Preparation of Vancomycin-Packed Capillaries

1. Dip on end of the fused silica capillary into a slurry of 2-μm silica particles and prepare the frit with a heated wire at a temperature of about 350°C for 60 s. Connect the opposite end of the capillary to a precolumn (2.1 mm × 5 cm) containing a slurry 5 μm diol silica/silica (3:1, w/w) in water. Connect the capillary to a liquid chromatography (LC) pump and pack for 15 cm, remove the slurry and flush with water for 30 min. The flushing was done at about 3000 p.s.i.; during the packing procedure capillary and precolumn were kept into an ultrasonic bath.

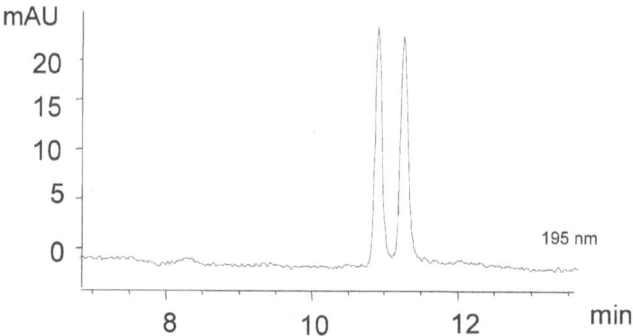

Fig. 4. Electrochromatogram of the enantiomeric separation of propranolol. Modified from **ref. *11***. Experimental conditions: capillary, 35 cm total length, 26.5 cm effective length; cm length of vancomycin stationary phase, 75 μm I.D.; mobile phase, 100 m*M* ammonium acetate, pH 6.0 mixed with 90% acetonitrile, 5 m*M* was the concentration of ammonium acetate. Sample, 0.2 mg/mL of racemic propranolol. Applied voltage, 25 kV, 20°C, pressurized at both sides at 10 bar. Injection, 12 bar × 0.5 min followed by a buffer plug at 12 bar × 0.2 min.

2. Pack an aqueous slurry (20 mg/mL composed by a mixture of vancomycin-silica/silica, 3:1, w/w) as described in **step 1** for 25 cm.
3. Repeat the procedure reported in **step 1** packing the capillary for 5 cm.
4. Flush the capillary with water for 30 min and prepare the two frits with the heating wire at a temperature of about 600°C for 60 s; cut the capillary close to the frits.
5. Remove the polyimide layer with a razor in the capillary zone where vancomycin is not present (8.4 cm from the end).

3.2.3. CEC Experiments Using Vancomycin CSP

1. Prepare stock solution of mobile phase 100 m*M* of acetic acid titrated with ammonia solution (33%) at pH 6.0.
2. Make the mobile phase mixing 1 mL of 100 m*M* ammonium acetate, pH 6.0/1 mL water and 8 mL of acetonitrile.
3. Flush the packed capillary with the mobile phase described in **step 2** using the LC pump at 3000 p.s.i.
4. Make stock standard racemic solutions (1 mg/mL) in methanol and daily diluted at 0.05–0.1 mg/mL with water for the injection.
5. Dip the ends of the capillary into the electrolyte compartments; flush the capillary at 12 bar for 30 min with the mobile phase.
6. Apply 25 kV pressurizing both inlet and outlet vials at 10 bar until a stable current and UV signal are observed (about 15 min).
7. Inject diluted standard racemic mixtures at 12 bar × 0.5 min; inject a mobile phase plug at 12 bar × 0.2 min.
8. Apply 25 kV, cartridge temperature at 20°C, inlet and outlet vials pressurized during the run at 10 bar and run the experiment (see **Fig. 4**).

4. Notes

1. Vancomycin is a potent antibiotic that can be harmful if inhaled; contact with skin must be also avoided.
2. The partial filling-counter current method is adopted in order to achieve good sensitivity. Selecting the appropriate background electrolyte pH (lower than the pI of the chiral selector, pI ≅ 7.2), analytes and vancomycin are moving in the opposite direction. Since the chiral selector is filling only part of the capillary, analyzed anions are detected with good sensitivity because no absorbing material is present in the path length.
3. The injection pressure is that used with a BioFocus® 3000 (Bio-Rad, Hercules, CA, USA). Employing other instrumentation preliminary experiments are necessary in order to find the optimum pressure as following: (i) flush with background electrolyte; (ii) inject the vancomycin-background electrolyte at a certain pressure measuring the time necessary to have the increase of the detector signal due to the vancomycin absorption; and (iii) select pressure/time where the vancomycin zone is not present at the path length of the detector.
4. The analytes are moving towards the anode as anions.

References

1. Innes, I. R. and Nickersen, M. (1970) *The Pharmacological Basis of Therapeutics*, in (Goodman, L. S. and Gilman, A., eds.), MacMillan Publishing, New York, p. 477.
2. Fanali, S. (1997) Controlling enantioselectivity in chiral capillary electrophoresis with inclusion-complexation. *J. Chromatogr. A* **792,** 227–267.
3. Fanali, S. (2000) Enantioselective determination by capillary electrophoresis with cyclodextrins as chiral selectors. *J. Chromatogr. A* **875,** 89–122.
4. Desiderio, C. and Fanali, S. (1998) Chiral analysis by capillary electrophoresis using antibiotics as chiral selector. *J. Chromatogr. A* **807,** 37–56.
5. Ward, T. J. and Farris, A. B. (2001) Chiral separations using the macrocyclic antibiotics: a review. *J. Chromatogr. A* **906,** 73–89.
6. Fanali, S., Catarcini, P., Blaschke, G., and Chankvetadze, B. (2001) Enantioseparations by capillary electrochromatography. *Electrophoresis* **22,** 3131–3151.
7. Armstrong, D. W., Tang, Y. B., Chen, S. S., Zhou, Y. W., Bagwill, C., and Chen, J. R. (1994) Macrocyclic antibiotics as a new class of chiral selectors for liquid chromatography. *Anal. Chem.* **66,** 1473–1484.
8. Armstrong, D. W., Rundlett, K. L., and Chen, J. R. (1994) Evaluation of the macrocyclic antibiotic vancomycin as a chiral selector for capillary electrophoresis. *Chirality* **6,** 496–509.
9. Bednar, P., Aturki, Z., Stransky, Z., and Fanali, S. (2001) Chiral analysis of UV nonabsorbing compounds by capillary electrophoresis using macrocyclic antibiotics: 1. Separation of aspartic and glutamic acid enantiomers. *Electrophoresis* **22,** 2129–2135.

10. Fanali, S., Desiderio, C., and Aturki, Z. (1997) Enantiomeric resolution study by capillary electrophoresis. Selection of the appropriate chiral selector. *J. Chromatogr. A* **772,** 185–194.
11. Desiderio, C., Aturki, Z., and Fanali, S. (2001) Use of vancomycin silica stationary phase in packed capillary electrochromatography. I. Enantiomer separation of basic compounds. *Electrophoresis* **22,** 535–543.

14

Separation of Enantiomers
by Capillary Electrophoresis Using Cyclodextrins

Wioleta Maruszak, Martin G. Schmid,
Gerald Gübitz, Elzbieta Ekiert, and Marek Trojanowicz

1. Introduction

Since the pioneering publication of first paper on application of capillary electrophoresis (CE) in separation of enantiomers in 1985 by Gassmann et al. *(1)* a large number of such applications with different chiral selectors have been developed *(2–8)*. Cyclodextrins (CDs) are the most frequently employed chiral selectors, since their first application in isotachophoresis *(9)*. The application of CDs for capillary zone electrophoresis was pioneered a one year later by Fanali *(10)* for chiral separation of symphatomimetic drugs.

CDs are cyclic oligosaccharides consisting of D-(+)-glucopyranose units connected with α-(1,4)-glucoside bonds. Although there are known CDs containing of 6 to 12 connected D-(+)-glucopyranoses, the most common used CDs are α, β, and γ composed of six, seven, and eight glucopyranose units, respectively. The connected glucopyranose units form a molecule in the form of a basket in the shape of a trucated cone providing a hydrophobic cavity for inclusion of various compounds. The enantioselectivity is based on formation of inclusion host-guest complexes where hydrophobic groups of analyte are included into hydrophobic cavity of CD, together with the secondary interactions between analyte and upper rim of CD, such as hydrogen bonding or dipole-dipole interaction. The principal physicochemical properties of native α-, β-, and γ-CDs are listed in **Table 1**.

The enantiomeric separation can be obtained when stabilities of complexes with CD differ sufficiently to provide various migration velocity. A theoretical model for CD-based separations was developed that relates the electrophoretic mobility differences of enantiomers to the CD concentration *(11,12)*. In this

From: *Methods in Molecular Biology, Vol. 243: Chiral Separations: Methods and Protocols*
Edited by: G. Gübitz and M. G. Schmid © Humana Press Inc., Totowa, NJ

Table 1
Physicochemical Properties of CDs

Parameter	Type of native CD		
	α	β	γ
Number of glucopyranose units	6	7	8
Number of hydroxyl groups	18	21	24
Number of 1st order OH groups	6	7	8
Number of 2nd order OH groups	12	14	16
Molecular weight	972	1135	1297
Internal cavity diameter (nm)	0.47–0.52	0.60–0.64	0.75–0.83
External cavity diameter (nm)	1.46 ± 0.05	1.54 ± 0.04	1.75 ± 0.04
Cavity capacity (nm^3)	0.176	0.346	0.510
Solubility in water at 25°C (g/100 mL)	14.50	1.82	23.20
pK$_a$ values for hydroxyl groups	12.1–12.6	12.1–12.6	12.1–12.6

approach, it was shown that for 1:1 CD-analyte complexes, the mobility difference goes past a maximum, and this maximum occurs when CD concentration is equal to the square root reciprocal of the product of the inclusion complex formation constants of each enantiomer with CD. This model was further extended to provide the resolution as a function of CD concentration in order to optimize the separation *(13)*. The degree of substitution in derivatized CDs was found to be a great importance for chiral separations *(14,15)*, but because usually CD derivatives represent mixtures of different products showing different substitution patterns, separations with the derivatized CDs are often difficult to reproduce. **Table 2** shows the most commonly used derivatized CDs in chiral separations.

A special variation is the use of CDs together with borate for the chiral resolution of diols *(16–18)*. It is assumed that mixed CD-borate-diol complexes are formed, involving the hydroxyls at C2 and C3 at the mouth of the cavity of the CD.

Another efficient approach is the combination of CDs with sodium dodecyl sulfate (SDS) making use of the principle of CD-modified micellar electrokinetic chromatography (CD-MEKC). This principle is subject of Chapter 20 in this book.

A wide usefulness of native and derivatized CD as chiral selectors in CE results from relative good solubility in aqueous background electrolytes (BGEs), a negligible absorptivity in commonly used UV range and usually sufficiently fast complexation rate, that leads to narrow peaks and large efficiency of separation *(7)*. Both efficiency and the resolution of chiral CE separation can be significantly influenced by addition of organic modifiers to BGE as well as though

Table 2
CDs Most Commonly Used in CE

Commonly used abbreviation	Type of CD
Native CDs	α-CD
	β-CD
	γ-CD
	α-CD
	β-CD
	γ-CD
Neutral substituted	Hydroxypropyl-α-CD
	Hydroxypropyl-β-CD
	Hydroxypropyl-γ-CD
	Methyl-β-CD
	Heptakis(di-*O*-methyl)-β-CD
	Heptakis(tri-*O*-methyl)-β-CD
	HP-α-CD
	HP-β-CD
	HP-γ-CD
	M-β-CD
	DM-β-CD
	TM-β-CD
Ionic substituted cyclodextrins	Carboxymethyl-β-CD
	Carboxyethyl-β-CD
	Carboxymethylethyl-β-CD
	Polymer carboxymethyl-β-CD
	Succinyl-β-CD
	Phosphated-β-CD
	Sulfated β-CD
	Sulfobuthyl-β-CD
	Sulfoethyl ether-β-CD
	Methylamino-β-CD
	Dimethylamino-β-CD
	6[(3-aminoethyl)amino]-6-deoxy-β-CD
	Mono(6-amino-6-deoxy)-β-CD
	CM-β-CD
	CE-β-CD
	CME-β-CD
	Polymer CM-β-CD
	Succ-β-CD
	p-β-CD
	S-β-CD
	SBE-β-CD
	SEE-β-CD
	MA-β-CD
	DMA-β-CD
	β-CD-NH$_2$

manipulation with the electroosmotic flow (EOF) by change of pH or addition of appropriate modifiers to BGE.

2. Materials

2.1. Apparatus

A commercially available CE apparatus with high voltage source up to 30 kV and UV or photodiode array detector should be used. The uncoated fused-silica capillaries (typically 50 or 75 μm inner diameter [I.D.] and effective length about 50 cm) were used for separations presented below.

2.2. Conditioning of the Capillary

1. 0.1 M NaOH, 10 mL.
2. 0.1 M HCl, 10 mL.

2.3. Preparation of BGE Solutions

1. NaH_2PO_4 or KH_2PO_4 of analytical grade.
2. Phosphoric acid of analytical grade.
3. Sodium tetraborate of analytical grade.
4. CDs (*see* **Notes 1–3**):
   a. Neutral: β-CD (Fluka, Buchs, Switzerland).
   b. Anionic: carboxymethyl-β-cyclodextrin (CM-β-CD) (Fluka), succinyl-β-CD (Wacker Chemie, Munich, Germany).
   c. Cationic: mono(6-amino-6-deoxy)-β-cyclodextrin (β-CD-NH$_2$) (Sigma, St. Louis, MO, USA).
5. Deionized water.
6. Syringe type membrane filters 0.45 μm.

2.4. Preparation of Analyte Solutions

1. Methanol of high-performance liquid chromatography (HPLC) grade.
2. Deionized water.

3. Methods

3.1. Preparation of BGEs

The advantage of application of CE for chiral separations is the possibility of the use of chiral selectors dissolved in BGE. The appropriate amount of compound used as chiral selector is dissolved in buffer solution and after adjustment of pH solution, filtered, and degassed by sonication (*see* **Notes 4–7**).

3.1.1. Example 1: Separation of Enantiomers of Basic Drug Using CE With Neutral, Unsubstituted CD

1. Prepare 20 mL 100 mM solution of NaH_2PO_4.
2. Adjust pH to 2.0 using 1:10 orthophosphoric acid.

3. Dissolve appropriate amount of solid β-CD in 25 mL phosphate buffer of pH 2.0 to obtain 15 m*M* β-CD solution.
4. Filter obtaining solution through 0.45-μm syringe filter and sonicate.
5. Fill the buffer reservoirs in CE setup with prepared BGE.

3.1.2. Example 2: Separation of Enantiomers of Neurotransmitters Using CE With Anionic CD

1. Prepare 50 mL 20 m*M* sodium tetraborate.
2. Dissolve carboxymethyl-β-CD in 25 mL borate buffer to obtain 20 m*M* solution.
3. Adjust pH of solution to 7.5 using by addition of boric acid (*see* **Note 6**).
4. Filter obtaining solution through 0.45-μm syringe filter and sonicate.
5. Fill the buffer reservoirs in CE setup with prepared BGE.

3.1.3. Example 3: Separation of Enantiomers of Phenyllactic Acid (DL-2-Hydroxy-3-Phenylpropanoic Acid) Using CE With Cationic CD

1. Prepare 5 mL 40 m*M* phosphoric acid, 18 m*M* ammediol (2-amino-2-methyl-1,3-propanediol) and dissolve mono(6-amino-6-deoxy)-β-CD to obtain 5 m*M* solution.
2. Adjust pH to 2.18 (*see* **Note 6**).
3. Filter obtained solution through 0.45-μm syringe filter and sonicate.
4. Fill the buffer reservoirs in CE setup with prepared BGE.

3.1.4. Example 4: Separation of Enantiomers of Vicinal Diols Using CE With CDs and Borate

1. Prepare 50 mL 50 m*M* sodium tetraborate.
2. Dissolve β-CD (or succinyl-β-CD) in 25 mL borate buffer to obtain 1.8% solution.
3. Adjust pH to 9.3 if necessary (*see* **Note 8**).
4. Filter obtaining solution through 0.45-μm syringe filter and sonicate.
5. Fill the buffer reservoirs in CE setup with prepared BGE.

3.2. Preparation of Analyte Solutions

Depending on the kind of analyte, the stock standard solutions are prepared from solid preparations by dissolution in methanol or water and then diluted with water to the required concentration prior to the injection in CE setup.

3.2.1. Example 1

Stock solution of 1 g/L was prepared in methanol and then, to obtain sample solution, it was diluted 1:10 with deionized water.

3.2.2. Example 2

10 m*M* Stock solution was prepared in deionized water and then diluted 1:10 with water to obtain sample solution.

3.2.3. Example 3

Sample solution was prepared by dissolving required amount of phenyllactic acid (DL-2-hydroxy-3-phenylpropanoic acid) in a mixture of methanol and water (1:1).

3.2.4. Example 4

Sample solution was prepared by dissolving required amount of vicinal diol in water or water/methanol mixtures, if necessary.

3.3. Conditioning of the Capillary

The conditioning of the capillary is an important step to obtain reproducible and reliable results of CE determination (*see* **Note 5**). The initial conditioning of brand new capillary should be performed according to the recommendation of manufacturer.

A typical everyday procedure should be as follows:

1. At the beginning of each series of measurements, rinse the capillary for 15 min with 0.5 *M* NaOH.
2. Rinse the capillary for 10 min with deionized water.
3. Rinse the capillary for 5 min with BGE.
4. Between measurements, rinse the capillary for 2 min with BGE.

3.4. Injection of Sample Solution

Inject sample solution using hydrodynamic or hydrostatic injection for 5 s or 4 s at 10 kV.

3.5. CE Analysis

After conditioning of the capillary and introduction of the analyte solution into the capillary, the CE measurement was carried out at 20–25 kV with positive polarization of high voltage electrodes. Depending on the analyte determined, the UV detection wavelength was set between 190 and 280 nm. The progress of separation was monitored by personal computer (PC) controlling the CE instrument.

3.6. Optimization of Separation Conditions

The choice of appropriate CDs as chiral selector and conditions of CE determination depends essentially on the molecular structure of analytes to be resolved. Among similar compounds, even minor structural differences may result in a change of selectivity of enantiomeric separation, which makes optimization of separation conditions quite a difficult task. In each case of chiral separation, the main optimized parameters are composition, concentration, and pH of BGE, as well as type, degree of substitution, and concentration of CD used *(19–21)*.

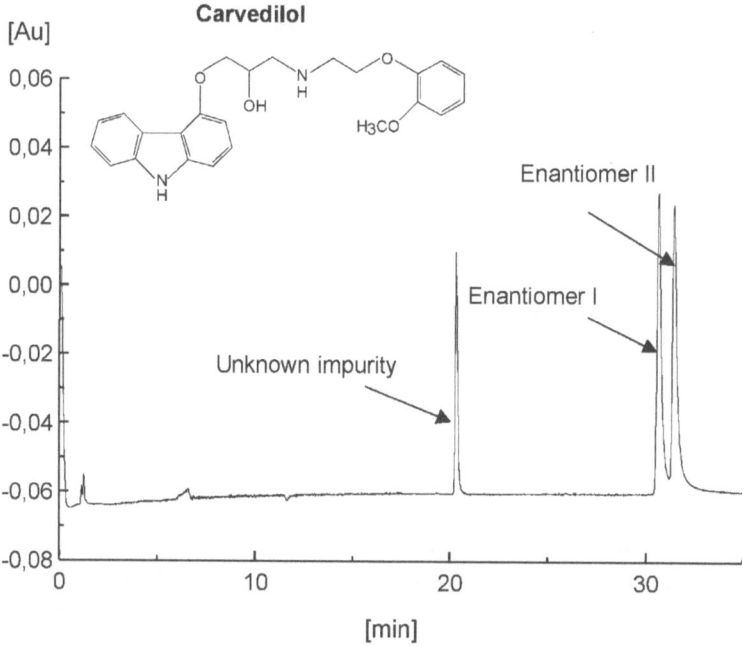

Fig. 1. Electropherogram obtained for separation of enantiomers of antihypertensive drug carvedilol with 100 m*M* phosphate buffer, pH 2.0, containing 15 m*M* β-CD. Positive polarization 14 kV, UV detection at 200 nm. Injected 0.1 g/L solution of racemate. Injection conditions: 8 kV for 6 s (capillary temperature 30°C). Fused-silica capillary 58.5 cm (49 cm to the detector) (×) 50 μm I.D.

The most often used buffers as BGE in chiral separation with CDs are phosphate, borate, mixed phosphate-borate, phosphate-citrate, and phosphate-Tris buffers, usually in concentration range from 10 to 100 m*M*. Separation of enantiomers of basic properties are usually carried out with electrolytes of low pH, where analytes are positively charged and EOF is low *(22,23)*. Enantiomers of analytes exhibiting acidic properties are separated with BGEs of high pH values *(23)*. The example of separation of enantiomers of an antihypertensive drug of basic properties, carvedilol, which is a nonselective β-adrenergic blocker with the use of neutral β-CD is shown in **Fig. 1** (example 1). The CE determination was carried out with phosphate BGE, pH 2.0, containing 15 m*M* β-CD at positive polarization 14 kV and UV detection at 200 nm.

Neutral CDs added to BGE at 5–30 m*M* concentrations are effective chiral selectors for separation of enantiomers of ionized compounds *(19,24–28)*. Their addition does not affect the ionic strength of BGE, but they change viscosity of BGE that may influence efficiency and chiral resolution. Neutral CDs

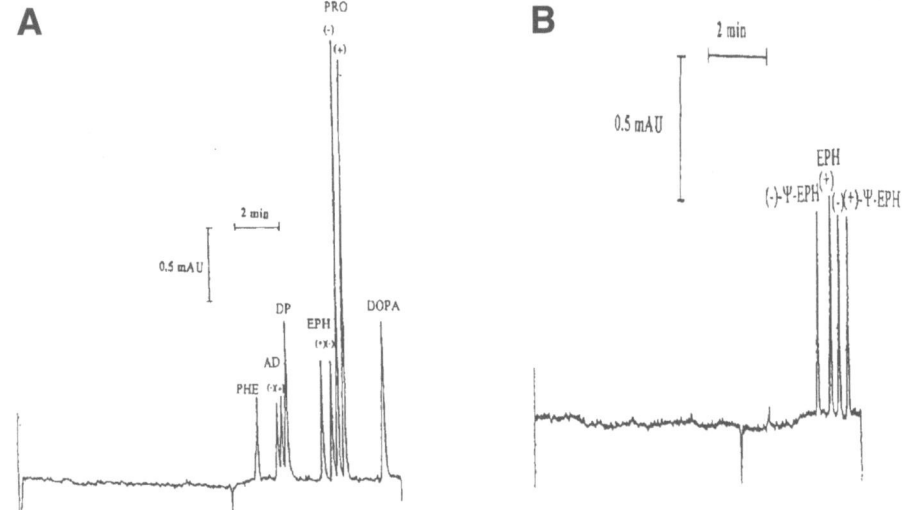

Fig. 2. Electropherograms obtained for separation of enantiomers of selected neuro-transmitters (**A**) and enantiomers of ephedrine (**B**) with 20 m*M* borate BGE, pH 7.5, containing 20 m*M* CM-β-CD *(34)*. Positive polarization 20 kV, UV detection at 214 nm. Hydrostatic injection 5 s. Fused-silica capillary 75 cm (65 cm to the detector) × 50 μm I.D. Abbreviations: PHE, DL-phenylalanine; AD, adrenaline; DP, dopamine; EPH, ephedrine; PRO, propranolol; DOPA, dihydroxyphenyloalanine.

together with ionic micelles or negatively or positively charged CDs may be successfully employed for CE separation of neutral hydrophobic analytes *(8)*.

CD derivatives with ionizable substituents exhibit electrophoretic mobility depending on the kind and number of ionizable groups. They are employed as chiral selectors, both for noncharged compounds and ionic species, where obtained chiral resolution is often better than that obtained with neutral CDs. Their own electrophoretic mobility increases the resolution of separation, especially when it is directed oppositely than electrophoretic mobility of analyte and/or EOF, which can result also from electrostatic interaction of CD and oppositely charged solution *(29,30)*. The electrostatic interactions are particularly significant in separations of solutes weakly interacting with neutral CDs. The most often used charged CDs in chiral separation are anionic CDs *(31–35)*. **Figure 2A** (example 2) shows electrophoretic separation of enantiomers of selected neurotransmitters obtained with the use of 20 m*M* anionic carboxymethyl-β-CD in 20 m*M* sodium tetraborate buffer, pH 7.5, with positive polarization 20 kV and UV detection at 214 nm *(35)*.

The analytes examined in this separation occurred both in neutral and anionic deprotonated form. The high pH value of BGE caused not only ionization of CD

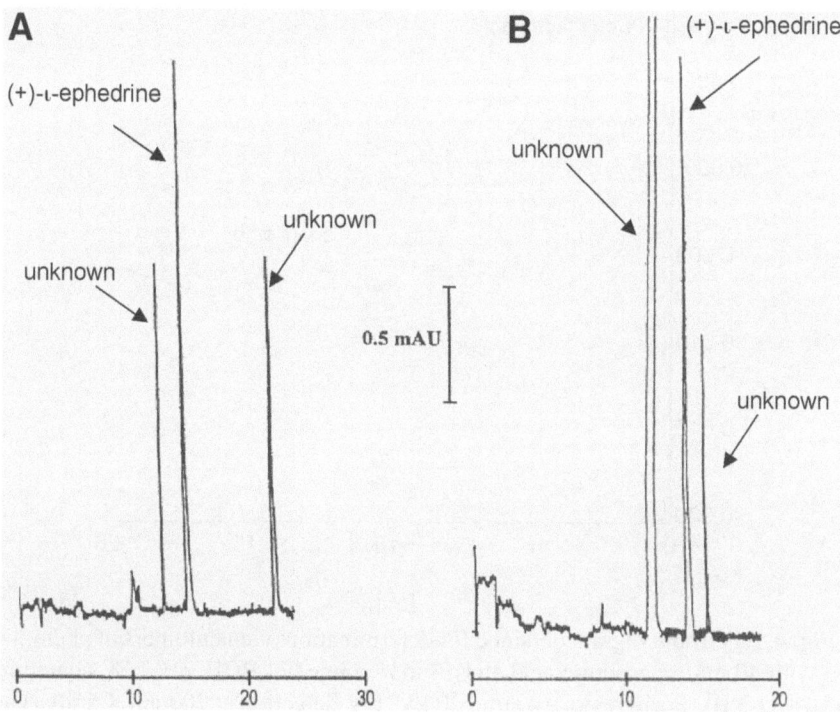

Fig. 3. Electropherogram obtained for determination of ephedrine enantiomers in pharmaceuticals Nurofen (**A**) and Gripex (**B**) *(34)*. Conditions are the same as in Fig. 2.

and most analytes, but also, due to strong EOF in such conditions, allows to obtain the migration of analytes to detector in a satisfactory period of time. CM-β-CD has been also shown to be a satisfactory chiral selector for separation of pairs of enantiomers having more than one chiral center (see separation of the ephedrine enantiomers in **Fig. 2B** *(35)*. Examples of determination of ephedrine in pharmaceutical preparations with this method are shown in **Fig. 3**.

The cationic derivatives of CDs were employed for separation of both neutral and ionized analytes. Separation of cationic enantiomers with quaternary ammonium β-CD was reported where weak binding of analytes form with neutral or anionic CDs was advantageous *(36)*. Such a separations were also demonstrated for several profens and amino acids in nonaqueous media *(37)*. Various cationic CDs were also employed for separation of a number of acidic and basic compounds *(38–40)*, including also the use of coated capillaries *(38)*. The chiral separation of anionic species, hydroxy acids, and carboxylic acids, was demonstrated with diamino-β-CD *(41)*. Using the mono(6-amino-6-deoxy)-β-CD (β-CD-NH$_2$), the separation of neutral and anionic enantiomers was shown

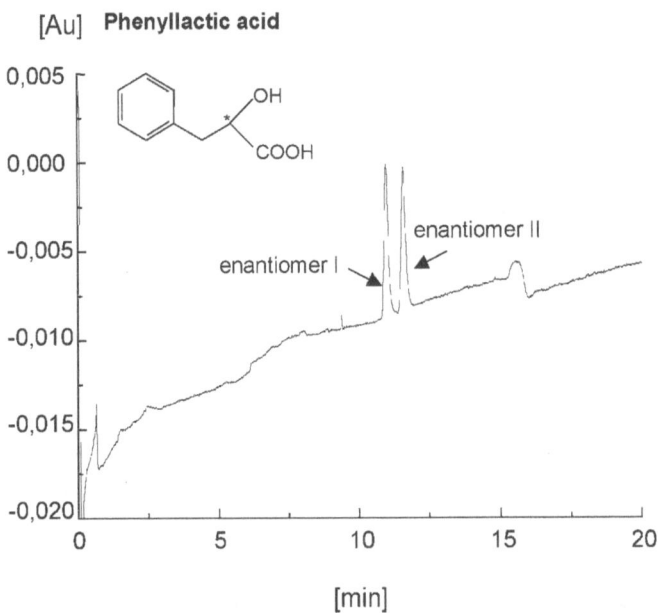

Fig. 4. Electropherogram obtained for determination of enantiomers of phenyllactic acid with 40 mM phosphoric acid and 18 mM ammediol BGE, pH 2.18, containing 5 mM β-CD-NH$_2$. Positive polarization 20 kV, UV detection at 200 nm. Capillary temperature 25°C. Injected 1.5 mM solution of phenyllactic acid (in MeOH/H$_2$O, 1:1). Injection conditions: 10 kV, 4 s. Fused-silica capillary 40.5 cm (30 cm to the detector) × 50 μm I.D.

together with demonstration of the intrinsic selectivity concept *(42)*. In this approach, selectivity corresponds to the concentration of the complexing agent. The separation of enantiomers of phenyllactic acid with 5 mM β-CD-NH$_2$ in 18 mM ammediol and 40 mM phosphoric acid BGE is shown in **Fig. 4** (example 3).

The use of low pH of BGE, assuring weak deprotonation of silanol groups at capillary walls, prevents adsorption of cationic CD. In BGE without CD in pH used a neutral analyte that migrated with velocity of EOF. Addition of β-CD-NH$_2$ does not affect EOF.

For the chiral separation of vicinal diols, which can not be resolved with CDs only, combination of CDs with borate was found to be useful *(16–18)*. Chiral separation of hydrobenzoin is shown in **Fig. 5** (example 4). Selectivity can be influenced by varying the concentration of CD and borate and by addition of organic modifiers (*see* **Note 9**).

The essential factor, which may affect efficiency and resolution of CE chiral separation, is the presence in BGE of organic modifiers that may change EOF, interactions with the capillary wall, solubility of CDs, and stability constants

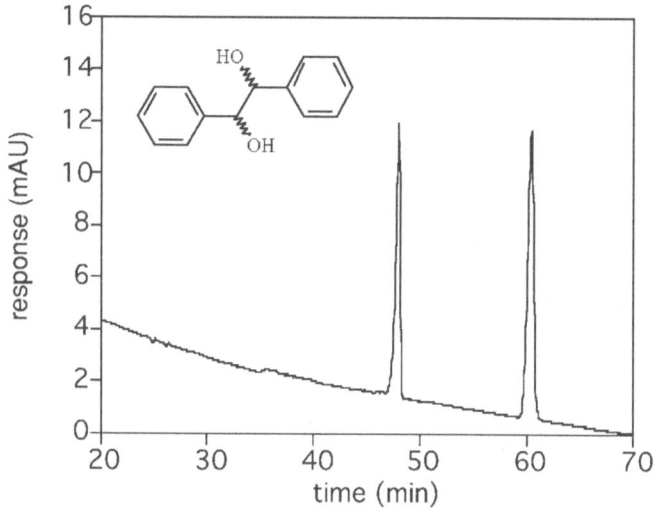

Fig. 5. Electropherogram of the chiral resolution of *RR*, *SS*-hydrobenzoin. Conditions: 1.8% β-CD, 50 m*M* borate, pH 9.3, and 20% (v/v) methanol, U = 18 kV, λ = 214 nm.

of complexes formed. Electrophoretic separation of enantiomers requires also careful control of temperature of the capillary, which affects of BGE viscosity and stability of inclusion complexes.

4. Notes

1. Due to appropriate dimensions of hydrophobic cavity, the β-CD is a suitable chiral selector forming inclusion complexes with numerous analytes. However, it is less soluble in aqueous solutions than α-CD and γ-CD, which can cause some limitations in CE separation.
2. In CE determinations, which are limited by solubility of native CDs, they can be replaced by better soluble substituted CDs, which may result in improvement of resolution *(28)*.
3. The electrophoretic mobility and complexation ability of substituted CDs essentially depends on degree of substitution. The use of neutral and ionic substituted CDs requires a careful control of degree of substitution, although there are no general rules to predict the optimal degree of substitution for a particular compound *(33)*.
4. The use of BGEs, where electrophoretic mobilities of charged CDs and analytes have opposite directions due to sample stacking effect, may cause an improvement of chiral resolution *(43)*. A simultaneously observed electrodispersion with increase of charge of CD, however, may result in pronounced broadening of peaks. This effect can be lowered by increase of ionic strength of BGE.
5. In the use of ionic derivatives of CDs, attention should be paid to the possibility of an increase of UV absorbance, ionic strength of BGE, and interaction of chiral selector with the capillary walls.

6. Because of the strong effect of pH on ionization of silanol groups of the capillary wall, ionization of ionic CDs and analytes, pH of BGE should be especially carefully optimized, besides the concentration of chiral selector. The changes of pH and CD concentration may result in changes of migration order of enantiomers.

7. The efficiency and enantiomeric resolution with CDs as chiral selectors can be additionally modified by: (i) addition of organic modifier to BGE; (ii) addition of another chiral selector to BGE; (iii) addition of polymer or surfactant to BGE; and (iv) changes of temperature of separation.

8. The pH value of the electrolyte is a crucial parameter, and pH 9.3 was found to be optimal. pH can be adjusted by the addition of boric acid or NaOH.

9. The addition of methanol results in an increase in resolution, however, connected with higher migration times.

References

1. Gassman, E., Kuo, J. E., and Zare, R. N. (1985) Electrokinetic separation of chiral compounds. *Science* **230,** 813–815.
2. Terabe, S., Otsuka, K., and Nishi, H. (1994) Separation of enantiomers by capillary electrophoretic techniques. *J. Chromatogr. A* **666,** 295–319.
3. Chanvetadze, B. (1997) *Capillary Electrophoresis in Chiral Analysis.* John Wiley & Sons, New York.
4. Gübitz, G. and Schmid, M. G. (1997) Chiral separation principles in capillary electrophoresis. *J. Chromatogr. A* **792,** 179–225.
5. Fanali, S. (1997) Controlling enantioselectivity in chiral capillary electrophoresis with inclusion-complexation. *J. Chromatogr. A* **792,** 227–267.
6. Gübitz, G. and Schmid, M. G. (2000) Recent progress in chiral separation principles in capillary electrophoresis. *Electrophoresis* **21,** 4112–4335.
7. Rizzi, A. (2001) Fundamental aspects of chiral separations by capillary electrophoresis. *Electrophoresis* **22,** 3079–3106.
8. Amini, A. (2001) Recent developments in chiral capillary electrophoresis and applications of this technique to pharmaceutical and biomedical analysis. *Electrophoresis* **22,** 3107–3130.
9. Snopek, J., Jelinek, I., and Smolkova-Keulemansova, E. (1988) Use of cyclodextrins in isotachophoresis; IV The influence of cyclodextrins on the chiral resolution of ephedrine alkaloid enantiomers. *J. Chromatogr. A* **438,** 211–218.
10. Fanali, S. (1989) Separation of optical isomers by capillary zone electrophoresis based on host-guest complexation with cyclodextrins. *J. Chromatogr. A* **474,** 441–446.
11. Wren, S. A. (1993) Theory of chiral separation in capillary electrophoresis. *J. Chromatogr.* **636,** 57–62.
12. Wren, S. A., Rowe, R. C., and Payne, R. S. (1994) A theoretical approach to clinical capillary electrophoresis with some practical implications. *Electrophoresis* **15,** 774–778.
13. Penn, S. G., Bergström, E. T., Goodall, D. M., and Loran, J. S. (1994) Capillary electrophoresis with chiral selectors: optimization of separation and determination

of thermodynamic parameters for binding of tioconazole enantiomers to cyclodextrins. *Anal. Chem.* **66,** 2866–2873.

14. Yoshinaga, M. and Tanaka, M. (1994) Use of selectively methylated-β-cyclodextrin derivatives in chiral separation of dansylamino acids by capillary zone electrophoresis. *J. Chromatogr. A* **679,** 359–365.

15. Fanali, S. and Aturki, Z. (1995) Use of cyclodextrins in capillary electrophoresis for the chiral resolution of some 2-arylpropionic acid non-steroidal anti-inflammatory drugs. *J. Chromatogr. A* **694,** 297–305.

16. Stefansson, M. and Novotny, M. (1993) Electrophoretic resolution of monosaccharide enantiomers in borate-oligosaccharide complexation media. *J. Am. Chem. Soc.* **115,** 11573–11580.

17. Jira, T., Bunke, A., Schmid, M. G., and Gübitz, G. (1997) Chiral resolution of diols by capillary electrophoresis using borate-cyclodextrin complexation. *J. Chromatogr. A* **761,** 269–276.

18. Schmid, M. G., Wirnsberger, K., Jira, T., Bunke, A., and Gübitz, G. (1997) Capillary electrophoretic chiral resolution of vicinal diols by complexation with borate and cyclodextrin—comparative studies on different cyclodextrin derivatives. *Chirality* **9,** 153–156.

19. Blanco, M., Coello, J., Iturriaga, H., Maspoch, S., and Pérez-Maseda, C. (1998) Separation of profen enantiomers by capillary electrophoresis using cyclodextrins as chiral selectors. *J. Chromatogr. A* **793,** 165–175.

20. Billiot, E., Thibodeaux, S., Shamsi, S., and Warner, I. M. (1999) Evaluating chiral separation interactions by use of diastereometric polymeric dipeptide surfactants. *Anal. Chem.* **71,** 4044–4049.

20a. Amini, A., Wiersma, B., Westerlund, D., and Paulsen-Sörman, U. (1999) Determination of the enantiomeric purity of S-ropivacaine by capillary electrophoresis with methyl-β-cyclodextrin as chiral selector using conventional and complete filling techniques. *Eur. J. Pharm. Sci.* **9,** 17–24.

21. Perrin, C., Vargas, M. G., Vander Heyden, Y., Maftouh, M., and Massart, D. L. (2000) Fast development of separation methods for the chiral analysis of amino acid derivatives using capillary eletrophoresis and experimental designs. *J. Chromatogr. A* **883,** 249–265.

22. Li, G., Lin, X., Zhu, Ch., Hao, A., and Guan, Y. (2000) New derivative of β-cyclodextrin as chiral selectors for capillary electrophoretic separation of chiral drugs. *Anal. Chim. Acta* **421,** 27–34.

23. Fischer, C., Schmidt, U., Dwars, T., and Oehme, G. (1999) Enantiomeric resolution of derivatives of α-aminophosphonic and α-aminophosphinic acids by high-performance liquid chromatography and capillary electrophoresis. *J. Chromatogr. A* **845,** 273–283.

24. Nielen, M. W. F. (1993) Chiral separation of basic drugs using cyclodextrin-modified capillary electrophoresis. *Anal. Chem.* **65,** 885–893.

25. Chankvetadze, B., Endresz, G., and Blaschke, G. (1995) Enantiomeric resolution of chiral imidazole derivatives using capillary electrophoresis with cyclodextrin-type buffer modifiers. *J. Chromatogr. A* **700,** 43–49.

26. Koppenhoefer, B., Epperlein, U., Christian, B., Lin, B., Ji, Y., and Chen, Y. (1996) Separation of enantiomers of drugs by capillary electrophoresis. β-cyclodextrin as chiral solvating agent. *J. Chromatogr. A* **735,** 333–343.

27. Koppenhoefer, B., Epperlein, U., Schlunk, R., Zhu, X., and Lin, B. (1998) Separation of enantiomers of drugs by capillary electrophoresis. Hydroxypropyl-α-cyclodextrin as chiral solvating agent. *J. Chromatogr. A* **793,** 153–164.

28. Pak, C., Marriott, P. J., Carpenter, P. D., and Amiet, R. G. (1998) Enantiomeric separation of propanolol and selected metabolites by using capillary electrophoresis with hydroxypropyl-β-cyclodextrin as chiral selector. *J. Chromatogr. A* **793,** 357–364.

29. Daali, Y., Cherkaoul, S., Christen, P., and Veuthey, J. L. (1999) Experimental design for enantioselective separation of celiprolol by capillary electrophoresis using sulfated β-cyclodextrin. *Electrophoresis* **20,** 3424–3431.

30. Ren, X., Dong, Y., Liu, J., Huang, A., Liu, H., and Sun, Z. (1999) Separation of chiral basic drugs with sulfobutyl-β-cyclodextrin in capillary electrophoresis. *Chromatographia* **50,** 363–368.

31. Chankvetadze, B., Endresz, G., and Blaschke, G. (1994) About some aspects of the use of charged cyclodextrins for capillary electrophoresis enatioseparation. *Electrophoresis* **15,** 804–807.

32. Schmitt, T. and Engelhardt, H. (1995) Optimization of enantiomeric separation in capillary electrophoresis by reversal of migration order and using different derivatized cyclodextrins. *J. Chromatogr. A* **697,** 561–570.

33. Francotte, E., Brandel, L., and Jung, M. (1997) Influence of the degree of substitution on cyclodextrin sulfobuthyl ether derivative on enantioselective separation by electrokinetic chromatography. *J. Chromatogr. A* **792,** 379–384.

34. Morin, Ph., Bellessort, D., Dreux, M., Troin, Y., and Gelas, J. (1998) Chiral resolution of functionalized piperidine enantiomers by capillary electrophoresis with native, alkylated and anionic β-cyclodextrin. *J. Chromatogr. A* **796,** 375–383.

35. Maruszak, W., Trojanowicz, M., Margasinska, M., and Engelhardt, H. (2001) Application of carboxymethyl-β-cyclodextrin as a chiral selector in capillary electrophoresis of enantiomer separation of selected neurotransmitters. *J. Chromatogr. A* **926,** 327–336.

36. Wang, F. and Khaledi, M. G. (1998) Capillary electrophoresis chiral separation of basic compounds using cationic cyclodextrin. *Electrophoresis* **19,** 2095–2100.

37. Wang, F. and Khaledi, M. G. (1998) Nonaqueous capillary electrophoresis chiral separations with quaternary ammonium β-cyclodextrin. *J. Chromatogr. A* **817,** 121–128.

38. Fanali, S. and Camera, E. (1996) Use of methyloamino-β-cyclodextrin in capillary electrophoresis. Resolution of acidic and basic enantiomers. *Chromatographia* **43,** 247–253.

39. Bunke, A. and Jira, Th. (1996) Chiral capillary electrophoresis using a cationic cyclodextrin. *Pharmazie* **51,** 672–673.

40. Bunke, A. and Jira, Th. (1998) Use of cationic cyclodextrin for enantioseparation by capillary electrophoresis. *J.Chromatogr. A* **798,** 275–280.

41. Galaverna, G., Paganuzzi, M. C., Corradini, R., Dossena, A., and Marchelli, R. (2001) Enantiomeric separation of hydroxy acids and carboxylic acids by diamino-β-cyclodextrins (AB, AC, AD) in capillary electrophoresis. *Electrophoresis* **22,** 3171–3177.
42. Lelievre, F., Gareil, P., and Jardy, A. (1997) Selectivity in capillary electrophoresis: application to chiral separations with cyclodextrins. *Anal. Chem.* **69,** 385–392.
43. Chien, R. L. and Burgi, D. D. (1992) On-column sample concentration using field amplification in CZE. *Anal. Chem.* **64,** 489A–496A.

15

Chiral Separations by Capillary Electrophoresis Using Proteins as Chiral Selectors

Jun Haginaka

1. Introduction

High-performance liquid chromatography (HPLC) chiral stationary phases based on a protein are of special interest because of their unique properties of stereoselectivity and because they are suited for separating a wide range of enantiomeric mixtures. These come from the multiple binding sites in a protein, and/or the multiple interactions between a solute and protein. Similarly, capillary electrophoresis (CE) methods using proteins as the immobilized or adsorbed ligands or running buffer additives have been developed for the separation of enantiomeric mixtures (1–4). Proteins used so far as chiral selectors have included albumins, such as bovine serum albumin (BSA), human serum albumin (HSA), and serum albumins from other species; glycoproteins such as α_1-acid glycoprotein (AGP), ovomucoid from chicken egg whites (OMCHI), ovoglycoprotein from chicken egg whites (OGCHI), avidin, and riboflavin-binding protein (or flavoprotein); enzymes, such as fungal cellulase from fungus *Aspergillus niger*, cellobiohydrolase I, pepsin, and lysozyme; and other proteins such as ovotransferrin (or conalbumin), β-lactoglobulin, casein, and human serum transferrin.

For chiral separations in protein-based CE, two methods were utilized. One is affinity capillary electrochromatography (CEC), and the other is affinity CE. In affinity CEC, protein-immobilized silica gels were packed into the capillary, or proteins were immobilized or adsorbed within the capillary. The applied electric fields result in solvent and solute flows through the system. The separation of enantiomers occurs by differences in interactions with an immobilized or adsorbed protein selector between enantiomers. This system is very similar to HPLC chiral stationary phase system, which is operated by the pressure-driven

From: *Methods in Molecular Biology, Vol. 243: Chiral Separations: Methods and Protocols*
Edited by: G. Gübitz and M. G. Schmid © Humana Press Inc., Totowa, NJ

flow. Thus, this technique is termed affinity CEC. There are several techniques for affinity CEC. A first way for affinity CEC technique is to use capillaries packed with protein-immobilized silica particles. HSA- *(5)* and AGP-immobilized *(6)* HPLC silica gels were packed into fused silica capillaries. A second way for affinity CEC technique is to immobilize chemically a protein to the inner surface of fused silica capillaries. **Figure 1** shows the procedure for the immobilization of BSA on the capillary wall *(7)*. The advantages of the methods are the small consumption of a chiral selector and the possibility of UV detection without limitations of the protein absorption. A third way is the use of capillaries filled with gels consisting of a protein crosslinked with glutaraldehyde. The method could have the potential to be applicable for all types of protein-based separations. However, since electroosmotic flow (EOF) is eliminated or negligible with these capillaries, the technique is not applicable to the separation of uncharged compounds. A fourth method employed in affinity CEC is to coat a protein dynamically in the capillary. The physically coated proteins were slowly desorbed in the presence of an electric field. However, the desorbed protein could automatically be replaced by adding a small amount of soluble protein to the running buffer. The advantages of the method are that it does not require the use of any packing material or the immobilization of a protein in the capillary and that the same capillary can be used for work with additional proteins. However, the enantioseparation of warfarin showing strong bindings to HSA ($Ka \geq 10^5$/M) was attained using the above method, but no enantioseparation of tryptophan showing weaker bindings to HSA ($Ka \leq 10^4$/M) was attained *(8)*. This is due to the fact that the effective protein concentration is low on the capillary wall.

The most common format using protein selectors in CE is to dissolve the protein in the running buffers. The technique is termed affinity CE. The advantages of affinity CE based on a protein are that no immobilization of a protein to packing materials or capillary walls is required and that packing procedures, which are needed for affinity CEC with a packed capillary, are not required. Further, since binding properties of an immobilized protein are rather different from those of the native protein, it is favorable to use soluble proteins. The disadvantages of the affinity CE method include: (i) use of the larger amount of a protein; (ii) adsoption of a protein to the capillary wall; (iii) absorption of UV light at the detection wavelength; and so on.

Uncoated and coated capillaries have been used in protein-based CE. With regard to adsorption of a protein to the capillary wall, some proteins (e.g., albumin) are relatively easy to use on uncoated capillaries, while others (e.g., AGP) are more difficult, because they quickly result in capillary blockages *(1)*. The adsorption of proteins on the wall will cause changes in the EOF, which can affect the reproducibility of migration times and peak area *(1,2)*. When uncoated

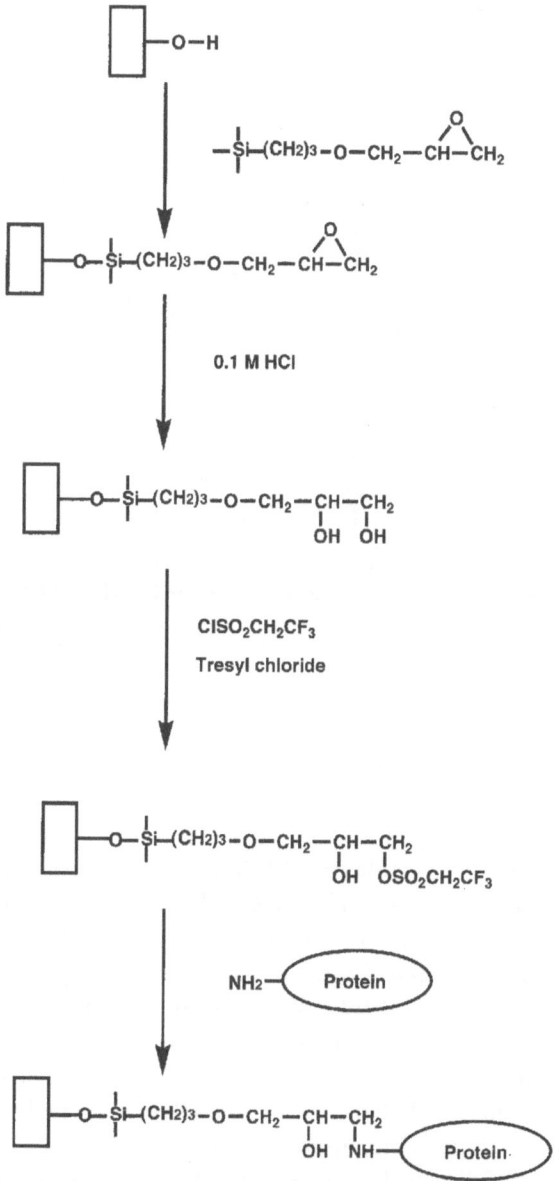

Fig. 1. Procedure for the immobilization of BSA on the capillary wall.

capillaries were used, it was important to wash the capillary between runs with sodium hydroxide or sodium dodecylsulfate to remove the adsorbed proteins completely. Two approaches to avoid the adsorption of proteins to the capillary wall are the use of coated capillaries and the use of additives to minimize

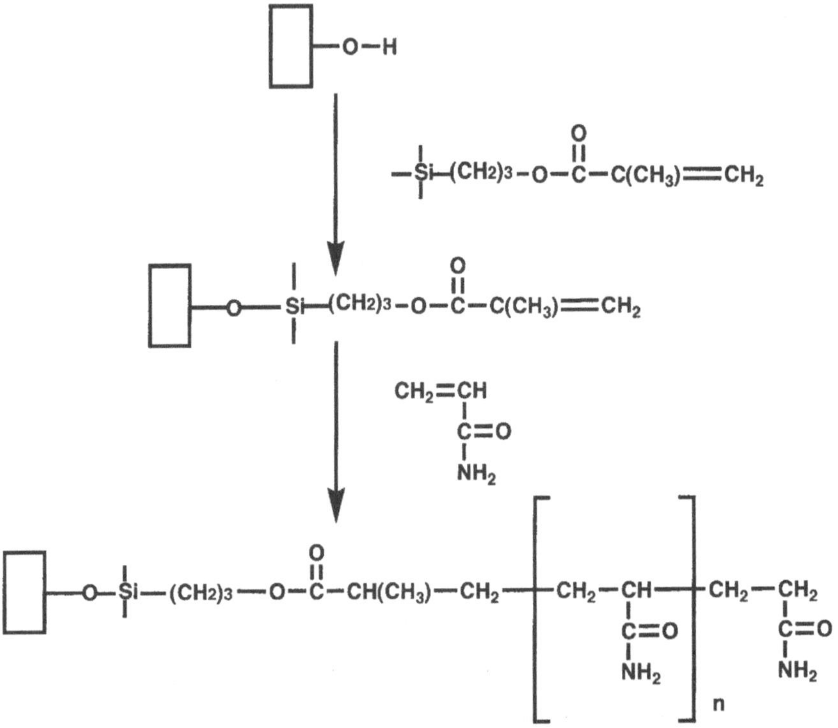

Fig. 2. Procedure for the preparation of linear polyacrylamide-coated capillary.

the protein-wall interaction. The most frequently used coating is linear poly-acrylamide developed by Hjertén *(9)*. **Figure 2** shows the preparation procedure of linear polyacrylamide-coated capillaries. On the other hand, the additives to minimize the protein-wall interaction include hydroxypropylcellulose, dextran, *o*-phosphorylethanolamine, 2-(cyclohexylamino)ethanesulfonic acid, and 3-[(3-chloramidopropyl)dimethylammonio]-2-hydroxy-1-propanesulfonate *(3)*.

When a protein is added in the running buffers, the background signals due to the protein interfered with detection of an analyte. Especially when high concentrations of a protein are being used, the problem is serious. To overcome this problem, the partial filling technique was developed. In the technique, the capillary was partially filled with a solution containing a protein, and the protein was not in the detector cell when the analyte reached that cell. **Figure 3** schematically illustrates the operating principle of the technique *(10)*. At the beginning of the separation, the capillary is partially filled with the solution containing an acidic protein such as BSA, AGP, or OGCHI (**Fig. 3A**). A sample solution of a cationic mixture is introduced at the end of capillary filled with the separation

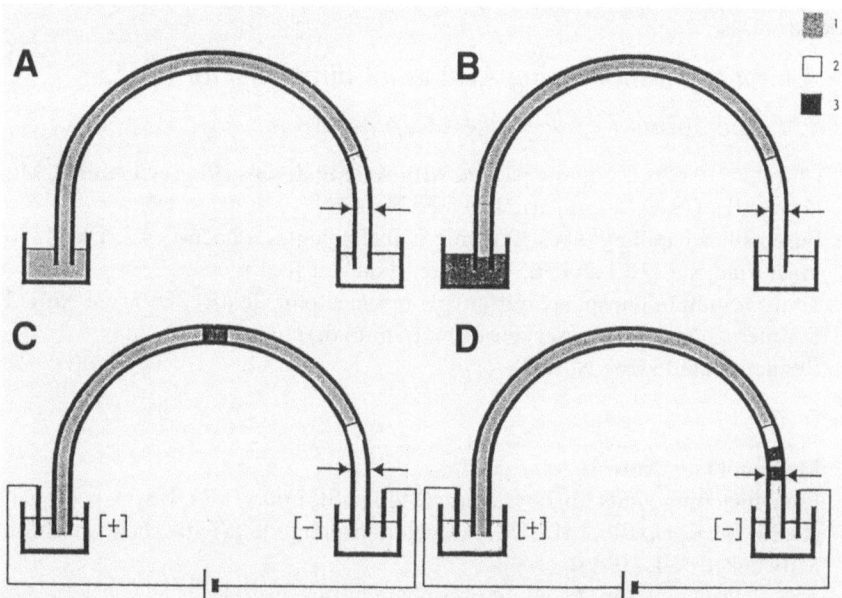

Fig. 3. Schematic illustration of the partial filling technique. Reproduced from **ref. 10** with permission. 1, separation zone; 2, running buffer solution; 3, sample solution; arrows indicate detection window. (**A**) The separation zone is introduced from the injection end to a point short of the detector cell. (**B**) The sample solution is introduced into the capillary. (**C**) A high voltage is applied between both ends of the capillary after both ends are dipped into the running buffer, and the analytes migrate toward the detector. (**D**) A separated zone reaches the detector cell, but the separation zone does not reach this cell.

solution (**Fig. 3B**). A cationic mixture migrates toward the cathode, while an acidic protein migrates in the opposite side. Since in this example, a coated capillary is used to eliminate the EOF, the separation zone or protein does not migrate significantly during the run. In the separation zone, enantiomer separations are attained (**Fig. 3C**), while the enantiomers migrate at identical velocities outside the separation zone and are detected in the absence of a protein (**Fig. 3D**). These procedures were automatically run using a commercial CE instrument. The partial filling techniques gave improved detection sensitivity and comparable reproducibilities of migration times and peak area to the conventional technique where the protein was completely filled in the separation capillary.

In the following sections, the affinity CEC based on packed AGP-immobilized silica gels *(6)* and immobilized BSA to fused silica capillaries *(7)* and the affinity CE based on OGCHI dissolved in the running buffer *(11,12)* will be precisely dealt with.

2. Materials

2.1. Chiral Separation Using AGP as a Chiral Selector (6)

2.1.1. Preparation of AGP-Packed Capillaries

1. Chiral-AGP HPLC column packed with AGP materials (Regis Chemical, Morton Grove, IL, USA) (5-µm particles) (*see* **Note 1**).
2. Fused silica capillary tubes (Polymicro Technologies, Phoenix, AZ, USA) (50-µm inner diameter [i.d.] and 365-µm outer diameter [o.d.]).
3. 10 mM Sodium phosphate buffer (pH 6.5)/acetonitrile (4:1, v/v) (*see* **Note 2**).
4. Stainless-steel tubing reservoir (40 × 6 mm i.d.).
5. Reducing union (*see* **Note 3**).

2.1.2. CE

1. Methanol (*see* **Note 4**).
2. Racemic compounds [disopyramide (Sigma, St. Louis, MO, USA), pentobarbital (U.S.P.C., Rockville, MD, USA), cyclophosphamide (Sigma) benzoin (Aldrich, Milwaukee, WI, USA)].
3. 15% 2-Propanol/4 mM sodium phosphate buffer, pH 6.8.
4. 2% 2-Propanol/2 mM sodium phosphate buffer, pH 5.5.
5. 3% 2-Propanol/2 mM sodium phosphate buffer, pH 6.5.
6. 5% 1-Propanol/5 mM sodium phosphate buffer, pH 6.5.
7. 0.45-µm Membrane filter.

2.2. Chiral Separation Using BSA as a Chiral Selector (7)

2.2.1. Preparation of BSA-Immobilized Capillaries

2.2.1.1. Etching of a Fused Silica Capillary

1. Fused silica capillary tubes (Ziemer, Mannheim, Germany) (60-cm effective length, 50-µm i.d. and 365-µm o.d.).
2. 0.1 M Sodium hydroxide solution.
3. 0.1 M Hydrochloric acid.

2.2.1.2. Epoxy-Diol-Coating of an Etched Fused Silica Capillary

1. A solution of 20% (v/v) of 3-glycidoxypropyltrimethoxysilane (Sigma) (*see* **Note 5**) in dry toluene (*see* **Note 6**).
2. 0.1 M Hydrochloric acid.

2.2.1.3. Activation of a Diol-Coated Capillary With 2,2,2-Trifluoroethanesulfonyl Chloride (Tresyl Chloride)

1. Acetone/water (9:1, v:v)
2. Acetone.
3. 17 µL Tresyl chloride (Fluka, Neu-Ulm, Germany) (*see* **Note 7**) in 1 mL dry toluene and 34 µL of pyridine.

2.2.1.4. COUPLING OF BSA TO THE ACTIVATED CAPILLARY

1. A solution of 20 mg/mL BSA (Sigma) in potassium phosphate buffer (50 mM, pH 7.4).
2. 1 M Sodium chloride solution.

2.2.2. CE

1. Racemic compounds (Sigma) (dinitrophenyl [DNP]-D,L-alanine and DNP-D,L-proline).
2. 50 mM Potassium phosphate buffer (pH 6.0).
3. 0.45-µm Membrane filter.

2.3. Chiral Separation Using OGCHI as a Chiral Selector (11,12)

2.3.1. Preparation of Linear Polyacrylamide-Coated Capillaries (9)

2.3.1.1. PREPARATION OF A 3-METHACRYLOXYPROPYLSILYLATED-FUSED SILICA CAPILLARY

1. Fused silica capillary tubes (GL Sciences, Tokyo, Japan) (30-cm effective length, 75-µm i.d. and 365-µm o.d.).
2. 0.1 M Sodium hydroxide solution.
3. 0.1 M Hydrochloric acid.
4. 3-Methacryloxypropyltrimethoxysilane (Sigma-Aldrich Japan, Tokyo, Japan).
5. Water (pH 3.5, adjusted with acetic acid).

2.3.1.2. COATING OF A 3-METHACRYLOXYPROPYLSILYLATED
CAPILLARY WITH LINEAR POLYACRYLAMIDE

1. Acrylamide (Nacalai Tesque, Kyoto, Japan) (*see* **Note 8**).
2. *N,N,N',N'*-Tetramethylethylenediamine (TEMED) (Nacalai Tesque) (*see* **Note 9**).
3. Ammonium peroxodisulfate (APS) (Nacalai Tesque) (*see* **Note 10**).

2.3.2. CE

1. OGCHI (*see* **Note 11**).
2. Racemic compounds [alimemazine (Daiichi Pharmaceutical, Tokyo, Japan) and eperisone (Eisai, Tokyo, Japan)].
3. Running buffer solutions: 50 mM sodium phosphate buffers (pH 5.0)/2-propanol (70:30, v/v) and 50 mM sodium phosphate buffers (pH 6.0)/2-propanol (90:10, v/v).
4. Separation solutions: running buffer solutions including 50 µM OGCHI as a chiral selector (*see* **Note 12**).
5. 0.45-µm Membrane filter.

3. Methods

3.1. Chiral Separation Using AGP as a Chiral Selector (6)

3.1.1. Preparation of AGP-Packed Capillaries

1. Obtain AGP packing material (5-µm particles) by emptying Chiral-AGP HPLC columns.

2. Cut the fused silica capillary into 42-cm length.
3. Make a frit at one end of the capillary (*see* **Note 13**).
4. Connect the lower end of a stainless-steel tubing reservoir (40 × 6 mm i.d.) to the inlet of the capillary, which is retained in a reducing union with a Vespel ferrule.
5. Make a slurry of 5-μm AGP packing by mixing approx 50 mg of packing material with 3 mL of 10 mM sodium phosphate buffer (pH 6.5)/acetonitrile (4:1, v/v) in an ultrasonic bath for approx 5 min (*see* **Note 14**).
6. Transfer the slurry to the stainless steel tubing reservoir and then pump the slurry into the capillary at a pressure of approx 5000 psi using a HPLC column slurry packer.
7. Check the capillary for blockages and voids in the packing using a 40× magnification microscope.
8. Switch off the pump after the desired length of capillary has been packed and wait for complete reduction of the residual pressure (*see* **Note 15**).
9. Make a retaining frit at approx 17 cm from the end frit (*see* **Note 16**).
10. Flush the capillary with the mobile phase of 10 mM sodium phosphate buffer (pH 6.5)/acetonitrile (4:1, v/v) from both ends using the column packer.
11. Burn away the polyamide coating of the capillary at 1 to 2 cm downstream of the retaining frit to make a detection window (*see* **Note 17**).

2.1.2. CE

1. Make sample solutions of racemic compounds (disopyramide, pentobarbital, cyclophosphamide, benzoin) in water or water-methanol at a concentration of approx 1 mg/mL and then filter through a 0.45-μm membrane filter (*see* **Note 18**).
2. Make a running buffer solution specified in **Subheading 2.1.2.3.–2.1.2.6.** and then filter through a 0.45-μm membrane filter.
3. Dip both ends of the capillary into the running buffer solution.
4. Inject the samples electrokinetically by applying a voltage of 5 kV for 1 s (*see* **Note 19**).
5. Carry out the CE separations at a constant voltage of 12–20 kV and detect the sample on UV absorbance measurements as shown in **Fig. 4 (6)**.

3.2. Chiral Separation Using BSA as a Chiral Selector (7)

3.2.1. Preparation of BSA-Immobilized Capillaries

3.2.1.1. Etching of a Fused Silica Capillary

1. Etch a fused silica capillary with 0.1 M sodium hydroxide solution and rinse with water for some minutes (*see* **Note 20**).
2. Flush the capillary with 0.1 M hydrochloric acid and then rinse with water (*see* **Note 21**).
3. Dry the capillary by flushing with nitrogen at 120°C.

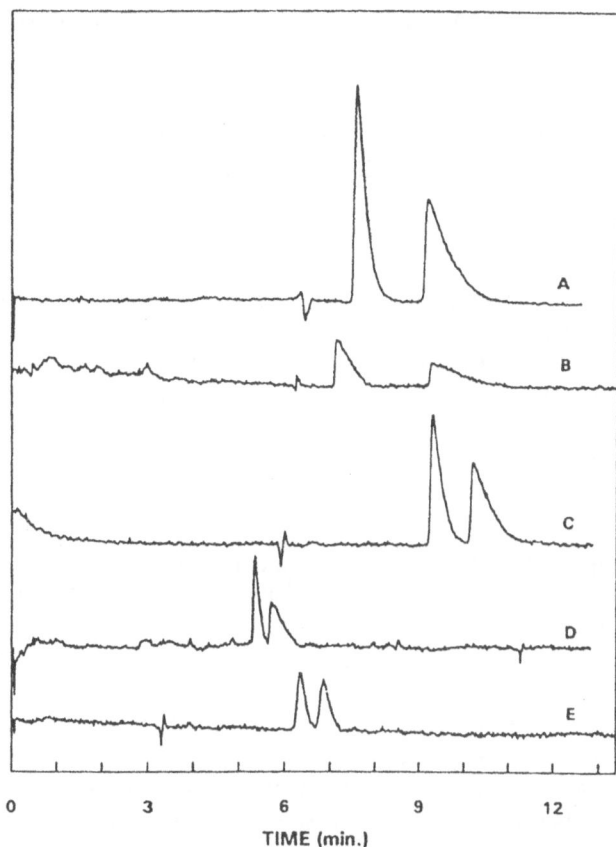

Fig. 4. Electrochrormatograms showing the enantiomeric separations of disopyramide (**A**), pentobarbital (**B**), hexobarbital (**C**), cyclophosphamide (**D**) and benzoin (**E**). Reproduced from **ref. 6** with permission. Conditions: (**A**) disopyramide (15% 2-propanol/4 m*M* sodium phosphate buffer, pH 6.8, applied voltage 12 kV, current 2 µA); (**B**) pentobarbital (2% 2-propanol/2 m*M* sodium phosphate buffer, pH 5.5, applied voltage 20 kV, current 2 µA); (**C**) hexobarbital (2% 2-propanol/2 m*M* sodium phosphate buffer, pH 5.5, applied voltage 18 kV, current 2 µA); (**D**) cyclophosphamide (3% 2-propanol/2 m*M* sodium phosphate buffer, pH 6.5, applied voltage 25 kV, current 2 µA); (**E**) benzoin (5% 1-propanol/5 m*M* sodium phosphate buffer, pH 6.5, applied voltage 15 kV, current 3 µA).

3.2.1.2. EPOXY-DIOL-COATING OF AN ETCHED FUSED SILICA CAPILLARY

1. Pump a solution of 20% (v/v) of 3-glycidoxypropyltrimethoxysilane in dry toluene through an etched fused silica capillary for 4 h at 110°C.
2. Rinse the capillary with toluene, methanol, and water, and then treat with 0.1 M hydrochloric acid for several hours at room temperature (*see* **Note 22**).

3.2.1.3. Activation of a Diol-Coated Capillary With Tresyl Chloride

1. Wash the diol-coated capillary with acetone/water (9:1, v/v) and then with dry acetone.
2. Pump a solution of 17 µL tresyl chloride in 1 mL dry acetone and 34 µL pyridine through the capillary (*see* **Note 23**). After 30 min, treat the capillary with acetone and flush with nitrogen.

3.2.1.4. BSA Coupling to the Activated Capillary

1. Pump a solution of 20 mg/mL BSA in potassium phosphate buffer (50 mM, pH 7.4) through the capillary for 10 min.
2. Seal the capillary and store at 4°C for 24 h (*see* **Note 24**).
3. Rinse the capillary with 1 M sodium chloride solution (*see* **Note 25**).
4. Burn away the polyamide coating of the capillary to make a detection window (*see* **Note 17**).

3.2.2. CE

1. Make sample solutions of racemic compounds (DNP-D,L-alanine and DNP-D,L-proline) in methanol at a concentration of approx 0.2 mg/mL and filter through a 0.45-µm membrane filter.
2. Make a running buffer solution of potassium phosphate buffer (50 mM, pH 6.0) and filter through a 0.45-µm membrane filter.
3. Dip both ends of the capillary into the running buffer solution.
4. Inject the sample by pressure, 3450 Pa for 1 s (*see* **Note 19**).
5. Carry out the CE separations at a constant voltage of 10 kV and detect the sample on UV absorbance measurements as shown in **Fig. 5** *(7)*.

3.3. Chiral Separation Using OGCHI as a Chiral Selector (11,12)

3.3.1. Preparation of Linear Polyacrylamide-Coated Capillaries (9)

3.3.1.1. Preparation of a 3-Methacryloxypropylsilylated Capillary
(*see* **Note 26**).

1. Wash successively a fused silica capillary for 30 min with 0.1 M sodium hydroxide solution, water, 0.1 M hydrochloric acid, and water (*see* **Notes 20** and **21**).
2. Mix 0.5 mL of 3-methacryloxypropyltrimethoxysilane with 5.0 mL of water, which is adjusted to pH 3.5 with acetic acid (*see* **Note 27**).
3. Suck up the silane solution into the capillary for 30 min and then allow the capillary to stand for 4 h at room temperature (*see* **Note 28**).
4. Wash the capillary with water by sucking up.

3.3.1.2. Coating of a 3-Methacryloxypropylsilylated
Capillary With Linear Polyacrylamide

1. Fill the silanized capillary with a degassed 3.5% acrylamide solution containing 2 µL TEMED and 2 mg APS/2 mL aqueous solution (*see* **Note 29**).

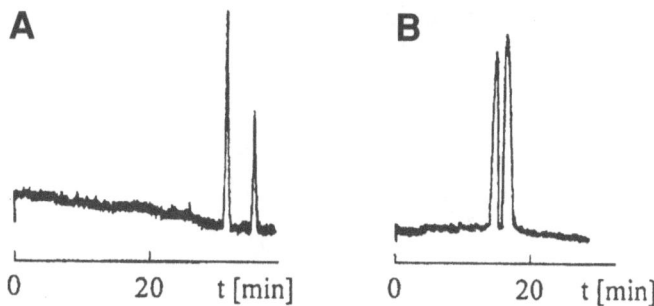

Fig. 5. Enantiomer separations of (**A**) DNP-D,L-alanine and (**B**) DNP-D,L-proline (nonracemic mixtures). Reproduced from **ref. 7** with permission. Conditions: running buffer solution, potassium phosphate buffer, pH 6.0, 50 mM; capillary, 60 cm effective length × 50 μm i.d. column.

2. Suck away the excess of polyacrylamide after 60 min and then rinse the capillary with water.
3. Burn away the polyamide coating of the capillary to make a detection window (*see* **Note 17**).

3.3.2. CE

1. Make sample solutions of racemic compounds (alimemazine and eperisone) in methanol at a concentration of approx 0.2 mg/mL, filter through a 0.45-μm membrane filter, and degas with an ultrasonic bath prior to use (*see* **Note 30**).
2. Make running buffer solutions and separation solutions (*see* **Note 31**), filter through a 0.45-μm membrane filter, and degas with an ultrasonic bath prior to use.
3. Rinse the capillary with water for 1 min, 50 mM sodium phosphate buffer (pH 2.5) for 3 min, water for 1 min, and running buffer for 3 min prior to the run (*see* **Note 32**).
4. Fill the capillary with the separation solution for 1 min.
5. Inject the sample by pressure, 3450 Pa for 1 s (*see* **Note 19**).
6. Dip both ends of the capillary into the running buffer solution.
7. Carry out the CE separations at a constant voltage of 12 kV and detect the sample on UV absorbance measurements as shown in **Fig. 6** (*12*).

4. Notes

1. AGP packing material (5-μm particles) is obtained by emptying commercially available Chiral-AGP HPLC columns. The packing material is taken from the outlet end of the column, and the first 1 to 2 cm of packing at the inlet end of the column is not used.
2. Generally, organic solvents and reagents used may be fatal or harmful if swallowed, inhaled, or absorbed through skin. They affect cardiovascular system, central nervous system, liver, and/or kidney. They may cause irritation to skin, eyes,

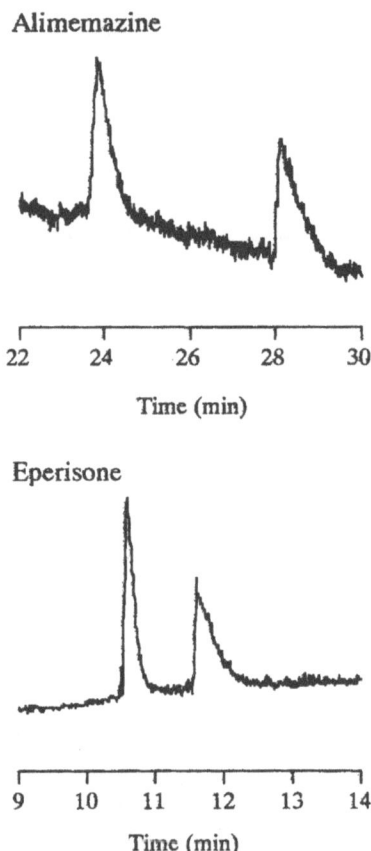

Fig. 6. Electropherograms of alimemazine and eperizone. Reproduced from **ref. *12*** with permission. Separation solutions: 50 m*M* sodium phosphate buffer (pH 5.0)/ 2-propanol (70:30, v/v) for alimemazine; 50 m*M* sodium phosphate buffer (pH 6.0)/ 2-propanol (95:5, v/v) for eperisone, which include 50 µ*M* OGCHI.

and/or respiratory tract. Therefore, they should be manipulated with extreme caution, using gloves, glasses, and so on. Store in a cool and dry well-ventilated location.
3. To connect the conventional HPLC line and capillary column, a reducing union is used.
4. It is a potent poison. It may be fatal or cause blindness if swallowed (*see* **Note 2**).
5. It is required to protect from moisture. It should be avoided to breath vapor (*see* **Note 2**).
6. It is a potent poison (*see* **Note 2**).
7. It is moisture-sensitive. It may cause severe and permanent damages to the digestive tract and cause irritation of the respiratory tract with burning pain (*see* **Note 2**).

8. It may polymerize explosively if heated to the melting point. Risk of cancer depends on level and duration of exposure (*see* **Note 2**).
9. It is combustible liquid and vapor (*see* **Note 2**).
10. It is a strong oxidizer. Contact with other material may cause fire (*see* **Note 2**).
11. It is found that OMCHI used in previous studies is crude *(13)*. In addition, a new glycoprotein from chicken egg whites is isolated, and it is termed OGCHI. Further, it is found that about 10% OGCHI is included in crude OMCHI preparations, and that chiral recognition ability of OMCHI reported previously *(14)* comes from OGCHI, and pure OMCHI has no chiral recognition ability *(13)*. For the isolation of OGCHI, *see* **ref. *13***.
12. The molecular mass of OGCHI is about 30,000. To prepare 50 μ*M* OGCHI, 1.5 mg of OGCHI is dissolved in 1 mL of a running buffer solution.
13. A small amount of silica gel (5-μm diameter) is moistened with deionized water to form a paste. The end of the capillary is then tapped into the paste until approx 2 mm of the tube is packed. The silica gel is sintered at the end of the capillary by gently heating with a small flame for approx 10 s.
14. The addition of a low concentration of electrolyte is useful to avoid clumping of the packing material. The rather low packing/liquid ratio also helps in reducing clumping and blockages during packing. In order to maintain the activity of immobilized proteins, it is better to avoid the use of balanced-density slurries and of other pure organic solvents. The use of an ultrasonic bath is effective for the preparation of slurries.
15. After switching off the pump, it takes 3 to 4 h before complete reduction of the residual pressure.
16. The capillary is first gently heated for a few seconds at the desired site for the frit to dry the packing. Rapid heating should be avoided, since this leads to a local disruption of the packing due to violent boiling of the buffer in the capillary. Then, the retaining frit is sintered by heating in the middle of the flame for approx 5 s. Localization of the heating is achieved by placing the capillary behind a 4-mm-diameter hole in an aluminum plate, mounted next to a Bunsen burner with a low flame. A low-pressure air jet is used to direct the flame through the hole in the plate and onto a localized region of the capillary.
17. To burn away the polyacrylamide coating, localization of the heating is achieved by placing the capillary behind a 5-mm-diameter hole in an aluminum plate as described in **Note 16**.
18. It is required to filter the sample to remove materials that might clog the capillary.
19. The commercially available CE instrument has two injection modes: one is an electrokinetic injection mode, and the other is a pressure mode.
20. The amounts of the silanol groups of a capillary surface are small without etching. Thus, by reaction with sodium hydroxide, the surface of a capillary wall is etched.
21. To remove Na$^+$ ions from the surface and to produce free silanol groups, the capillary is flushed with 0.1 *M* hydrochloric acid.
22. By reaction with hydrochloric acid, the epoxide is hydrolyzed to a diol.

23. Pyridine is added to remove hydrochloric acid, which yields as a reaction product, as pyridinium chloride. Without removing hydrochloric acid, the reaction does not proceed completely.
24. Both ends of the capillary are sealed by heating.
25. The capillary is washed with 1 *M* sodium chloride solution to remove noncovalently adsorbed BSA.
26. The method is based on the use of a bifunctional compound, in which one group specifically reacts with the capillary wall and the other reacts with a monomer taking part in polymerization process. In this case, 3-methacryloxypropyltrimethoxysilane is used. This procedure gives a thin, well-defined monomolecular layer of a polymer covalently bound to the capillary wall.
27. The reaction of a silane reagent is generally performed in nonaqueous solvent. The reaction is performed in acidic, aqueous solution, too. In this case, the latter is used.
28. The suction is performed using an aspirator.
29. Degassing is performed using an ultrasonic bath.
30. To avoid the generation of a bubble, the running buffer solution is degassed using an ultrasonic bath.
31. The running buffer solution used does not include OGCHI, but the separation solution, whose composition is the same as the running buffer solution, includes OGCHI.
32. The polyacrylamide-coated capillary is only stable at pH ranges 2.0–8.0. The capillary should be washed with special care.

References

1. Lloyd, D. K., Aubry, A.-F., and De Lorenzi, E. (1995) Selectivity in capillary electrophoresis: the use of proteins. *J. Chromatogr. A* **792,** 349–369.
2. Hage, D. S. (1997) Chiral separation on capillary electrophoresis using proteins as stereoselective binding agents. *Electrophoresis* **18,** 2311–2321.
3. Haginaka, J. (2000) Enantiomer separation of drugs by capillary electrophoresis using proteins as chiral selectors. *J. Chromatogr. A* **875,** 235–254.
4. Tanaka, Y. and Terabe, S. (2001) Recent advances in enantiomeric separations by affinity capillary electrophoresis using proteins and peptides. *J. Biochem. Biophys. Methods* **48,** 103–116.
5. Lloyd, D. K., Li, S., and Ryan, P. (1995) Protein chiral selector in free-solution capillary electrophoresis and packed-capillary electrochromatography. *J. Chromatogr. A* **694,** 285–296.
6. Li, S. and Lloyd, D. K. (1993) Direct chiral separations by capillary electrophoresis using capillaries packed with an α_1-acid glycoprotein chiral stationary phase. *Anal. Chem.* **65,** 3684–3690.
7. Hofstetter, H., Hofstetter, O., and Schurig, V. (1998) Enantiomer separation using BSA as chiral stationary phase in affinity OTEC and OTLC. *J. Microcol. Sep.* **10,** 287–291.

8. Hage, D. S. and Yang, J. (1994) Chiral separations in capillary electrophoresis using human serum albumin as buffer additive. *Anal. Chem.* **66,** 2719–2725.

9. Hjertén, S. (1985) High-performance electrophoresis. Elimination of electroosmosis and solute interaction. *J. Chromatogr.* **347,** 191–198.

10. Tanaka, Y. and Terabe, S. (1995) Partial separation zone technique for the separation of enantiomers by affinity electrokinetic chromatography with proteins as chiral pseudostationary phases. *J. Chromatogr. A* **694,** 277–284.

11. Haginaka, J. and Kanasugi, N. (1997) Separation of basic drug enantiomers by capillary electrophoresis using ovoglycoprotein as a chiral selector. *J. Chromatogr. A* **782,** 281–288.

12. Matsunaga, H. and Haginaka, J. (2001) Separation of basic drug enantiomers by capillary electrophoresis using ovoglycoprotein as a chiral selector: Comparison of chiral resolution ability of ovoglycoprotein and completely deglycosylated ovoglycoprotein. *Electrophoresis* **22,** 3252–3256.

13. Haginaka, J., Seyama, C., and Kanasugi, N. (1995) The absence of chiral recognition ability in ovomucoid. Ovoglycoprotein-bonded HPLC stationary phases for chiral recognition. *Anal. Chem.* **67,** 2539–2547.

14. Miwa, T., Ichikawa, M., Tsuno, M., et al. (1987) Direct liquid chromatographic resolution of racemic compounds. Use of ovomucoid as a column ligand. *Chem. Pharm. Bull.* **35,** 682–686.

16

Cellulases as Chiral Selectors in Capillary Electrophoresis

Gunnar Johansson, Roland Isaksson, and Göran Pettersson

1. Introduction

The vast majority of proteins exercise their action by means of interaction with small molecules. These interactions are in most cases stereospecific, which is not surprising since the proteins themselves are chiral. As a useful consequence, proteins could have a potential as enantioselectors. This is immediately evident in the case of carrier/transport proteins, such as serum albumin *(1)* or α-1 acid glycoprotein (orosomucoid) *(2)*, but also enzymes *(3,4)* bind their substrates and other ligands in a stereospecific way.

This phenomenon has been exploited, and chiral stationary phases have been created by coupling selected proteins to a solid support. Chromatographic enantioselective columns with a very broad applicability based on α-1 acid glycoprotein *(1)* are commercially available. The filamentous fungus *Trichoderma reesei* secretes a set of cellulose hydrolyzing enzymes, cellulases *(5)*, in large quantities. Among these, the two cellobiohydrolases Cel 7A *(3,6)* and Cel 6A *(7)* have been documented as useful chiral selectors. These two enzymes degrade the cellulose chains sequentially from one end. As an adaptation to this kind of action, the active sites of both enzymes have evolved into a 20–50 Å long tunnel structure *(8)*, which is believed to be of vital importance for their enantioselective behavior. A strongly homologous sibling to Cel 7A, the endoglucanase Cel 7B, which has all interacting groups positioned in an identical configuration but lacks the closed tunnel structure, displays poor enantioselective ability *(9)*.

When Cel 7A was immobilized to silica to form a stationary phase, it displayed excellent enantioselectivity for separation of β-adrenergic blocking agents

From: *Methods in Molecular Biology, Vol. 243: Chiral Separations: Methods and Protocols*
Edited by: G. Gübitz and M. G. Schmid © Humana Press Inc., Totowa, NJ

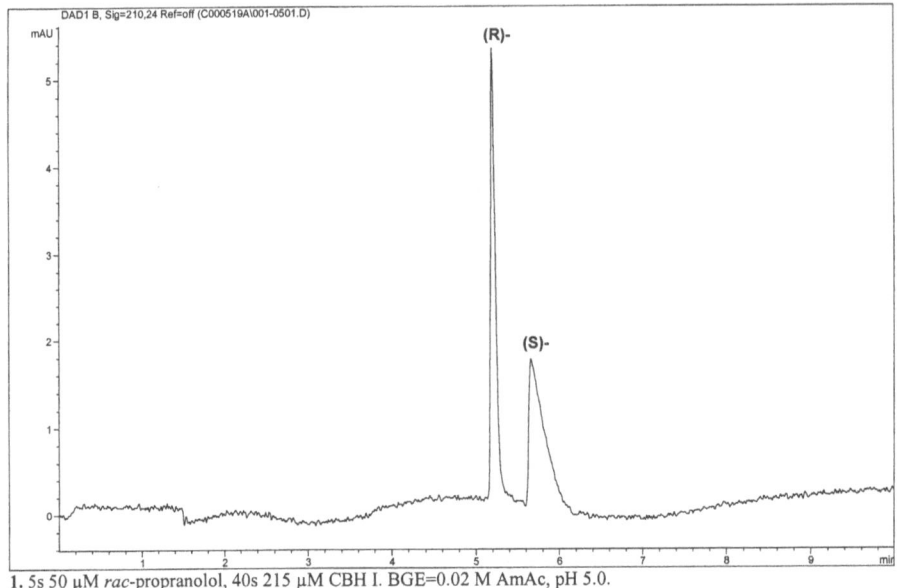

DAD1 B, Sig=210,24 Ref=off (C000519A\001-0501.D)

(R)-

(S)-

1. 5s 50 μM *rac*-propranolol, 40s 215 μM CBH I. BGE=0.02 M AmAc, pH 5.0.

Fig. 1. 5 s 50 μ*M rac*-propranolol, 40 s 215 μ*M* CBH I. BGE = 0.02 *M* AmAc, pH 5.0.

(1,6). Some acidic and ampholytic compounds could be resolved as well by this phase. The enantioselective properties of Cel 7A-based columns have been extensively studied, and commercial columns on which a variety of compounds can be separated are now on the market.

The research has in recent years been focused on the molecular mechanism forming the basis for the enantioselectivity, and crystal structures for the complexes between the enzymes and the chiral compounds have been described *(8)*. As can be expected, the binding site is situated in the tunnel, and the enantiomers of propranolol are bound at the same site with only minor differences in the interaction pattern.

A logical extension of the use of cellulases as imobilized chiral selectors in chromatography was to add cellulases to the background electrolytes (BGE) in capillary electrophoresis (CE) *(10–12)*. In this case, the critical procedure to couple the protein to a carrier matrix is omitted, and the disturbing nonselective interaction expected to take place between matrix and analyte disappears. The experimental data obtained demonstrated that the cellulase was equally efficient as an additive to the buffer in electrophoresis as was earlier found in chromatography. A typical result is shown in **Fig. 1** *(13)*. The high solubility and good stability of the enzyme contributed to the usefulness (*see* **Notes 1–6**).

Table 1
Some Examples of Enantiomer Separation of Drugs
Using Cellulases as Chiral Selector in the Background Electrolyte

Substances	Type of cellulase	References
Alprenolol	**Cel7A**	*11*
Bambuterol analogues	CBH58	*12*
Mexiletine	CBH58	*12*
Norephedrine	CBH58	*12*
Oxprenolol	Cel7A	*11*
Propranolol	Cel7A, CBH 58	*11,12*
Warfarin	Cel7A	*11*
Di-*p*-toluoyltartaric acid	Cel7A	*11*
Pindolol	Cellulase (*Aspergillus niger*)	*14*
Trimipramine	CBH 58	*12*

1.1. Area of Use:
Documented Applications of Celluloses as Chiral Selectors in CE

As shown in several reports, cellulases are powerful chiral selectors in capillary electrophoretic enantiomer separations of both acidic and basic drugs. Some representative examples of such separations are shown in **Table 1**. The first separations were carried out using Cel7A *(1)* as selector, but lately, successful separation were achieved also by use of *Phanerochaete chrysosporium* Cel 7D *(2)* and *Aspergillus niger* cellulase *(3)*.

As were found in the liquid chromatographic applications of cellulases, the best candidates for separations are basic drugs, especially the aromatic aminoalcohols serving as β-blocking agents.

In the majority of these reported CE-separations, the partial filling technique has been adopted. Like other proteins, Cel 7A displays strong UV absorption in a wide wavelength range ($\varepsilon = 78800/(M)$ (cm) at 280 nm and >10,000 still at 300 nm). Including the protein in the BGE may thus cause severe background absorption and reduce the sensitivity of detection for analytes with main absorption bands below 300 nm. To circumvent this problem Valtcheva et al. *(10)* and later Tanaka and Terabe *(15)* introduced the partial filling technique (plug technique). This technique makes it possible to prevent the selector (protein) from reaching the detection window, and thus, an improved detectability of the analyte is obtained. Enantiomer separations of aminoalcohol can illustrate this method. If the pH of the BGE is higher than the isoelectric point (pI) of the selector (protein) but lower than the pK_a-value of the aminoalcohol, the selector and the analyte will migrate in the opposite direction, and only the analyte will reach the detector window.

However, even in such cases, when both the selector and the analyte pass the detection window, it is still possible to determine the enantiomeric composition without having a problem with the high UV absorption of the protein. An absolute minimum condition necessary for a successful detection is, of course, that the selector and the analyte have distinctively different mobilities, thus preventing them from comigrating to the detector region. This was recently demonstrated nicely by the separation of the warfarin enantiomers *(11)*. In this case, both the selector and the acid (analyte) were detected at the anodic site, but the analyte migrated considerably faster, leading to the observation of two sharp separated analyte peaks followed by a plateau signal originating from the selector zone.

The system has also been subject to optimization by chemometric studies *(16)*.

2. Materials

2.1. Choice of Cellulase

1. The trade name "cellulase" cannot be used as a reliable search parameter for material suitable here, since most cellulase products marketed are really a mixture of components secreted in parallel by cellulolytic organisms in which only a few, if any, components display the desired separation capability (*see* **Note 7**).
2. Some crude preparations, at least those from *Trichoderma* strains may indeed display some separation power, since the most useful enzyme is a dominating component, but both performance and reproducibility are deemed to be poor. A successful result can only be expected for highly purified enzyme preparations. As evident from the introduction, a limited number of cellulases are documented for use in this field so far.
3. The most studied is definitely Cel7A (CBH1) from *T. reesei*, and since it has several attractive features, including good stability, high solubility, strong ligand binding, and a relative ease of preparation, it will form the focus for this article.
4. Other cellulases that could function, and which to some extent are complementary to the former one, include Cel 6A (CBH-2) from *T. reesei* and Cel 7D (CBH-58) from *P. chrysosporium*.

3. Methods

3.1. Preparation of Cellulase

The Cel 7A can be prepared in good yield from several commercially available crude enzyme mixtures from *Trichoderma* strains, including Celluclast from Novo Nordisk AS (Copenhagen, Denmark) and Sigma (cat. nos. C 8546 or C 9422). *T. reesei/viride* material from other suppliers should generally give similar result.

Note: Preparations that are explicitly described as endoglucanases should be avoided, since they probably have a poorer content of the Cel 7A enzyme.

The following is the suggested purification protocol starting from lyophilized enzyme powder *(17)*:

1. Dissolve in 10 m*M* ammonium acetate to an estimated protein concentration of 50 mg/mL.
2. Desalt to the same buffer using a desalting column packed with, e.g., Sephadex® G-25 (Amersham Biosciences, Piscataway, NJ, USA) or Bio-Gel® P-6 DG (Bio-Rad, Hercules, CA, USA) according to the instructions from the gel supplier.
3. Apply the desalted sample to an ion exchange column, e.g., diethylaminoethyl (DEAE)-Sepharose® from Amersham Biosciences, which has been equilibrated with 10 m*M* ammonium acetate, pH 5.0 (*see* **Note 8**). Elute with a linear gradient from 10–500 m*M* ammonium acetate, pH 5.0, using 3-column vol of start and end buffer, respectively. The crude Cel 7A should represent a large peak at the end of the chromatogram, which displays a good activity against *p*-nitrophenyl lactoside *(18)*, but virtually no activity towards carboxymethyl cellulose. The material in this peak should be useful for many purposes, but a preparation which meets high demands requires an additional purification step. Here, the ion exchange chromatography is repeated on the same type of column using a gradient of 50–300 m*M* ammonium acetate, pH 3.7. In this step, the Cel 7A probably represents the first eluting main peak. Activity criteria are as before. The highly purified enzyme should appear as a slightly fuzzy band with an apparent molecular weight of approx 65 kDa in sodium dodecyl sulfate-polyacrylamide gel electrophoresis (SDS-PAGE) with 12% polyacrylamide and a pI in the range 3.9–4.3.

The purified enzyme is generally very stable. Lyophilized powder or sterile filtered solutions in pH 5.0 buffers can be expected to have shelf lives of several months in the refrigerator and longer for frozen material.

3.2. Standard Experimental Procedures

In order to suppress electroosmosis and protein wall absorption, the capillaries coated with, e.g., polyacrylamide or polyvinyl alcohol (PVA) are preferred, except for the case of noncharged analytes *(10)*. Mesityloxide can be utilized to monitor the extent of osmosis. As a consequence, only charged compounds can be analyzed under these conditions. First, the selector dissolved in the buffer (background electrolyte) was introduced to the capillary by pressure, and then the analytes also dissolved in the buffer. Prior to the introduction, however, the analyte samples were diluted up to 100 times in order to utilize the stacking effect. In **Fig. 1** a typical separation of a β-blocking agent is shown *(13)*. Frequently, an improved efficiency is obtained by addition of a minor amount of water-miscible organic solvent, such as isopropanol, acetonitrile, etc., to the BGE.

In cases when a high enantioselectivity is obtained, very short selector plugs, even shorter than the analyte plug, can be used. Caution must be taken in cases

A

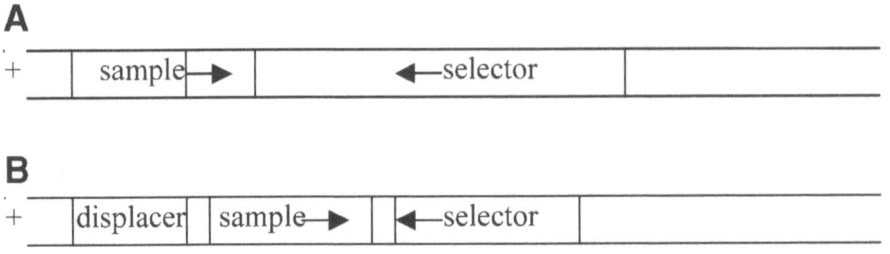

B

Fig. 2. (A) Capillary loading scheme for a cationic sample and an anionic selector. (B) Capillary loading scheme when a displacer is employed.

where the analytes show high affinity to the protein, as there is a risk that one or even both of the enantiomers never reach the detector window *(12)*. To eliminate this risk, a displacer plug containing, for instance, cellobiose can be introduced at the capillary inlet end (**Fig. 2B**). A similar displacing effect can also be used by other modifications of the BGE in the "starting" end of the capillary.

Chemometrical methods are powerful tools to optimize CE separations. In a recent study using Trichoderma CBH-1 as a chiral selector *(16)*, a good separation with high efficiency and good peak symmetry was achieved using pH 6.5, in the BGE at 0.015 *M* ionic strength, and 17% (v/v) of acetonitrile.

4. Notes

1. Useful parameters. The protein-ligand interaction and, thus, the chiral selectivity is influenced by pH, as outlined below. Increased ionic strength may generally weaken the interaction for charged analytes, but so will the addition of 5–20% of an organic solvent such as 2-propanol or acetonitrile, stressing that a larger number of analytes interact with the protein via both ionic and hydrophobic interactions *(8)*. As a consequence, experiments can also be optimized by varying the temperature in the interval at approx 50°C.

2. pH-dependence. The chiral resolution power of Cel7A in chromatography and electrophoresis is strongly pH-dependent. For positively charged analytes that, up to date, are most studied, we find an improved effect as pH is increased, mostly due to the combined effects of stronger interaction at the chiral site *(8)* and a higher countermigration velocity of the protein. Acidic or neutral analytes, on the other hand, are more likely to be resolved at pH 3.0 to 4.0.

3. The following is a summary of sample-selector application sequences in partial filling mode for Cel 7A:

   a. Analyte positively charged, pH 4.0–7.0, or analyte negatively charged pH 3.0 to 4.0. Apply first the desired volume of selector, followed by the analyte sample. Here, the sample and selector zones will "collide" and thus interact. The selector will not reach the detector zone and interfere with the monitoring.

b. Analyte negatively charged (acidic) pH 4.0–7.0 and migrating faster than the selector. Apply first the selector zone followed by the analyte. The analyte will here "catch up" to the selector zone and experience retardation due to the interaction, but a careful tuning of the selector zone, in terms of selector concentration and zone length, will permit the sample components to "run free" of the selector zone before reaching the detector, as demonstrated for the enantioresolution of warfarin *(11)*.

c. Analyte acidic, but migrating slower than the selector, pH 4.0–7.0. In this tentative situation, we have conditions opposite to the previous case. The analyte should be applied before the selector zone, which will then overtake the analyte and speed up its migration. Similarly to above, the experiment has to be designed to make sure that analytes and selector have separated completely before they reach the detector.

d. Extension to noncharged analytes. In this case, we must rely on electroosmosis to transport the free sample molecules. We can, however, employ the procedures outlined above as follows: (i) if sample and selector *de facto* migrate in opposite directions, design the experiment according to case; (ii) if the sample migrates in the same direction as the selector but faster, choose scheme; and (iii) if the sample migrates in the same direction as the selector but slower, choose scheme.

4. Warning for backwards migration. The Cel 7A is a strongly acidic protein and will quickly acquire a high negative charge as the pH goes above its pI of approx 3.9. If a cationic analyte is analyzed at conditions where [Cel7A] is higher than the K_d for the protein-analyte interaction, there is a risk that the most retarded component will follow the selector backwards and thus get partially lost, resulting in an incorrect estimate of analyte composition. For remedies, *see* **Notes 5** and **6**.

5. Spread-out selector zone. It is important to note that the retention of a certain analyte by a selector in partial filling mode is basically dependent on the total amount of selector acting, but virtually independent on its distribution *(19)*. Thus, in order to avoid the problems mentioned above, it advantageous to present the selector in the largest possible portion of the capillary, allowing it to have a moderate influence on the migration velocity in a large portion of the travelled distance rather than a very dramatic effect, even reversal, in a short section. Furthermore, the relative errors in the amount of selector introduced will be minimized and the spread of the selector in a larger volume will tend to diminish the migration artifacts due to the influence from the selector on the conductivity.

6. Integrator automation. In chromatography, the quantitative determination of analytes, in particular within enantiomer pairs, is easily achieved by integration of the signal peaks, e.g., from a UV-detector. The integration here is straightforward, since all components pass the detector with the velocity of the liquid flow. The same achievement in CE is more complicated, since the detector here monitors the components during the separation process. The integral observed in time domain here will not only depend on concentration and absorbance properties of the components but also on the migration velocity of the components as they pass

the detector. In conventional zone electrophoresis, this can be corrected if a weighing function is applied. Assuming that all components migrate with steady velocities in the systems, it is appropriate to use $1/t_{mig}$ as weighing parameter, since t_{mig} is proportional to their residence time in the detector. When a selector is employed in partial filling mode, or in other cases where the migration velocity of the components varies during the experiment, this correction mode is not valid, since the t_{mig} is no longer proportional to the time in the detector region. For separated enantiomers in particular, their velocity in the detector area should be equal, despite the difference in migration time achieved in the selector region. Here, it is recommended to disable the correction function in order to obtain a correct ratio for the pair of enantiomers. In this case, the integral ratio will not be correct for components with different mobilities in the BGE.

7. Choice of cellulase preparations. Cellulase preparations explicitly described as "endoglucanase" can generally be expected to be poor with respect to enantioselective components and should be avoided.

8. Faster equilibration of column. The equilibration may be quicker if you allow 0.5 column vol of 0.5 M buffer of the same kind to pass the column before you introduce the starting buffer. Check equilibration carefully with respect to pH and conductivity before applying the sample.

9. The use of a displacer or dissociating conditions. Another experimental design that avoids the loss of components due to pronounced backward migration of the analyte-selector complex was reported by Hedeland et al. *(12)*. Here, a zone of cellobiose, a natural inhibitor to the Cel 7A enzyme, is introduced after the analyte sample at the inlet end of the capillary. Furthermore, one uses a selector solution of sufficient concentration to provide a transient backward migration of the analytes. This backward migration into the competing cellobiose zone results in a release of the analytes from the selector. A considerable concentration of dilute sample zones can here be achieved together with a remaining enantioselectivity.

References

1. Allenmark, S., Bomgren, B., and Borén, H. (1984) Direct liquid chromatographic separation of enantiomers on immobilized protein stationary phases: IV. Molecular interaction forces and retention behaviour in chromatography on bovine serum albumin as a stationary phase. *J. Chromatogr.* **316,** 617–624.
2. Hermansson, J. and Eriksson, M. (1986) Direct liquid chromatographic resolution of acidic drugs using a chiral α_1-acidic glycoprotein column (Enantiopac®). *J. Liq Chromatogr.* **9,** 621–639.
3. Erlandsson, P., Marle, I., Hansson, L., Isaksson, R., Pettersson, C., and Pettersson, G. (1990) Immobilised cellulase (CBH1) as a chiral stationary phase for direct resolution of enantiomers. *J. Am. Chem. Soc.* **112,** 4573–4574.
4. Jadaud, P., Thelohan, S., Schonbaum, G. R., and Wainer, I. W. (1989) The Stereochemical resolution of enantiomeric free and derivatized amino acids using an HPLC chiral stationary phase based on immobilized α chymotrypsin: chiral separation due to solute structure or enzyme activity. *Chirality* **1,** 38–44.

5. Ilmen, M., Saloheimi, A., Onnela, M. L., and Penttilä, M. E. (1997) Regulation of cellulase gene expression in the filamentous fungus *Trichoderma reesei*. *Appl. Environ. Microbiol.* **63**, 1298–1306.
6. Marle, I., Erlandsson, P., Hansson, L., Isaksson, R., Pettersson, C., and Pettersson, G. (1991) Separation of enantiomers using cellulase (CBH I) silica as a chiral stationary phase. *J. Chromatogr.* **586**, 233–248.
7. Henriksson, H., Petersson, G., and Johansson, G. (1999) Discrimination between enantioselective and non-selective binding sites on cellobiohydrolase-based stationary phases by selective displacers. *J. Chromatogr. A* **857**, 107–115.
8. Ståhlberg, J., Henriksson, H., Divne, C., et al. (2001) Structural basis for enantiomer binding and separation of a common β-blocker: crystal structure of cellobiohydrolase 1 with bound (S)-propranolol at 1.9 Å resolution. *J. Mol. Biol.* **305**, 79–93.
9. Henriksson, H., Stålberg, J., Isaksson, R., and Pettersson, G. (1996) The active sites of cellulases are involved in chiral recognition: a comparison of cellobiohydrolase I and endoglucanase I. *FEBS Lett.* **390**, 339–344.
10. Valtcheva, L., Mohammad, J., Pettersson, G., and Hjerten, S. (1993) Chiral separation of β-blockers by high performance capillary electrophoresis based on non-immobilised cellulase as enantioselective protein. *J. Chromatogr.* **638**, 263–267.
11. Hedeland, M., Isaksson, R., and Pettersson, C. (1998) Cellobiohydrolase I as a chiral additive in capillary electrophoresis and liquid chromatography. *J. Chromatogr. A* **807**, 297–305.
12. Hedeland, M., Nygard, M., Isaksson, R., and Pettersson, C. (2000) Cellulases from the fungi *Phanerochaete chrysosporium* and *Trichoderma reesei* as chiral selectors in capillary electrophoresis; applications with displacer plugs and sample preconcentration. *Electrophoresis* **21**, 1587–1596.
13. Lindberg, K. (2000) Undergraduate thesis. Department of Biochemistry, Uppsala University, Uppsala, Sweden.
14. Busch, S., Kraak, J. C., and Poppe, H. (1993) Chiral separations by complexation with proteins in capillary zone electrophoresis. *J. Chromatogr.* **635**, 119–126
15. Tanaka, Y. and Terabe, S. (1995) Partial separation zone technique for the separation of enantiomers by affinity electrokinetic chromatography with proteins as chiral pseudo-stationary phases. *J. Chromatogr. A* **694**, 277–284.
16. Harang, V., Tysk, M., Westerlund, D., Isaksson, R., and Johansson, G. (2002) A statistical experimental design to study factors affecting enantioseparation of propranolol by capillary electrophoresis with cellobiohydrolase (Cel7A) as chiral selector. *Electrophoresis* **23**, 2306–2319.
17. Bhikhabhai, R., Johansson, G., and Pettersson, G. (1984) Isolation of cellulolytic enzymes from *Trichoderma Reesei* QM 9414. *J. Appl. Biochem.* **5**, 336–345.
18. Deshpande, M. V., Eriksson, K.-E. L., and Pettersson, L. G. (1984) An assay for selective determination of exo-1,4,-β-glucanases in a mixture of cellulolytic enzymes. *Anal. Biochem.* **138**, 481–487.
19. Johansson, G., Isaksson, R., and Harang, V. (2003) Migration time and peak area artifacts caused by systemic effects in voltage controlled capillary electrophoresis. *J. Chromatogr A* **1004**, 91–98.

17

Use of Chiral Crown Ethers in Capillary Electrophoresis

Martin G. Schmid and Gerald Gübitz

1. Introduction

Chiral crown ethers are known to include stereoselectively compounds containing primary amino groups. This principle has successfully been applied in liquid chromatography (LC) using crown ether-based stationary phases *(1–10)*. Kuhn et al. *(11)* transferred this basic principle to capillary electrophoresis (CE) using (+)-18-crown-6-tetracarboxylic acid (**Fig. 1**) as a chiral selector added to the electrolyte for the chiral separation of amino acids. This crown ether was shown to be applicable also to the chiral separation of dipeptides *(12,13)* (**Fig. 2**), sympathomimetics *(14)* (**Fig. 3**), and various other drugs containing primary amino groups *(15)* by CE.

The chiral recognition mechanism is based on the formation of hydrogen bonds between the three amine hydrogens and the oxygens of the macrocyclic ether. The substituents of the crown ether, which are arranged perpendicular to the plane of the ring, form a chiral barrier dividing the cavity into two domains. Thereby two diastereomeric inclusion complexes are formed with the analyte enantiomers.

In this chapter, procedures are given for the chiral separation of amino acids, dipeptides, and other compounds containing primary amino groups by CE.

2. Materials

2.1. Apparatus

1. CE instrument equipped with a UV detector.
2. Personal computer (PC) for data aquisition.
3. Fused silica capillaries, e.g., from Microquartz (Munich, Germany).
4. Special capillary cutting blade.

From: *Methods in Molecular Biology, Vol. 243: Chiral Separations: Methods and Protocols*
Edited by: G. Gübitz and M. G. Schmid © Humana Press Inc., Totowa, NJ

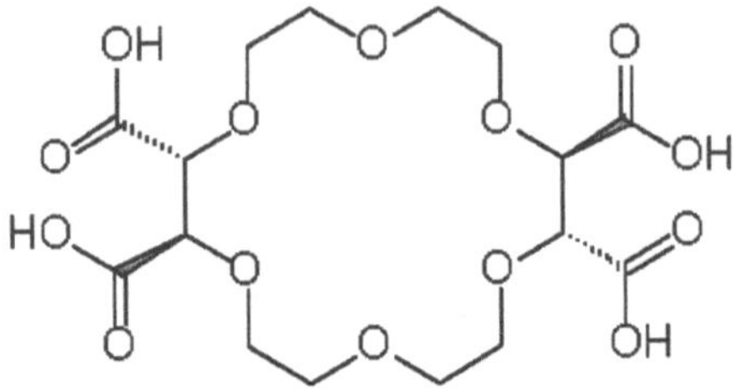

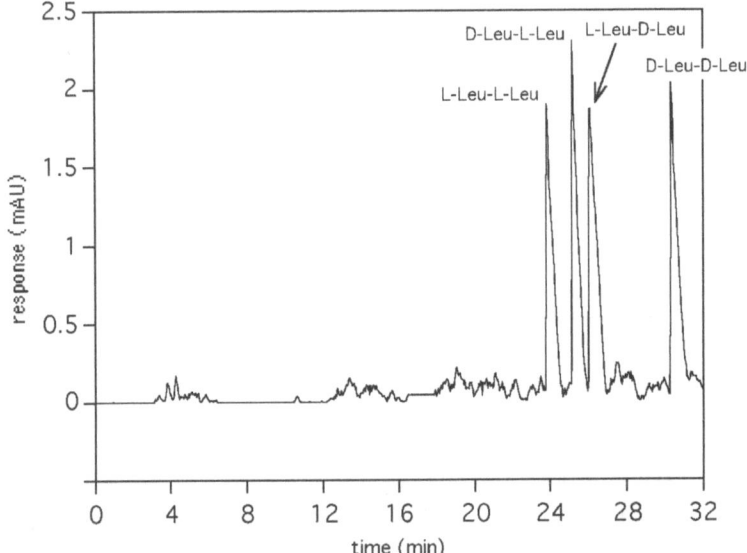

Fig. 1. Chemical structure of (+)-18-crown-6-tetracarboxylic acid.

Fig. 2. Electropherogram of the chiral resolution of leucyl-leucine (electrolyte: 15 mM 18C$_6$H$_4$, 10 mM Tris-citrate, 20% [v/v] methanol, U = 30 kV). From **ref.** *13* with permission.

2.2. Conditioning of the Capillary

1. 0.1 M NaOH, 10 mL.
2. 0.1 M HCl, 10 mL.

2.3. Preparation of Background Electrolyte Solutions

1. Citric acid.
2. Tris base.
3. 18-Crown-6-tetracarboxylic acid.
4. Formamide.

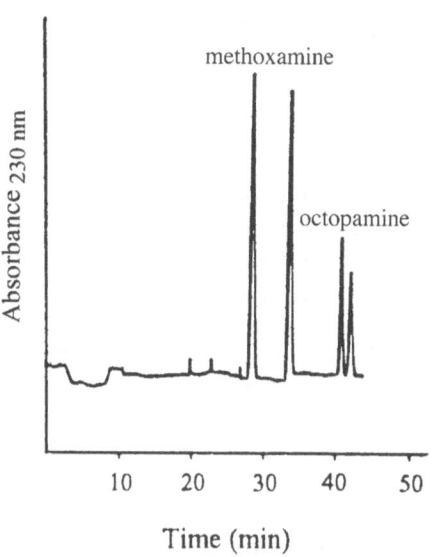

Fig. 3. Electropherogram of the enantiomer separation of methoxamine and octopamine using 30 m*M* 18C₆H₄, pH 2.07, U = 15 kV. From **ref. *14*** with permission.

5. Tetra-*n*-butyl ammonium perchlorate.
6. Triethylamine (TEA).
7. Methanol of analytical grade.
8. Double-distilled water.
9. Syringe type membrane filters 0.20 or 0.45 μm.

3. Methods

3.1. Aqueous CE

3.1.2. Separation Conditions

1. Dissolve up to 30 m*M* 18-crown-6-tetracarboxylic acid in water and adjust pH to 2.0–2.2 (*see* **Notes 1** and **2**).
2. After degassing and filtration through a syringe filter, electrolyte is ready for use.
3. Dissolve primary amines in electrolyte or water (1 mg/mL).
4. Observe enantioseparation and verify enantiomeric elution order by injecting the pure enantiomers at equal conditions.
5. To enhance separation, up to 20% methanol (v/v) may be added (*see* **Note 3**).

3.1. Nonaqueous CE (16)

3.1.1. Separation Conditions

1. Dissolve up to 50 m*M* 18-crown-6-tetracarboxylic acid in formamide (*see* **Note 4**).
2. After degassing and filtration through a syringe filter, electrolyte is ready for use.

3. Dissolve primary amines in electrolyte (1 mg/mL).
4. Observe enantioseparation and verify enantiomeric elution order by injecting the pure enantiomers at equal conditions.

4. Notes

1. Avoid cations such as potassium or ammonium, because they compete with the enantiomers for the crown ether's cavity.
2. In some cases, electrolyte may consist of the chiral crown ether dissolved in water only, but in other cases, Tris-citric acid may be useful as a background electrolyte system
3. Addition of TEA may improve resolution in some cases by increasing migration time drastically and reducing efficiency.
4. Separation efficiency may be improved by adding up to 100 mM tetra-n-butyl ammonium perchlorate as a supporting electrolyte *(16)*.

References

1. Sogah, G. D. Y. and Cram, D. J. (1979) Host-guest complexation.14. Host covalently bound to polystyrene resin for chromatographic resolution of enantiomers of amino acids and ester salts. *J. Am. Chem. Soc.* **101,** 3035–3042.
2. Shinbo, T., Nishimura, K., Sugiura, M., and Yamaguchi, T. (1987) Chromatographic separation of racemic amino-acids by use of chiral crown ether-coated reversed-phase packings. *J. Chromatogr.* **405,** 145–153.
3. Lee, W. and Hong, C. Y. (2000) Direct liquid chromatographic enantiomer separation of new fluoroquinolones including gemifloxacin. *J. Chromatogr. A* **879,** 113–120.
4. Machida, Y., Nishi, H., and Nakamura, K. (1999) Enantiomer separation of hydrophobic amino compounds by HPLC using crown ether dynamically coated chiral stationary phase. *J. Chromatogr. A* **830,** 311–320.
5. Péter, A., Fülöp, F., and Tourwé, D. (1995) High-performance liquid chromatographic method for the separation of isomers of cis- and trans-2-amino-cyclopentane-1-carboxylic acid. *J. Chromatogr. A* **715,** 219–226.
6. Péter, A., Tóth, G., Török, G., and Tourwé, D. (1998) Separation of enantiomeric beta-methyl amino acids and of beta-methyl amino acid containing peptides. *J. Chromatogr. A* **728,** 455–465.
7. Machida, Y., Nishi, H., Nakamura, K., Nakai, H., and Sato, T. (1998) Enantiomer separation of amino compounds by a novel chiral stationary phase derived from crown ether. *J. Chromatogr. A* **805,** 85–92.
8. Hyun, M. H., Jin, J. S., and Lee, W. (1998) Liquid chromatographic resolution of racemic amino acids and their derivatives on a new chiral stationary phase based on crown ether. *J. Chromatogr. A* **822,** 155–161.
9. Hyun, M. H., Jin, J. S., Koo, H. J., and Lee, W. (1999) Liquid chromatographic resolution of racemic amines and amino alcohols on a chiral stationary phase derived from crown ether. *J. Chromatogr. A* **837,** 75–82.

10. Hyun, M. H., Han, S. C., Lipshutz, B. H., Shin, Y.-J., and Welch, C. J. (2001) New chiral crown ether stationary phase for the liquid chromatographic resolution of α-amino acid enantiomers. *J. Chromatogr. A* **910,** 359–365.

11. Kuhn, R., Erni, F., Bereuter, T., and Häusler, J. (1992) Chiral recognition and enantiomeric resolution based on host guest complexation with crown ethers in capillary zone electrophoresis. *Anal. Chem.* **64,** 2815–2820.

12. Kuhn, R., Riester, D., Fleckenstein, B., and Wiesmüller, K.-H. (1995) Evaluation of an optically-active crown-ether for the chiral separation of dipeptides and tripeptides. *J. Chromatogr. A* **716,** 371–379.

13. Schmid, M. G. and Gübitz, G. (1995) Capillary zone electrophoretic separation of the enantiomers of dipeptides based on host-guest complexation with a chiral crown-ether. J. *Chromatogr. A* **709,** 81–88.

14. Höhne, E., Krauss, G.-J., and Gübitz, G. (1992) Capillary zone electrophoresis of the enantiomers of aminoalcohols based on host-guest complexation with a chiral crown-ether. *J. High Resol. Chromatogr.* **15,** 698–700.

15. Nishi, H., Nakamura, K., Nakai, H., and Sato, T. (1997) Separation of enantiomers and isomers of amino-compounds by capillary electrophoresis and high-performance liquid-chromatography utilizing crown-ethers. *J. Chromatogr. A* **757,** 225–235.

16. Mori, Y., Ueno, K., and Umeda, T. (1997) Enantiomeric separations of primary amino-compounds by nonaqueous capillary zone electrophoresis with a chiral crown-ether. *J. Chromatogr. A* **757,** 328–332.

18

Chiral Separations by Capillary Electrophoresis Using Cinchona Alkaloid Derivatives as Chiral Counter-Ions

Michael Lämmerhofer and Wolfgang Lindner

1. Introduction

Cinchona alkaloids and derivatives thereof, obtained by dedicated structural modifications, have a long tradition in various stereochemical methods. Among them, their stereoselective ion-pairing capabilities for acidic compounds have been exploited for capillary electrophoretic enantiomer separations of chiral acids *(1–14)*.

The native cinchona alkaloids, quinine and quinidine (**Fig. 1A**), possess 5 stereogenic centers both with (1*S*, 3*R*, 4*S*)-configuration and opposite configurations at the carbons 8 and 9, which are (8*S*, 9*R*) for quinine and (8*R*, 9*S*) for quinidine. Since the latter two configurations usually exert stereocontrol, both the alkaloids exhibit pseudo-enantiomeric behavior. This means that, in separation technologies, they show reversed affinity towards the enantiomers of an acidic analyte, which then translates into reversed elution orders (*vide supra*). These semi-rigid molecules feature various functionalities, which are quinuclidine group, quinoline ring, and hydroxyl group, representing the binding sites for intermolecular interaction with complementary moieties of analytes. The pK values of the basic quinuclidine group are 9.7 (at 18°C) for quinine and 10.0 (at 20°C) for quinidine *(15)*, and corresponding figures for the aromatic quinoline are 5.07 and 5.4, respectively. The protonated tertiary amine represents the primary ionic interaction site for the ion-pairing mechanism, while the quinoline that is largely undissociated under operating conditions may support ion-pair formation through π-π-interaction with corresponding complementary groups of the analyte.

The native cinchona alkaloid quinine has first been suggested as chiral counter-ion for nonaqueous ion-pair capillary electrophoresis (CE) with methanolic background electrolytes by Stalcup and Gahm *(1)* and was employed in

From: *Methods in Molecular Biology, Vol. 243: Chiral Separations: Methods and Protocols*
Edited by: G. Gübitz and M. G. Schmid © Humana Press Inc., Totowa, NJ

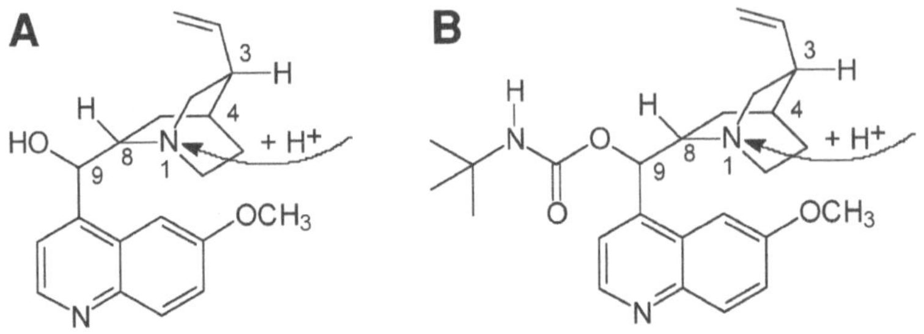

Fig. 1. Structure of cinchona alkaloid-derived chiral counter-ions. (**A**) Native cinchona alkaloids quinine (1*S*, 3*R*, 4*S*, 8*S*, 9*R*) and quinidine (1*S*, 3*R*, 4*S*, 8*R*, 9*S*), and (**B**) corresponding *tert*-butyl carbamates.

a dual selector system in combination with cyclodextrins in aqueous acetate buffer to improve the resolution in CE enantiomer separation of α-arylcarboxylic acids and aromatic hydroxy carboxylic acids like mandelic acid derivatives *(2)*. Piette et al. compared the enantiomer separation capability of quinine and quinidine as well as cinchonidine and cinchonine that have replaced the methoxy group of the quinoline ring by a hydrogen *(5)*. Throughout, the latter were slightly less enantioselective for the enantiomer separation of *N*-derivatized amino acids. The overall separation factors achieved with these native cinchona alkaloids for chiral acids by all the aforementioned methods were quite moderate.

On the contrary, we could show that dedicated modification of the native cinchona alkaloids, e.g., the introduction of a carbamate group in combination with a bulky carbamate residue, at the hydroxyl at carbon 9 leads to an enormous gain of the enantiomer discrimination capability of the cinchonan selector. For example, *N*-benzoyl-β-phenylalanine enantiomers could not be resolved with native quinine as selector, while they were well separated with an R_S value of 5.1 employing the *tert*-butylcarbamate of quinine (**Fig. 1B**) as chiral counter-ion in the background electrolyte (BGE) (Note: The R_s value denotes the resolution between the separated enantiomer peaks and is calculated by the peak width method at half height, $R_s = 1.18 (tr_2 - tr_1) / (w_{1/2\,1} + w_{1/2\,2})$; wherein $tr_{1,2}$ are the migration times of first and second eluted enantiomer and $w_{1/2\,1,2}$ are the peak widths at half height of first and second migrating enantiomers.) *(5)*. For *N*-3,5-dinitrobenzoyl-leucine, R_S could be enhanced from 5.5 (quinine as counter-ion) to 64.3 (*tert*-butylcarbamate of quinine). This trend of significant improvement of enantioselectivities with carbamate selectors compared to corresponding native cinchona alkaloids was observed throughout the investigated analyte set of *N*-derivatized amino acids and may be mainly attributed to the more rigid selector structure with better defined binding pocket as well as the favorable hydrogen donor-acceptor properties of the carbamate group *(5)*. A variety of other carbamate derivatives such as cyclohexyl, 1-adamantyl, 3,4-dichlorophenyl, 3,5-dinitrophenyl carbamates, and bis-(carbamoylquinine) derivatives were then synthesized and showed partly even higher enantio-selectivity (e.g., 1-adamantyl carbamate) or to some extent complementary stereodiscrimination potential (e.g., aromatic carbamates) compared to the *tert*-butyl carbamate *(3–6,11,12)*. Due to its effectiveness and easy accessibility, we suggest the *tert*-butyl carbamates of quinine and quinidine as standard chiral counter-ions among the cinchona alkaloid derivatives for enantioselective ion-pair CE of chiral acids. Herein, we therefore focus our attention in the following description and depicted separation examples on these carbamate derivatives (**Fig. 1B**). Again, like underivatized quinine and quinidine also, the correspond-ing *tert*-butyl carbamates exhibit pseudo-enantiomeric behavior, which is manifested in reversed elution orders of (*R*) and (*S*) enantiomers as exempli-fied in **Fig. 2**.

The spectrum of applicability of *tert*-butyl carbamoylated quinine and qui-nidine counter-ions comprises all kinds of acidic chiral compounds and include in particular the following compounds:

1. *N*-Derivatized amino acids: virtually all of the protection groups or labeling reagents commonly employed in amino acid and peptide chemistry may be utilized for *N*-derivatization. The group of derivatives include 9-fluorenylmethoxycarbonyl (FMOC), benzyloxycarbonyl (Z), 2,4-dinitrophenyl (DNP), 4-(*N,N*-dimethylamino-sulfonyl)-2,1,3-benzoxadiazole-7-yl (DBD), 5-dimethylaminonaphthalin-1-sulfo-nyl (DNS), 6-nitroveratrylmethoxycarbonyl (NVOC), 4-nitrobenzyloxycarbonyl (PNZ), 3,5-dinitrobenzyloxycarbonyl (DNZ), carbazole-9-carbonyl (CC), 3,5-dinitrobenzoyl (DNB), benzoyl (Bz), acetyl (Ac), and many others (*see* **Note 1**).
2. *N*-Derivatized peptide stereoisomers (**Fig. 3**).
3. *N*-Derivatized amino phosphonic acids (**Fig. 2**) and phosphinic acids.
4. *N*-Derivatized amino sulfonic acids.
5. α-Aryloxy carboxylic acids and α-aryl carboxylic acids.
6. Chiral acidic drugs (including carboxylic, sulfonic, and phosphonic acid derivatives).

As already pointed out, the separation mechanism is based on stereoselective ion-pairing that creates a difference of net migration velocities of the both enan-tiomers. Thus, the basic cinchona alkaloid derivative is added as chiral counter-ion [chiral selector, in the following denoted as $SO_{(R)}$] to the BGE. Under the chosen acidic conditions of the BGE the positively charged counter-ion [$SO_{(R)}^+$] associates with the acidic chiral analytes [selectands, in the following denoted as $SA_{(S)}$ and $SA_{(R)}$] usually with 1:1 stoichiometry to form electrically neutral ion-pairs, driven mostly by electrostatic ion-ion interactions and supported by other intermolecular interactions, like hydrogen-bonding, π-π-interaction, and steric interactions, according to the following ion-pair equilibrium reactions:

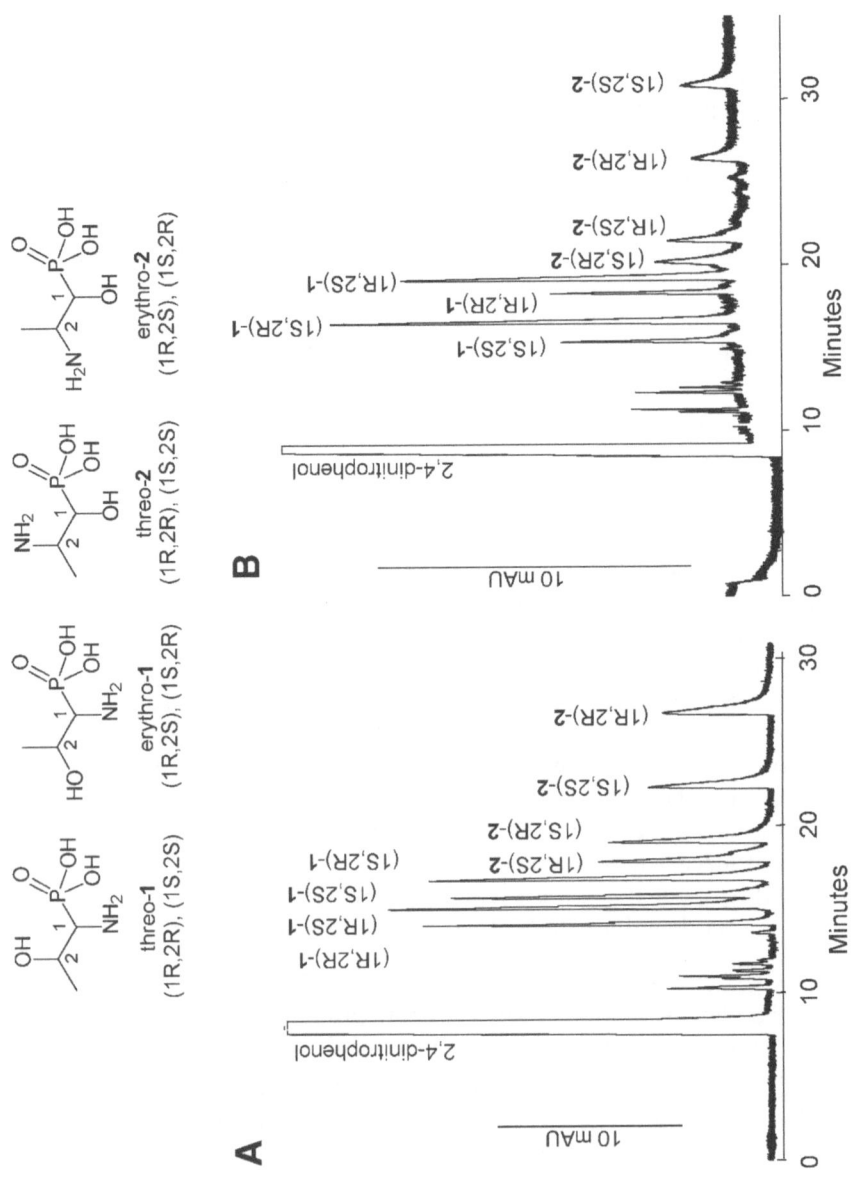

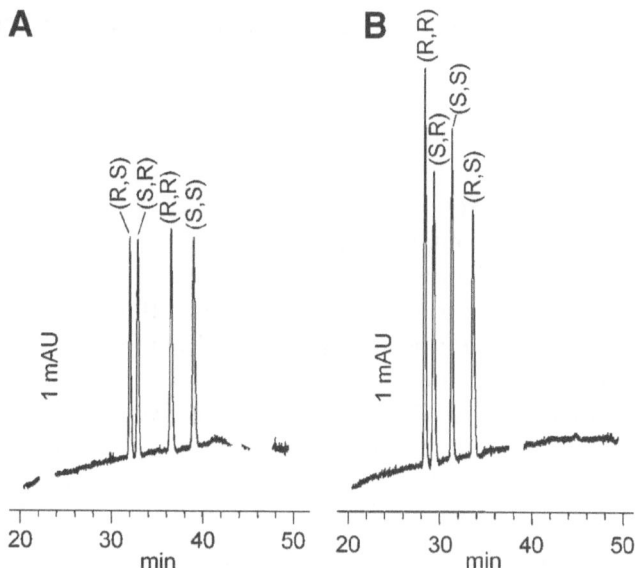

Fig. 3. Separation of all 4 stereoisomers of (**A**) DNZ-Ala-Ala and (**B**) DNP-Ala-Ala (reprinted with permission from **ref. *14***). Experimental conditions: polyvinyl alcohol (PVA)-coated capillary, 50 μm i.d., 64.5 cm total length, 56.0 cm effective length; BGE, methanol containing 100 mM acetic acid, and 12.5 mM triethylamine; selector solution, 10 mM O-9-(*tert*-butylcarbamoyl) quinine in BGE; partial filling, applied plug length: 50.4 cm; applied voltage, –25 kV (plain BGE at both inlet and outlet electrode vessels during run); temperature, 15°C.

$$SO_{(R)}^{\ominus} + SA_{(R)}^{\ominus} \overset{K_{ip,(R)}}{\rightleftharpoons} [SO_{(R)} - SA_{(R)}]^0 \qquad [\text{Eq. 1a}]$$

$$SO_{(R)}^{\ominus} + SA_{(S)}^{\ominus} \overset{K_{ip,(S)}}{\rightleftharpoons} [SO_{(R)} - SA_{(S)}]^0 \qquad [\text{Eq. 1b}]$$

Fig. 2. (*opposite page*) Separation of the stereoisomers of 1-amino-2-hydroxypropane phosphonic acid **1** and 2-amino-1-hydroxypropane phosphonic acid **2** as *N*-2,4-dinitrophenyl derivatives by nonaqueous CE with O-9-(*tert*-butylcarbamoyl) quinine (**a**) and O-9-(*tert*-butylcarbamoyl) quinidine (**b**) as counter-ions illustrating the reversal of elution orders of enantiomers that can be obtained with the both pseudo-enantiomeric counter-ions (with modifications from **ref. *8***). Experimental conditions: fused-silica capillary, 50 μm i.d., 45.5 cm total length, 37 cm to detection window; BGE, 100 mM acetic acid and 12.5 mM triethylamine in ethanol/methanol (60:40, v/v); selector solution, 10 mM counter-ion in BGE; partial filling technique, filling of the selector solution with 50 mbar for 5 min (corresponds to approx 30-cm selector plug length); injection, 50 mbar for 5 s; applied voltage, –25 kV (plain BGE at both inlet and outlet electrode vessels during run); temperature, 15°C.

These equilibrium reactions are characterized by the ion-pair formation constant $K_{ip,(R,S)}$:

$$K_{ip,(R)} = \frac{[SO_{(R)} - SA_{(R)}]}{[SO_{(R)}] \cdot [SA_{(R)}]} \quad \text{and} \quad K_{ip,(S)} = \frac{[SO_{(R)} - SA_{(S)}]}{[SO_{(R)}] \cdot [SA_{(S)}]} \quad \text{[Eq. 2]}$$

where $[SO_{(R)} - SA_{(R,S)}]$, $[SO_{(R)}]$, and $[SA_{(R,S)}]$ are the equilibrium concentrations of the respective SO-SA complex, the free selector, and the respective free selectand. Since intermolecular SO-SA interactions may evolve stereoselectively, ion-pair formation constants $K_{ip,(R)}$ and $K_{ip,(S)}$ may differ for (R)- and (S)-selectand enantiomers and the formed $[SO_{(R)} - SA_{(R)}]^0$ and $[SO_{(R)} - SA_{(S)}]^0$ ion-pairs are diastereomeric to each other.

Under the influence of an electric field, the effective mobility (μ_{eff}) of the SA will be comprised by its mobilities in free (μ_{free}) and complexed forms (μ_{ip}), weighted by the corresponding molar fractions the SA is present in the two states:

$$\mu_{eff} = \mu_{app} - \mu_{EOF} = \left(\frac{[SA]}{[SA] + [SO - SA]}\right) \cdot \mu_{free} + \left(\frac{[SO - SA]}{[SA] + [SO - SA]}\right) \cdot \mu_{ip} \quad \text{[Eq. 3]}$$

where μ_{app} is the apparent mobility calculated from the migration times as specified in **Table 1** (footnote a) and μ_{EOF} is the electroosmotic mobility (*see* **Notes 2 and 3**). Using Eq. 2, $[SO - SA]$ can be replaced by $K_{ip} \cdot [SO] \cdot [SA]$ so that, according to Wren and Rowe, Eq. 3 can be rewritten as:

$$\mu_{eff,(R)} = \frac{\mu_{free,(R)} + \mu_{ip,(R)} \cdot K_{ip,(R)} \cdot [SO_{(R)}]}{1 + K_{ip(R)} \cdot [SO_{(R)}]} \quad \text{[Eq. 4a]}$$

$$\mu_{eff,(S)} = \frac{\mu_{free,(S)} + \mu_{ip,(S)} \cdot K_{ip,(S)} \cdot [SO_{(R)}]}{1 + K_{ip(S)} \cdot [SO_{(R)}]} \quad \text{[Eq. 4b]}$$

In ion-pair CE, the SO-SA-complex (ion-pair) is supposed to have a net charge of zero, and therefore the electrophoretic mobility of the ion-pair μ_{ip} is zero as well, which therefore, moves solely with the velocity of the electroosmotic flow (EOF). Mobilities of the both enantiomers in the free state are equal. Hence, Eq. 4 can be reduced to:

$$\mu_{eff,(R)} = \frac{\mu_{free}}{1 + K_{ip(R)} \cdot [SO_{(R)}]} \quad \text{and} \quad \mu_{eff,(S)} = \frac{\mu_{free}}{1 + K_{ip,(S)} \cdot [SO_{(R)}]} \quad \text{[Eq. 5]}$$

Thus, the dependence of the separation factor (α) on the SO concentration for the present ion-pair CE system can be written as:

Table 1
Selected Enantiomer Separation Data of Chiral Acids
by CE Employing O-9-(*tert*-Butylcarbamoyl) Quinine as Chiral Counter-Ion

Compound	$\mu_{app,1}{}^a$	$\mu_{app,2}{}^a$	α^b	R_S	e.o.^{c}
	[× 10⁻⁵ cm²/(V) (s)]				
Total filling technique^d					
DNB-Leu	−9.51	−4.77	1.99	55.8	(R)
DNZ-Leu	−4.52	−3.23	1.40	14.8	(R)
Bz-β-Phe	−3.61	−3.33	1.08	3.8	(S)
Bz-Leu	−5.26	−4.39	1.20	9.3	(R)
Bz-Tle	−8.75	−7.65	1.15	1.8	(R)
FMOC-Leu	−3.83	−3.31	1.16	7.5	(R)
1,1'-Binaphthyl-2,2'-diyl hydrogenphosphate	−9.09	−8.61	1.06	3.0	(R)
Counter-current technique^e					
DNB-Leu	−6.80	−3.00	2.26	64.3	(R)
DNZ-Leu	−3.77	−2.79	1.35	20.6	(R)
DNZ-β-Abu	−3.86	−3.29	1.17	10.1	(R)
Bz-β-Phe	−2.93	−2.68	1.09	5.1	(S)
Bz-Phe	−5.21	−4.62	1.13	10.1	(R)
Bz-Leu	−4.42	−3.73	1.19	12.4	(R)
Partial filling technique^f					
DNZ-Leu	−4.35	−3.45	1.26	12.3	(R)
Bz-β-Phe	−4.95	-4.35	1.14	7.5	(S)
Bz-Leu	−3.48	−3.29	1.06	3.3	(R)
FMOC-Leu	−3.61	−3.29	1.10	5.6	(R)
DNP-Leu	−9.31	−8.70	1.07	5.5	(S)
DNP-Pro	−8.35	−7.76	1.08	6.6	(S)
DNZ-2-aminopropane sulfonic acid	−10.53	−9.52	1.11	2.0	(R)
DNZ-2-aminobutane sulfonic acid	−10.04	−8.78	1.14	2.8	(R)
DNZ-2-amino-3,3-dimethylbutane sulfonic acid	−8.65	−6.12	1.41	7.8	(R)

$^a\mu_{app} = (L_{eff} \cdot L_{tot})/(t_r \cdot V)$ where L_{eff} is length from injection end to detection point, L_{tot} is total length of capillary, and t_r is migration time, V is applied voltage.

$^b\alpha = \mu_{app,1}/\mu_{app,2}$.

^{c}e.o., configuration of first eluted enantiomer.

^{d}BGE, 100 mM octanoic acid and 12.5 mM triethylamine in ethanol/methanol (60:40, v/v); selector solution, 10 mM in BGE; applied voltage, −25 kV (selector solution in both inlet and outlet electrode vessel); temperature, 15°C.

^{e}BGE, 100 mM octanoic acid and 12.5 mM ammonia in ethanol/methanol (60:40, v/v); selector solution, 10 mM in BGE; applied voltage, −25 kV (selector solution in inlet and BGE in electrode vessel); temperature, 15°C.

^{f}BGE, 100 mM acetic acid and 12.5 mM triethylamine in ethanol/methanol (60:40, v/v); selector solution, 10 mM in BGE; partial filling, 50 mbar for 5 min (approx 30 cm plug length); applied voltage, −25 kV (BGE in both inlet and outlet electrode vessel); temperature, 15°C.

$$\alpha = \frac{\mu_{eff,(R)}}{\mu_{eff,(S)}} = \frac{1 + K_{ip,(S)} \cdot [SO]}{1 + K_{ip,(R)} \cdot [SO]} \qquad \text{[Eq. 6]}$$

where $K_{ip,(S)} > K_{ip,(R)}$ and $\mu_{eff,(R)} > \mu_{eff,(S)}$. It is noted that due to opposite charge of free SO and uncomplexed analytes, their migrations will take place in opposite directions, thus leading to a counter-current type separation process with the stronger bound enantiomer migrating slower than the other one. A counter-current migration does also exist for free (anodic direction) and complexed solute species (cathodic with EOF), which is favorable in terms of separation.

From Eq. 6, it can be derived that inequality of thermodynamic binding constants of the both enantiomers ($K_{ip,(R)}$ and $K_{ip,(S)}$) is the only source for enantioselectivity in the present systems (binding selectivity term), while selectivity contributions arising from mobility differences of diastereomeric associates are supposed to be negligible. It is also seen that the selector concentration plays a major role (see **Note 4**). With enhancement of the chiral counter-ion concentration in the BGE, separation selectivity of the system increases (**Fig. 4**). However, this adversely affects migration velocities (Eq. 5), so that analysis takes longer at higher counter-ion concentrations.

Ion-pair CE with cinchona alkaloid-type chiral counter-ions has been performed mostly with nonaqueous media that consisted of methanol or methanol-ethanol mixtures containing organic acids, such as acetic acid or octanoic acid, and bases, such as ammonia and triethylamine, as electrolytes (see **Note 5**). The reason for preference of nonaqueous solvents over aqueous BGE may be essentially explained by 2 factors. One is the much better solubility of the relatively lipophilic cinchonan derivatives that need to be added to the BGE at concentrations between 2 and 100 mM. On the other hand, the nonaqueous media strengthen electrostatic, e.g., ion-ion interactions being in favor for ion-pairing interactions. In contrast, solvophobic effects, which may also exist and be active between hydrophobic molecule parts of the analyte and the *tert*-butyl group of the counter-ion, are negatively affected in such solvents. Here, it must be emphasized that nonetheless aqueous-based BGEs are also applicable, e.g., methanol/ammonium acetate buffer (80:20, v/v), and should be employed if analyte solubility is poor in nonaqueous BGE. Other variables like apparent pH or acid-base ratio, type of solvents, concentration of electrolytes, and temperature represent influential parameters to optimize separations (see **Note 5**) (3,14).

In the standard experimental set-up of CE enantiomer separation methods, the chiral selector is incorporated to both inlet and outlet buffer reservoir (henceforth termed total filling technique), and the standard detection scheme is UV detection. Since cinchonan derivatives have a high molar absorptivity at the detection wavelengths of most analytes, originating from the electron-rich aromatic quinoline moiety, they produce a strong background signal, which deterio-

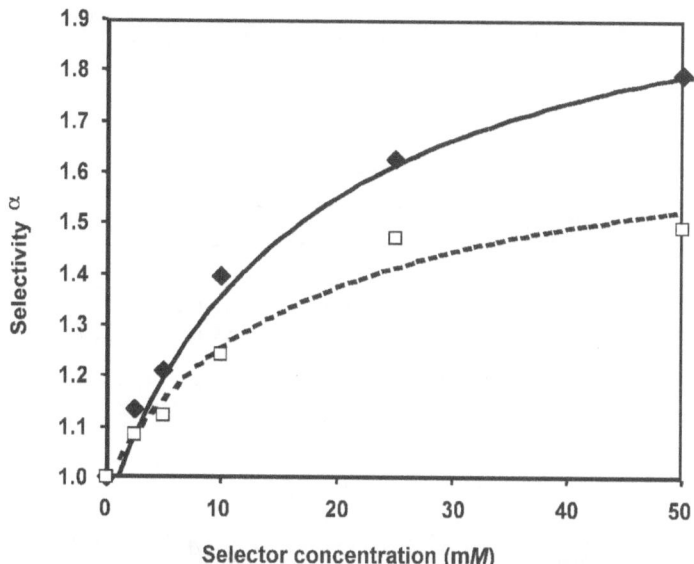

Fig. 4. Plot of enantioselectivity α of DNB-Ala (◆) and DNB-Ala-Ala (□) vs selector concentration (selector: O-*tert*-butylcarbamoyl quinine) (the mobilities used for calculation of α were EOF-corrected) (reprinted with permission from **ref. *14***). The experimental data points and the fitted lines according to Eq. 6 show good agreement. Experimental conditions: PVA-coated capillary, 50 μm i.d., 64.5 cm total length, 56 cm effective length; BGE, 100 mM acetic acid, and 12.5 mM triethylamine in methanol/ethanol (80:20, v/v); partial filling technique, applied plug length of selector solution, 50 cm; injection, 50 mbar for 5 s; temperature, 15°C; applied voltage, −25 kV (plain BGE at both inlet and outlet electrode vessels during run).

rates detection sensitivity considerably. Therefore, this standard experimental set-up (total filling technique) is not very useful for most practical applications with cinchonan-derived counter-ions. Instead, other methodologies, such as counter-current technique, and partial filling technique have been employed that circumvent this drawback of UV absorbing selectors.

The steps of the counter-current technique that may be adopted, because the chiral selector and analytes possess opposite charges and therefore migrate in the capillary in opposite directions, are schematically outlined in **Fig. 5A**. First, the capillary is filled with the BGE containing the chiral counter-ion over the entire length, as in the total filling technique. After injection of the sample, the separation is carried out with BGE-selector solution only in the inlet reservoir (cathodic end) and plain BGE devoid of selector in the outlet electrode vessel (anodic end). The selector zone migrates toward the injection end of the capillary (with its self-electrophoretic velocity and EOF) so that the detection window

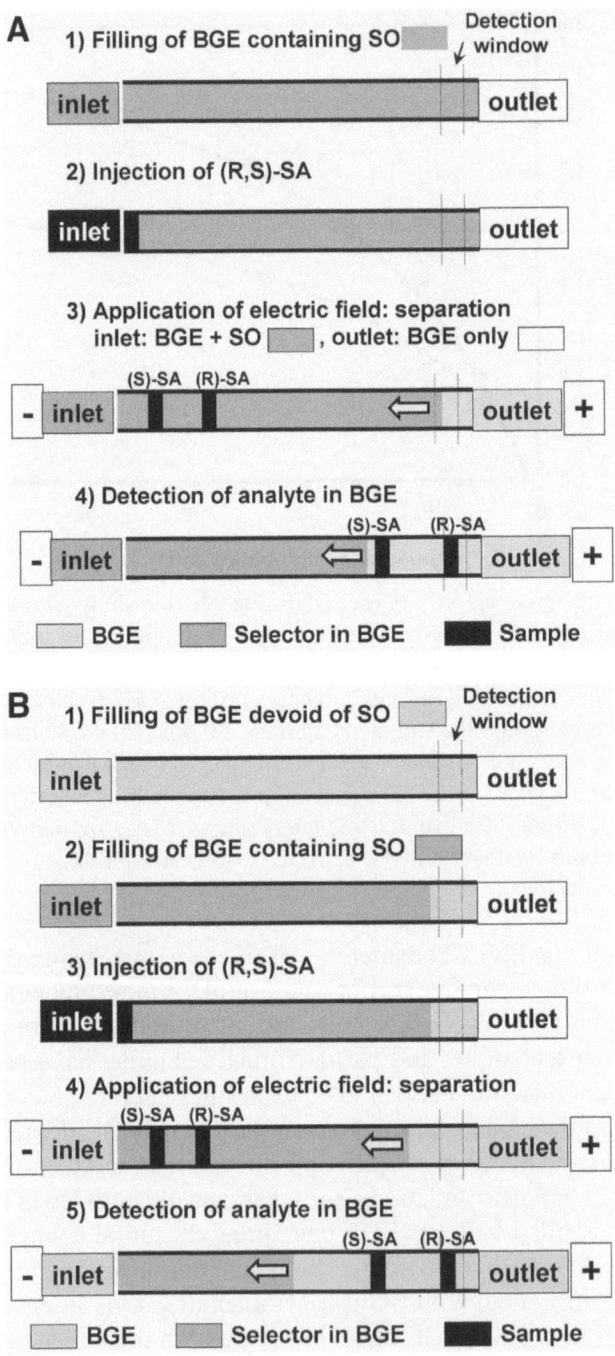

Fig. 5. Schematic illustration of counter-current technique (**A**) and partial filling technique (**B**).

is cleared from the strong UV background after a certain time (depending on EOF velocity and mobility of the selector). This is readily visible in the electropherogram by a stepwise drop of the baseline. The analytes that are separated in the selector zone on their way to the detector reach the detection window after the selector has passed the detection cell and thus can be detected with high sensitivity in a portion of selector-free BGE. Breakthrough times were, under the conditions specified in the experimental section, 7.7 min for quinine, 7.9 min for quinidine, 7.3 min for *tert*-butyl carbamoylated quinine, and 7.4 min for the corresponding quinidine derivative, while analytes such as *N*-derivatized amino acids showed migration times longer than 12 min *(5)*.

In the partial filling technique (**Fig. 5B**) *(16)*, only a part of the capillary shorter than the effective length (distance from injection end to detection point, L_{eff}) is filled with the selector-BGE solution before applying the solute to the capillary (*see* **Note 6**). The separation is carried out with plain BGE solution in both inlet and outlet electrode vessels. During the run, the selector zone moves toward the inlet end of the capillary and away from the detection window. Thus, the selector is precluded from the detection cell all the time, and the migration of the selector zone toward the inlet end prevents the risk that the selector enters the detection window avoiding any problems of detection interferences. **Figure 6** depicts separations of *N*-benzoyl-leucine enantiomers employing the total filling technique (**Fig. 6A**), the counter-current technique (**Fig. 6B**), and the partial filling technique (**Fig. 6C**) under comparable conditions. It is seen that all three methods separate the enantiomers adequately, with the partial filling technique providing slightly smaller separation factors and resolutions due to the reduced contact time caused by the shorter selector plug length in the capillary (*see* **Notes 6, 7,** and **8**).

Both the counter-current technique and the partial filling technique are also the methods of choice when CE enantiomer separation systems are to be coupled to mass spectrometers. The presence of selector in the effluent would have a deleterious effect on ionization efficiency so that relative abundances are usually decreased with increasing selector concentrations.

Although the precision of counter-current, and in particular the partial filling techniques, may be worse than that of the total filling technique, run-to-run repeatabilities with cinchonan derivatives as selectors were typically below 1% relative standard deviation (RSD), which is quite acceptable for practical applications.

2. Materials

2.1. Derivatization of Amino Acids, Peptides, Amino Sulfonic Acids, and Phosphonic Acids

1. 0.1 *M* Carbonate buffer, obtained by mixing 0.1 *M* sodium bicarbonate and 0.1 *M* sodium carbonate in ratio of 2:1 (v/v).

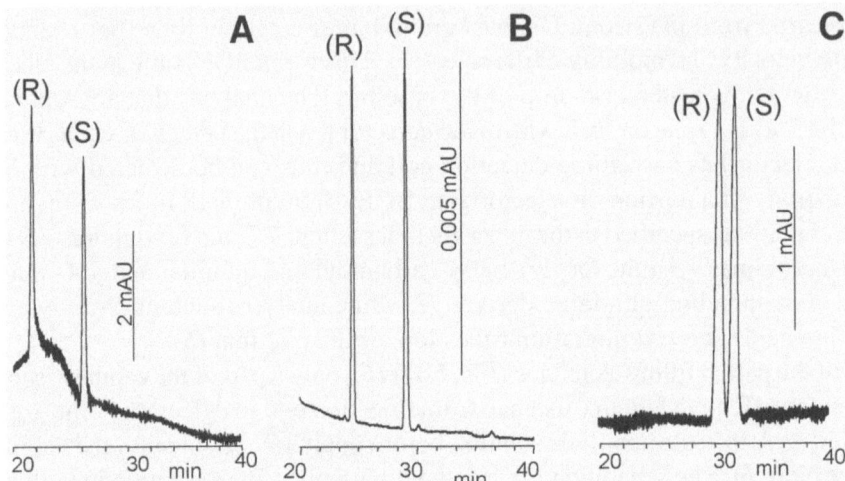

Fig. 6. CE enantiomer separations of *N*-benzoyl-leucine with O-9-(*tert*-butylcarba-moyl) quinine as chiral counter-ion by (**A**) total filling technique, (**B**) counter-current technique, and (**C**) partial filling technique (with modifications from **ref. 10**) (experimental conditions as specified in **Subheadings 2.** and **3.**).

2. 0.1 *M* Borate buffer: 0.1 *M* sodium tetraborate ($Na_2B_4O_7$) (for *N*-carbazole-9-carbonyl [CC] derivatives).
3. 3,5-Dinitrobenzoyloxy succinimide (DNB-OSu) in acetontrile (1.4%, w/v).
4. 2,4-Dinitrofluorobenzene in acetonitrile (Sanger's reagent; 2.5%, w/v).
5. *N*-(9-Fluorenylmethoxycarbonyl)succinimide (FMOC-OSu) in acetonitrile (2.5%, w/v).
6. *N*-3,5-Dinitrobenzyloxycarbonyloxy succinimide (DNZ-OSu) in 1,4-dioxane (2.5%, w/v).
7. *N*-Carbazole-9-carbonyl chloride in acetonitrile (0.5%, w/v) (freshly prepared).
8. Sample solution (e.g., approx 5 μ*M*) of the amino acid, peptide, amino sulfonic acid, or amino phosphonic acid to be analyzed in carbonate buffer (DNB, DNP, FMOC, and DNZ derivatives) or borate buffer (CC derivatives).

2.2. Synthesis of O-9-(tert-Butylcarbamoyl) Quinine and Quinidine Counter-Ions (17)

1. Quinine or quinidine (as free base) (Fluka, Buchs, Switzerland).
2. Dry toluene (for quinine derivative) or dry 1,4-dioxane (for quinidine derivative).
3. *tert*-Butyl isocyanate (Sigma, St. Louis, MO, USA).
4. Dibutyl tin dilaurate (catalyst) (Sigma).
5. *n*-Hexane.
6. Cyclohexane.

2.3. Preconditioning of Fused-Silica Capillary

1. 0.1 *M* aqueous sodium hydroxyde solution.
2. Water (double-distilled).
3. Methanol (p.a.).
4. BGE: ethanol/methanol (60:40, v/v) containing 100 m*M* acetic acid or octanoic acid and 12.5 m*M* triethylamine or ammonia.

2.4. CE Experiments: Total Filling Technique, Counter-Current Technique, and Partial Filling Technique

1. CE instrument: for example, an Agilent HP[3D] capillary electrophoresis instrument (Agilent Technologies, Waldbronn, Germany) equipped with a diode array detector was utilized for the experiments shown herein. UV detection wavelengths were set at 215, 230, 250, and 280 nm. The capillary was kept at constant temperature of 15°C.
2. Bare fused-silica capillary from Polymicro (Phoenix, AZ, USA) with 50 µm inner diameter (i.d.) and a total length (L_{tot}) of 45.5 cm. A detection window is fabricated at a distance of 37 cm from the inlet end (effective length, L_{eff}) by removing the polyimide coating with a razor blade or by burning it off.
3. The nonaqueous BGE, daily fresh prepared, is composed of an ethanol/methanol (60:40, v/v) mixture containing 100 m*M* octanoic acid and 12.5 m*M* triethylamine (total filling technique), or 100 m*M* octanoic acid and 12.5 m*M* ammonia (counter-current technique), or 100 m*M* acetic acid and 12.5 m*M* triethylamine (partial filling technique).
4. Selector solution (daily fresh prepared): the chiral counter-ion, O-9-(*tert*-butyl-carbamoyl) quinine or corresponding quinidine derivative, is dissolved in BGE at a concentration of 10 m*M*.
5. Sample solution: the corresponding sample is dissolved in BGE (approx 1 mg/mL). If derivatization according to any one of the protocols described above was carried out, an aliquot of 100 µL of the derivatization reaction mixture is diluted with BGE to 1 mL.

3. Methods

3.1. Derivatization of Amino Acids, Peptides, Amino Sulfonic Acids, and Phosphonic Acids

1. DNB derivatives: DNB derivatization is carried out by adding 300 µL of the DNB-OSu reagent to the sample solution in carbonate buffer (200 µL). The reaction is allowed to proceed at 50°C overnight.
2. *N*-2,4-Dinitrophenyl (DNP) derivatives: 300 µL of Sanger's reagent are admixed to 200 µL sample solution in carbonate buffer, followed by reaction for 2 h at room temperature.
3. FMOC derivatives: 300 µL of the FMOC-OSu reagent solution are mixed with 200 µL of the sample solution in carbonate buffer and allowed to react for 2 h at room temperature. Excess of the reagent is extracted twice with diethylether.

4. DNZ derivatives: DNZ derivatives are obtained by adding 300 µL of a solution of DNZ-OSu in 1,4-dioxane (2.5%, w/v) to 200 µL sample solution in carbonate buffer. The mixture is then shaken vigorously at room temperature for 3 h. Excess of the reagent is extracted twice with diethylether.
5. CC derivatives: for the preparation of the CC derivatives 300 µL of reagent solution are combined with 200 µL sample solution in borate buffer. The resulting mixture is shaken vigorously. After reaction for 2 h at room temperature, the reaction mixture is extracted twice with n-heptane in order to remove excess of reagent.

3.2. Synthesis of O-9-(tert-Butylcarbamoyl) Quinine and Quinidine Counter-Ions (17)

1. Dissolve 6 mmol of quinine (as free base) in dry toluene. For the synthesis of the quinidine derivative, quinidine base is dissolved in dry 1,4-dioxane.
2. 6.6 mmol *tert*-Butyl isocyanate and 1 drop of dibutyl tin dilaurate as catalyst are added.
3. The mixture is refluxed for 4 h.
4. The solvent is evaporated, and the remaining raw material is washed with n-hexane.
5. The crude white solid is crystallized with cyclohexane.
6. Yield: 70%.
7. Physical properties of O-9-(*tert*-butylcarbamoyl)quinine:

   m.p.: 122°C; $[\alpha]^{23}_{Na589} = -10.9°$, $[\alpha]^{23}_{Hg546} = -15.8°$ (C = 1.01; MeOH)

   IR (KBr): 1718, 1622, 1593, 1532, 1508, 1267, 1035 cm^{-1}
   1H-NMR (200 MHz, dMeOD): 8.68 (d, 1H), 7.95 (d, 1H), 7.57 (m, 2H), 7.45 (dd, 1H), 6.50 (d, 1H), 5.80 (m, 1H), 4.9 – 5.1 (m, 2H), 4.01 (s, 3H), 3.2 – 3.4 (m, 3H), 3.0 – 3.3 (m, 1H), 2.5 – 2.8 (m, 2H), 2.25 – 2.45 (m, 1H), 1.7 – 2.0 (m, 3H), 1.5 – 1.70 (m, 2H), 1.2 – 1.4 (s, 9H) ppm.
8. Physical properties of O-9-(*tert*-butylcarbamoyl)quinidine:

   m.p.: 161°C; $[\alpha]^{25}_{Na589} = + 0.30°$, $[\alpha]^{25}_{Hg546} = + 0.57°$ (C = 1.03; MeOH)

   IR (KBr): 1725, 1621, 1592, 1506, 1245, 1031 cm^{-1}
   1H-NMR (200 MHz, dMeOD): 8.65 (d, 1H), 7.9 – 8.05 (d, 1H), 7.5 – 7.65 (m, 2H), 7.35 – 7.5 (m, 1H), 6.65 (d, 1H), 6.1 – 6.3 (m, 1H), 5.05 – .25 (m, 2H), 4.0 (s, 3H), 3.2 – 3.45 (m, 4H), 2.65 – 3.15 (m, 5H), 2.25 – 2.45 (m, 1H), 2.0 – 2.2 (m, 1H), 1.7 – 1.90 (m, 1H), 1.2 – 1.4 (s, 9H) ppm.

3.3. Preconditioning of Fused-Silica Capillary

1. Before each series, the capillary is rinsed with 0.1 *M* aqueous sodium hydroxide solution by application of a pressure of 50 mbar for 10 min (*see* **Note 2**).
2. Wash the capillary first with water and then with methanol, each for 10 min at an inlet pressure of 50 mbar.
3. Afterwards, the capillary is flushed with BGE at 50 mbar for 30 min.
4. Between the runs, the capillary is rinsed with BGE at 50 mbar for 5 min.

3.4. Total Filling Technique

1. After rinsing with BGE, the capillary is equilibrated for 10 min with selector solution applying a pressure of 50 mbar at the inlet end.
2. The sample is injected hydrodynamically by application of a pressure of 50 mbar for 5 s.
3. The separations are carried out with an applied voltage of –25 kV with both inlet and outlet reservoirs being filled with selector solution.

3.5. Counter-Current Technique

1. After rinsing the capillary with plain BGE, the capillary is filled with selector solution over the entire length applying a pressure of 50 mbar for 10 min at the inlet end of the capillary.
2. Sample injections are made in the hydrodynamic mode (50 mbar) for a period of 5 s.
3. During the separations, the inlet reservoir (at the cathodic side) is filled with the selector solution, while the outlet reservoir (at the anodic side) contains the same electrolyte solution devoid of selector, i.e., plain BGE.
4. The separations are performed in the reversed polarity mode with an applied voltage of –25 kV.

3.6. Partial Filling Technique

3.6.1. Determination of Filling Time and Selector Zone Length

1. Initially, the filling time needs to be determined once for each new conditions by monitoring a breakthrough curve. Thus, after rinsing the capillary with plain BGE, the inlet reservoir is filled with selector (counter-ion) solution.
2. A pressure of 50 mbar is applied to the inlet electrode vessel and the breakthrough of the selector solution monitored, which is visible by a stepwise change of the UV signal.
3. From the breakthrough time, the applied plug length (PL_{app}) of the filling step can be calculated by:

$$PL_{app} = v_{app}\, t_{app} \qquad \text{[Eq. 7]}$$

where v_{app} is the linear velocity of the SO zone in the filling step, which is obtained by the effective length of the capillary divided by the breakthrough time, and t_{app} is the filling time, i.e., the time pressure is applied to the inlet in the filling step (*see* **Note 6**).

3.6.2. Partial Filling Experiment

1. After rinsing the capillary with plain BGE, the capillary is filled with the selector solution by applying a pressure of 50 mbar to the injection end of the capillary over a period of 5 min, which results in a selector plug length of approx 30 cm with the above specified selector solution.

2. Then, the sample dissolved in BGE is injected hydrodynamically (50 mbar for 5 s).
3. Finally, the analytical run is performed applying a constant voltage of –25 kV using plain BGE without chiral selector as running buffer at both inlet and outlet home vials.

4. Notes

1. The choice of protection group and derivatizing agent, respectively, for amino acid, amino sulfonic acid, and amino phosphonic acid analysis is essentially based on two criteria: one is related to separability of the respective derivatives, and the other to detection issues. Upon derivatization or labeling, favorable interactive groups for binding with complementary sites of the selector may be introduced like an electron-poor aromatic moiety (π-acid) such as DNP, DNB, DNZ, and PNZ group. N-acyl derivatives such as DNB, Bz, DNZ, Z, and FMOC also provide a rigid hydrogen donor-acceptor system for hydrogen bonding with the carbamate of the selector and therefore, in case of primary amino acids, are usually resolved with high selectivity. In case of secondary amino acids, the N-H hydrogen donor is missing in the corresponding N-acylated derivatives, which is associated with a significant drop of separation factors or even leads to a loss of enantioselectivity. For analysis of secondary amino acids like proline, N-aryl derivatives such as DNP and DBD are recommended, since they discriminate between enantiomers by a different chiral recognition mechanism and do not depend on the hydrogen donor qualities of the nitrogen. In conclusion, for primary amino acids, separation factors roughly follow the order: DNB > DNZ > Bz ~ DNP ~ PNZ ~ FMOC > Z (see **Table 1**), while for secondary amino acids DNP derivatives show similar enantioselectivity as corresponding primary amino acid derivatives. Overall, aside from detection issues, through the N-protection group separation selectivity may be fine-tuned and enable free selection of the elution order (see **Fig. 3**).

   DNP derivatives, obtained by derivatization with Sanger's reagent, are often a good selection for UV detection, due to their good separability, also for secondary amino acids, and strong UV absorptivities with favorable UV maximum at 360 nm. DBD (λ_{ex} 450 nm, λ_{em} 590 nm), DNS (λ_{ex} 310 nm, λ_{em} 540 nm), and FMOC (λ_{ex} 266 nm, λ_{em} 305 nm) derivatives can be used in combination with fluorescence detectors.

2. It is strongly recommended to carry out a washing step with 0.1 M sodium hydroxide solution before each series (with untreated fused-silica capillaries only), because the basic cinchonan derivatives show a tendency to adhere to the acidic fused-silica capillary wall by electrostatic forces. As a result of selector adsorption, the surface charge and ζ-potential of the fused-silica capillary wall will vary. The concomitant change of EOF velocity will lead to a deterioration of run-to-run repeatabilities. If such changes in run-to-run repeatabilities are noticed, the washing step with sodium hydroxide should be performed more frequently. Coated capillaries should be rinsed as specified by the supplier.

3. A serious difficulty inhibits the precise and accurate determination of the electro-osmotic mobility of the present separation systems. The problems arise (i) from the strong UV background of the selector; (ii) the counter-current separation process, which makes it impossible to measure the EOF with a neutral marker together with the analytes in a single run, but demands a separate and eventually different experiment; (iii) in case of partial filling as well as counter-current techniques, from discontinuous separation zones with different EOF velocities and actual field strengths in both the distinct zones. For the total filling technique, a strong UV absorbing neutral compound that unfortunately may also interact with the quinoline moiety of the selector by π-π-interactions, thus slightly falsifying EOF determination, may be used, or alternatively, one can appy the indirect detection method with a non-UV absorbing neutral compound (normal polarity mode applying +25 kV). Benzylic alcohol was used as neutral flow marker for EOF determination in the counter-current experimental set-up, and the normal polarity (+25 kV) was used for this purpose. In the partial filling technique, EOF determinations were carried out with acetone. First, the capillary was filled with selector solution to a length of about 30 cm. Then the EOF marker was injected and voltage with normal polarity (+25 kV) applied. The break-through of the selector zone was observed after a few minutes and the EOF marker eluted several minutes after the entire selector zone had passed the detection window. The electroosmotic mobility may then be calculated from the peak of the neutral EOF marker by the formula described in **Table 1, footnote a**. However, it represents only an approximated average value.

4. If poor separation selectivity and inadequate resolution between enantiomers is observed, it is proposed to increase the concentration of the chiral counter-ion in the BGE from 10 mM up to 50 or 100 mM (**Fig. 4**). This is usually the most effective means to enhance resolution. However, a fine-tuning of other conditions like acid-base ratio, type and percentage of solvents (e.g., acetonitrile instead of ethanol), total electrolyte concentrations, and temperature may also lead to considerably better resolution. In particular, phosphonic and sulfonic acids may require adaption of conditions, because those described in the experimental part and **Table 1** are specifically optimized for amino carboxylic acid derivatives.

5. Mobilities of the anionic co-ion in the BGE should match the mobilities of the analyte anion to avoid serious peak broadening contributions arising from electrokinetic dispersion. Accordingly, the peak performance strongly depends on the type of co-ion used. For example, mobility matching was accomplished in the analysis of DNB-amino acid derivatives by replacement of acetic acid by octanoic acid *(3)*. Such considerations, however, must always take into account the mobility of the analyte ion, and thus the optimal co-ion may vary from analyte to analyte. No generalized recommendation can, therefore, be given here.

6. Remarks on the partial filling technique (for details see recent review on this subject in **ref. 16**): It is always advisable to determine the applied plug length (PL_{app}) under the conditions employed (see Eq. 7). It is affected by the applied pressure (ΔP), the diameter (d), the length (L_t) of the capillary, and the viscosity of the solution (η) according to the relationship:

$$PL_{app} = \Delta P\ d^2/32\ \eta\ L_t \qquad\qquad\qquad \text{[Eq. 8]}$$

Thus, a change of the solvent system or of the selector concentration may affect the applied length of the selector plug, requiring readjustment of the plug length during optimization of separations and method development, respectively. As described, in the course of the separation, the selector zone moves with its electrophoretic mobility and EOF toward the injection end, so that the final selector plug length (effective selector plug length, PL_{eff}), when the analyte leaves the plug, is actually shorter than the applied one. The effective plug length depends on the applied selector plug length (PL_{app}), the velocity of the selector plug (v_{PL}), and the residence time of the analyte in the plug (t_{PL}), i.e., the mobility of the analyte in the selector zone:

$$PL_{eff} = v_{PL}\ t_{PL} + PL_{app} \qquad\qquad\qquad \text{[Eq. 9]}$$

Separation factors and resolutions are smaller the shorter the applied and effective selector plug lengths (7), i.e., the higher mobilities of selector zone and analyte. As a matter of fact, resolutions are typically lower in the partial filling technique than the counter-current or total filling techniques where the analyte resides for a longer time in the selector zone and thus longer separation zones are active (see **Fig. 6**).

7. Another disturbing effect of the partial filling technique concerns the EOF behavior. In the two distinct zones, EOF velocities are different, as pointed out above. This mismatch of EOF inside and outside the separation plug (selector zone) may create laminar flow that impairs the efficiency of the system. Moreover, faster migration of the analyte in the BGE in relation to the selector zone brings about additional band spreading at the interface of the both distinct zones. Both effects become striking, particularly at higher selector concentrations. Thus, the partial filling technique usually provides lower plate counts, as can be seen from **Fig. 6**. One of the main advantages besides detection issues is the significant minimization of selector consumption in the partial filling technique over both counter-current and total filling technique. The vol of selector solution applied to the capillary amounts to <1 µL per run, so that milligrams of the cinchonan derivatives last for several series of injections.

8. The question as to which of the three techniques, i.e., total filling, counter-current, or partial filling, to apply should take the above considerations into account. As already pointed out, the counter-current and partial filling techniques are certainly better suited for practical applications owing to elimination of background noise and better detection sensitivity. If selectivity is critical, the counter-current technique may be preferred, because it yields higher separation factors and, often, better efficiency. However, one must ensure that the analytes do not reach the detector before the selector zone is purged from the detection compartment. This may sometimes be difficult for analytes with high mobility. With the partial filling technique such problems do not exist; therefore, we suggest this as method of choice for CE enantiomer separations with cinchonan-derived selectors when separation selectivities and resolutions are large enough with the slightly shorter effective separation zones.

Acknowledgments

Financial support of the Austrian Christian-Doppler Research Society and the Austrian Science Fund (project no. 13965-CHE) is gratefully acknowledged.

References

1. Stalcup, A. M. and Gahm, K. H. (1996) Quinine as a chiral additive in nonaqueous capillary zone electrophoresis. *J. Microcol. Sep.* **8**, 145–150.
2. Jira, T., Bunke, A., and Karbaum, A. (1998) Use of chiral and achiral ion-pairing reagents in combination with cyclodextrins in capillary electrophoresis. *J. Chromatogr. A* **798**, 281–288.
3. Piette, V., Lämmerhofer, M., Lindner, W., and Crommen, J. (1999) Enantiomeric separation of N-protected amino acids by non-aqueous capillary electrophoresis using quinine or tert-butylcarbamoylated quinine as chiral additive. *Chirality* **11**, 622–630.
4. Piette, V., Lindner, W., and Crommen, J. (2000) Enantioseparation of anionic analytes by non-aqueous capillary electrophoresis using quinine and quinidine derivatives as chiral counter-ions. *J. Chromatogr. A* **894**, 63–71.
5. Piette, V., Fillet, M., Lindner, W., and Crommen, J. (2000) Non-aqueous capillary electrophoretic enantioseparation of N-derivatized amino acids using cinchona alkaloids and derivatives as chiral counter-ions. *J. Chromatogr. A* **875**, 353–360.
6. Piette, V., Fillet, M., Lindner, W., and Crommen, J. (2000) Enantiomeric separation of amino acid derivatives by non-aqueous capillary electrophoresis using quinine and related compounds as chiral additives. *Biomed. Chromatogr.* **14**, 19–21.
7. Lämmerhofer, M., Zarbl, E., and Lindner, W. (2000) tert.-Butylcarbamoylquinine as chiral ion-pair agent in non-aqueous enantioselective capillary electrophoresis applying the partial filling technique. *J. Chromatogr. A* **892**, 509–521.
8. Lämmerhofer, M., Zarbl, E., Lindner, W., Peric Simov, B., and Hammerschmidt, F. (2001) Simultaneous separation of the stereoisomers of 1-amino-2-hydroxy and 2-amino-1-hydroxypropane phosphonic acids by stereoselective capillary electrophoresis employing a quinine carbamate type chiral selector. *Electrophoresis* **22**, 1182–1187.
9. Hammerschmidt, F., Lindner, W., Wuggenig, F., and Zarbl, E. (2000) Enzymes in organic chemistry. Part 10: chemo-enzymatic synthesis of L-phosphaserine and L-phosphaisoserine and enantioseparation of amino-hydroxyethylphosphonic acids by non-aqueous capillary electrophoresis with quinine carbamate as chiral ion pair agent. *Tetrahedron: Asymmetry* **11**, 2955–2964.
10. Lämmerhofer, M., Zarbl, E., Piette, V., Crommen, J., and Lindner, W. (2001) Evaluation of enantioselective nonaqueous ion-pair capillary electrophoresis as screening assay in the development of new ion exchange type chiral stationary phases. *J. Sep. Sci.* **24**, 706–716.
11. Franco, P., Klaus, P. M., Minguillón, C., and Lindner, W. (2001) Evaluation of the contribution to enantioselectivity of quinine and quinidine scaffolds in chemically and physically mixed chiral selectors. *Chirality* **13**, 177–186.

12. Piette, V., Lindner, W., and Crommen, J. (2002) Enantiomeric separation of N-protected amino acids by non-aqueous capillary electrophoresis with dimeric forms of quinine and quinidine derivatives serving as chiral selectors. *J. Chromatogr. A* **948,** 295–302.

13. Peric Simov, B., Wuggenig, F., Lämmerhofer, M., Lindner, W., Zarbl, E., and Hammerschmidt, F. (2002) Indirect evidence for the biosynthesis of (1S,2S)-1,2-epoxypropylphosphonic acid as a Co-metabolite of fosfomycin [(1R,2S)-1,2-epoxy-propylphosphonic acid] by Streptomyces fradiae. *Eur. J. Org. Chem.* 1139–1142.

14. Czerwenka, C., Lämmerhofer, M., and Lindner, W. (2002) Electrolyte and additive effects on enantiomer separation of peptides by non-aqueous ion-pair capillary electrophoresis using tert-butylcarbamoyl quinine as chiral counter-ion. *Electrophoresis* **23,** 1887–1899

15. Budavari, S. (ed.) (1989) *The Merck Index.* Merck & Co., Rahway, NJ.

16. Amini, A., Paulsen-Sörman, U., and Westerlund, D. (1999) Principle and applications of the partial filling technique in capillary electrophoresis. *Chromatographia* **50,** 497–506.

17. Mandl, A., Nicoletti, L., Lämmerhofer, M., and Lindner, W. (1999) Quinine versus carbamoylated quinine-based chiral anion exchangers. A comparison regarding enantioselectivity for N-protected amino acids and other chiral acids. *J. Chromatogr. A* **858,** 1–11.

19

Chiral Separation by Capillary Electrophoresis Using Polysaccharides

Hiroyuki Nishi

1. Introduction

Capillary electrophoresis (CE) offers numerous advantages, such as high separation efficiency, analysis speed, instrumentation simplicity, and reduced operating costs, compared with other separation analytical methods. One of the successful application areas of CE is in drug analysis, where the focus is on relatively small synthetic drugs. Various CE methods (modes) such as micellar electrokinetic chromatography (MEKC) and capillary gel electrophoresis (CGE) have been developed within CE techniques for the analysis of drugs. Among them, electrokinetic chromatography (EKC), including MEKC, and capillary zone electrophoresis (CZE) have been used for the separation of enantiomers.

Typical enantiomer separation by CE (CZE or EKC) is performed by simply employing chiral additives to the running buffer solutions (homogeneous solution). Various chiral additives, such as cyclodextrins (CDs), polysaccharides, proteins, crown ethers, and chiral surfactants, which are also effective chiral moieties in high-performance liquid chromatography (HPLC) except chiral surfactants, have been found to be useful for the CE enantiomer separation as chiral selectors *(1–8)*. In CE, the separation solution can be easily altered to find the optimum separation medium.

Polysaccharides are one of the promising chiral additives in the CE enantiomer separation as in the HPLC enantiomer separation. It is well known that polysaccharide derivatives are effective and useful chiral moieties in chromatography, especially in HPLC. Many ionic polysaccharides, such as heparin *(9–12)*, chondroitin sulfates *(12–16)*, dextran sulfate *(12,17)*, and λ-carrageenan *(18)*, and neutral polysaccharides, such as dextran *(14,19)*, dextrin *(20–26)*, laminaran *(27)*, and pullulan *(27)*, have been successfully employed for the CE

From: *Methods in Molecular Biology, Vol. 243: Chiral Separations: Methods and Protocols*
Edited by: G. Gübitz and M. G. Schmid © Humana Press Inc., Totowa, NJ

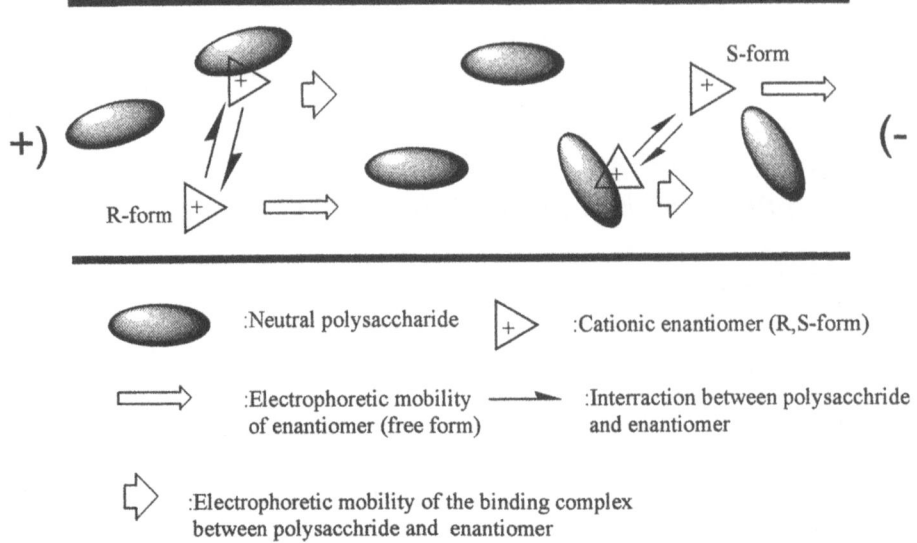

Fig. 1. A schematic of the separation principle of CZE, in case of acidic medium with electrically neutral chiral selector and positively charged analyte (enantiomers).

enantiomer separation (*see* **Note 1**). However, it is not easy to use positively charged polysaccharides in the CE enantiomer separation because of its low solubility and adsorption to the inner surface of the capillary.

In the case of electrically neutral polysaccharides, the separation mode belongs to CZE, where basic and acidic enantiomeric compounds can be enantioseparated. On the other hand, charged polysaccharides are used in EKC mode, where electrically neutral enantiomeric compounds can be separated as well as charged compounds. A schematic of the separation principle of CZE is shown in **Fig. 1**, where an acidic medium with electrically neutral chiral selector such as dextrin is assumed. When enantiomers of positively charged drugs (e.g., diltiazem) are injected from the positive end, and a high voltage is applied between the both electrodes, solutes electrically migrate toward the negative electrode. Under the acidic conditions mentioned above, the electroosmotic flow (EOF) generated in use of uncoated capillary tube is very weak (neglected in **Fig. 1**) compared with the neutral and alkaline conditions. Consequently, solutes mainly migrate with their electrophoretic mobilities. During the migration, solutes interact with chiral selectors added to the buffers, leading to the different mobilities of solutes, i.e., successful enantiomer separation. In the enantiomer separation of diltiazem (Ca-channel blocker) with dextrin, 2*S*,3*S*-form migrated faster than the corresponding 2*R*,3*R*-form. This means that the slowly migrated 2*R*,3*R*-form strongly interacted with dextrin (longer arrow in **Fig. 1**). In other words, separa-

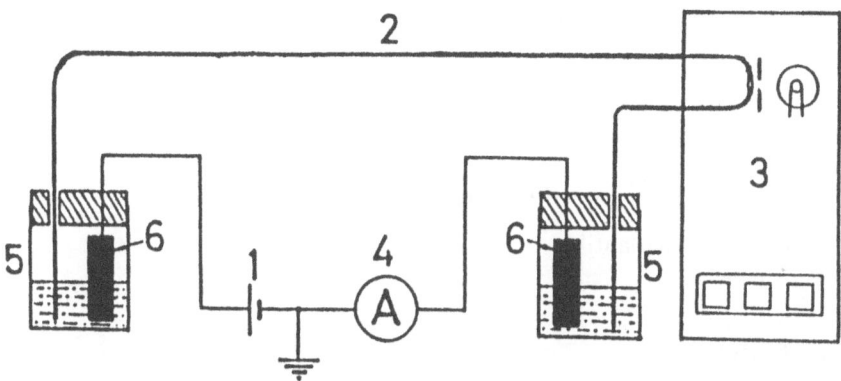

Fig. 2. A schematic illustration of of the apparatus. (1) Regulated high voltage DC power supply, (2) fused silica tube, (3) UV detector, (4) ammeter, (5) electrode vessel (buffer reservoir), and (6) electrode.

tion in CE is achieved as a result of the different mobilities of analytes. In the usual CZE or EKC, without chiral selectors, mobilities of enantiomers do not differ, which means an unsuccessful separation. However, through the addition of chiral compounds to the buffer solutions in CE, many enantiomer separations have been achieved, owing to the difference in binding constants between each enantiomer and a chiral selector.

2. Materials

2.1. Apparatus

A commercially available CE apparatus equipped with UV or photodiode array detector, uncoated fused-silica capillary tubes (typically 50 or 75 µm inner diameter [ID], effective length around 50 cm), and a personal computer for the instrument control and data collections. A schematic illustration of the apparatus is shown in **Fig. 2**. Homemade apparatus also can be set up with low costs.

2.2. Preparation of Buffer Solutions

1. KH_2PO_4 or NaH_2PO_4 (of analytical reagent grade).
2. Phosphoric acid (of analytical reagent grade).
3. Polysaccharides: heparin, chondroitin sulfate A, chondroitin sulfate B (dermatan sulfate), chondroitin sulfate C, dextran sulfate, carboxymethyl-dextran, dextran, and dextrin. Most of charged polysaccharides are obtained as salt form (typically sodium salt) (*see* **Notes 2–6**).
4. Membrane filter (0.45 µm) (for syringe type).
5. Measuring cylinder (25 mL) and beaker (100 mL).

2.3. Conditioning of Capillary

1. 0.1 M NaOH solution, 10 mL.
2. Purified water, 20 mL.
3. Methanol, 20 mL.

2.4. Preparation of Sample Solutions

1. Methanol or acetonitrile.
2. Purified water.

3. Methods

3.1. Preparation of Buffer Solutions

A 10 or 20 mM phosphate buffer solution (100 mL) is prepared, and the pH of the buffer is adjusted at the specified value by adding a diluted phosphoric acid (1 in 10) (one-tenth dilution of phosphoric acid with water). Running buffer solutions (20 mL) are prepared by adding appropriate concentration of polysaccharides to the above phosphate buffer. Then the solution is filtered with a membrane filter of pore size 0.45 µm and degassed by sonication with an ultrasonic wave bath before use.

3.1.1. Example 1:
Enantioseparations of Basic Drugs
by CE With Electrically Neutral Polysaccharides

1. Prepare 100 mL of 20 mM KH_2PO_4 or NaH_2PO_4 solution.
2. Adjust the pH of the solution to 2.5 by adding a diluted phosphoric acid (1 in 10).
3. Dissolve electrically neutral polysacchrides (for example, dextran or dextrin) (*see* **Note 3**) in 20 mL of the phosphate buffer solution of pH 2.5 to the concentration 3 or 6% (solubility of polysaccharide employed should be checked in advance). In case of poor solubility, warming may be effective (*see* **Note 4**).
4. Filter the solutions through a 0.45-µm membrane filter (syringe type).
5. Degas the solutions by sonication (ultrasonic wave bath).
6. Fill the buffer reservoirs (vials) with the running buffer solutions prepared (each 10 mL).

3.1.2. Example 2:
Enantioseparations by CE With Anionic Polysaccharides

1. Prepare 100 mL of 20 mM KH_2PO_4 or NaH_2PO_4 solution.
2. Adjust the pH of the solution to 3.0 or 5.0 or 7.0.
3. Dissolve anionic polysaccharides (for example, heparin, chondroitin sulfate C, etc., as salt form) in 20 mL of the phosphate buffer solution (pH 3.0, 5.0, or 7.0) to the concentration 3 or 6% (solubility of polysaccharide employed should be

checked in advance. pH of buffers after addition of polysaccharides also should be monitored) (*see* **Note 4**).

4. Filter the solutions through a 0.45-μm membrane filter (syringe type).
5. Degas the solution by sonication (ultrasonic wave bath).

3.2. Preparation of Sample Solutions

1. Stock standard solutions: the stock standard solutions for the enantiomers are prepared in methanol or acetonitrile with an approximate concentration of 1.0 mg/mL.
2. Sample solutions: the stock standard solutions are diluted with water to a concentration of approx 0.1 mg/mL for the CE injection.

3.3. Conditioning of Capillary

Conditioning of capillary is important to obtain the reproducible migration time. The following procedures are a typical example. The conditioning procedures can be performed automatically in the commercially available CE apparatus. When the capillary is contaminated with hydrophilic compounds, washing with an organic solvent such as MeOH, following with 0.1 *M* NaOH and water, is effective.

1. Rinse with alkaline solution: at the beginning of each experiment, the capillary is washed with a 0.1 *M* NaOH solution for 5 min.
2. Rinse with water: after washing with alkaline solution, the capillary is washed with water for 10 min.
3. Rinse with the buffer: finally the capillary is washed with the running buffer solution for 10 min.
4. Rinse before CE run: before each injection, the capillary is washed with the buffer for 1 min.

3.4. Injection of Sample Solution

The sample solutions are injected from one of the capillary ends by the pressure mode of the CE apparatus. In case of **Subheadings 3.1.1.** and **3.1.2.**, samples are injected from the positive end of the capillary. Typically, injection time is set at around 5 s.

3.5. CE Analysis

After conditioning of the capillary, CE analysis is performed. The capillary is usually thermostated at 25°C, and typical applied voltage is set at around 15–30 kV. Typical detection wavelength is in the range 190–280 nm, depending to the character of analytes. General speaking, the low detection wavelength (i.e., 190 nm) can be adaptable in the CE because of good transparency of the solutions employed. All CE analysis and data collection are controlled by a personal computer in the commercial CE apparatus as in HPLC apparatus.

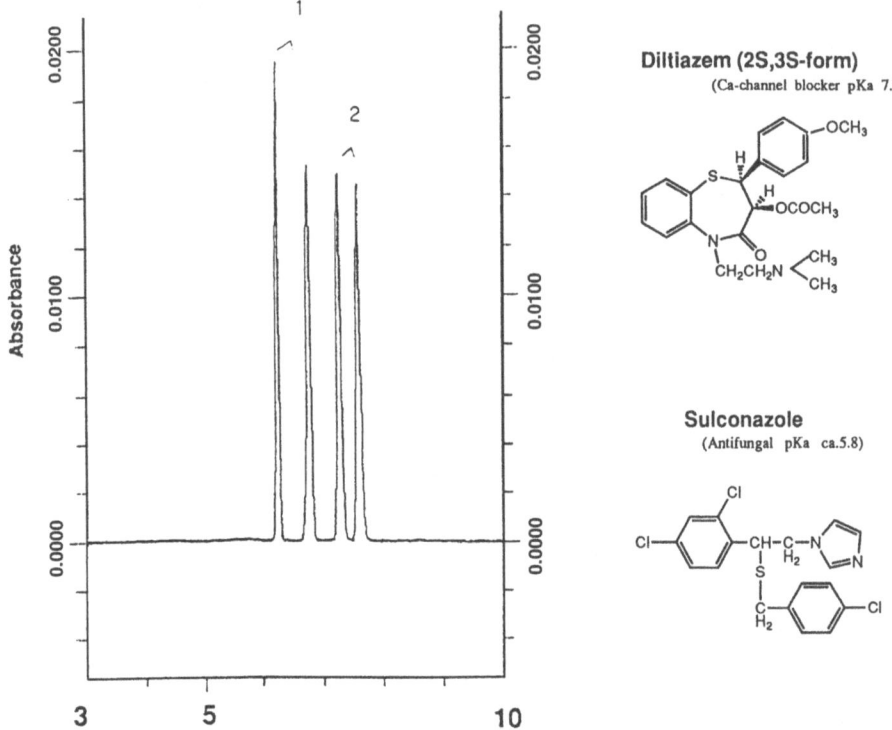

Fig. 3. Separation of enantiomers of (1) diltiazem and (2) sulconazole by CZE with dextrin. Conditions: 6% dextrin (JP grade) in 20 m*M* phosphate buffer, pH 2.5, diameter of 75 µm, and effective length of 40 cm fused silica capillary, applied voltage +30 kV, detection 220 nm, and capillary temperature 20°C.

3.5.1. Example 1:
Enantioseparations of Basic Drugs
by CE With Electrically Neutral Polysaccharides

In use of electrically neutral polysaccharides, the separation mode is CZE. Usually acidic conditions are employed to obtain low EOF. Basic compounds, i.e., positively charged analytes, are injected from the positive end and migrate toward the negative end according to their mobilities. On the other hand, electrically neutral polysaccharides do not migrate because of the low EOF. One example is shown in **Fig. 3**, where enantiomers of diltiazem (2*S*,3*S*-form is diltiazem, and 2*R*,3*R*-form) and sulconazole (racemic drug, antifungal) were simultaneously separated by CZE with dextrin *(19)*. The buffer employed was 6% dextrin (Japanese Pharmacopoeia grade) in 20 m*M* phosphate buffer, pH 2.5.

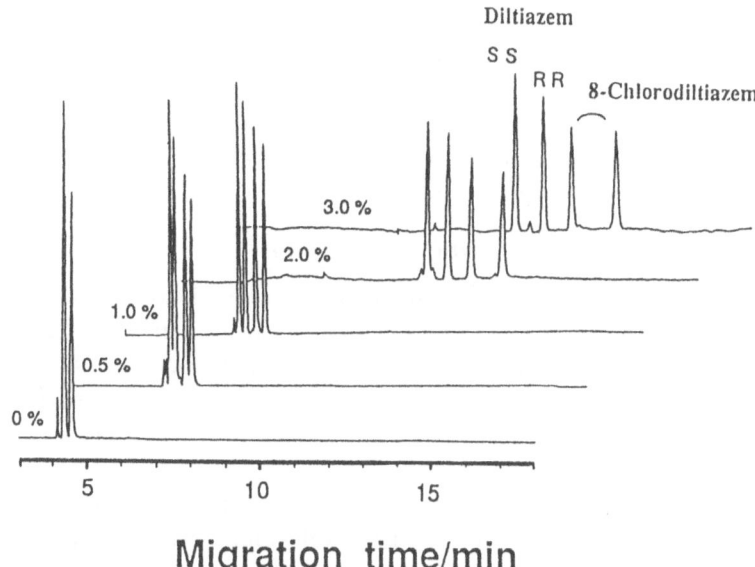

Migration time/min

Fig. 4. Separation of enantiomers of diltiazem and its 8-chloroderivative by EKC with chondroitin sulfate C. Conditions: 0–3% chondroitin sulfate C in 20 mM phosphate borate buffer, pH 2.8, diameter of 75 μm, and effective length of 50 cm fused silica capillary, applied voltage +20 kV, detection 235 nm, and capillary temperature 23°C.

For negatively charged analytes, samples are injected from the negative end, and negative voltage is applied (reverse mode of **Fig. 1**).

3.5.2. Example 2:
Enantioseparations by CE With Anionic Polysaccharides (EKC)

Anionic polysaccharides are effective for the enantiomer separation both of electrically neutral and charged analytes. The separation mode can be classified as EKC. Typically, approximate acidic neutral conditions (approx pH 3.0–7.0), depending upon the degree of substitution of ionic residues such as SO_4^{2-} or COO^-, are employed (*see* **Note 7**). One example is shown in **Fig. 4**. Enantiomers of diltiazem and its 8-chloro derivative were simultaneously separated by EKC with chondroitin sulfate C. The conditions employed were 0–3% chondroitin sufate C in 20 mM phosphate-borate buffer of pH 2.8 *(12)*. Judging from the literatures, heparin, chondroitin sulfates (A, B, and C) and dextran sulfate seem to have relatively wide capability of enantiomer separation. Among these ionic polysaccharides, ionic character of chondroitin sulfates *(12–16)* is small, leading to the adaptability to acidic conditions as shown in **Fig. 4**. On the other hand, dextran sulfate *(17,21)* or heparin *(9–12)* should be employed under the higher pH buffers around 5.0 because of their strong ionic character, leading to the long migration times of analytes. Other than the ionic character, the species of the monosaccharide residues of ionic polysaccharides may be important factors affecting enantioselectivity (*see* **Note 8**).

3.6. Optimization of Enantiomer Separation

Once enantiomer separation is obtained under the above-mentioned conditions, the conditions should be optimized. One of the important factors is the concentration of the chiral selector employed, although there may be a limit due to high viscosity of polysaccharides. Solubility of each polysaccharide should be investigated before the CE analysis. The optimum concentration $[C]_{opt}$ of the selector is mathematically given by Wren et al. as $[C]_{opt} = (K_1K_2)^{-1/2}$, where K_1 and K_2 are formation constants between the enantiomer and the selector *(28)*. It is noted that the optimum concentration of the chiral selector exists in the CE enantiomer separation. In the typical concentration range employed for polysaccharides (1 to approx 5%), the higher the concentration of the selector, the better the separation of enantiomers as shown in **Fig. 4**. Buffer pH is also important factor. Migration time of enantiomers depends on both electrophoretic mobility and EOF (*see* **Note 9**). If migration times are too long, buffer pH should be changed to higher values, leading to strong EOF. Sometimes addition of organic modifiers improves enantioselectivity, probably owing to the change of formation constants.

4. Notes

1. CE is established as a viable option for the separation of enantiomers. Recent review papers on the application and status of CE enantiomer separations confirmed that CE has become established as a routine technique. The CE enantiomer separation can have benefits in terms of method robustness, cost, and time. Various chiral selectors including polysaccharides have been employed for the CE enantiomer separation. Although not quite as versatile and powerful as CDs, inexpensive polysaccharides showed remarkable enantioselectivity in some cases.

2. Polysaccharides are typically complex mixtures of homologs and isomers, which can vary greatly from lot to lot. Therefore, especially in the case of natural (ionic) polysaccharides or derivatized polysaccharides, different enantioselectivity may be obtained from "the same" polysaccharides, due to its wide molecular mass range and different ratios of the unit components.

3. Dextrans are storage polysaccharides in yeasts and bacteria. Dextrans are polymers in which the D-glucose units are joined almost extensively by α-(1,6)-linkages. Hydrolysis of starch, which consists of amyloses (linear polysaccharides with molecular mass approx $5 \times 10^5 - 2 \times 10^6$) and amylopectins (nonlinear polysac-

charides with molecular mass approx $15 \times 10^6 - 400 \times 10^6$), yields a mixture of dextrins where D-glucose units are joined by α-(1,4)-linkages. These two polysaccharides can be soluble in water. Among these neutral polysaccharides, dextrins in general show a wide enantioselectivity, probably due to their helical structure. A full turn in the helix requires at least six glucose units.

4. For the strict comparison of the capability of enantioselectivity, the concentration of polysaccharides should be adjusted at the same percentage. Furthermore, molecular mass, molecular mass distribution, degree of substitution residue, etc., must be controlled. However, preparation of solution at the same concentration is sometimes difficult due to the high viscosity. The practically usable maximum concentration for polysaccharides seems to be around 5%.

5. Polysaccharides with a molecular mass above a certain level may show the same enantioselectivity at the same percentage concentrations, although there was a relationship between the enantioselectivity and molecular mass in much smaller oligosaccharides, such as maltodextrins.

6. Various kinds of polysaccharides can be employed for the CE enantiomer separation. Solubility and transparency of solutions are important factors as chiral selectors in CE.

7. In the case of charged polysaccharides as chiral selectors, the ionic character is important. Therefore, the choice of pH and the concentration of the selector are very principal for the improvement in selectivity. The optimization of the separations could be achieved through the proper choice of the experimental conditions.

8. As in CDs, where various ionic or nonionic derivatives have been developed, small modification or introduction of some residues to polysaccharides may lead to the change of enantioselectivity. One example is diltiazem, which was enantioseparated by dextrin and not by dextran, but a dextran derivative: carboxymethyldextran *(29)*.

9. In the CE enantiomer separation, manipulation of the migration order of enantiomers can be easily performed by changing chiral additives or buffer pH (with or without use of EOF).

References

1. Terabe, S., Otsuka, K., and Nishi, H. (1994) Separation of enantiomers by capillary electrophoretic techniques. *J. Chromatogr. A* **666,** 295–319.
2. Chankvetadze, B. (1997) *Capillary Electrophoresis in Chiral Analysis.* John Wiley & Sons, New York.
3. Gubitz, G. and Schmid, M. G. (1997) Chiral separation principles in capillary electrophoresis. *J. Chromatogr. A* **792,** 179–225.
4. Fanali, S. (1997) Controlling enantioselectivity in chiral capillary electrophoresis with inclusion-complexation. *J. Chromatogr. A* **792,** 227–267.
5. Verleysen, K. and Sandra, P. (1998) Separation of chiral compounds by capillary electrophoresis. *Electrophoresis* **19,** 2798–2833.
6. Nishi, H. and Terabe, S., eds. (2000) *Applications of Chiral Capillary Electrophoresis. Special Vol., J. Chromatogr. A* **875,** 3–484.

7. Nishi, H. (1997) Enantioselectivity in chiral capillary electrophoresis with poly-saccharides. *J. Chromatogr. A* **792,** 327–347.
8. Nishi, H. and Kuwahara, Y. (2001) Enantiomer separation by capillary electro-phoresis utilizing noncyclic mono-, oligo- and polysaccharides as chiral selec-tors. *J. Biochem. Biophys. Methods* **48,** 89–102.
9. Stalcup, A. M. and Agyei, N. M. (1994) Heparin: a chiral mobile-phase additive for capillary zone electrophoresis. *Anal. Chem.* **66,** 3054–3059.
10. Agyei, N. M., Gahm, K. H., and Stalcup, A. M. (1995) Chiral separation using heparin and dextran sulfate in capillary zone electrophoresis. *Anal. Chim. Acta* **307,** 185–191.
11. Jin, Y. and Stalcup, A. M. (1998) Application of heparin to chiral separations of antihistamines by capillary electrophoresis. *Electrophoresis* **19,** 2119–2123.
12. Nishi, H., Nakamura, K., Nakai, H., and Sato, T. (1995) Enantiomer separation of drugs by mucopolysacchride-mediated electrokinetic chromatography. *Anal. Chem.* **67,** 2334–2341.
13. Nishi, H. and Terabe, S. (1995) Enantiomeric separation of diltiazem, clentiazem, and its related compounds by capillary electrophoresis using polysaccharides. *J. Chromatogr. Sci.* **33,** 698–402.
14. Nishi, H. (1996) Enantiomer separation of basic drugs by capillary electrophore-sis using ionic and neutral polysaccharides as chiral selectors. *J. Chromatogr. A* **735,** 345–351.
15. Gotti, R., Cavrini, V., Andrisano, V., and Mascellani, G. (1998) Dermatan sulfate as useful chiral selector in capillary electrophoresis. *J. Chromatogr. A* **814,** 205–211.
16. Gotti, R., Furlanetto, S., Andrisano, V., Cavrini, V., and Pinzauti, S. (2000) Design of experiments for capillary electrophoretic enantioresolution of salbutamol using dermatan sulfate. *J. Chromatogr. A* **875,** 411–422.
17. Nishi, H., Nakamura, K., Nakai, H., Sato, T., and Terabe, S. (1994) Enantiomeric separation of drugs by affinity electrokinetic chromatography using dextran sul-fate. *Electrophoresis* **15,** 1335–1340.
18. Beck, G. and Neau, S. (1996) Evaluation of quantitative structure property rela-tionships necessary for enantioresolution with lambda- and sulfobutylether lambda-carrageenan in capillary electrophoresis. *Chirality* **8,** 503–510.
19. Nishi, H., Izumoto, S., Nakamura, K., Nakai, H., and Sato, T. (1996) Dextran and dextrin as chiral selectors in capillary zone electrophoresis. *Chromatographia* **42,** 617–630.
20. D'Hulst, A. and Verbeke, N. (1992) Chiral separation by capillary electrophoresis with oligosaccharides. *J. Chromatogr.* **608,** 275–287.
21. D'Hulst, A. and Verbeke, N. (1994) Separation of the enantiomers of coumarinic anticoagulant by capillary electrophoresis using maltodextrins as chiral modifi-ers. *Chirality* **6,** 225–229.
22. D'Hulst, A. and Verbeke, N. (1996) Chiral analysis of basic drugs by oligosaccha-ride-mediated capillary electrophoresis. *J. Chromatogr. A* **735,** 283–293.
23. D'Hulst, A. and Verbeke, N. (1994) Quantitation in chiral capillary electrophore-sis: theoretical and practical considerations. *Electrophoresis* **15,** 854–863.

24. Soini, H., Stefansson, M., Riekkola, M.-L., and Novotny, M. (1994) Maltooligo-saccharides as chiral selectors for the separation of pharmaceuticals by capillary electrophoresis. *Anal. Chem.* **66,** 3477–3484.
25. Kuwahara, Y. and Nishi, H. (1998) Fast drug analysis by capillary electrophore-sis-content uniformity and assays for diltiazem tablets and trimetoquinol tablets. *Yakugaku Zasshi (Japanese)* **118,** 456–463.
26. Watanabe, T., Takahashi, K., Horiuchi, M., et al. (1999) Chiral separation and quan-titation of pentazocine enantiomers in pharmaceuticals by capillary zone electro-phoresis using maltodextrins. *J. Pharm. Biomed. Anal.* **21,** 75–81.
27. Chankvetadze, B., Saito, M., Yashima, E., and Okamoto, Y. (1997) Enantiosepa-ration using selected polysaccharides as chiral buffer additives in capillary elec-trophoresis. *J. Chromatogr. A* **773,** 331–338.
28. Wren, S. A. C. and Rowe, R. C. (1992) Theoretical aspects of chiral separation in capillary electrophoresis—Initial evaluation of a model. *J. Chromatogr.* **603,** 234–241.
29. Nishi, H. and Kuwahara, Y. (2002) Enantiomer separation by capillary electro-phoresis utilizing carboxymethyl derivatives of polysaccharides as chiral selec-tors. *J. Pharm. Biomed. Anal.* **27,** 577–585.

20

Chiral Micellar Electrokinetic Chromatography

Koji Otsuka and Shigeru Terabe

1. Introduction

Micellar electrokinetic chromatography (MEKC), which was first introduced in 1984 *(1)*, has become one of major separation modes in capillary electrophoresis (CE), owing to its applicability to the separation of neutral compounds as well as charged ones *(2,3)*. The separation principle of MEKC with an anionic micelle, such as sodium dodecyl sulfate (SDS), is schematically shown in **Fig. 1**. A separation capillary is filled with an SDS micellar solution. Under neutral or basic conditions, the entire solution migrates toward the cathode by the electroosmotic flow (EOF) when a high voltage is applied, while the micelle is forced toward the anode by electrophoresis. Normally the EOF is stronger than the electrophoretic migration of the SDS micelle, and hence, the micelle migrates toward the cathode at a slower velocity than the aqueous phase. When a neutral analyte is injected into the micellar solution at the anodic end, it will be distributed between the micelle and the aqueous phase. An analyte that is not incorporated into the micelle at all migrates at the same velocity as the EOF toward the cathode, whereas an analyte that is totally incorporated into the micelle migrates at the lowest velocity or the same velocity as the micelle toward the cathode. The more the analyte is incorporated into the micelle, the slower the analyte will migrate. As long as the analyte is electrically neutral, it migrates at a velocity between the two extremes or between the velocity of the EOF and that of the micelle. The analytes are detected in an increasing order of the distribution coefficients at the cathodic end.

Chiral separation is one of major objectives of CE, and a number of successful reports on enantiomer separations by CE has been published *(4–7)*, including enantioseparations of drugs by MEKC *(8)*. In chiral separations by MEKC, MEKC using chiral micelles, which is called chiral MEKC, and cyclodextrin (CD)-modified MEKC (CD-MEKC) are commonly employed. In this chapter, however, only a chiral MEKC mode is briefly introduced.

From: *Methods in Molecular Biology, Vol. 243: Chiral Separations: Methods and Protocols*
Edited by: G. Gübitz and M. G. Schmid © Humana Press Inc., Totowa, NJ

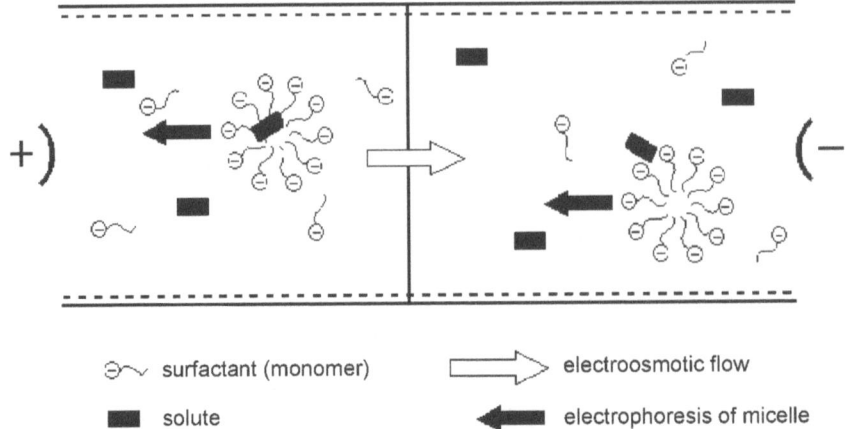

Fig. 1. Schematic illustration of the separation principle of MEKC.

2. Materials

2.1. Apparatus

The use of a commercially available automated CE instrument is recommended for reproducible results. Usually a UV detector is included in the CE system. A laser-induced fluorescence detector, however, is also available from some suppliers. Untreated or uncoated fused silica capillaries (50–75 μm inner diameter [i.d.], 375 μm outer diameter [o.d.]) are frequently employed as separation capillaries as well as some kinds of coated ones, where the effective length normally ranges 30–50 cm.

2.2. Chiral Micelles

In chiral MEKC, an anionic or cationic chiral micelle is used as a pseudostationary phase, and it works as a chiral selector. Bile salts, e.g., sodium cholate (SC), sodium deoxycholate (SDC), sodium taurocholate (STC), and sodium taurodeoxycholate (STDC), are the most popular chiral surfactants in chiral MEKC (*see* **Note 1**).

Digitonin, which is a glycoside of digitogenin and used for the determination of cholesterol, is a neutral surfactant and is also employed as a chiral pseudostationary phase together with an ionic surfactant, e.g., SDS.

2.3. Buffer Components

1. Sodium dihydrogenphosphate (NaH_2PO_4), disodium hydrogenphosphate (Na_2HPO_4), and/or sodium tetraborate ($Na_2B_4O_7$).

2. Phosphoric acid and boric acid.
3. Urea.
4. Organic modifiers, *e.g.*, methanol, acetonitrile, and 2-propanol.

3. Methods

3.1. Preparation of Buffer Solutions

A phosphate buffer solution or a borate-phosphate solution (100 mL) is prepared by mixing 10 or 20 mM NaH$_2$PO$_4$ and Na$_2$HPO$_4$ or NaH$_2$PO$_4$ and Na$_2$B$_4$O$_7$, of which pH is adjusted at the adequate value. If necessary, 10 or 20 mM phosphoric acid or boric acid is added for the pH adjustment. A background solution (BGS) or separation solution (10 mL) is prepared by dissolving an appropriate amount of a chiral surfactant with the buffer solution. The BGS is filtered through a membrane filter (0.45 μm) and degassed by ultrasonication.

If an acidic condition is required, the buffer solution is made with 20 mM sodium dihydrogenphosphate and 20 mM phosphoric acid (*see* **Note 2**).

3.2. Additives Other Than Micelles

Sometimes several additives other than chiral micelles are used to improve separation or resolution and/or peak shapes. Urea is a most commonly used additive in MEKC. Relatively high concentration, e.g., 1–5 M, of urea is added to the micellar solution. Organic modifiers, such as methanol, acetonitrile, and 2-propanol, are also employed as additives to the micellar solution. Typically, 10–30% (v/v) content of an organic modifier is added after preparing the micellar solution.

3.3. Sample Preparation

Sample preparation is carried out the same way as in conventional MEKC: an appropriate amount of the sample is weighed and then dissolved in the separation solution or micellar solution to be used. However, the sample solvent is sometimes replaced with the buffer solution without the micelle or a phosphate buffer only. Moreover, sometimes water, organic solvents, e.g., methanol and acetonitrile, or aqueous organic solvents are used instead of the aforementioned buffer solutions. The concentration of the sample is 100–1000 ppm.

3.4. Operation Conditions

Most commonly, the separation capillary is an untreated or bare fused silica capillary, and the effective length is 30–50 cm, and i.d. is 50–75 μm. Detection wavelength (if employed UV detection), applied voltage, and temperature are dependent on samples to be analyzed.

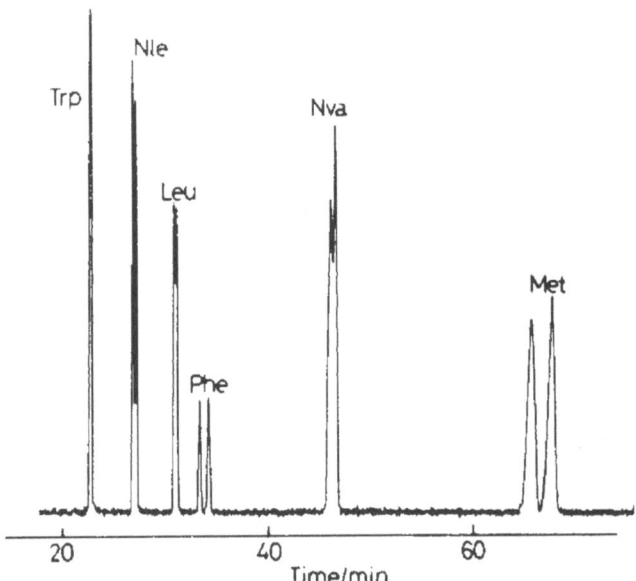

Fig. 2. Chiral separation of Dns-DL-AAs with 50 m*M* STDC in 50 m*M* phosphate buffer (pH 3.0) *(9)*. Separation capillary, 50 μm i.d. × 70 cm, 50 cm effective; current, 50 μA; temperature, 40°C.

3.5. Enantiomer Separations by MEKC Using Chiral Micelles

3.5.1. Using Bile Salts

Typically, 10–50 m*M* bile salts are employed as chiral pseudostationary phases.

3.5.1.1. EXAMPLE 1:
SEPARATION OF DANSYLATED-DL-AMINO ACIDS (DNS-DL-AAS) (**FIG. 2**) *(9)*

1. A separation capillary (50 μm i.d. × 50 cm in effective) is filled with 50 m*M* STDC (pH 3.0).
2. Each sample Dns-DL-AA is dissolved in methanol/water (1:1) at 0.1 mg/mL.
3. Injection is carried out hydrodynamically (pressurized method), for 1 s at 50 mbar.
4. Detection wavelength is 210 nm.
5. Applied voltage is 25 kV. However, if the reversed polarity is used: the injection end is cathodic and the detection end is anodic.

3.5.1.2. EXAMPLE 2:
SEPARATION OF DRUG COMPONENTS (**FIG. 3**) *(10)*

1. The BGS is 50 m*M* STDC (pH 7.0).
2. Samples are several drug components, such as diltiazem and trimetoquinol.

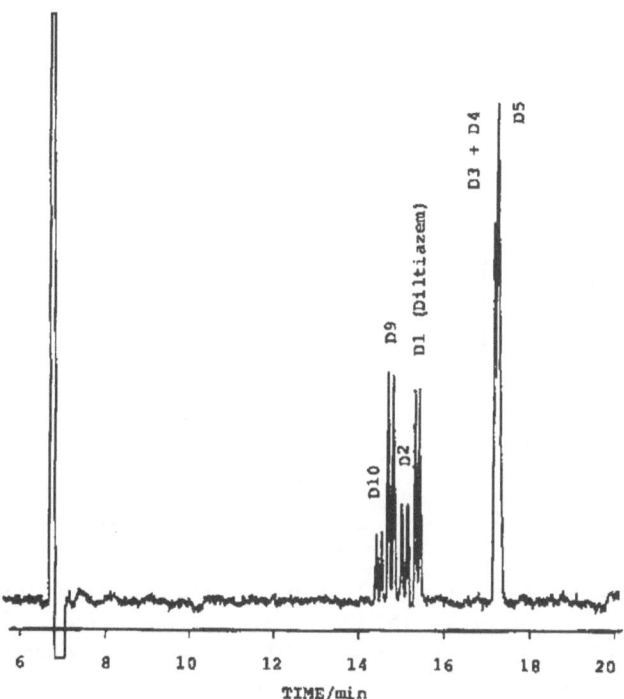

Fig. 3. Chiral separation of diltiazem hydrochloride and related compounds by MEKC using STDC *(10)*. Separation solution, 50 m*M* STDC in 20 m*M* phosphate-borate buffer (pH 7.0); separation capillary, 50 μm i.d. × 65 cm, 50 cm effective; applied voltage, 20 kV; detection wavelength, 210 nm; temperature, ambient.

3. Detection wavelength is 210 nm.
4. Applied polarity is normal: the injection end is anodic.

3.5.2. Using Digitonin

Since digitonin is a neutral surfactant, an ionic micelle is necessary to form a mixed micelle that is effective as a pseudostationary phase in MEKC. As an anionic achiral micelle, SDS is used (*see* **Note 3**).

3.5.2.1. EXAMPLE 3:
ENANTIOSEPARATION OF PHENYLTHIOHYDANTOIN (PTH)-DL-AAS (**FIG. 4**) *(11)*

1. The BGS is 10 m*M* digitonin together with 20 m*M* SDS (pH 3.0).
2. Each sample PTH-DL-AA is dissolved in acetonitrile-water (1:1).
3. The reversal polarity (25 kV) is applied.

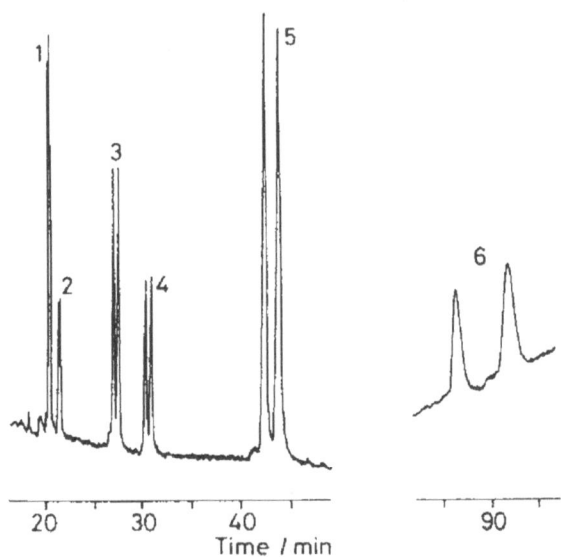

Fig. 4. Chiral separation of six PTH-DL-AAs by MEKC *(11)*. Corresponding AAs: (1) Trp, (2) Nle, (3) Nva, (4) Val, (5) Aba, (6) Ala. Micellar solution, 25 mM digitonin, 50 mM SDS (pH 3.0); separation capillary, 50 μm i.d. × 63 cm, 49 cm effective; total applied voltage, 20 kV; current, 34 μA; detection wavelength, 260 nm; temperature, ambient.

3.5.3. Using Synthetic Chiral Surfactants

Other than aforementioned natural occurring surfactants or bile salts and digitonin, several synthetic chiral micelles have been also employed in chiral MEKC separations.

As for such chiral micelles, various *N*-alkanoyl-L-amino acids, e.g., sodium *N*-dodecanoyl-L-amino acids, have been examined to use in chiral MEKC, and several enantiomers were successfully separated (**Fig. 5**) *(12)*. In each case, the addition of SDS, urea, and organic modifiers such as methanol or 2-propanol were essential to obtain improved peak shapes and enhanced enantioselectivity. Unfortunately, these surfactants are not commercially available.

Other amino acid chiral micelles, such as *S*- and *R-N*-dodecoxycarbonyl-amino acids, are reported to be effective for enantioseparations, and these are obtainable from a supplier.

Anionic alkylglucoside chiral surfactants, such as dodecyl β-D-glucopyra-noside monophosphate and monosulfate, can also be used as chiral pseudosta-tionary phases in MEKC. A synthesized chiral surfactant based on *(R,R)*-tartaric

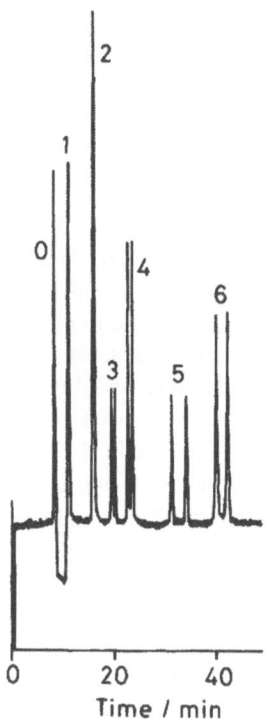

Fig. 5. Chiral separation of six PTH-DL-AAs by MEKC *(12)*. Corresponding AAs: (1) Ser, (2) Aba, (3) Nva, (4) Val, (5) Trp, (6) Nle; (0) acetonitrile. Micellar solution, 50 mM sodium N-dodecanoyl-L-valinate, 30 mM SDS, 0.5 M urea (pH 9.0) containing 10% (v/v) methanol; separation capillary, 50 μm i.d. × 65 cm, 50 cm effective; total applied voltage, 20 kV; current, 17 μA; detection wavelength, 260 nm; temperature, ambient.

acid was used for the enantiomer separation in MEKC, where enantiomers having fused polyaromatic rings were separated easier than having only single aryl group. Neutral steroidal glucoside surfactants, e.g., N,N-bis-(3-D-gluconamido-propyl)-cholamide (Big CHAP), has been introduced for the use as chiral pseudo-stationary phases in MEKC.

Recently, the use of a high-molecular-mass surfactant, such as poly(sodium N-undecylenyl-L-valinate), has been investigated for enantioseparations by MEKC. Polymerized dipeptide surfactants, derived from sodium N-undecylenyl-L-valine-L-leucine, sodium N-undecylenyl-L-leucine-L-valine, and so on, were employed.

Anyhow, the synthetic chiral micelles have not been popular to date in using as chiral pseudostationary phases in MEKC.

4. Notes

1. Bile salts are the most popular chiral micelles employed in MEKC, not only for enantioseparations, but also for usual separations of rather hydrophobic compounds. As mentioned before, several types of bile salts are commercially available, so that by changing the bile salt micelle from one to another, a different selectivity may be obtained.

2. In MEKC, the migration order of the samples might be reversed. This is sometimes quite important for enantiomeric quantitation, e.g., optically purity test. In some cases, the reversal of the migration order can be easily attained by the suppression of the EOF through the decrease in pH or the use of a capillary of which the inside wall is coated with neutral material such as polyacrylamide

3. Saponins, e.g., glycyrrhizic acid (GRA) and β-escin, are also used as chiral pseudostationary phases in MEKC *(13)*. Some Dns-DL-AAs were optically resolved by a 30 mM GRA, 50 mM octyl-β-D-glucoside, 10 mM SDS solution (pH 7.0), but the reproducibility in migration times were not good. A β-escin, SDS system under acidic conditions was effective for enantioseparations of PTH-DL-AAs, and while under neutral conditions insufficient separations were obtained.

4. As another important and effective mode for enantioseparations, CD-MEKC is widely used especially for the separation of neutral drug components. In CD-MEKC, various types of CDs are employed as chiral selectors. In exact meanings, only neutral CDs can be used in CD-MEKC, that is, if an ionic CD is employed, the system is no longer CD-MEKC, but is called cyclodextrin electrokinetic chromatography (CDEKC). Practically, however, one can use both neutral and ionic CDs as chiral selectors as additives in MEKC. As for neutral CDs, nonderivatized CDs, e.g., α-, β-, and γ-CDs, and derivatized CDs, e.g., 2,6-di-*O*-methyl-β-CD, 2,3,6-tri-*O*-methyl-β-CD, and hydroxypropyl-β-CD are frequently used. Some reviews are available on the use of CDs as chiral selectors in CE as well as MEKC *(4,14)*.

References

1. Terabe, S., Otsuka, K., Ichikawa, K., Tsuchiya, A., and Ando, T. (1984) Electrokinetic separations with micellar solutions and open-tubular capillaries. *Anal. Chem.* **56,** 111–113.
2. Terabe, S., Otsuka, K., and Ando, T. (1985) Electrokinetic chromatography with micellar solution and open-tubular capillary. *Anal. Chem.* **57,** 834–841.
3. Otsuka, K. and Terabe, S. (1998) Micellar electrokinetic chromatography. *Bull. Chem. Soc. Jpn.* **71,** 2465–2481.
4. Terabe, S., Otsuka, K., and Nishi, H. (1994) Separation of enantiomers by capillary electrophoretic techniques. *J. Chromatogr. A* **666,** 295–319.
5. Gübitz, G. and Schmid, M. G. (1997) Chiral separation principles in capillary electrophoresis. *J. Chromatogr. A* **792,** 179–225.
6. Chankvetadze, B. (1999) *Capillary Electrophoresis in Chiral Analysis.* John Wiley & Sons, New York.
7. Verleysen, K. and Sandra, P. (1998) Separation of chiral compounds by capillary electrophoresis. *Electrophoresis* **19,** 2798–2833.

8. Otsuka, K. and Terabe, S. (2000) Enantiomer separation of drugs by micellar electrokinetic chromatography using chiral surfactants. *J. Chromatogr. A* **875,** 163–178.
9. Terabe, S., Shibata, M., and Miyashita, Y. (1989) Chiral separation by electrokinetic chromatography using bile salt micelles. *J. Chromatogr.* **480,** 403–411.
10. Nishi, H., Fukuyama, T., Matsuo, M., and Terabe, S. (1990) Chiral separation of diltiazem, trimetoquinol and related compounds by micellar electrokinetic chromatography with bile salts. *J. Chromatogr.* **515,** 233–243.
11. Otsuka, K. and Terabe, S. (1990) Enantiomeric resolution by micellar electrokinetic chromatography with chiral surfactants. *J. Chromatogr.* **515,** 221–226.
12. Otsuka, K., Kawahara, J., Tatekawa, K., and Terabe, S. (1991) Chiral separations by micellar electrokinetic chromatography with sodium N-dodecanoyl-L-valinate. *J. Chromatogr.* **559,** 209–214.
13. Ishihama, Y. and Terabe, S. (1993) Enantiomeric separation by micellar electrokinetic chromatography using saponins. *J. Liq. Chromatogr.* **16,** 933–944.
14. Fanali, S. (2000) Enantioselective determination by capillary electrophoresis with cyclodextrins as chiral selectors. *J. Chromatogr. A* **875,** 89–122.

21

Chiral Separation by Capillary Electrophoresis in Nonaqueous Medium

Marja-Liisa Riekkola and Heli Sirén

1. Introduction

In the last few years, nonaqueous background electrolyte (BGE) solutions have attracted increasing interest, not only in capillary zone electrophoresis (CZE) but also in other capillary electrophoretic techniques (CE) *(1)*. Nonaqueous capillary electrophoresis (NACE) exploits the vastly different physicochemical properties of organic solvents. With mere adjustment of BGE composition, parameters such as resolution, analysis time, and selectivity can be fine-tuned or even altered dramatically *(2–9)*. The solubility of the analytes and additives can be enhanced, and the possible adsorption of the analytes on the capillary wall can be reduced. Because solvation of the analytes differs considerably from that in aqueous solutions, a strong effect on the separation selectivity can be seen. In addition, low currents make the use of higher electric field strengths possible in separations. The interactions, which hardly take place in water, can be utilized in nonaqueous solvents permitting capillary electrophoretic separation of even uncharged compounds after the formation of complexes with ionic additives *(6,10)*. Moreover, the solutes labile in aqueous solutions can be stable in nonaqueous media.

The organic solvents applied in capillary electrophoretic separations must, nevertheless, meet specific requirements. They should dissolve the analytes and the electrolytes used in the separation; they should not be inflammable, toxic, or reactive; they should be available as pure substances and at low cost; and they should exhibit relatively low viscosity, moderately high relative permittivity, and low vapor pressure *(1)*. Relative permittivity ε (dielectric constant) (**Table 1**) of the solvent is an important parameter, because it describes the strength

From: *Methods in Molecular Biology, Vol. 243: Chiral Separations: Methods and Protocols*
Edited by: G. Gübitz and M. G. Schmid © Humana Press Inc., Totowa, NJ

Table 1
Properties of Some Solvents Used in CE[a]

Solvent	T_m	T_b	P	γ	ϵ	μ	η	pK_{auto}
MeOH	-97.7	64.5	16.9	22.3	32.66	2.87	0.551	16.91
EtOH	-114.5	78.2	7.89	21.9	24.55	1.66	1.083	19.1
1-PrOH	-126.2	97.1	2.73	23.1	20.45	3.09	1.943	19.4
2-PrOH	-88.0	82.2	6.03	21.2	19.92	1.66	2.044	21.08
1-BuOH	-88.7	117.6	0.82	24.2	17.51	1.75	2.571	20.89
ACN	-43.9	81.6	12.2	28.3	35.94	3.92	0.341	32.2
PC	-55.0	241.7	0.00616	41.4	64.92	4.94	2.53	—
FA	2.5	210.5	0.00882	58.2	109.5	3.37	3.302	16.8
NMF	-3.8	199.5	0.0338	39.5	182.4	3.86	1.65	10.74
DMF	-60.5	153.0	0.53	36.4	36.71	3.82	0.802	23.1
DMSO	18.5	189.0	0.077	43.0	46.45	4.06	1.991	31.8
H_2O	0	100.0	3.17	71.8	78.36	1.85	0.8903	17.51

[a]Temperature-dependent properties are at 25°C (12).

Units: melting point (T_m) and boiling point (T_b) in °C; vapor pressure p in kPa; surface tension γ in 10^{-3} N/m; relative permittivity ϵ is dimensionless; dipole moment μ in D (1 D = 3.33564 10^{-30} Cm); dynamic viscosity η in mPa s (numerically equal to cP); autoprotolysis constant pK_{auto} is dimensionless (note that concentration of bulk solvent is taken into account). (—) value not reported.

of the interactions between ions in the solvent. Most of the solvents used in NACE studies have $\varepsilon \geq 30$, where ionic dissociation of electrolytes can be considered to take place to a large or full extent. Ethanol and 1-butanol belong to solvents with $10 < \varepsilon < 30$, where ionic dissociation takes place to some extent, but the dominant effect is extensive ion-pair formation. Low dynamic viscosity is preferred in HCE, because it allows high mobilities of the solvated analyte ions. Also important is the compatibility of the organic solvents with specific detection techniques. When UV detection is applied, the solvent should be transparent to UV light. It can be concluded that the variety of solvent properties, the low currents, and the wide range of detection systems suitable for NACE offer a good basis for finding attractive conditions and for devising tailor-made solutions for a variety of analytical problems.

Methanol and acetonitrile and their mixtures, as well as methanol-ethanol mixtures, are the most frequently used organic solvents in enantioseparations by NACE *(2,5–9,11–15)*. In addition, amide-type solvents have been found useful. Because of relatively low surface tension of methanol, it has been especially popular in studies with mass spectrometer as detector. Most common buffering systems in organic solvents consist of acids and their ammonium salts, of which acetic acid and ammonium acetate have been clearly the most common ones to adjust the pH. The traditional organic solvent compositions in nonaqueous capillary electrochromatography (NACEC) have been acetonitrile, tetrahydrofuran (THF), and methanol.

Uncoated fused silica capillaries have been mostly used in NACE separations, because adsorption of analytes on the capillary wall may be reduced in nonaqueous media, with the benefit that coating of the capillary causes suppression of analyte-wall interactions and is not necessarily important in most studies. However, polyvinyl alcohol-coated capillaries have successfully been used in enantioseparations in ethanol-methanol BGE solutions *(15,16)*.

NACE may be the method of choice for separations that are not easily accomplished in aqueous media. The inherent feasibility and ease of adjusting the composition of the BGE solution are significant advantages of the CE technique. A wide variety of chiral selectors in nonaqueous solvents have been investigated over the past few years, extending the applicability of the NACE technique to chiral separations also. Even though most CE chiral applications to date have been carried out in aqueous–organic mobile phase systems, nonaqueous media have opened and will certainly open new avenues in chiral separations by CE. Hydrophobic chiral selectors, not sufficiently soluble in aqueous medium, are easily applied in the enantiomer separations by NACE. Even though hydrophobic interactions are generally weak in organic solvents, nonaqueous conditions offer the possibility to utilize different ion-ion interactions, like ion-pair formation in separations. Furthermore, under nonaqueous media, the shifts in the pKa and pH scales, the changed dissociation behavior, stronger electrostatic interactions, and/or distinct solvational effects in organic solvents widen the basis for chiral separations *(14)*. The mobilities of analytes can be faster, slower, or equal compared to aqueous solvents depending on the organic solvent used as the BGE. Quinine and quinidine derivatives, as well as derivatized β- and γ-cyclodextrins, have been actively utilized as the chiral selectors to facilitate ion-pair formation or ion-dipole interactions in NACE *(2,17,18)*.

In NACEC various types of silica materials, such as porous native silica gel with unmodified silanol groups, aminopropylsilanized silica, and octadecyl-silica, have mainly been used as supports of a chiral selector for enantiosep-arations. These supports have been modified on silica by covalent bondings of tert-butylcarbamoylquinine via a 3-mercaptopropyltriethoxysilane spacer or by coating with poly(diphenyl-2-pyridylmethylmethacrylate) and polysaccharide derivatives, such as cellulose and amylose *tris*-3,5-dimethylphenylcarbamates, cellulose tris-3,5-dichlorophenylcarbamate, cellulose tris-4-methylbenzoate, amylose tris-*S*-phenylethylcarbamate, or a mixer of two polysaccharide phenyl-carbamates. Macrocyclic antibiotics, such as vancomycin and tecoplanin, chem-ically immobilized on silica, and organic methacrylate monoliths containing quinine derivatives under NACEC conditions have also been reported. Anion and cation exchange sorbents for chiral acids and bases have also been used with acetonitrile-methanol electrolytes *(17)*.

The complimentary potential of NACE to aqueous CE for chiral separations has been realized by several scientists *(2,13,14,17,18)*. In this chapter, common procedures to separate chiral compounds with NACE and NACEC are given. Especially, fast and repeatable separations can be obtained for the enatiomers of a large number of hydrophobilic and hydrophobic chiral bases.

2. Materials

2.1. Reagents

1. Organic solvents such as methanol, ethanol, propanols, formamide, and acetoni-trile should be of high-performance liquid chromatography (HPLC)-grade.
2. In nonaqueous enantiomerseparations, chiral selectors, such as β-cyclodextrin, γ-cyclodextrin, or their sulfonated, methylated, or hydroxylated products, should be of analytical grade. Derivatives of quinine and quinidine as well as β- and γ-cyclo-dextrins are applicable to NACE. In addition to chiral selectors, chiral packing materials, i.e., silica modified with vancomycin or cellulosa/amylose-carbamate are useful in NACEC.
3. Nitromethane is used as the electroosmosis marker.
4. Other chemicals used for chiral separations should be of analytical grade or of the highest purity available.

2.2. Instruments

1. A CE instrument for NACE analyses or a CE instrument with the possibility to pressurize (5–50 bar, normally 8 bar constant pressure) both ends of the capillary for NACEC analyses can be employed.
2. A photodiode array (PDA) or a UV detector with wavelength filters of, e.g., 200, 214, and 254 nm can be used.
3. Temperature control units for capillary, electrolyte solutions, and samples. The temperatures of capillary and sample trays should be optimized (approx 15–40°C) and maintained stable during the separations. As the temperature of the buffer trays is not adjustable in the instrument, the largest buffer vials available should be used in order to minimize the vol change in the electrolyte composition caused by evaporation (*see* **Note 1**).
4. Glass or polymer vials provided with lids for electrolytes and samples (*see* **Note 1**).
5. Hydrodynamic injections should be preferred in NACE. If different organic electrolyte compositions are used, sample injections must be adjusted in different solvents. For example, injections of 3.1–8.9 s using alcoholic BGE solutions can be made with pressures of 0.4–0.5 p.s.i. (1 p.s.i. = 6894.76 Pa). Electrokinetic injections may be preferred in NACEC analyses.
6. The voltage applied in alcoholic BGE separations containing 20 mM ammonium acetate (99:1, v/v) may be 20–30 kV corresponding to an electric field of 500 V/cm.
7. Polytetra-fluoroethylene (PTFE) or nylon filters (0.45 μm) for sample and BGE filtering.
8. Uncoated fused silica capillaries or packed CE capillaries are used for NACE and NACEC analyses, respectively. In NACE the detection lengths (L_{det}) of the capillaries are below 60 cm (usually 20–50 cm when injection is made from inlet, and approx 7–10 cm when injection is done from outlet) with 25–75 μm inner diameter (i.d.) and 150–365 μm outer diameter (o.d.) *(18)*. In NACEC, the capillary dimensions may be: particle size 5 μm, 180 μm i.d., 350 μm o.d., L_{tot} 30 cm. The packed columns should be shorter than 60 cm. For example, in order to use a 50-cm long capillary packed with 1.5 μm material at 2.0 mm/s with pressure-driven flow, a pressure drop to 3500 bar is recommended *(20,21)*.

The packed columns can be obtained from manufacturers or they can be homemade by using, for example, supercritical CO_2 *(19)*. Pellicular packings or porous layer beads having particle diameters of 30–55 μm can be used to fill a fused silica capillary for NACEC separations. Also, nonporous particles of 1–5 μm may be used.

Monolithic packings can be fabricated with polymerization of a monomer mixture, fusion of individual particles, and the application of an external field. Sol-gel-bonded continuous-bed columns can be homemade as an alternative to packed columns. They have nearly the same properties as packed columns, but without end-frits *(20)*.

The sorbent materials have been activated for enantiomer separations with native bondings to silica, if no chiral selector is added to BGE *(2,14,21)*. For example, cation exchange type chiral stationary phases (CSPs) based on 3,5-dichlorobenzoyl amino as chiral selectors and silica as the sorbent may be applied to enantiomer separations of chiral bases by NACEC. Also, silica phase stationary phases have been used after derivatization with vancomycin *(20)*. The organic solvent contained 13 mM ammonium acetate in a methanol-acetonitrile mixture.

3. Methods

3.1. Preparation of Electrolyte Solutions for Analysis

1. Prepare an electrolyte solution (BGE) by weighting ammonium, potassium or sodium acetate, formate, or hydroxide to make a 10–30 mM stock electrolyte solution for NACE separations. Dilute it with alcohol (methanol, ethanol, 1-propanol, 1-butanol)/acetonitrile-acetic acid/formic acid (100%) in 50:49:1 (v/v/v), alcohol/acetonitrile (40:60, v/v), or in pure organic solvent only (acetonitrile) to the final vol needed. As an example, CE-UV analyses can be made with a 20 mM ammonium acetate dissolved in alcohol/acetic acid (99:1,v/v) mixture. Weigh the chiral selector, e.g., β-cyclodextrin (*see* Chapter 2) to make a BGE solution containing 100 mM of the chiral selector for the analyses (*see* **Note 1**).

    An example of the BGE in NACE: 100 mM β-cyclodextrin in 150 mM citric acid, 100 mM Tris in formamide (pH* adjustment to 5.1).
2. Prepare an electrolyte solution for NACEC analysis using 1–5 mM lithium, potassium, sodium, or ammonium salt in DMF, methanol, or ethanol or use the electrolytes described in **Subheading 3.1.**

    Examples of the BGE in NACEC with packed capillaries: (i) 10 mM Tris in methanol, or (ii) 10 mM ammonium acetate in methanol both containing, e.g., 50 mM β-cyclodextrin. Instrumental conditions: capillary length 32.5 cm, pressure at the both capillary ends 12 bar, separation voltages 5–10 kV, and injection with 10 bar for 3 s.
3. Filtrate the electrolyte solution, e.g., with 0.45 μm PTFE or nylon membranes and degas by sonication prior to use.

3.2. Prepare the Sample Mixture

Analyte mixtures can be made, e.g., of β-blockers, anti-inflammatory drugs, stimulants, aromatic amines, metanephrines, and catecholamines, by dissolving them into pure organic solvents or into nonaqueous electrolyte solutions used for separations.

3.3. Capillary Conditioning

1. Prepare the capillary for NACE analysis. For nonaqueous analyses, the flushing is performed with pure solvent for 90 min and with the electrolyte solution for 60 min. Between analyses the capillary is flushed with the electrolyte solution for 2 min.

2. Prepare the capillary for NACEC analysis. Purge the capillary with 1-propanol/methanol with acid or base (99:1, v/v), or salts (ammonium acetate), which should be of analytical grade.
3. Conditioning between runs: organic solvent mixture at 40°C for 1 h using voltage, conditioning with voltage gradient from 100 V/cm to 650 V/cm, followed by a steady state at 650 V/cm.

3.4. Checking the Performance of the Analyses and Sequences

If small (vol of 2 mL) vials are used, the inlet and outlet vials should be changed every h to avoid electrolyte depletion *(19)*. Temperatures of the BGE and sample solutions should be low, but high enough to keep the chemicals soluble.

4. Notes

1. Temperature control keeps the mobilities of the analytes repeatable and eliminates evaporation of the organic solutions.
2. Conditioning of the capillaries with the organic solvent is important.
3. The polyimide coating should be removed at a length of 1–1.5 cm from the capillary outer wall before use.
4. The randomly sulfated cyclodextrins are not sufficiently soluble in pure methanol or acetonitrile.
5. Much higher concentrations of β-cyclodextrin are needed in the nonaqueous solvents to achieve similar separations to those in water.
6. Because the packing and the sorbent in CEC capillaries may cause unrepeatable migration times for the analytes, internal standards should be used to correct the values. Also, mobilities can be used to correct the peak values calculated from the electropherograms.
7. Short injection times (1 s) must be avoided due to the inaccurate vol feeding.

References

1. Riekkola, M.-L. (2002) Recent advances in non-aqueous capillary electrophoresis. *Electrophoresis* **23,** 3868–3883.
2. Chankvetadze, B. and Blaschke, G. (2000) Enantioseparations using capillary electromigration techniques in nonaqueous buffers. *Electrophoresis* **21,** 4159–4178.
3. Cherkaoui, S., Geiser, L., and Veuthey, J.-L. (2000) Rapid separation of basic drugs by nonaqueous capillary electrophoresis. *Chromatographia* **52,** 403–407.
4. Ward, V. L. and Khaledi, M. G. (1999) Efficiency studies in nonaqueous capillary electrophoresis. *J. Chromatogr. A* **859,** 203–219.
5. Porras, S. P., Riekkola, M.-L., and Kenndler, E. (2001) Capillary zone electrophoresis of basic analytes in methanol as non-aqueous solvent. Mobility and ionisation constant. *J. Chromatogr. A* **905,** 259–268.

6. Steiner, F. and Hassel, M. (2000) Nonaqueous capillary electrophoresis: a versatile completion of electrophoretic separation techniques. *Electrophoresis* **21,** 3994–4016.
7. Porras, S. P., Jyske, P., Riekkola, M.-L., and Kenndler, E. (2001) Mobility and ionization constant of basic drugs in methanol. Application of nonaqueous background electrolytes for capillary zone electrophoresis based on a conventional pH scale. *J. Microcolumn Sep.* **13,** 149–155.
8. Porras, S. P., Riekkola, M.-L., and Kenndler, E. (2000) Capillary zone electrophoresis of basic drugs in non-aqueous acetonitrile with buffers based on a conventional pH scale. *Chromatographia* **52,** 290–294.
9. Porras, S. P., Wiedmer, S. K., Strandman, S., Tenhu, H., and Riekkola, M.-L. (2001) Novel dynamic polymer coating for capillary electrophoresis in nonaqueous methanolic background electrolytes. *Electrophoresis* **22,** 3805–3812.
10. Burger, K. (1983) Solvation, in *Ionic and Complex Formation Reactions in Non-Aqueous Solvents*, Elsevier, Amsterdam.
11. Tobler, E., Lämmerhofer, M., Wuggening, F., Hammerschmidt, F., and Lindner, W. (2002) Low-molecular-weight chiral cation exchangers: novel chiral stationary phases and their application for enentioseparation of chiral bases by nonaqueous capillary electrochromatography. *Electrophoresis* **23,** 462–476.
11a. Marcus, Y. (1998) *The Properties of Solvents, Wiley Series in Solution Chemistry, Vol. 4.* Wiley & Sons, Chichester, pp. 136–140.
12. Porras, S. P., Jussila, M., Sinervo, K., and Riekkola, M.-L. (1999) Alcohols and wide-bore capillaries in nonaqueous capillary electrophoresis. *Electrophoresis* **20,** 2510–2518.
13. Cherkaoui, S., Varesio, E., Christen, P., and Veuthey, J.-L. (1998) Selectivity manipulation using nonaqueous capillary electrophoresis. Application to tropane alkaloids and amphetamine derivatives. *Electrophoresis* **19,** 2900–2906.
14. Sahota, R. S. and Khaledi, M. G. (1994) Nonaqueous capillary electrophoresis. *Anal. Chem.* **66,** 1141–1146.
15. Czerwenka, C., Lämmerhofer, M., and Linder, W. (2002) Electrolyte and additive effects on enantiomer separation of peptides by nonaqueous ion-pair capillary electrophoresis using tert.-butylcarbamoylquinine as chiral counterion. *Electrophoresis* **23,** 1887–1899.
16. Zarbl, E., Lämmerhofer, M., Franco, P., Petracs, M., and Linder, W. (2001) Development of stereoselective nonaqueous capillary electrophoresis system for the resolution of cationic and amphoteric analytes. *Electrophoresis* **22,** 3297–3307.
17. Chankvetadze, B. and Blaschke, G. (2001) Enantioseparations in capillary electromigration techniques: recent developments and future trends. *J. Chromatogr. A* **906,** 309–363.
18. Busby, M. B., Maldonado, O., and Vigh, G. (2002) Nonaqueous capillary electrophoretic separation of basic enatiomers using octakis(2,3-O-dimethyl-6-O-sulfo)-γ-cyclodextrin, a new, single-isomer chiral resolving agent. *Electrophoresis* **23,** 456–461.
19. Roed, L., Lundanes, E., and Greibrokk, T. (1999) Nonaqueous electrochromatography on C_{30} columns: separation of retinyl esters. *Electrophoresis* **20,** 2373–2378.

20. Fanali, S., Catarcini, P., and Quaglia, M. G. (2002) Use of vancomycin silica stationary phase in packed capillary electrochromatography: III. Enantiomeric separation of basic compounds with the polar organic mobile phase. *Electrophoresis* **23,** 477–485.
21. Karlsson, C., Wikstrom, H., Armstrong, D. W., and Owens, P. K. (2000) Enantioselective reversed-phase and non-aqueous capillary electrochromatography using a teicoplanin chiral stationary phase. *J. Chromatogr. A* **897,** 349–363.

22

Chiral Ligand-Exchange Capillary Electrophoresis and Capillary Electrochromatography

Martin G. Schmid and Gerald Gübitz

1. Introduction

The principle of chiral ligand-exchange chromatography was introduced in the early 1970s by Davankov and Rogozhin *(1)*. This basic principle was successfully transferred also to capillary electrophoresis (CE). A survey of applications of ligand-exchange capillary electrophoresis (LECE) is given in a recent review *(2)*.

The first application of this principle in CE was reported by Zare's group using L-histidine *(3)* or aspartame-Cu(II) complexes *(4)* as additives to the electrolyte for the chiral separation of Dns-amino acids. Another selector used in combination with sodium dodecyl sulfate (SDS) for the separation of Dns-amino acids is *N,N*-didecyl-L-alanine/Cu(II) *(5,6)*. Desiderio et al. *(7)* used L-Pro-, L-4-hydroxyproline (L-Hypro)- or aspartame-Cu(II) complexes for the chiral resolution of α-hydroxy acids. Schmid and Gübitz *(8)* succeeded in direct resolution of underivatized amino acids using L-Pro- or L-Hypro-Cu(II) complexes. L-Hypro was found to show a higher resolution power compared to L-Pro.

N-alkyl derivatives of L-Hypro, such as *N*-(2-hydroxyoctyl)-L-4-hydroxyproline (HO-L-4-Hypro) or *N*-(2-hydroxypropyl)-L-4-hydroxyproline (HP-L-4-Hypro) were found to show significantly enhanced chiral recognition ability for amino acids compared to L-Hypro *(9,10)* and were also shown to be applicable to the chiral resolution of dipeptides *(10)*, α-hydroxy acids *(11)*, and amino alcohols such as sympathomimetics *(12)* and β-blockers *(11)*.

The chiral recognition mechanism is based on the formation of diastereomeric ternary mixed metal complexes between a chiral selector ligand and an

From: *Methods in Molecular Biology, Vol. 243: Chiral Separations: Methods and Protocols*
Edited by: G. Gübitz and M. G. Schmid © Humana Press Inc., Totowa, NJ

Table 1
Separation Data for Racemic Amino Acids

Compound	t_1 (min)	α (t_2/t_1)	Rs
β-Phenylserine	17.83	1.017	0.97
α-Phenylglycine	18.30	1.000	—
β-(3,4-Dihydroxyphenyl)serine	19.53	1.016	0.91
Dopamine (DOPA)	19.81	1.027	1.19
α-Methylphenylalanine	20.62	1.017	1.00
m-Tyrosine	20.74	1.025	1.65
p-Tyrosine	21.14	1.033	2.76
Phenylalanine	21.46	1.019	1.19
o-Tyrosine	21.51	1.023	1.22
Tryptophan	22.61	1.031	1.92
α-Methyl DOPA	22.85	1.040	2.36

Electrolyte composition: 80 mM L-Hypro, 40 mM CuSO$_4$, adjusted with NH$_3$ to pH 4.0. Migration time of the faster enantiomer (t_1), separation factor (α) and resolution (Rs) are given.

analyte ligand. Resolution occurs due to the difference in complex stability constants of the two diastereomeric mixed complexes with the analyte enantiomers:

$$Cu\ (L\text{-Sel})_2 + S\text{-A} \rightleftharpoons Cu\ (L\text{-Sel})(S\text{-A}) + L\text{-Sel} \quad Sel: selector$$

$$Cu\ (L\text{-Sel})_2 + R\text{-A} \rightleftharpoons Cu\ (L\text{-Sel})(R\text{-A}) + L\text{-Sel} \quad A: analyte$$

Generally, the migration order for underivatized amino acids using L-Pro, L-Hypro or their *N*-alkyl derivatives as chiral selectors is D before L indicating that the complexes with the L-enantiomers show the higher stability (**Fig. 1A**). By addition of SDS above (**Table 1**), the critical micelle concentration hydrophobic micelles are formed creating a pseudostationary phase (micellar electrokinetic chromatography; [MEKC]). A significant improvement in resolution is observed connected with a reversal of the enantiomer migration order (EMO) (**Fig. 1B**) (*see* **Note 1**). This behavior was also noticed by Chen et al. *(13)* who investigated the influence of different surfactants *(14)*. The bidentate selector complex is neutral and migrates with the same velocity as the electroosmotic flow (EOF). At the applied pH , the amino acids are positively charged and migrate faster than the EOF, whereby the migration order is D before L. Hydroxy acids that tend to migrate to the anode, however, are transported by the EOF to the cathode, but more slowly than the EOF. When SDS is added, the amino acids are distributed between the bulk phase and the hydrophobic micelles. The

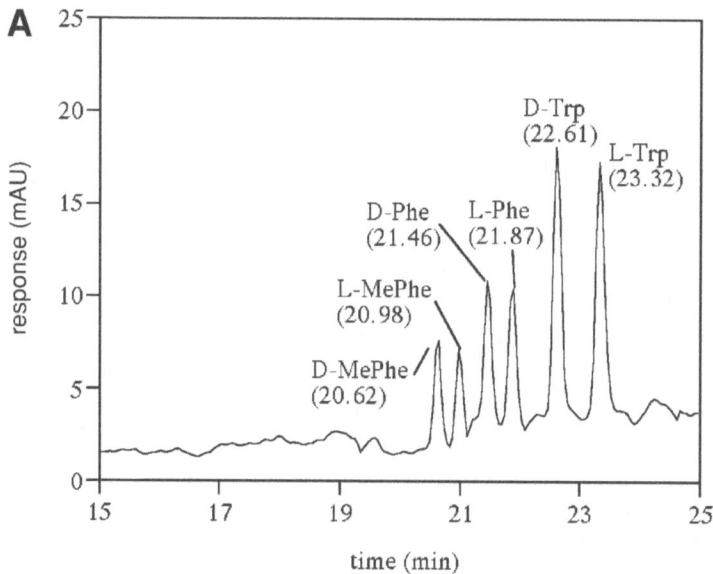

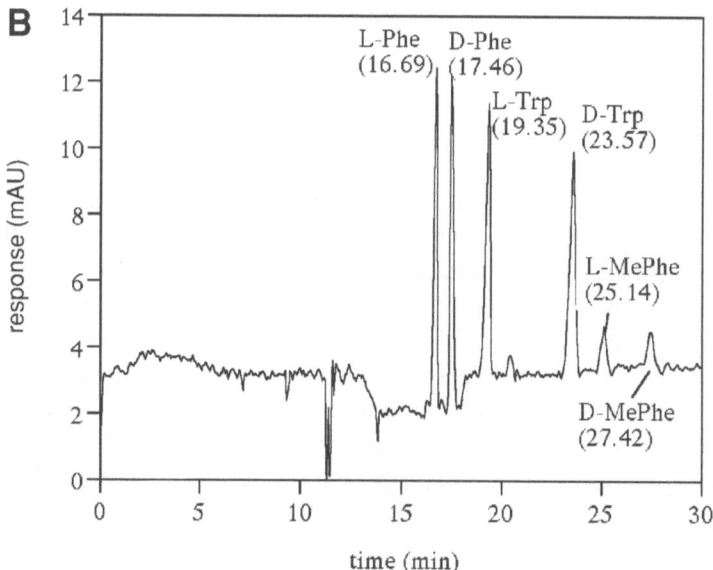

Fig. 1. (**A**) Electropherogram of the enantiomer separation of α-methylphenylala-nine (MePhe), phenylalanine, and tryptophan without using SDS (8). Buffer: 80 m*M* L-Hypro, 40 m*M* Cu^{2+}, adjusted with NH$_3$ to pH 4.0. (**B**) Electropherogram of the enantiomer separation of Phe, Trp, and MePhe using SDS (8). Buffer: 50 m*M* L-Hypro, 25 m*M* Cu^{2+}, 15 m*M* SDS, and 3 *M* urea, adjusted with NH$_3$ to pH 4.0.

negatively charged micelles have negative mobility and thereby retain the free and complexed amino acids that now show negative effective mobilities. The more strongly complexing L-enantiomers are now transported more rapidly than the D-enantiomers to the cathode (**Fig. 1B**).

Two approaches were reported for the preparation of monolithic ligand exchange capillary electrochromatography (LECEC) phases. One is based on the preparation of a continuous bed by *in situ* polymerization of methacrylamide, piperazine diacrylamide, vinylsulfonic acid, and N-(2-hydroxy-3-allyloxypropyl)-L-4-hydroxyproline *(15)*. This phase was applied to the chiral separation of underivatized amino acids *(15)* and hydroxy acids *(16)*. The other approach is based on the preparation of a monolithic silica phase starting from tetramethoxysilane and subsequent derivatization of the monolith with L-proline amide *(17)*. On this phase, Dns-amino acids and some hydroxy acids were resolved.

This chapter presents procedures for chiral LECE and LECEC developed in the laboratory of the authors. LECE separations are carried out using the copper (II)complexes of L-Pro, L-Hypro, N-(2-hydroxyoctyl)-L-4-hydroxyproline and N-(2-hydroxypropyl)-L-4-hydroxyproline. Application examples are given for underivatized amino acids, hydroxy acids, dipeptides, and amino alcohols.

For LECEC, the preparation of chiral phases on continuous bed basis is described. Procedures for the chiral separation of amino acids and hydroxy acids are given.

2. Materials

2.1. Apparatus

To perform CE experiments, a CE instrument equipped with a UV-detector and a facility to collect data is needed. Fused silica capillaries can be purchased at various suppliers, e.g., MicroQuartz (Munich, Germany). For CEC experiments, the CE apparatus should contain an additional pressure device for flushing the chiral stationary phase (CSP) between runs, moreover, a simple high-performance liquid chromatography (HPLC) pump providing constant pressure mode and a connection to fused silica capillaries is useful for conditioning CSPs. All reagents and analytes should be of analytical grade. A special capillary cutting blade is available for precise and proper cutting fused silica capillaries.

2.2. CE Experiments

2.2.1. Conditioning of the Capillary

1. 0.1 *M* NaOH, 10 mL.
2. 0.1 *M* HCl, 10 mL.

2.2.2. Preparation of Background Electrolyte Solutions

1. Cu(II) sulfate.
2. Phosphoric acid.
3. Ammonia solution.
4. L-Proline or L-4-hydroxyproline (Fluka, Buchs, Switzerland).
5. SDS of analytical grade.
6. Triethyl amine (TEA) (Fluka).
7. 2 *M* NaOH.
8. Double-distilled water.
9. Syringe type membrane filters 0.20 or 0.45 μm.

2.2.3. Synthesis of N-(2-Hydroxyoctyl)-
L-4-Hydroxyproline and N-(2-Hydroxypropyl)-L-4-Hydroxyproline

1. Sodium hydroxide.
2. Double-distilled methanol.
3. L-Hypro (Fluka).
4. 1,2-Epoxyoctane or 1,2-epoxypropane (Fluka).
5. Sodium hydroxyde.
6. Double-distilled water.
7. Diethyl ether.
8. Dry toluene.

2.3. CEC Experiments

2.3.1. Conditioning of the Capillary

1. Before first run: 10% aqueous copper(II) sulfate solution.
2. Between runs: use mobile phase only.

2.3.2. Preparation of Mobile Phase

1. Cu(II) sulfate.
2. NaH$_2$PO$_4$ or KH$_2$PO$_4$.
3. Double-distilled water.
4. Syringe type membrane filters 0.20 or 0.45 μm.

2.3.3. Preparation of the Ligand-Exchange
Chiral Continuous Bed (LE-CCB)

2.3.3.1. PRETREATMENT OF THE CAPILLARY

1. Acetone.
2. Double-distilled water.
3. 0.2 *M* NaOH, 10 mL.
4. 0.2 *M* HCl, 10 mL.
5. γ-Methacryloxypropyltrimethoxysilane (Sigma, St. Louis, MO, USA).

2.3.3.2. Preparation of the Monomer Solution and Polymerization Procedure

1. L-Hypro sodium salt.
2. Double-distilled water.
3. Allylglycidyl ether of analytical grade (Sigma).
4. Methacrylamide (Sigma).
5. Piperazine diacrylamide (Bio-Rad, Hercules, CA, USA).
6. Vinylsulfonic acid (30%) (Sigma).
7. Ammonium sulfate of analytical grade.
8. NaH_2PO_4 and Na_2HPO_4 of analytical grade.
9. Ammonium peroxodisulfate (Bio-Rad).
10. Tetramethylethylenediamine (TEMED) (Bio-Rad).
11. Fat.

3. Methods

3.1. CE

3.1.1. Synthesis of N-(2-Hydroxyoctyl)-
L-4-Hydroxyproline and N-(2-Hydroxypropyl)-L-4-Hydroxyproline

1. Dissolve sodium hydroxide (0.8 g, 0.02 mol) in 30 mL distilled methanol.
2. Add 2.62 g (0.02 mol) L-Hypro and 12.28 mL (0.08 mol) 1,2-epoxyoctane or 0.08 mol 1,2-epoxypropane.
3. Stir the mixture for 3 d.
4. Evaporate the organic solvent and obtain a highly viscose product.
5. Dissolve this crude product in 10 mL bidistilled water and titrate with NaOH to a pH > 10.0.
6. Shake the aqueous solution with six portions (50 mL) diethylether to remove excess epoxide.
7. Evaporate the aqueous phase and remove traces of water azeotropically by evaporating the resulting residue twice with 60 mL dry toluene.
8. Obtain the sodium salt as a colorless powder (*see* **Note 2**).

3.1.2. Separation Conditions
for Amino Acids Using L-Pro or L-4-Hypro

1. Dissolve up to 80 m*M* L-pro or L-Hypro with half molarity of copper(II) sulfate and adjust pH to 4.0 by adding diluted ammonia.
2. After degassing and filtration through a syringe filter, electrolyte is ready for use.
3. Dissolve underivatized amino acid racemates in electrolyte (1 mg/mL).
4. Observe enantioseparation (*see* **Note 3**) and verify enantiomeric elution order by injecting the pure enantiomers at equal conditions.
5. To change elution order of the enantiomers an electrolyte consisting of 50 m*M* L-Hypro, 25 m*M* Cu(II) and 20 m*M* SDS pH 4.0 can be used instead of 80 m*M* L-Hypro, 40 m*M* Cu(II) pH 4.0 (*see* **Note 1**).

3.1.3. Separation Conditions for Amino Acids Using N-(2-Hydroxyoctyl)-L-4-Hydroxyproline or N-(2-Hydroxypropyl)-L-4-Hydroxyproline

1. Dissolve up to 15 mM N-(2-hydroxyoctyl)-L-4-hydroxyproline or 20 mM N-(2-hydroxypropyl)-L-4-hydroxyproline in 10 mM phosphoric acid and add half molarity of copper(II) sulfate; adjust pH to 4.3.
2. After degassing and filtration through a syringe filter, electrolyte is ready for use.
3. Dissolve underivatized amino acid racemates in electrolyte (1 mg/mL).
4. Observe enantioseparation and verify enantiomeric elution order by injecting the pure enantiomers at equal conditions.

3.1.4. Separation Conditions for Hydroxy Acids

Generally, the same electrolytes that are shown in **Subheading 3.2.** can be used. Note that hydroxy acids tend to the anode and thus are transported by the EOF much slower than compared to amino acids.

3.1.5. Separation Conditions for Dipeptides

1. Dissolve up to 15 mM N-(2-hydroxyoctyl)-L-4-hydroxyproline or 20 mM N-(2-hydroxypropyl)-L-4-hydroxyproline in 10 mM phosphoric acid and add half molarity of copper(II) sulfate; adjust pH to 6.0.
2. After degassing and filtration through a syringe filter, electrolyte is ready for use.
3. Dissolve dipeptide racemates in electrolyte (1 mg/mL).
4. Observe enantioseparation and verify enantiomeric elution order by injecting the pure enantiomers at equal conditions.

3.1.6. Separation Conditions for Amino Alcohols

1. Dissolve up to 20 mM N-(2-hydroxyoctyl)-L-4-hydroxyproline or 20 mM N-(2-hydroxypropyl)-L-4-hydroxyproline in 10 mM copper(II) sulfate, add 0.1 M TEA and adjust pH to 12.0.
2. After degassing and filtration through a syringe filter, electrolyte is ready for use (*see* **Note 4**).
3. Dissolve amino alcohol racemates in electrolyte (1 mg/mL).
4. Observe enantioseparation and verify enantiomeric elution order by injecting the pure enantiomers at equal conditions.

3.2. CEC

3.2.1. Preparation of the LE-CCB

3.2.1.1. Pretreatment of the Capillary

1. Wash fused silica tubing (35–70 cm length, 75 μm inner diameter [I.D.] successively with acetone (5 min), purified water (5 min), hydrochloric acid 0.2 M (5 min),

purified water (5 min), sodium hydroxide 0.2 *M* (15 min), purified water (10 min), acetone (5 min), and finally air to activate the silica innerwall of the capillary (*see* **Note 5**).

2. Silanize capillary by sucking in a solution of 50% γ-methacryloxypropyltrimeth-oxysilane in acetone, thus introducing a double-bond function for subsequent co-polymerization at the capillary inner surface.
3. After 1 min, seal ends of the capillary with fat.
4. After 12 h, cut off 1 cm of capillary ends (*see* **Note 6**).
5. Flush capillary with acetone (for 5 min) and water, respectively, prior to fill in the monomer solution.

3.2.1.2. PREPARATION OF THE MONOMER SOLUTION AND POLYMERIZATION PROCEDURE

1. Synthesize *N*-(2-hydroxy-3-allyloxypropyl)-L-4-hydroxyproline by stirring an aqueous solution (10 mL) of the sodium salt of L-Hypro (1 g) with an equimolar amount of allylglycidyl ether overnight.
2. Store this solution at 4°C under nitrogen and use it without purification.
3. Mix methacrylamide (18 mg) with 22 mg piperazine diacrylamide (crosslinker), 3 μL vinylsulfonic acid (30%), 25 μl *N*-(2-hydroxy-3-allyloxypropyl)-L-4-hydrox-yproline solution and 10 mg ammonium sulfate in 175 μL 0.05 *M* phosphate buffer (pH 7.0–8.0).
4. Degas this solution for 10 min.
5. Initiate polymerization by addition of ammonium peroxodisulfate (10% aqueous solution) and TEMED (10% aqueous solution) (*see* **Note 7**).
6. Immediately suck the solution into the capillary using vacuum. Stop sucking before the zone reaches the detection window.
7. Seal ends of the capillary with fat.
8. After 12 h, the polymerization is virtually complete.
9. Prior to use, cut off 1 cm of capillary ends (*see* **Note 6**) and connect the bed to a HPLC pump providing constant pressure mode (*see* **Notes 8** and **9**).
10. Set constant pressure to 60 bar and flush the CSP with double-distilled water (for 30 min) (*see* **Note 8**), 10% aqueous copper(II) sulfate solution (for 30 min), again with double-distilled water (for 30 min), and mobile phase.

3.2.2. Separation Conditions for Underivatized Amino Acids *(Table 2)*

1. Set capillary into the CEC device.
2. Flush capillary (ext. pressure device of instrument) with 50 m*M* sodium dihydro-gen phosphate, pH 4.3, 0.1 m*M* Cu(II) as a mobile phase.
3. Prepare samples by dissolving amino acid racemates in mobile phase (1 mg/mL).
4. There are two possibilities to run a measurement: (i) CEC: set high voltage to 15 kV without pressure support (*see* **Note 9**); and (ii) pressurized CEC: set ext. pressure device to pressurize inlet vial. CEC provides better resolution, whereas pressur-ized CEC shows faster results (**Fig. 2**).
5. Observe enantioseparation and verify enantiomeric elution order by injecting the pure enantiomers at equal conditions.

Table 2
Chiral Separation Data of Amino Acids

Amino acid	t_1	t_2	α	Rs
Asparagine	6.564	7.612	1.210	0.976
DOPA	9.149	13.963	1.629	1.239
α-Methyl-DOPA	10.935	26.50	2.649	1.721
α-Methylphenylalanine	8.297.	10.535	1.329	1.416
p-Tyrosine	8.204	12.281	1.608	2.708
Phenylalanine	7.958	11.012	1.473	1.417
Serine	6.844	7.771	1.185	1.033
Threonine	7.021	8.305	1.257	1.271
Tryptophan	12.320	20.378	1.744	1.239

Conditions: CCB, 51.5 (8.5) cm × 0.075 mm; mobile phase: 50 mM sodium dihydrogen phosphate, 0.1 mM Cu(II), pH 4.6, 7 kV (20 µA), 12 bar pressure on inlet; injection: 2 kV × 5 s, detection wavelength was 223 nm.

3.2.3. Separation Conditions for Underivatized Hydroxy Acids

Conditions are equal to **Subheading 3.2.2.**, but polarity must be changed when applying an electric field, since hydroxy acids migrate to the anode. When preparing the CCB, no vinyl sulfonic acid should be added to the monomer solution.

4. Notes

1. Reversal of EMO is sometimes desirable, e.g., for purity checks of enantiomers. The unwanted enantiomer (distomer) present as impurity in samples of the active enantiomer (eutomer) should always appear as the first peak to avoid overlapping with the tailing of the large peak of the eutomer. (Otherwise it would be covered by the tailing of the peak of the eutomer.) Means of reversing the EMO in LECE are:
   a. Change of the sense of chirality of the selector *(14,18,19)*.
   b. Change from *trans*-L-Hypro to *cis*-L-Hypro *(14,18,19)*.
   c. Addition of SDS (not the case with all selectors) *(14,18,19)*.
   d. Addition of cetyl trimethyl ammonium bromide (CTAB) (reversal of the EOF) *(14,18,19)*.
2. The purity of the product may be checked by thin-layer chromatography (TLC):
   a. Stationary phase: precoated TLC plastic sheets silica gel 60 F_{254}.
   b. Mobile phase: ethanol/isopropanol/water (50:30:20).
   c. Rf L-Hypro: 0.47.
   d. Rf *N*-(2-Hydroxy-octyl)-L-4-hydroxyproline: 0.70.
   e. Detection: iodine vapor.
3. Using this electrolyte, the detection of aromatic amino acids is possible, however, detection problems occur using aliphatic amino acids as analytes. In this case,

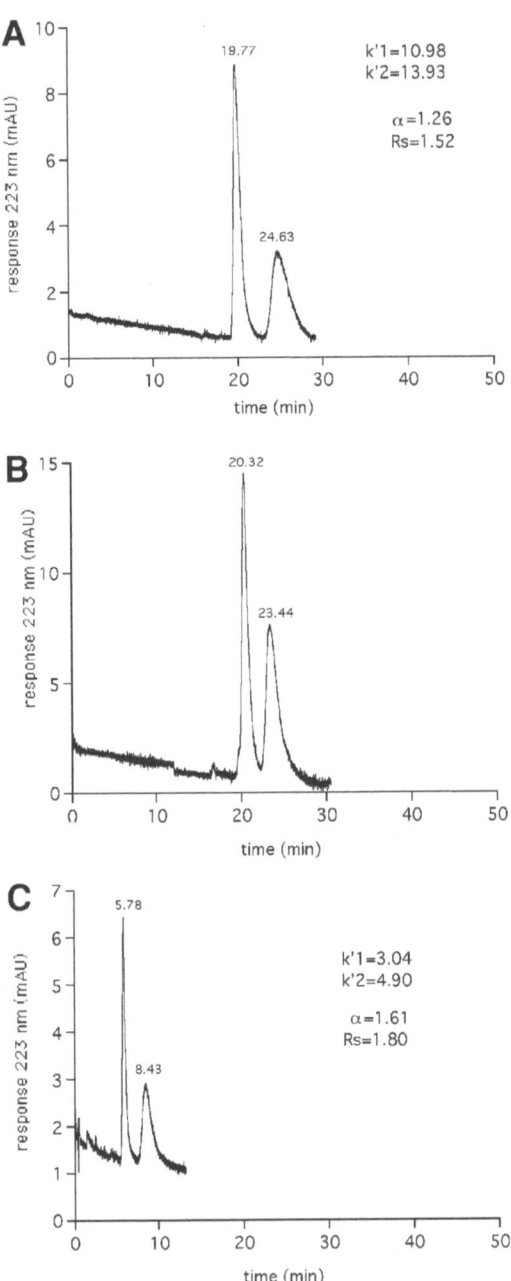

Fig. 2. Chiral separation of DL-Phe comparing CEC (**A**), pressure-driven micro-HPLC (**B**), and pressure-supported CEC (**C**). Conditions: mobile phase: 50 m*M* sodium dihydrogen phosphate, 0.1 m*M* Cu(II), pH 4.6; stationary phase: LE-CCB (26 cm × 75 μm); injection: 10 kV × 6 s; (A) 30 kV, (B) 12 bar, (C) 30 kV and 12 bar.

HO-L-Hypro or HP-L-Hypro cause less problems, since a lower selector concentration is sufficient for enantioseparation.

4. Electrolytes with pH 12.0 should be used immediately and are stable for about 1 h.
5. The most simple way is to connect the capillary to a water vacuum pump and suck in successively the washing reagents. This method can also be used to fill the capillary with monomer solution, however, this can also be done by a syringe including capillary connector.
6. Capillaries can be cut with a special available capillary cutter or a cutter blade. If you are not sure whether the capillary is cut correctly and plane, you can check it with an ordinary microscope.
7. To be sure that the solution will polymerize in the capillary after a period that is sufficient for sucking in the monomer solution, take some aliquots and measure the time from starting the polymerization until the first white polymer zone appears. This can be done in batch experiments.
8. Using this method, within 10 min, a small drop of liquid should appear at the capillary outlet. If not, increase pressure. If many bubbles appear immediately, the CCB did not polymerize, and the monomer solution is washed out again.
9. If air bubbles occur generated by Joule' heating, the baseline signal will become very noisy. In this case, the CCB should be pressurized from the capillary inlet for a while until the baseline signal is stable again.

References

1. Davankov, V. A. and Rogozhin, S. V. (1971) Ligand chromatography as a novel method for the investigation of mixed complexes: stereoselective effects in α-amino acid copper(II) complexes. *J. Chromatogr.* **60,** 280–283.
2. Schmid, M. G., Grobuschek, N., Lecnik, O., and Gübitz, G. (2001) Chiral ligand-exchange capillary electrophoresis. *J. Biochem. Biophys. Methods* **48,** 143–154.
3. Gassmann, E., Kuo, J. E., and Zare, R. N. (1985) Electrokinetic separation of chiral compounds. *Science* **230,** 813–814.
4. Gozel, P., Gassman, E., Michelsen, H., and Zare, R. N. (1987) Electrokinetic resolution of amino acid enantiomers with copper(II)-aspartame support electrolyte. *Anal. Chem.* **59,** 44–49.
5. Cohen, A. S., Paulus, A., and Karger, B. L. (1987) High-performance capillary electrophoresis using open tubes and gels. *Chromatographia* **24,** 15–24.
6. Sundin, N. G., Dowling, T. M., Grindberg, N., and Bicker, G. (1996) Enantiomeric separation of dansyl amino acids using MECC with a ligand exchange mechanism. *J. Microcolumn. Sep.* **8,** 323–329.
7. Desiderio, C., Aturki, Z., and Fanali, S. (1994) Separation of alpha-hydroxy acid enantiomers by high-performance capillary electrophoresis using copper(II)-L-amino acid and copper(II)-aspartame complexes as chiral selectors in the background electrolyte. *Electrophoresis* **15,** 864–869.
8. Schmid, M. G. and Gübitz, G. (1996) Direct resolution of underivatized amino acids by capillary zone electrophoresis based on ligand-exchange. *Enantiomer* **1,** 23–27.

9. Végvári, Á., Schmid, M. G., Kilár, F., and Gübitz, G. (1998) Chiral separation of alpha-amino acids by ligand-exchange capillary electrophoresis using N-(2-hydroxyoctyl)-L-4-hydroxyproline as a selector. *Electrophoresis* **19,** 2109–2112.

10. Schmid, M. G., Rinaldi, R., Dreveny, D., and Gübitz, G. (1999) Enantioseparation of alpha-amino acids and dipeptides by ligand-exchange capillary electrophoresis using different L-4-hydroxyproline derivatives. *J. Chromatogr. A* **846,** 157–163.

11. Schmid, M. G., Lecnik, O., Sitte, U., and Gübitz, G. (2000) Application of ligand-exchange capillary electrophoresis to the chiral separation of α-hydroxy acids and beta-blockers. *J. Chromatogr. A* **875,** 307–314.

12. Schmid, M. G., Laffranchini, M., Dreveny, D., and Gübitz, G. (1999) Chiral separation of sympathomimetics by ligand-exchange capillary electrophoresis. *Electrophoresis* **20,** 2458–2461.

13. Chen, Z., Lin, J. M., Uchiyama, K., and Hobo, T. (1998) Simultaneous separation of o-, m-, p-fluoro-DL-phenylalanine and o-, m-, p-DL-tyrosine by ligand-exchange micellar electrokinetic capillary chromatography. *J. Chromatogr. A* **813,** 369–378.

14. Chen, Z., Lin, J. M., Uchiyama, K., and Hobo, T. (1999) Simultaneous separation of sixteen positional and optical isomers of the tryptophan family by ligand-exchange micellar electrokinetic chromatography. *Chromatographia* **49,** 436–443.

15. Schmid, M. G., Grobuschek, N., Tuscher, C., et al. (2000) Chiral separation of amino acids by ligand-exchange capillary electrochromatography using continuous beds. *Electrophoresis* **21,** 3141–3144.

16. Schmid, M. G., Grobuschek, N., Lecnik, O., Gübitz, G., Végvári, Á., and Hjertén, S. (2001) Enantioseparation of hydroxy acids on easy-to-prepare continuous beds for capillary electrochromatography. *Electrophoresis* **22,** 2616–2619.

17. Chen, Z. and Hobo, T. (2001) Chemically L-prolinamide-modified monolithic silica column for enantiomeric separation of dansyl amino acids and hydroxy acids by capillary electrochromatography and µ-high performance liquid chromatography. *Electrophoresis* **22,** 3339–3346.

18. Chen, Z., Lin, J. M., Uchiyama, K., and Hobo, T. (2000) Reversal behaviors of the enantiomer migration order and the stereo-selectivity of Cu(II) complex with amino acid enantiomers in ligand exchange-micellar electrokinetic chromatography. *Anal. Sci.* **16,** 131–137.

19. Lecnik, O., Schmid, M. G., Presser, A., and Gübitz, G. (2002) Influence of structure and chirality of the selector on the chiral recognition of amino acids using ligand-exchange capillary electrophoresis. *Electrophoresis* **23,** 3006–3012.

23

Enantioseparation in Capillary Chromatography and Capillary Electrochromatography Using Polysaccharide-Type Chiral Stationary Phases

Bezhan Chankvetadze

1. Introduction

Polysaccharide esters and phenylcarbamates represent powerful chiral stationary phases (CSPs) for enantioseparations in high-performance liquid chromatography (HPLC) (*1*). Within the last few years, these materials have been increasingly used for enantioseparations in capillary liquid chromatography (CLC) and capillary electrochromatograpy (CEC) (*2,3*).

Polymeric-type components may be either supports for a chiral selector or chiral selectors themselves. Both kinds of materials have been reported as useful CSPs for enantioseparations in CLC and CEC. The CSPs containing achiral polymers as supports for chiral selectors have been described by Peters and Svec et al. (*4*), Lämmerhofer et al. (*5*), and Koide and Ueno (*6*). The CSPs, containing chiral polymers as selectors have been studied by Chankvetadze et al. (*2,3, 7*), Mayer, Francotte and coworkers (*8*), and Otsuka and Terabe's group (*9*). Two different types of chiral polymers have been used as chiral selectors in CLC and CEC enantioseparations—in particular, helically chiral synthetic polymethacrylates (*10*) and ester and phenylcarbamate derivatives of the most abundant natural chiral polymers cellulose (*2,3,7–9*) and amylose (*2,3*). This chapter focuses on polysaccharide derivatives. The application of polysaccharide-type chiral selectors has been reported in the following formats in CLC and CEC: (i) coated (*11*) and (ii) covalently immobilized (*12*) on the fused-silica capillary wall; as well as (iii) coated (*2,3,7,9*); or (iv) covalently immobilized (*8*) on the surface of silica gel, which is then packed in a fused-silica capillary. The immobilization can be performed by radical polymerization (*12*) or photopolymerization (*8*). Numerous factors, such as the selection of a chiral selector, its

From: *Methods in Molecular Biology, Vol. 243: Chiral Separations: Methods and Protocols*
Edited by: G. Gübitz and M. G. Schmid © Humana Press Inc., Totowa, NJ

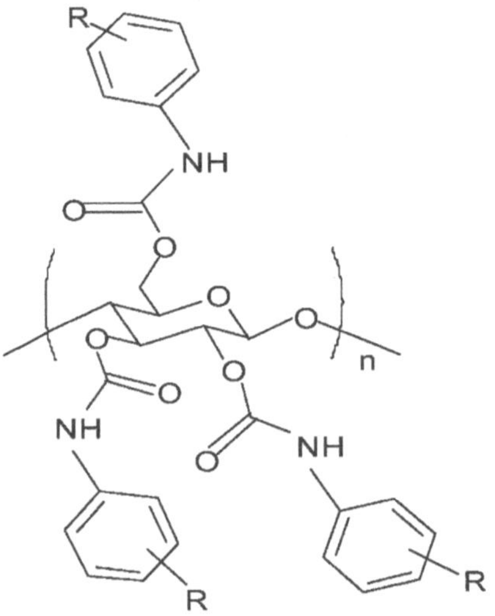

R

1. 4-CH$_3$O	21. 3,5-(CH$_3$)$_2$
2. 4-C$_2$H$_5$O	22. 2,6-(CH$_3$)$_2$
3. 4-(CH$_3$)$_2$CHO	23. 3,4,5-(CH$_3$)$_3$
4. 4-(CH$_3$)$_2$CHCH$_2$O	24. 3,5-Cl$_2$
5. 4-(CH$_3$)$_3$Si	25. 3,4-Cl$_2$
6. 4-CH$_3$	26. 2,6-Cl$_2$
7. 4-CH$_3$CH$_2$	27. 3,5-F$_2$
8. 4-(CH$_3$)$_2$CH	28. 3.5-(CF$_3$)$_2$
9. 4-(CH$_3$)$_3$C	29. 2-Cl-4-CH$_3$
10. 3-CH$_3$	30. 2-Cl-5-CH$_3$
11. 2-CH$_3$	31. 2-Cl-6-CH$_3$
12. H	32. 3-Cl-2-CH$_3$
13. 4-F	33. 3-Cl-4-CH$_3$
14. 4-Cl	34. 4-Cl-2-CH$_3$
15. 2-Cl	35. 4-Cl-3-CH$_3$
16. 3-Cl	36. 3-F-4-CH$_3$
17. 4-Br	37. 5-F-2-CH$_3$
18. 4-CF$_3$	38. 3-F-5-CH$_3$
19. 4-NO$_2$	39. 3-Cl-5-CH$_3$
20. 3,4-(CH$_3$)$_2$	40. 3-Br-5-CH$_3$

Fig. 1. Cellulose derivatives used as chiral selectors in HPLC, CLC, and CEC.

nature and amount, as well as pore diameter, particle size and surface area of the silica gel, the packing of the capillaries, and performing experiments, affect peak resolution in CLC and CEC. The limited size of this chapter does not allow adequate coverage of these aspects, so the reader is referred also to current review papers on the subject *(2,3,13,14)*. Here, we cover some methodological aspects of CSP preparation, capillary fabrication, and performance of CLC and CEC experiments with polysaccharide-type CSPs.

2. Materials

2.1. Preparation of Cellulose Trisphenylcarbamate Derivatives (15)

Cellulose trisphenylcarbamate derivatives (**Fig. 1**) may be prepared by the reaction of microcrystalline cellulose (Avicel; Merck, Darmstadt, Germany) with an excess of the corresponding isocyanates and isolation of the reaction product as the methanol-insoluble fraction. In a typical preparation, 0.5 g of cellulose is placed in a 25-mL three-neck round-bottom flask and dried in oil bath at 90°C for 2 h under vacuum. Thereafter, 10 mL dry pyridin is added under dry nitrogen flow in order to protect the dried cellulose and the solvent from contact with air and moisture (*see* **Note 1**). When the cellulose is completely dissolved, the corresponding isocyanate is added in a four-fold molar excess (1 mol of cellulose reacts with 3 mol of isocyanate) and left under stirring at 90°C for 12 h (*see* **Note 2**). Cellulose trisphenylcarbamate derivatives are isolated by drop-wise addition of the reaction mixture to 250 mL of methanol under stirring. The precipitate is filtrated and dried in vacuum at 60°C for 3 h.

For the preparation of ester derivatives of cellulose, the corresponding acid chlorides can be used instead of isocyanates *(1,5,16)*. This technique can be extended to other polysaccharides, such as amylose *(1)*, chitosan, curdlan, etc. *(1)* (*see* **Note 3**).

2.2. Preparation of Aminopropylsilanized Silica Gel

Aminopropylsilanized silica gel can be prepared in the common way by reaction of silica gel dried under vacuum at 160°C for 2 to 3 h with 3-aminopropyltriethoxysilane in dry benzene (*see* **Note 4**) at 80°C for 12 h (*see* **Note 5**). The product is isolated by filtration and dried in vacuum at 120°C for 3 h (*see* **Note 6**).

2.3. Preparation of Coated Type CSP

A given amount of the polysaccharide derivative is dissolved in a suitable solvent (commonly tetrahydrofuran [THF]) and added dropwise to native or aminopropylsilanized silica gel with a given particle and pore size (*see* **Notes 7** and **8**). *N,N*-Dimethylacetamide, chloroform, pyridine, and other solvents can be used for polysaccharide derivatives, which are insoluble in tetrahydrofurane. The solvent is evaporated, and the CSP is dried in a vacuum at 60°C for 3 h.

2.4. Preparation of Covalently Immobilized Polysaccharide Derivatives

Coated-type CSPs are instable in some organic solvents, which may be of interest as additives to the mobile phase in CLC and CEC experiments. In order to overcome this problem, several techniques have been proposed for a covalent immobilization of polysaccharide derivatives on silica gel *(1,17–22)*. The size limitation does not allow discussing of each of these techniques separately. Therefore, some general characteristics of each method are given below. The method proposed by Okamoto and coworkers in **ref.** *17* involves covalent immobilization by using diisocyanate-type spacers between the polysaccharide derivatives containing a few free hydroxyl groups and aminopropyl silica gel. As the authors noted, the enantiomer-resolving ability of these materials is somewhat lower compared to the corresponding coated-type CSPs. The method described by Kimata et al. *(18)* involves radical co-polymerization between accordingly derivatized silica gel and polysaccharide derivatives. The method proposed by

Oliveros and co-workers *(19)* and Kubota, Okamoto and co-workers *(20)* is also based on a radical co-polymerization of accordingly derivatized polysaccharides onto silica gel. In contrast to these methods, Francotte and Huynh *(21)* have reported the application of photopolymerization techniques for obtaining polysaccharide-based CSP stable in various solvents. One additional technique proposed by Enomoto and Okamoto's group involves covalent immobilization of amylose derivatives with nonreducing terminal groups *(22)*. This technique allows high flexibility of the polymer chain of immobilized polysaccharide and is advantageous for achieving good separation characteristics. However, it cannot be used for cellulose derivatives.

2.5. Preparation of Packed Capillary Columns (7)

Fused-silica capillaries of 50–100 μm inner diameter (I.D.) can be used for the fabrication of the capillary columns. In a typical preparation procedure described below, fused-silica capillaries of 100 μm I.D. from Polymicro Technologies (Phoenix, AZ, USA) were used for the preparation of packed capillaries. The inlet end of the capillary was connected to an HPLC precolumn (4.6 × 50 mm), which served as a reservoir for the slurry of the packing material in methanol or water. A commercially available HPLC column frit was connected to the outlet end of the capillary in order to retain the packing material (*see* **Note 9**). The slurry of the packing material was sonicated in a water bath (for 5 min) and transferred into the reservoir. The system was closed tightly, and a pressure of up to 400 bar was applied using an HPLC pump from Jasco (Tokyo, Japan) and maintained for 1 h. After complete reduction of the residual pressure (3 to 4 h), bi-distilled water was pumped through the packed bed for approx 1 h. The outlet and inlet frits were sintered by local heating of the capillaries at approx 750–800°C for 5–10 s. The capillaries prepared according to this technique can be used for CLC and CEC separations.

2.6. Preparation of Wall-Coated Open Tubular Capillary Columns (11)

According the procedure described by Francotte and Jung *(11)*, wall-coated open tubular (WCOT) capillaries can be prepared by the so-called static method involving the following steps: The cellulose derivative is dissolved in dichloromethane or THF at a concentration of 0.2% (w/v). This solution was carefully filtered and pumped with a pressure of 0.3 mbar at 35°–40°C into the untreated fused-silica capillary of 50 μm I.D. The solution was slowly evaporated and produced a film with a thickness of about 0.025 μm on the inner wall of the capillary. As the authors noted, such a film without covalent attachment on the capillary inner wall may be unstable.

2.7. Preparation of Wall Immobilized
Open Tubular Capillary Columns (12)

Fabrication of fused-silica capillary columns containing covalently immo-bilized polysaccharide derivative on the inner capillary surface involves the pretreatment of the capillary inner wall, covalent attachment of polymerizable functionalities onto the capillary wall, and co-polymerization of the latter with the polysaccharide derivative also containing a certain amount of polymeriz-able groups. In a typical preparation described below *(12)*, a fused-silica capil-lary of 75 μm I.D. was filled with 0.1 *M* sodium hydroxide and heated to 100°C for 2 h. Thereafter, the capillary was washed successively with 0.1 *M* HCl for 5 min, deionized water for 10 min, acetone for 15 min, and purged with dry nitrogen for 15 min. In order to attach methacryloyl groups onto the inner sur-face of the capillary, it was filled with a solution of 1,1-diphenyl-2-picryl-hydroxide (0.37 mg) and 3-trimethoxysilylpropyl methacrylate (0.9 mL) and dry *N,N*-dimethylformamide (2.1 mL), sealed at the both ends, and heated in an oven at 120°C for 6 h. The capillary was washed for 2 h with dimethylform-amide (DMF), for 1.5 h with methanol, and 1.5 h with dichloromethane, fol-lowed by purging with dry nitrogen for 15 min. The cellulose 3,5-dichlorophe-nylcarbamate derivative containing 30% 4-vinylphenylcarbamate residues (CVDCPC) was prepared according to the procedure reported for the 3,5-dimethylphenylcarbamate derivative of cellulose *(20)* for further immobiliza-tion on the silica gel particles. The above dried capillary was filled to the length of 50 cm with a solution containing CVDCPC (64 mg), styrene (6.4 mg), dry THF (0.9 mL), and 0.1 mL of a THF solution of 2,2-azobisisobutyronitrile (2.78 mg in 1 mL) and heated to perform the co-polymerization at 60°C for 20 h. After finishing the polymerization, the capillary was dried at 60°C for 12 h, washed with *n*-hexane/2-propanol (95/5, v/v) for 3 h, and purged with dry nitrogen at 60°C for 12 h (*see* **Note 10**).

3. Methods

3.1. CLC and CEC Experiment Using Packed Capillary Columns (7)

When working with capillary columns, the injection of the minute samples and the efficient detection of the sample components becomes a critical issue. Most of the commercially available capillary liquid chromatography (LC) equip-ment is based on splitting systems of the mobile phase. This facilitates an exact injection of the nanoliter size samples and maintenance of linear flows in the nL/min range. Some commercial capillary electrophoresis equipment (e.g., Agilent HP-3D) is suitable for performing capillary LC experiments in the constant pressure mode up to a pressure of 12 bar without splitting of the flow. In our laboratory, homemade variable pressure equipment is used for performing CLC experiments in the constant pressure mode. The equipment can be used in the range of an applied pressure of 5–50 bar *(7)*.

CEC experiments in packed capillary columns can be performed in various modular and automatic commercial capillary electrophoresis equipment, but technically the most suitable appears to be the HP-3D capillary electrophoresis equipment from Agilent (Waldbronn, Germany). In this instrument, a pressure of up to 12 bar can be applied at both the inlet and the outlet buffer vials. Air circulation is used for cooling the separation capillary. This may be associated with certain problems, with regard to the cooling efficiency, but facilitates the easy mounting of the capillary column into a cassette. Sample injection is possible from the both inlet and outlet vials. This allows performance of the so called "short end injection" of the sample and, if the separation factor enables it, significant shortening of the analysis time *(3)*. In a typical CEC enantioseparation with a capillary column packed with a polysaccharide containing CSP, 5 m*M* ammonium acetate in methanol is used as mobile phase (buffer solution). The UV-absorbing sample, dissolved in the same buffer at a concentration of 0.1 mg/mL, may be injected into the capillary by pressure (commonly 10 bar for 3 s) or by voltage. The enantioseparation may be performed with the applied voltage in the range of 5–30 kV. In order to avoid bubble formation, pressurization of both the inlet and the outlet ends of the capillary column may appear useful.

Aqueous-organic buffers can be used instead of pure organic mobile phases as described above (*see* **Note 11**). The detection is commonly performed at 214 nm, although other UV wavelengths, as well as alternative detection systems, can also be applied.

3.2. CLC and CEC Experiment Using WCOT Capillary Columns *(11,12)*

CEC experiments with WCOT capillary columns can be performed using almost all commercially available capillary electrophoresis equipment. For CLC experiments, it is required that the CE instrument is equipped with a pumping system in the range up to 1000 mbar. In a typical experiment, as described in **ref. *11***, sample injection was performed for 2 to 5 s in the low pressure mode (35 mbar, concentration of the sample 0.2–1.0 mg/mL). The enantioseparations were performed in the same low pressure mode (P/ACE™ 5510 capillary electrophoresis instrument, Beckman Coulter, Fullerton, CA, USA) maintaining pneumatically a constant nitrogen pressure of 35 mbar on the mobile phase reservoir. The detection type and detection wavelength needs to be selected depending on the analytes (*see* **Note 12**).

The fabrication of open tubular capillaries and performing a CLC and CEC experiment in these capillaries is easier compared to packed capillary columns. However, WCOT capillary columns commonly contain only small amounts of a CSP. Therefore, the sample loading capacity of this type of capillary columns is very low. This may be associated with severe detection sensitivity problems as well as a drastic decrease of separation efficiency due to overloading of the WCOT capillary columns. These issues must be considered when selecting the type of capillary column.

3.3. Comparison Between CLC and CEC in Packed Capillaries (7,23)

A CEC experiment is more demanding to the properties of the stationary phases (generation of electroosmotic flow [EOF]) and buffers (electric conductivity and support of the EOF) compared to a CLC experiment. However, CEC may offer significant advantages from the viewpoint of separation characteristics. The advantages of CEC enantioseparations compared to CLC are illustrated in **Fig. 2**. This figure shows van Deemter curves for the enantiomers of piprozolin in the CLC and CEC modes. As these data indicate, CEC offers (i) at least a two times higher separation efficiency at the optimal flow rate compared to CLC; and (ii) CEC tolerates high linear flow rates much better than CLC. The impact of both of these advantages on the enantioseparations is illustrated in **Figs. 3** and **4** *(7)*. Thus, as shown in **Fig. 3**, no separation can be observed in the pressure-driven mode for the enantiomers of piprozolin, whereas almost a baseline enantioseparation can be observed in the electrokinetically-driven mode under otherwise similar conditions. No sharp decrease of the plate numbers with increasing linear flow rates of the mobile phase shown in **Fig. 2** facilitates ultrafast enantioseparations in the CEC mode (**Fig. 4**). Thus, a CEC experiment, although more demanding compared to CLC, may offer significant advantages from the viewpoint of separation characteristics.

3.4. Examples of Applications

The advantages of miniaturization in separation science has been recognized for a long time. However, the technical problems associated with injection, separation, and detection of minute samples did not allow for several decades the practical realization of the obvious potential advantages. The most recent developments in CLC, CEC, and mass spectrometric techniques as well as with other detection technologies bring the advantages of miniaturized separation techniques close to being fully realized.

As shown in **Fig. 5**, these techniques may be applied at present not only to simple standard samples, but also to rather complex mixtures of compounds of biomedical interest. Some additional examples of CEC applications are summarized in **refs.** *2* and *3*.

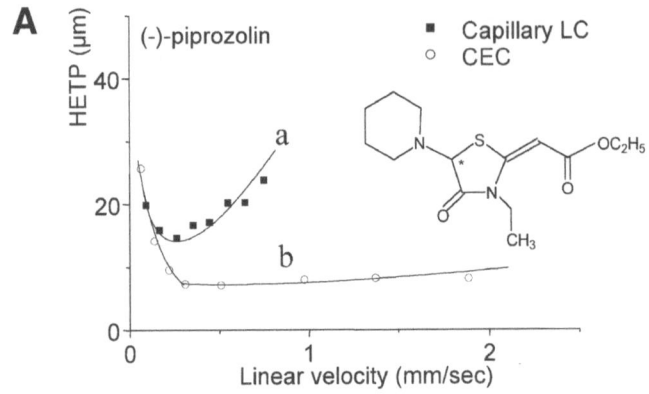

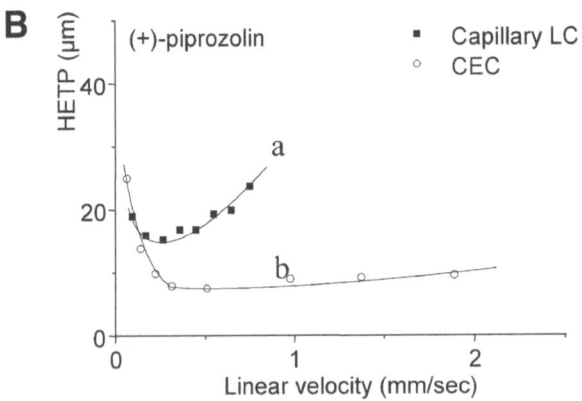

Fig. 2. Van Deemter curves (**A**) in the CLC and (**B**) the CEC mode for the enanti-omers of piprozolin. The separation capillary was prepared by coating cellulose tris (3,5-dichlorophenylcarbamate) (5%, w/w) onto aminopropylsilanized silica gel (5 μm, 100 nm). The separation medium was 2.5 mM ammonium acetate in methanol. Applied pressure and voltage in the CLC and CEC modes were varied. Reproduced from **ref. 7** with permission.

In conclusion, both CLC and CEC appear to be very promising techniques for enantioseparations. However, still significant progress must be achieved in the theory and methods of miniaturized enantioseparation technologies, such as CLC and especially CEC, in order to gain a status comparable to those of HPLC, gas chromatography (GC), and capillary electrophoresis (CE) in the field of enantioseparations.

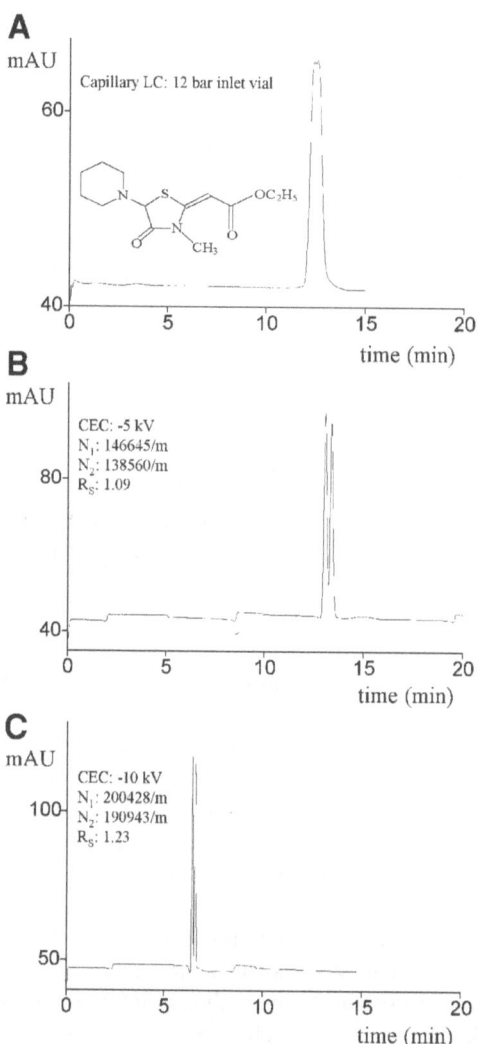

Fig. 3. Enantioseparation of etozilin in (**A**) the CLC and (**B** and **C**) the CEC mode in the same capillary. The CSP was prepared by coating cellulose tris(3,5-dichlorophenylcarbamate) (1.0%, w/w) on aminopropylsilanized silica gel (5 μm, 100 nm). The applied pressure in (**A**) was 12 bar; the applied voltage was in (**B**) –5 kV and (**C**) –10 kV.

4. Notes

1. Isocyanates are toxic compounds and need to be handled carefully.
2. The reaction between polysaccharides and isocyanates can be monitored using infrared (IR) spectrometry. The disappearance of the absorbance band close to 2270 cm^{-1}, characteristic for the N-C=O bond in isocyanates, indicates that there

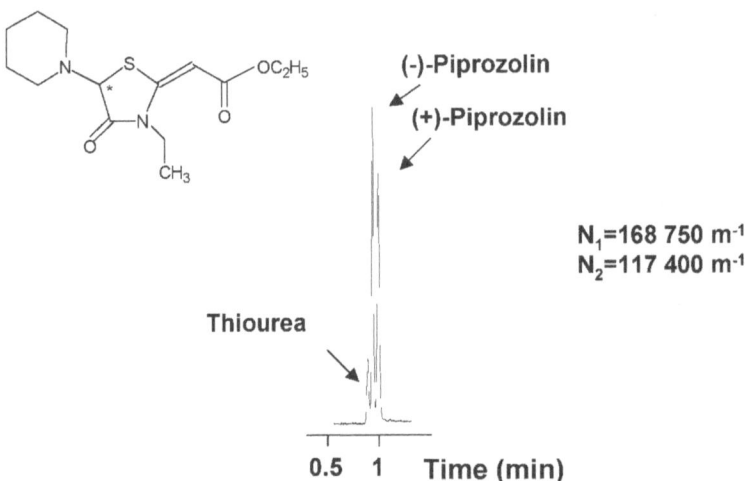

Fig. 4. Fast CEC enantioseparation of piprozolin on a cellulose tris(3,5-dichloro-phenylcarbamate) containing CSP. Reproduced from **ref. 7** with permission.

is no unreacted isocynate present in the reaction mixture. An appearance of an absorbance band in the range of 1650–1780 cm^{-1} is a characteristic for a carbonyl group in the carbamate moiety and can be used in order to examine its presence in the reaction product (also when the reaction still proceeds).

3. Due to low solubility of some polysaccharides and reaction products *N,N*-di-methyl-acetamide in combination with lithium chloride can be used as a solvent.

4. In laboratories where benzene is not allowed to be used, it may be exchanged with toluene.

5. Mechanical mixing/stirring leading to breakage and formation of fines should be avoided in all reactions with porous silica gels. Therefore, it is advised to perform the reaction as close as possible to the boiling point of the solvent used and to equip the reaction flask with an efficient glass condenser with dry nitrogen reservoir.

6. Aminopropylsilanized silica gels are recommended as support for polysaccharide type chiral selectors in HPLC. However, as recent studies indicated, native silica gel can be used with the same success as aminopropylsilanized materials in both CLC and CEC experiments.

7. It is advised to use wide pore silica gel, especially in CEC experiments. This allows generation of a stronger EOF, and the achievement of higher peak efficiency (*see* **refs. 2, 3,** and **7**).

8. The particle size of silica gel and its uniformity are important characteristics, and the materials should be selected carefully. Smaller particle size materials are favorable from the viewpoint of achieving high peak efficiencies. However, with decreasing particle size of the packing material, the backpressure increases significantly in HPLC (also in CLC) experiments. The backpressure is not a problem in CEC,

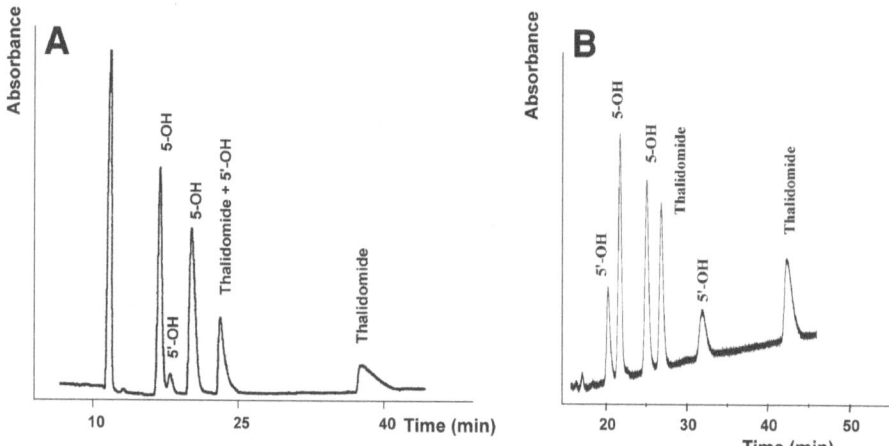

Fig. 5. Simultaneous enantioseparation of thalidomide and its 5- and 5'-hydroxy metabolites in (**A**) the CLC and (**B**) the CEC mode. The separation capillary was prepared by coating 12.4% cellulose tris(3,5-dimethylphenylcarbamate) and 12.4% amylose tris(3,5-dimethylphenylcarbamate) onto native silica gel (5 µm, 200 nm). The mobile phase was 10 mM ammonium acetate dissolved in 55% methanol + 45% ethanol (v/v). The applied pressure in the CLC mode was 13 bar. The applied voltage in the CEC mode was 20 kV.

but problems with small particles are associated with packing of the capillaries and frit fabrication. Commonly, spherical silica gels with 5 µm particle size are used in CLC and with 3 and 5 µm in CEC experiments.

9. Many laboratories use preliminary frits prepared with native silica materials before packing the capillary with a desired stationary phase. According to our experience, the application of commercially available HPLC frits instead of silica-based preliminary frits significantly simplifies the fabrication of capillary columns.

10. The concentration of CVDCPC should be carefully optimized in order to avoid clogging of the capillary.

11. Comparative characteristics of organic-aqueous and polar organic mobile phases in CEC experiments are described in **ref. 2**.

12. CE equipments with variable low pressure facility such as Agilent's HP 3D, Beckman Coulter's P/ACE MDQ, Prin CE (Emmen, The Netherlands), etc., may appear favorable for performing CLC experiments with WCOT capillaries.

References

1. Okamoto, Y. and Yashima, E. (1998) Polysaccharide derivatives for separation of enantiomers. *Angew. Chem.* **37**, 1021–1043.
2. Chankvetadze, B. and Blaschke, G. (2000) Enantioseparations using capillary electromigration techniques in nonaqueous buffers. *Electrophoresis* **21**, 4159–4178.

3. Fanali, S., Catarcini, P., Blaschke, G., and Chankvetadze, B. (2001) Enantiosepa-rations by capillary electrochromatography. *Electrophoresis* **22,** 3131–3151.

4. Peters, E. C., Lewandowski, K., Petro, M., Svec, F., and Frechet, J. M. (1998) Chiral electrochromatography with a "moulded" rigid monolithic capillary col-umn. *Anal. Commun.* **35,** 83–86.

5. Lämmerhofer, M., Svec, F., Frechet, J. M. J., and Lindner, W. (2000) Chiral mono-lithic columns for enantioselective capillary electrochromatography prepared by copolymerization of a monomer with quinidine functionality. 2. Effect of chro-matographic conditions on the chiral separations. *Anal. Chem.* **72,** 4623–4628.

6. Koide, T. and Ueno, K. (2000) Enantiomer separations of cationic and neutral com-pounds by capillary electrochromatography with monolithic chiral stationary phases of β-cyclodextrin bounded negative polyacrylamide gels. *J. Chromatogr. A* **893,** 177–187.

7. Chankvetadze, B., Kartozia, I., Breitkreutz, J., et al. (2001) Comparative capillary chromatographic and capillary electrochromatographic enantioseparations using cellulose tris(3,5-dichlorophenylcarbamate) as chiral stationary phase. *J. Sep. Sci.* **24,** 251–257.

8. Mayer, S., Briand, X., and Francotte, E. (2000) Separation of enantiomers by packed capillary electrochromatography on a cellulose-based stationary phase. *J. Chro-matogr. A* **875,** 331–339.

9. Otsuka, K., Mikami, M., and Terabe, S. (2000) Enantiomer separation by capil-lary electrochromatography using chiral stationary phases. *J. Chromatogr. A* **887,** 457–463.

10. Krause, K., Chankvetadze, B., Okamoto, Y., and Blaschke, G. (1999) Chiral sepa-rations in nonaqueous capillary electrochromatography using helically chiral poly (diphenyl-2-pyridylmethyl methacrylate) as chiral stationary phase. *Electrophore-sis* **20,** 2772–2778.

11. Francotte, E. and Jung, M. (1996) Enantiomer separation by open-tubular liquid chromatography and electrochromatography in cellulose-coated capillaries. *Chro-matographia* **42,** 521–527.

12. Wakita, T., Chankvetadze, B., Yamamoto, C., and Okamoto, Y. (2001) Chromato-graphic enantioseparations on capillary column containing covalently bound cel-lulose (3,5-dichlorophenylcarbamate) as chiral stationary phase. *J. Sep. Sci.* **25,** 167–169.

13. Gübitz, G. and Schmid, M. G. (2000) Chiral separation by capillary electrochro-matography. *Enantiomer* **5,** 5–11.

14. Lämmerhofer, M., Svec, F., Frechet, J. M., and Lindner, W. (2000) Separation of enantiomers by capillary electrochromatography. *Trends Anal. Chem.* **19,** 676–698.

15. Okamoto, Y., Kawashima, M., Yamamoto, K., and Hatada, K. (1984) Chromato-graphic resolution. 6. Useful chiral packing materials for high-performance liquid chromatographic resolution. Cellulose triacetate and tribenzoate coated on macro-porous silica gel. *Chem. Lett.* 739–742.

16. Ichida, A., Shibata, T., Okamoto, I., Yuki, Y., Namikoshi, H., and Toga, Y. (1984) Resolution of enantiomers by HPLC on cellulose derivatives. *Chromatographia* **19,** 280–284.
17. Okamoto, Y., Aburatani, R., Miura, S., and Hatada, K. (1987) Chiral stationary phases for HPLC: cellulose tris(3,5-dimethylphenylcarbamate) and tris(3,5-dichlorophenylcarbamate) chemically bonded to silica gel. *J. Liq. Chromatogr.* **10,** 1613–1628.
18. Kimata, K., Tsuboi, R., Hosoya, K., and Tanaka, N. (1993) Chemically bonded chiral stationary phase prepared by the polymerization of cellulose p-vinylbenzoate. *Anal. Methods Instrum.* **1,** 23–29.
19. Oliveras, L., Lopez, P., Minguillon, C., and Franco, P. (1995) Chiral chromatographic discrimination ability of a cellulose 3,5-dimethylphenylcarbamate/10-undecenoate mixed derivative fixed on several chromatographic matrices. *J. Liq. Chromatogr.* **18,** 1521–1532.
20. Kubota, T., Kusano, T., Yamamoto, C., Yashima, E., and Okamoto, Y. (2001) Cellulose 3,5-dimethylphenylcarbamate immobilized onto silica gel via copolymerization with a vinyl monomer and its chiral recognition ability as a chiral stationary phase for HPLC. *Chem. Lett.* 724–725.
21. Francotte, E. and Huynh, D. (2002) Immobilized halogenylphenylcarbamate derivatives of celulose as novel stationary phases for enantioselective drug analysis. *J. Biomed. Pharm. Anal.* **27,** 421–429.
22. Enomoto, N., Furukawa, S., Ogasawara, Y., et al. (1996) Preparation of silica gel-bonded amylose through enzyme-catalyzed polymerization and chiral recognition ability of its phenylcarbamate derivative in HPLC. *Anal. Chem.* **68,** 2798–2804.
23. Girod, M., Chankvetadze, B., Okamoto, Y., and Blaschke, G. (2001) Highly efficient enantioseparations in non-aqueous capillary electrochromatography using cellulose tris(3,5-dichlorophenylcarbamate) as chiral stationary phase. *J. Sep. Sci.* **24,** 27–34.

24

Chiral Separation by Capillary Electrochromatography Using Cyclodextrin Phases

Dorothee Wistuba, Jingwu Kang, and Volker Schurig

1. Indroduction

Cyclodextrin and its derivatives are well known chiral selectors in gas chromatography (GC), high-performance liquid chromatography (HPLC), capillary electrophoresis (CE), and also in capillary electrochromatography (CEC), a hybrid method of CE and liquid chromatography (LC). The separation of enantiomers by CEC on cyclodextrin stationary phases (1–4) can be achieved by three different methods: (i) open tubular capillary electrochromatography (o-CEC); (ii) packed capillary electrochromatography (p-CEC); and (iii) rod-CEC. In o-CEC, the internal capillary wall is coated with the chiral stationary phase (*see* **Fig. 1**), e.g., with Chirasil-Dex (*see* **Fig. 2**), which is a permethyl-β-cyclodextrin covalently linked via a spacer to dimethylpolysiloxane (4–6). For p-CEC, using capillaries filled with typical chiral HPLC packing materials, two different techniques for fixing cyclodextrins or its derivatives to the silica are described: (i) the cyclodextrin or its derivatives are covalently bound via a spacer to the silica particles (4,7) (e.g., Chira-Dex-silica [*see* **Fig. 2**]); or (ii) Chirasil-Dex is immobilized on the silica material (4,8). The packing bed is held in position by frits, and detection occurs in the empty part of the capillary (*see* **Fig. 1**). In rod-CEC, columns consisting of a porous solid prepared by an *in situ* sol-gel-process (9) or by sintering of silica (10) were used. For enantiomer separation, the resulting monolith is derivatized with Chirasil-Dex (9,10) (*see* **Fig. 1**).

Advantages of o-CEC are the simple preparation of the capillaries, the simple instrumental handling, and the short conditioning time. But the sample capacity is very low and the electroosmotic flow (EOF) is reduced by the coating of the

From: *Methods in Molecular Biology, Vol. 243: Chiral Separations: Methods and Protocols*
Edited by: G. Gübitz and M. G. Schmid © Humana Press Inc., Totowa, NJ

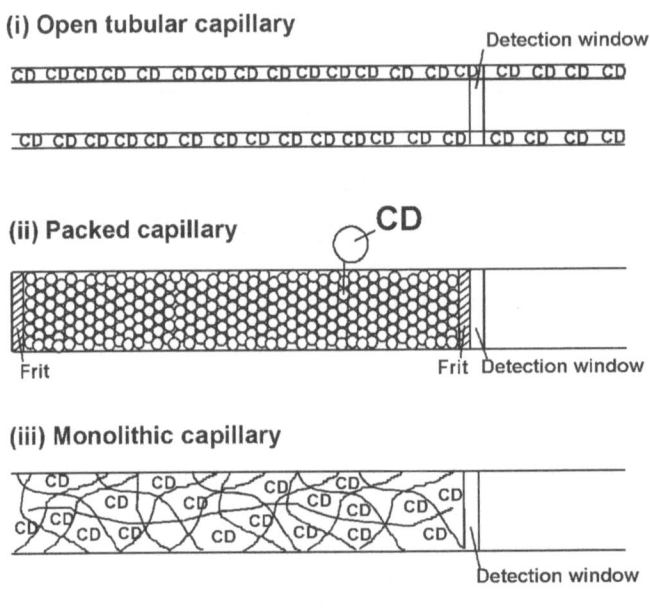

(i) Open tubular capillary

Detection window

(ii) Packed capillary   CD

Frit   Frit Detection window

(iii) Monolithic capillary

Detection window

CD = permethylated β-cyclodextrin

Fig. 1. Capillaries for CEC. (**i**) Open tubular capillary coated with Chirasil-Dex for o-CEC. (**ii**) Packed capillary filled with Chira-Dex-silica or with Chirasil-Dex-modified silica particles for p-CEC. (**iii**) Capillary with a Chirasil-Dex derivatized monolith for rod-CEC.

internal capillary wall. While for p-CEC the sample capacity and the sensitivity is higher and the migration time is normally shorter, the presence of frits can lead to bubble formation and long conditioning times. To minimize the risk of bubble formation, pressurization of the separation capillaries are required. With monolithic capillaries, problems associated with frit preparation are irrelevant. Disadvantages are the time-consuming and difficult preparation and the reproducibility.

The enantiomer separation of various chiral compounds, such as mephobarbital, hexobarbital, pentobarbital, 5-ethyl-1-methyl-5-(*n*-propyl)-barbituric acid, 1-methyl-5-(2-propyl)-5-(*n*-propyl)-barbituric acid, benzoin, α-methyl-α-phenylsuccinimide, glutethimide, methylthiohydantoin (MTH)-proline, methyl mandelate, chlorinated alkyl phenoxypropanoates, γ-phenyl-γ-butyrolactone, 1-(2-)naphthylethanol, 1-phenylethanol, ibuprofen, flurbiprofen, cicloprofen, etodolac, carprofen, and 1,1'-binaphthyl-2,2'-diylhydrogenphosphate, are feasible by CEC with above-mentioned capillaries.

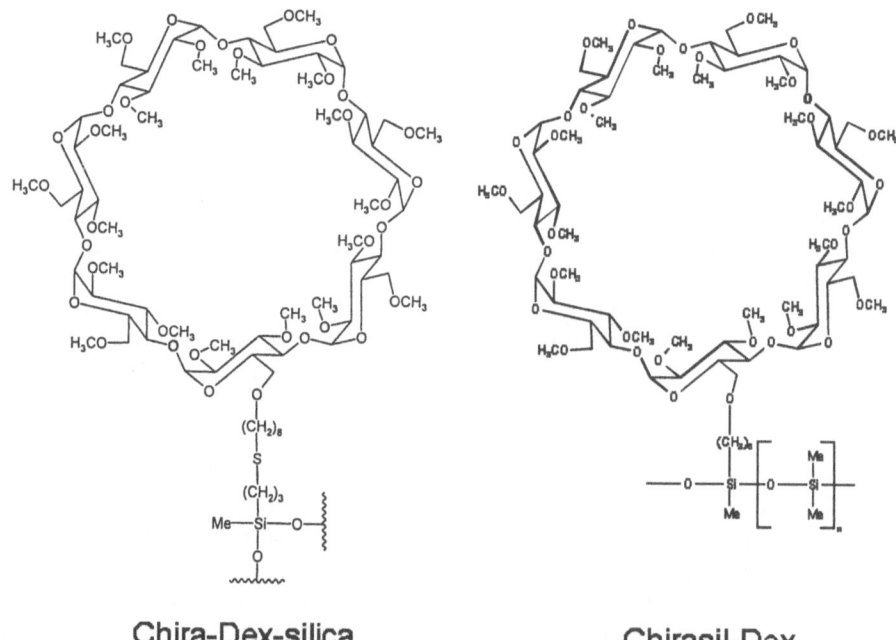

Chira-Dex-silica **Chirasil-Dex**

Fig. 2. Schematic representation of Chira-Dex-silica and Chirasil-Dex.

2. Materials

2.1. Preparation of Capillaries for o-CEC

1. Capillaries: 1 m × 50 µm.
2. Synthesis of Chirasil-Dex *(11,12)*.
   a. Synthesis of monokis-6-(octen-7-yl)-β-cyclodextrin: in a three-necked round-bottomed flask equipped with a dropping funnel and a reflux condenser 18 g (16 mM) β-cyclodextrin were dissolved under a nitrogen atmosphere in 400 mL dry dimethyl sulfoxide. To the solution 1.9 g (47 mM) sodium hydroxide powder were added. After being stirred for half an hour, the solution gets clear. Stirring was continued an additional hour, then 7.58 mL (40 mM) 8-bromo-octene dissolved in 50 mL dry dimethyl sulfoxide were slowly added. After 48 h, sodium bromide and sodium hydroxide were filtered, and the solvent was removed at the evaporater until a vol of about 30 mL. The residue was diluted with 30 mL metanol. Diethyl ether (400 mL) was added, and the white precipitate was quickly filtered with a Büchner funnel. The product was isolated by repeated flush chromatography (ethanol/toluene, 2:1). The white crystals was dried at 60°C and 0.05 torr overnight. Yield: 5.3 g (27%) (*see* **Notes 1, 2,** and **3**).

  b. Synthesis of permethyl-monokis-6-(oct-7-enyl)-β-cyclodextrin: in an atmo-
     sphere of nitrogen 3.7 g (145 mM) of sodium hydride (95% in paraffin) were
     transferred into an ice-cooled 0.25-L four-necked round-bottomed flask equipped
     with a nitrogen inlet, two dropping funnels, and reflux condenser fitted with a
     mercury valve. Monokis-6-(octen-7-yl)-β-cyclodextrin (3.5 g) was dissolved
     in 115 mL of anhydrous dimethyl formamide, and half of it was added via the
     dropping funnel to the sodium hydride in a nitrogen atmosphere, whereupon a
     vigorous reaction with evolution of hydrogen (Caution!) started. After the vig-
     orous reaction has ceased, half of the amount of 13.6 mL (218 mM) of methyl
     iodide was added slowly at a bath temperature of 20°C. After stirring for 30
     min, the second half of the reagent was added to the reaction mixture. After
     being stirred for 1 h, the reaction mixture was decanted from unreacted sodium
     hydride and was carefully poured into 250 mL ice water. The aqueous phase
     was extracted three times with 150 mL diethyl ether. The combined diethyl
     ether layers were washed three times with 20 mL water to remove residual di-
     methyl formamide, and it was subsequently dried over anhydrous sodium sul-
     fate. The solvent was removed, and the white product was dried at 60°C and
     0.01 torr. Yield: 3.4 g (80%) (*see* **Notes 1** and **4**).
  c. Synthesis of Chirasil-Dex: in an atmosphere of nitrogen, 1.02 g (approx 0.34
     mM) of dimethylpolysiloxane containing 10% Si-H-groups and 0.4 g (0.262
     mM) of permethyl-monokis-6-(oct-7-enyl)-β-cyclodextrin, dried in a vacuum
     (10^{-2} torr) at 40°C over P_4O_{10} for 72 h and 40 mL of dry toluene were placed
     into a 0.1-L three-necked round-bottomed flask equipped with a nitrogen inlet
     and reflux condenser fitted with a mercury valve. To the refluxing reaction
     mixture were added a few droplets of a semiconcentrated solution of the cata-
     lyst H_2PtCl_8 in anhydrous tetrahydrofurane at some portions during the reac-
     tion time (approx 0.5 mg). After 24 h, the solvent was evaporated in a vacuum
     with a rotary evaporator, and the residue was taken up in 50 mL anhydrous
     methanol. The turbid methanolic phase was separated from a blackish phase,
     and the methanol was evaporated. The residue was extracted with petroleum
     ether. After filtration the solution was concentrated, the polysiloxane was dried
     in a vacuum. Yield: 0.7 g (49%).
 3. Chirasil-Dex solution for coating of the capillaries: 16 mg/1 mL ether (leads to a
    film thickness of 200 μm) (*see* **Note 5**).

2.2. Preparation of Capillaries for p-CEC

2.2.1. Cyclodextrin Covalently Bound to Silica (7)

 1. Synthesis of permethyl-monokis-6-(oct-7-enyl)-β-cyclodextrin (*see* **Subheading
    2.1.**).
 2. Synthesis of (3-mercaptopropyl)-methyl-silica (*7*). Silica (2 g) was dried by azeo-
    tropic destillation with toluene. To the silica, 40 mL dry toluene, 2.5 mL dry pyri-
    dine, and 4 mL (3-mercaptopropyl)methyl dimethoxysilane were added and heated
    for 48 h at 90°C under a nitrogen atmosphere. After filtration, the modified silica

was washed with toluene, acetone, diethyl ether, pentane, and diethyl ether and dried in vacuum (*see* **Notes 6** and **7**).

2.2.2. Chirasil-Dex Immobilized on Silica (8,13)

1. Synthesis of Chirasil-Dex (*see* **Subheading 2.1.**).
2. Coating solution: dissolve 45 mg Chirasil-Dex and 5 mg of an acidic co-polymer in about 750 μL dry toluene.

2.2.3. Packing Procedure

1. Make a slurry of 20 mg modified silica in 75 μL methanol (*see* **Note 8**).

2.3. Preparation of Capillaries for Rod-CEC (9,10)

2.3.1. Preparation of a Silica Monolith by Sintering (10,14)

1. Fused silica capillary: 100 μm × 0.5 m.
2. Make a slurry of 10 mg bare silica (5 μm, 300 Å) in 100 μL acetone/toluene (1:1).
3. Dissolve 105.99 mg (1 mM) sodium carbonate in 10 mL water.

2.3.2. Preparation of a Silica Monolith by Sol-Gel Technique (9,15)

1. Fused silica capillary (50 μm inner diameter [I.D.]): treat the capillary by flushing with 0.1 mol/L sodium hydroxide solution for 2 h, followed by washing with 0.1 mol/L hydrochloric acid, water and acetone for 0.5 h, respectively. Finally, the capillary is purged with dry nitrogen for 0.5 h.
2. Freshly prepared acetic acid solution: 0.001 mol/L.
3. Freshly prepared polyethylene glycol (PEG) 10,000 solution: 1.10 g PEG in 10 mL of acetic acid solution (0.001 mol/L).
4. Sol solution: mix 4 mL of tetramethoxysilane (*see* **Note 1**) in 10 mL of PEG 10,000 solution with vigorous stirring by a magnetic stirrer under the cooling of an ice bath. Keep stirring for 45 min until the mixture becomes homogeneous.

2.3.3. Surface Modification With Chirasil-Dex (9,10)

1. Make a solution of 5% Chirasil-Dex in acetone (w/v).

3. Methods

3.1. Preparation of Capillaries for o-CEC (5,6)

1. Heat the capillary under slow hydrogen stream (in a GC oven) for 2 h at 250°C.
2. Fill the capillary with the Chirasil-Dex solution by a nitrogen pressure. Close one end of the capillary by dipping about 3 cm of the capillary in a Pasteur pipet filled with silicon paste (*see* **Note 9**). Place the capillary in a water bath (25°C). Apply vacuum (10^{-3}–10^{-4} bar) at the other end of the capillary and evaporate the solvent.
3. For immobilization, heat the coated capillary under a slow hydrogen stream for 24 h at 190°C.

4. Install the capillary in a GC. For conditioning, heat the capillary 12 h under a hydrogen stream (1 bar) at 100°C. Analyze *n*-dodecane and 1-phenylethanol. Calculate the retention factor k (*see* **Note 10**).
5. Rinse the capillary successively with 5 mL methanol, mixtures of methanol/dichloromethane (9:1, 7:3, 1:1, 3:7, 1:9) and 5 mL dichloromethane.
6. For calculating the degree of immobilization (quotient of the retention factor after and before washing), install the capillary in a GC again and determine the retention factor k of *n*-dodecane and 1-phenylethanol.
7. Flush the capillary with water (for 30 min) and with the mobile phase (for 120 min) by applying pressure.
8. For preparing a detection window, burn off a section of about 3 mm of the polyimide outer coating with a hot wire.
9. Install the capillary into the electrophoresis instrument.
10. Condition the capillary at 30 kV.

3.2. Preparation of Capillaries for p-CEC (7,8)

3.2.1. Cyclodextrin Covalently Bound to Silica (7)

1. Dissolve 53.4 mg monokis-6-(octen-7-yl)-permethyl-β-cyclodextrin in 5 mL dry toluene under nitrogen (*see* **Note 7**).
2. Add 50 mg (3-mercaptopropyl)methyl-silica at 0°C under nitrogen.
3. Add 25 µL 9-borabicyclo[3.3.1.]nonane (9-BBN) (0.5 *M* in tetrahydrofurane) (*see* **Note 11**) and allow the mixture to warm up to room temperature.
4. After 24 h, add an additional amount of 9-BBN (25 µL).
5. After 3 h, filter the mixture and wash the modified silica successively with diethyl ether, methanol, *n*-hexane, methanol, and diethyl ether.

3.2.2. Chirasil-Dex Immobilized on Silica (8,13)

1. Dry 200 mg silica (5 µm, 300 Å) by azeotropic destillation with toluene (*see* **Note 6**).
2. Coat the dried silica with a solution of the Chirasil-Dex and an acidic co-polymer in toluene.
3. Slowly remove the solvent during rotation, with the rotation evaporator under reduced pressure (*see* **Note 12**).
4. Heat the coated silica for thermal immobilization under high vacuum at 190°C for 24 h (*see* **Note 13**).
5. Wash the silica particles with methanol, dichloromethane, and diethyl ether.

3.2.3. Packing Procedure (7,8)

1. Dip one end of a capillary (0.5 m × 100 µm I.D.) into silica (10 µm) wetted with water. After drying for 24 h, sinter the silica plug with a hot wire by a homemade heater.
2. Sonicate the slurry of modified silica in methanol for about 3 min.
3. Fill the slurry quickly into the capillary (placed in an ultrasonic bath) with an HPLC pump.

4. Pack the capillary with methanol by applying a pressure of 380 bar for 1 h.
5. Flush the capillary with water.
6. Make the ultimate outlet frit by sintering a small section of the packing material (approx 600°C) about 15 cm from the end of the capillary (depended from the CE instrument). Flush the capillary with water during the sintering.
7. Cut off the temporary frit and empty the capillary up to the final frit by flushing with water.
8. Make the inlet frit by sintering the packing material at the other end of the capillary.
9. Prepare the detection window by the side of the outlet frit in the empty part of the capillary by placing one drop of concentrated nitric acid at the capillary for about 5–10 min. Then scratch carefully the polyimide outer coating.
10. Install the capillary into the CE instrument and flush the capillary with the mobile phase.
11. Condition the capillary by applying voltage (*see* **Notes 14** and **15**).

3.3. Preparation of Capillaries for Rod-CEC (9,10,14,15)

3.3.1. Preparation of a Silica Monolith by Sintering (10,14)

1. Prepare a temporary outlet frit (*see* **Subheading 3.2.3.**)
2. Sonicate a slurry of bare silica in acetone/toluene (1:1) for about 3 min.
3. Fill about 20–25 cm of the capillary, which is placed in a ultrasonic bath with the slurry by pumping first acetone/toluene (1:1), then acetone, and then finally water, through the capillary with an HPLC pump (380 bar) for 1 h.
4. Rinse the silica bed with a 0.1 *M* sodium carbonate solution (with about the vol of one capillary filling), then with water, and afterwards with acetone.
5. Put the capillary into a GC oven and install the capillary with the inlet end into the injector.
6. Apply a pressure of 2 bar hydrogen first at room temperature, 4 h at 120°C, and then 10 h at 380°C.
7. Cut off the temporary outlet frit.

3.3.2. Preparation of a Silica Monolith by Sol-Gel Technique (9,15)

1. Fill the capillary with the sol solution using a syringe. Seal both ends of the capillary with a piece of soft rubber.
2. Keep the capillary in a GC oven at 40°C for 24 h.
3. Heat the column in the GC oven using a temperature program from 40°C to 100°C at a rate of 0.5°C/min. Keep the final temperature for 20 h.
4. Wash the capillary with ethanol for 2 h to remove PEG from the silica matrix.
5. Heat again using a temperature program from 40° to 190°C at a rate of 0.5°C/min. Keep the final temperature for 2 h.

3.3.3. Surface Modification With Chirasil-Dex

1. Coat the monolith with a solution of Chirasil-Dex in acetone by pushing the solution through the monolithic capillary by pressure.

2. Dry the capillary at room temperature in a GC oven under a hydrogen pressure of 2 bar overnight. Heat the capillary for three days at 235°C under a hydrogen pressure of 2 bar.
3. Wash the capillary with acetone and water.
4. Prepare the detection window as described in **Subheading 3.2.3.**
5. Install the capillary into the CE instrument and condition the monolith by flushing with the mobile phase.

4. Notes

1. Dimethyl formamide, methyl iodide, pyridine, methanol, tetramethoxysilane, and dichloromethane are hazardous to health and should be handled with caution.
2. Pulverize the sodium hydroxide immediately before starting the reaction.
3. Control the temperature during the drying step of monokis-6-(octen-7-yl)-β-cyclodextrin. The temperature should not be higher than 60°C.
4. Add the monokis-6-(octen-7yl)-β-cyclodextrin dissolved in dimethy formamide very carefully to the sodium hydride because hydrogen evolve.
5. Calculation of the film thickness: film thickness (μm) = 2.5 × I.D. of the capillary (mm) × concentration of the coating solution (%).
6. Dry the silica particles by azeotropic destillation in rotary evaporator. Repeat this procedure three times.
7. For modification of silica avoid stirring with a magnetic stirrer. Use a rotary evaporator (without vacuum).
8. To enhance the EOF of capillaries packed with Chirasil-Dex silica, add 10–20% bare silica to the slurry of the packing material.
9. Be sure that there is no air bubble inside the filled capillary. Then the capillary should be empted by applying vacuum.
10. Calculation of the retention factor: $k = t_R/t_0$ (t_R = retention time, t_0 = breakthrough time).
11. 9-BBN is moisture-sensitive.
12. Use a rotary evaporator. Remove the solvent very slowly (during 2 h) under slight vacuum.
13. Use a rotating-strip column.
14. For conditioning, first apply 5 kV, then enhance the voltage very slowly (during 4–6 h) up to 30 kV.
15. To avoid bubble formation apply pressure (10–12 bar) at the inlet side or at both sides of the capillary. Degas the mobile phase.

References

1. Chankvetadze, B. and Blaschke, G. (2001) Enantioseparations in capillary electromigration techniques: recent developments and future trends. *J. Chromatogr. A* **906**, 309–363.
2. Wistuba, D. and Schurig, V. (2000) Recent progress in enantiomer separation by capillary electrochromatography. *Electrophoresis* **21**, 4136–41158.

3. Gübitz, G. and Schmid, M. G. (2000) Recent progress in chiral separation principles in capillary electrophoresis. *Electrophoresis* **21,** 4112–4135.

4. Schurig, V. and Wistuba, D. (1999) Recent innovations in enantiomer separation by electrochromatography utilizing modified cyclodextrins as stationary phases. *Electrophoresis* **20,** 2313–2328.

5. Mayer, S. and Schurig, V. (1993) Enantiomer separation by electrochromatography in open tubular columns coated with Chirasil-Dex. *J. Liq. Chromatogr.* **16,** 915–931.

6. Mayer, S. and Schurig, V. (1992) Enantiomer separation by electrochromatography on capillaries coated with Chirasil-Dex. *J. High Resolut. Chromatogr.* **15,** 129–131.

7. Wistuba, D., Czesla, H., Roeder, M., and Schurig, V. (1998) Enantiomer separation by pressure-supported electrochromatography using capillaries packed with a permethyl-β-cyclodextrin stationary phase. *J. Chromatogr. A* **815,** 183–188.

8. Wistuba, D. and Schurig, V. (1999) Enantiomer separation by pressure-supported electrochromatography using capillaries packed with Chirasil-Dex polymer-coated silica. *Electrophoresis* **20,** 2779–2785.

9. Kang, J.-W., Wistuba, D., and Schurig, V. (2002) A silica monolithic column prepared by the sol-gel process for enantiomeric separation by capillary electrochromatography. *Electrophoresis* **23,** 1116–1120.

10. Wistuba, D. and Schurig, V. (2000) Enantiomer separation by capillary electrochromatography on a cyclodextrin-modified monolith. *Electrophoresis* **21,** 3152–3159.

11. Schurig, V., Schmalzing, D., Mühleck, U., et al. (1990) Gas chromatographic enantiomer separation on polysiloxane-anchored permethyl-β-cyclodextrin (Chirasil-Dex). *J. High Resolut. Chromatogr.* **13,** 713–717.

12. Jung, M. and Schurig, V. (1993) Enantiomeric separation by GC on Chirasil-Dex: systematic study of cyclodextrin concentration, polarity, immobilization, and column stability. *J. Microcol. Sep.* **5,** 11–22.

13. Schurig, V., Negura, S., Mayer, S., and Reich, S. (1996) Enantiomer separation on a Chirasil-Dex-polymer-coated stationary phase by conventional and micro-packed high-performance liquid chromatography. *J. Chromatogr. A* **755,** 299–307.

14. Asaie, R., Huang, X., Farnan, D., and Horváth, C. (1998) Sintered octadecylsilica as monolithic column packing in capillary electrochromatography and micro high-performance liquid chromatography. *J. Chromatogr. A* **806,** 251–263.

15. Tanaka, N., Nagayama, H., Kobayashi, H., et al. (2000) Monolithic silica columns for HPLC, micro-HPLC, and CEC. *J. High Resolut. Chromatogr.* **23,** 111–116.

25

Chiral Separations
by Capillary Electrochromatography
Using Molecularly Imprinted Polymers

Peter Spégel, Jakob Nilsson, and Staffan Nilsson

1. Introduction

It is strongly recommended for the reader to first study the introduction in Chapter 9, "Chiral Separations by HPLC Using Molecularly Imprinted Polymers," to extract the basic features off molecularly imprinted polymer (MIP) synthesis. Also, it can be useful to read the Methods section and to study the notes where practical information can be found that complements the information given in this chapter.

The capillary columns used in capillary electrochromatography (CEC) and capillary liquid chromatography offer several advantages over conventional columns, including lower sample consumption and, often, faster analysis times. However, packing a thin (20–100 μm) capillary with an ordinary particle-based material can be tricky and time-consuming (1). When preparing packed capillary columns, there is a need for retaining frits, and these might cause, e.g., band broadening and sample adsorption (2). Recently, much effort has been put into developing new types of stationary phase materials for the capillary format (1). One of the more successful approaches is the monolithic format (**Fig. 1**) (3). Using this technique, the stationary phase is prepared *in situ*, i.e., inside the capillary. The monolith can either be synthesized de novo or constructed by entrapment of prefabricated particles. When synthesizing the monolith de novo, particle packing procedures and the need for retaining frits are circumvented. The monolith has a sponge-like structure, which can be designed to provide a high porosity, stability, and flexibility. Also, stationary phases coated on the inner surface of the capillary for open tubular (OT) applications have been identified as promising stationary phase types for the capillary format (**Fig. 2**)

From: *Methods in Molecular Biology, Vol. 243: Chiral Separations: Methods and Protocols*
Edited by: G. Gübitz and M. G. Schmid © Humana Press Inc., Totowa, NJ

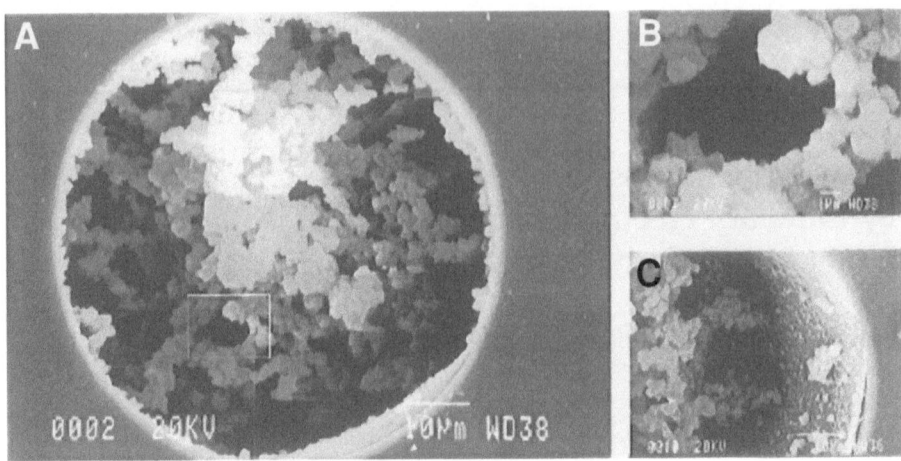

Fig. 1. Scanning electron micrograph of an MIP monolith inside a capillary. (**A**) Cross-section of the capillary. (**B**) Closer view of the square in panel A showing the super-pour network. (**C**) Picture showing the covalent attachment of the monolith to the inner surface of the capillary. Reprinted with permission from **ref. 8**.

(4). The synthesis and use of MIP monoliths and coatings for CEC will be covered in this chapter.

CEC has become known as a powerful tool in analytical chemistry *(5)*, and it has been identified as a technique able to improve separations based on MIP stationary phases *(6)*. Synthesis of MIPs has been described previously (Chapter 9). In short, MIPs are synthesized utilizing a template-assisted polymerization, where the monomers assembled around the template are polymerized, together with cross-linking monomers, to create a cavity in the polymer that is complementary to the template in shape, size, and chemical functionality. This cavity is able to recognize and rebind the template (analyte) with high selectivity.

One advantage of the MIP, when compared to immobilized biological selectors, is that the MIP can be used under very harsh conditions. The MIP can be stored for several years and still retain its initial affinity for the template molecule, and it can also withstand extreme conditions such as treatment with strong acids and bases as well as temperatures up to around 150°C *(7)*.

1.1. Monolithic MIP Preparation

The monolithic MIP is synthesized *in situ*, i.e., inside the capillary, and retaining frits can be avoided due to the continuous structure of the monolith (**Fig. 3**).

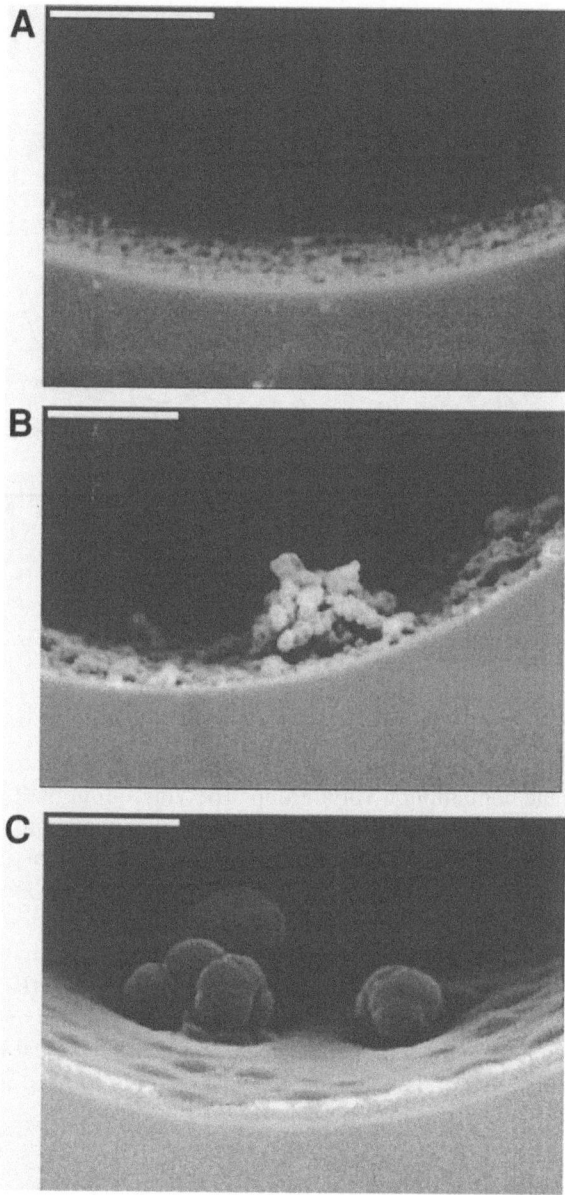

Fig. 2. Scanning electron micrographs of MIP surface coatings on the inner surface of a capillary. The MIP coatings are prepared with (**A**) toluene, (**B**) dichloromethane, and (**C**) acetonitrile as the polymerization solvent. Reprinted with permission from **ref. *13***.

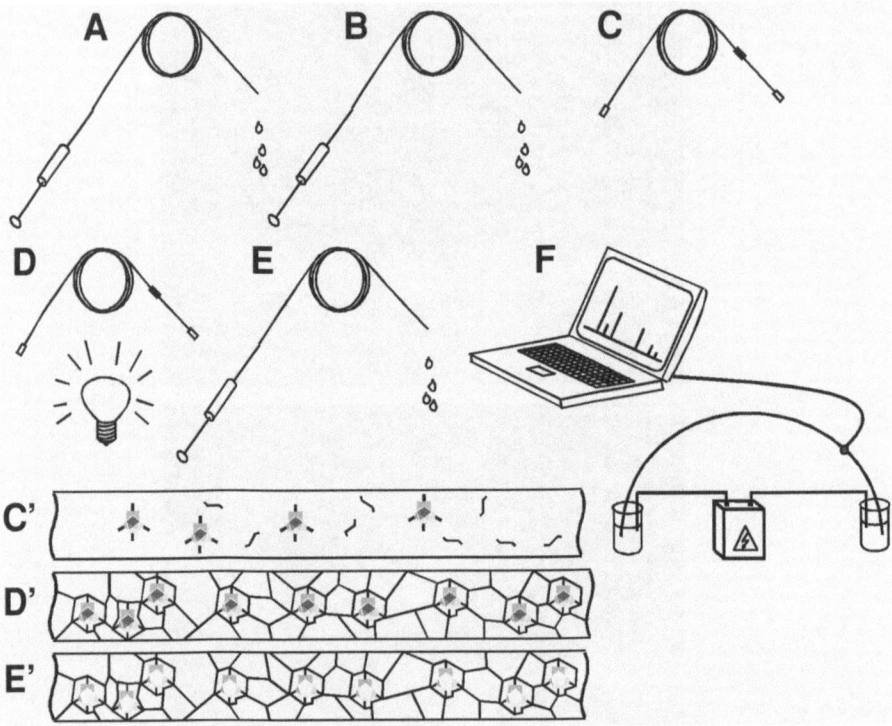

Fig. 3. Schematics of *in situ* MIP monolith preparation. (**A**) The capillary is derivatized with a silane containing a vinyl group. The vinyl group will participate in the polymerization, thus covalently linking the monolith to the capillary. (**B**) The prepolymerization mixture is filled into the capillary. (**C**) The capillary is sealed, and the detection window is shielded. (**C**) Inside the capillary, complexes are formed between the template molecule and functional monomers. (**D**) Polymerization is initiated using UV light. (**D**) Inside the capillary, the monomers, in complex with the template, react with crosslinking monomers to form a rigid polymer network. (**E**) The capillary is washed. (**E**) The template molecule is extracted from the MIP, leaving behind a cavity that is complementary to the template in shape, size, and chemical functionality. (**F**) The capillary is installed in the CE apparatus and ready to be used.

The monolith appears as a sponge-like polymer rod (**Fig. 1**), which contains charged groups, due to the most often charged functional monomers used to complex/recognize the template/analyte. Owing to these charges on the monolith, it can become mobile when exposed to an electric field. This might result in elution of parts or the entirety of the monolith during electrochromatography. To prevent MIP elution, the inner surface of the capillary can be derivatized prior to monolith synthesis (**Fig. 3**). The most common approach is silanization using ([methacryloxy]propyl)trimethoxy silane that contains vinyl groups able to participate in the polymerization reaction *(8)*. Thus, the MIP monolith will become covalently bound to the capillary inner surface during polymerization (**Fig. 1**), and stationary phase elution will be prevented.

After having synthesized the monolith, the remaining polymerization solution is exchanged for a washing solution, able to extract the template, and followed by conditioning in electrolyte (**Fig. 3**). For some MIP monolith systems, this can be achieved electrophoretically *(9)*. This approach is time-consuming and requires special considerations when preparing the prepolymerization mixture. Another approach to enable solvent exchange in the monolith is to synthesize a monolith containing super-pores. This approach yields good flow properties, easy solvent exchange, and column regeneration. One technique used in order to introduce super-pores in the monolith is to use interrupted polymerization *(8)*. When precipitation polymerization is used, particles are initially formed inside the capillary. If the monomer concentration is high enough, these particles become crosslinked, and the volume in between them is gradually filled with polymer. In order to produce super-pores using the interrupted polymerization approach, the capillary is flushed to get rid of monomers before the interparticle volumes are completely filled with polymer. This approach requires careful timing, but allows the use of a great number of polymerization solvents. An alternative to this approach is to use a pore-forming solvent in the prepolymerization mixture. For instance, iso-octane has successfully been used with toluene as polymerization solvent, where a higher amount of iso-octane yields a more porous polymer *(10)*. However, the application of this approach is limited in terms of solubility of the template and monomers in the polymerization solvent. If a pore-forming solvent cannot be found that is able to solubilize the monomers and the template molecule, the interrupted polymerization approach might be the method of choice.

Polymerization can be initiated using either heat or UV light (**Fig. 3**). When performing UV light-initiated polymerization, an UV light-permeable capillary coating is needed. These capillaries are commercially available. It is advantageous to use UV light for a couple of reasons. First, polymerization can be initiated at low temperature, which is beneficial in terms of imprint quality *(11)*. It is believed that the complexes between functional monomers and the template molecule are more stable at a lower temperature due to a more favorable entropy term. Second, a photochemical masking procedure can be used, which prevents polymer growth in the detection window (if on column detection is used) *(8)*. This is easily achieved by covering the detection window during polymerization with a piece of UV light-impermeable paper.

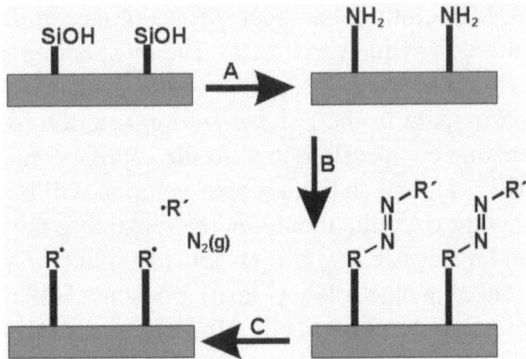

Fig. 4. Schematics of immobilization of radical initiator to the inner surface of the capillary for MIP surface synthesis. (**A**) The capillary is derivatized to contain amine groups. (**B**) A carboxylic acid group containing radical initiator is linked to the capillary bound amine using a carbodiimide reagens. (**C**) Upon UV light irradiation, the bound radical initiator decomposes, and polymerization begins at the capillary inner surface.

1.2. Open Tubular MIP Preparation

The monolithic MIP offers a high binding site density, and separations can thus be performed on very short monoliths *(12)*. However, it can be difficult to optimize the polymerization procedure to achieve super-pores in the resultant MIP monolith. Furthermore, due to the possible presence of irregularities in the monolith structure, eddy diffusion might affect the efficiency disadvantageously. An interesting approach to circumvent optimization of super-pore formation and efficiency losses due to eddy diffusion is to synthesize an MIP coating on the inner surface of the capillary *(13,14)*. The polymerization procedure can be easily adapted to yield a polymer of the desired thickness, e.g., some few micrometers, which is limited in its presence to the inner walls of the capillary (**Fig. 2**). The resultant capillary will thus possess good flow-through properties, offering fast flushing, stationary phase conditioning, electrolyte regeneration, and almost no problems with column clogging.

One technique that successfully has been used to synthesize MIP coatings is to immobilize the radical initiator on the inner surface of the capillary (**Fig. 4**) *(13)*. The polymer will thus start to grow from the capillary surface and propagate into the center of the capillary. However, there is a need for timing of the polymerization, as the capillary eventually will be completely filled with polymer if the polymerization is allowed to propagate for too long a time. The MIP coatings for OT-CEC can easily be fabricated using solvents impossible to use in the synthesis of super-porous monolithic MIPs *(13)*.

1.3. Separations Using MIP

One opinion when choosing the eluent for MIP chromatography is to choose one that has the same composition as the polymerization solvent (*see* Chapter 9). This approach might be useful in MIP-high-performance liquid chromatography (HPLC), but it is not as readily transferred to MIP-CEC. The reason for this is the need for the eluent in CEC to contain an electrolyte able to accept or transfer protons or to undergo an eventual autoproteolysis. The polymerization solvents used for MIP monolith or surface preparation do not normally possess these properties, i.e., they are inert in this aspect. The most frequently used electrolytes in MIP-CEC contain mixtures of acetonitrile with different amounts of an aqueous buffer *(6)*. The electrochromatography is fine-tuned by optimizing the amount of acetonitrile and the apparent pH of the electrolyte *(10,15)*.

1.4. New Approaches in MIP-CEC

Recently, a new type of MIP has been developed and is used in CEC for partial filling applications *(15–17)*. This involves injection of the MIP phase as a plug in front of the sample. When the electric field is applied, the sample is allowed to pass the plug, due to the opposite charges of the MIP and the analyte, and reach the detection window prior to the MIP plug. The MIP used in this approach is in the form of particles of submicrometer size that is prepared utilizing precipitation polymerization of dilute monomer solutions. The advantages of this technique include the use of an entirely new MIP phase in every new separation, the ability to use particle-based MIP without the use of retaining frits and packing procedures, and the ability to design the MIP phase for every new application using MIP particles with differing selectivity.

2. Materials

2.1. Preparation of Capillaries for Monolithic MIP

1. Fused silica capillary, 100 μm inner diameter, 375 μm outer diameter, and 35 cm of total length with a UV-transparent capillary coating (e.g., from Composite Metal Services, West Yorkshire, UK; cat. no. TSU100375), 0.1 *M* NaOH, 1 *M* HCl, distilled water, acetone, nitrogen gas, and toluene (HPLC-grade) dried over molecular sieve. Syringe (outer diameter of the needle 330 μm), PVC tubing (inner diameter 127 μm).
2. Silanization solution consisting of dry toluene (HPLC-grade)/(methacryloxy)propyltrimethoxysilane (85:15, v/v). Syringe and PVC tubing (*see* **step 1.**), and soft plastic rubber. The silanization solution should be prepared fresh as required. The (methacryloxy)propyltrimethoxysilane should be stored below 8°C.
3. Dry toluene (HPLC-grade) and nitrogen gas. Syringe and PVC tubing (*see* **step 1.**).

2.2. Preparation of Capillaries for OT-MIP

1. Fused silica capillary, 50 μm inner diameter, 375 μm outer diameter, and 35 cm of total length with a UV-transparent capillary coating (e.g., from Composite Metal Services; cat. no. TSU050375), 0.1 *M* NaOH, 1 *M* HCl, distilled water, acetone, nitrogen gas, and toluene (HPLC-grade) dried over molecular sieve. Syringe (outer diameter of the needle 330 μm), PVC tubing (inner diameter 127 μm).
2. Dry toluene (HPLC-grade)/3-aminopropyltriethoxysilane (85:15, v/v) (amino derivatization solution), syringe and PVC tubing (*see* **step 1**) and soft plastic rubber. The silanization solution should be prepared fresh as required. 3-Amino-propyltriethoxysilane should be stored below 8°C.
3. Dry toluene (HPLC-grade) and nitrogen gas. Syringe and PVC tubing (*see* **step 1**).
4. 20 m*M* 4,4'-Azobis(4-cyanopentanoic acid) (ACPA) (radical initiator) in methanol, 60 m*M* 1-ethyl-3-(3-dimethylaminopropyl)carbodiimide hydrochloride (EDAC) (coupling reagense) in 50 m*M* 2-(*N*-morpholino)ethanesulfonic acid (MES) buffer adjusted to pH 5.5 with 50 m*M* NaOH. Syringe and PVC tubing (*see* **step 1**) and soft plastic rubber. The ACPA and EDAC solutions should be prepared fresh as required. EDAC should be stored dry below 0°C, and ACPA should be stored dry at 4–8°C.
5. Methanol/distilled water (1:1, v/v) and nitrogen gas. Syringe and PVC tubing (*see* **step 1**).

2.3. Preparation of Monolithic MIP

1. Dry toluene (HPLC-grade)/iso-octane (99:1, v/v) (polymerization solvent), methacrylic acid (MAA) (functional monomer), trimethylolpropane trimethacrylate (TRIM) (crosslinking monomer), free base *S*-propranolol (template molecule), 2,2'-azobisisobutyronitrile (AIBN) (radical initiator). Glass vial (1 mL) with cover. Be aware of the risk of explosion by shock, friction, fire, or other sources of ignition when handling AIBN. Be aware of that all monomers are toxic and potential skin sensitizers.
2. Syringe (outer diameter of the needle 330 μm), PVC tubing (inner diameter 127 μm), soft plastic rubber.
3. Freezer with the ability to house a UV lamp that emits light at 350 nm (polymerization initiation). The UV-irradiation might harm the eyes and should be handled carefully.
4. Methanol and methanol/acetic acid (9:1, v/v) (MIP washing solutions), syringe and PVC tubing (*see* **step 2**).

2.4. Preparation of OT-MIP

1. Dichloromethane (polymerization solvent), MAA (functional monomer), TRIM (crosslinking monomer), free base *S*-propranolol (template molecule). Glass vial (1 mL) with cover. Be aware of that all monomers are toxic and potential skin sensitizers.

2. Syringe (outer diameter of the needle 330 μm), PVC tubing (inner diameter 127 μm), soft plastic rubber.
3. Freezer with the ability to house a UV lamp emitting light at 350 nm (polymerization initiation). The UV light might harm the eyes and should be handled carefully.
4. Methanol and methanol/acetic acid (9:1, v/v) (MIP washing solutions). Syringe and PVC tubing (*see* **step 2**).

2.5. Enantiomer Separation Using MIP in CEC

1. Capillary electrophoresis (CE) equipment preferably with the ability to apply pressure on both the inlet and the outlet vials simultaneously. Electrolyte consisting of acetonitrile/25 mM phosphoric acid adjusted to pH 3.0 with triethanolamine (80:20, v/v). Stock solutions of 10 mM of rac-propranolol, S-propranolol and R-propranolol, respectively in distilled water.

3. Methods

3.1. Preparation of Capillaries for Monolithic MIP

1. In order to achieve a stable and even coating, the capillary needs to be extensively washed with water, 0.1 M NaOH, water, 1 M HCl, water, acetone, and toluene, respectively. The washing solutions can easily be introduced into the capillary using a syringe connected to the capillary via PVC tubing. The capillary is subsequently dried under a flow of nitrogen gas.
2. The silanization solution is introduced into the capillary using a syringe and PVC tubing. Allow a small vol to pass through the capillary (10–15 droplets), seal the capillary at both ends using soft rubber, and leave the capillary at room temperature over night (*see* **Note 1**).
3. Rinse the capillary with toluene, using a syringe and PVC tubing, followed by drying with a flow of nitrogen. The capillaries used are UV-transparent, however, a detection window needs to be created in order to allow a broader range of wavelengths to be transmitted. A detection window is created on the capillary by masking the capillary outside the detection area with aluminium foil and burning of the part of the outer polymer coating that, after installation in the CE apparatus, will be present in the detector (*see* **Note 2**).

3.2. Preparation of Capillaries for OT-MIP

1. In order to achieve a stable and even coating, the capillary needs to be extensively washed with water, 0.1 M NaOH, water, 1 M HCl, water, acetone, and toluene, respectively. The washing solutions can easily be introduced into the capillary using a syringe connected to the capillary via PVC tubing. The capillary is subsequently dried under a flow of nitrogen.
2. The silanization solution is introduced into the capillary using a syringe and PVC tubing. Allow a small vol to pass through the capillary (10–15 droplets), seal the

capillary at both ends using soft rubber, and leave the capillary at room temperature overnight (*see* **Note 1**).

3. The capillary is rinsed by toluene, using a syringe and PVC tubing, followed by drying with a flow of nitrogen. The capillaries used are UV-transparent, however, a detection window needs to be created in order to allow a broader range of wavelengths to be transmitted. A detection window is created on the capillary by masking the capillary outside the detection area with aluminium foil and the burning of the part of the outer polymer coating present in the detection window (*see* **Note 2**).

4. The ACPA solution is mixed with the EDAC solution (1:1, v/v) carefully by sonication for 5–10 s. The solution is filled into a syringe, and the syringe is connected to the capillary via PVC tubing. The solution is introduced into the capillary while allowing a small vol to pass through the capillary (10–15 droplets). The capillary is sealed at both ends using soft plastic rubber and left at room temperature overnight (*see* **Note 1**).

5. Wash the capillary extensively, using a syringe connected to the capillary via PVC tubing, with methanol/distilled water (1:1, v/v) to remove unbound ACPA and dry the capillary under a flow of nitrogen gas. These modified capillaries should be prepared as required (*see* **Note 3**).

3.3. Preparation of Monolithic MIP

1. The prepolymerization mixture is prepared by mixing functional monomer (MAA) 0.24 M, crosslinking monomer (TRIM) 0.24 M, template molecule (S-propranolol) 0.03 M, and radical initiator (AIBN) 3.6 g/L in the polymerization solvent (toluene/iso-octane [99:1, v/v]) to give an appropriate working volume of 200 µL. This solution is degassed by sonication for 5 min. Prepolymerization solutions should be prepared fresh as required (*see* **Notes 4–7**).

2. The prepolymerization solution is filled into a syringe, and the syringe is connected to the capillary via PVC tubing. Allow a small portion of the prepolymerization solution to flow through the capillary (10–15 droplets), then seal the capillary at both ends using small pieces of soft plastic rubber. Protect the detection window by masking with a piece of UV-blocking paper in order to avoid polymer growth in the detection area (*see* **Note 1**).

3. Place the capillary under a UV lamp in a freezer and allow polymerization to propagate overnight (*see* **Notes 8–11**).

4. Remove the template molecule by extensive washing with methanol/acetic acid (9:1, v/v) and methanol, respectively, using a syringe and a PVC tubing (*see* **Note 12**).

5. The MIP capillaries can be stored at room temperature shielded from light until use (*see* **Note 13**).

3.4. Preparation of OT-MIP

1. The prepolymerization mixture is prepared by mixing functional monomer (MAA) 0.48 M, crosslinking monomer (TRIM) 0.48 M, template molecule (S-propranolol) 0.06 M in the polymerization solvent (dichloromethane) to give an appropriate working volume of 200 µL. This solution is degassed by sonication for 5 min. Pre-

polymerization solutions should be prepared fresh as required (*see* **Notes 4, 5, 7, and 14**).

2. The prepolymerization solution is filled into a syringe, and the syringe is connected to the capillary via PVC tubing. Allow a small portion of the prepolymerization solution to flow through the capillary (10–15 droplets) and seal the capillary at both ends using small pieces of soft plastic rubber. Protect the detection window by masking it with a piece of UV-blocking paper in order to avoid polymer growth in the detection area (*see* **Note 1**).

3. The capillary is placed under a UV lamp at room temperature, and polymerization is interrupted after 2 to 3 h (*see* **Notes 10, 11, 15,** and **16**).

4. Remove the template molecule by extensive washing with methanol/acetic acid (9:1, v/v) and methanol, respectively, using a syringe and a PVC tubing (*see* **Note 12**).

5. The MIP capillaries can be stored at room temperature shielded from light until use (*see* **Note 13**).

3.5. Enantiomer Separations Using MIP in CEC

1. Install the capillary in the CE apparatus. Condition the MIP phase by flushing the capillary with electrolyte (100–400 µL). Dilute the sample stock solutions with electrolyte to achieve samples with analyte concentrations of 25 µ*M*. Injection is performed electrokinetically at 5 kV for 3 s. The capillary can be thermostated to 60°C to improve the interaction kinetics. Separation is performed at 15 kV (429 V/cm) with detection at 215 nm (*see* **Notes 17** and **18**).

4. Notes

1. Use a piece of plastic rubber large enough to be able to stick in the capillary and completely seal the ends of the capillary. Do not use to large pieces, as there is a risk of breaking the capillary if there is too much weight on the capillary ends. It is especially important to handle the capillaries with UV-transparent coating carefully, as they are far more brittle than the standard polyimide-coated capillaries.

2. The on-column detection window can be prepared prior to surface derivatization of the capillary in order not to destroy parts of it when using heat. Also, the window can be created by exposing the detection area to fuming sulfuric acid or by gently scratching the protecting polymer layer using a razor blade. However, it is very easy to break the capillary in the detection window, and it is recommended to prepare the window as late as possible in the MIP synthesis procedure.

3. It is important to dry the capillary thoroughly, as even very small amounts of protic solvents present in the capillary during MIP synthesis can be disastrous.

4. The functional monomer can, in part, be exchanged for 2-vinylpyridine (2-VPy) or 4-vinylpyridine (4-VPy). The combination of MAA with VPy has been shown to improve imprinting in some systems. Also, weakly interacting monomers can be used to improve electrochromatography on the resultant MIP. Improvement in resolution and efficiency can be governed by exchanging 50% of MAA for, e.g., methyl methacrylate (MMA) or butyl methacrylate (BMA).

5. The crosslinker TRIM can be exchanged for, e.g., ethylene glycol dimethacrylate (EDMA). As this monomer has only two vinyl groups (compared to three for TRIM), the crosslinking ratio needs to be increased, e.g., 75% of the total monomer content.

6. The amount of iso-octane in the polymerization solvent can be adjusted to alter the porosity of the resulting MIP. A higher iso-octane content yields a more porous polymer. Also, if other types of monomers are used, e.g., 2-VPy, 4-VPy, MMA, or BMA, there might be a need for a higher iso-octane content in order to obtain appropriate super-pores.

7. It is extremely important to carefully degas the prepolymerization solution, as small amounts of dissolved oxygen will inhibit polymerization.

8. Polymerization can also be initiated at room temperature or in a fridge. However, at these higher temperatures, the polymer might grow into the detection area even though the masking procedure has been performed carefully.

9. When performing polymerization in a freezer, the UV lamp and/or the capillary might become covered with ice. The presence of ice in between the UV source and the polymerization solution might decrease the intensity of the UV light reaching the polymerization solution (due to light scattering). In order to decrease these effects, it can be beneficial to fix the capillary as close to the lamp as possible or directly fasten the capillary on the UV lamp with tape.

10. As the polymer scatters and in some cases absorbs light, there might be a lower polymerization yield in the parts of the capillary that are not closest to the lamp. These problems can be avoided by turning the capillary every few hours to make sure that all parts of the capillary are exposed the same UV light intensity.

11. Heat-initiated polymerization can be used with AIBN at 65°C. However, it is then extremely hard to avoid MIP growth in the detection area. To remove polymer from the detection area, the capillary is flushed with water while quickly burning of the MIP present in the detection area.

12. A stronger template eluent is achieved if the amount of acetic acid in the wash solution is increased.

13. The MIP might turn brittle upon prolonged light exposure.

14. The polymerization solvent can be exchanged for, e.g., acetonitrile, chloroform, or toluene. A change of polymerization solvent will affect the imprinting process as well as the morphology of the resultant polymer.

15. Polymerization can be performed in a freezer at lower temperature to improve the imprinting.

16. The polymerization time needs to be optimized, as it is dependent on the intensity of the UV source. Thus, it is appropriate in a first experiment to use three or four capillaries and interrupt the polymerization after, e.g., 1, 2, 3, and 4 h, respectively.

17. The electrolyte might be varied in order to alter the electrochromatography. Parameters that can be changed include the amount of acetonitrile, the buffer system, and the pH. The use of triethanol amine in the buffer is advantageous, as it tends to decrease the EOF.

18. The MIP often shows some cross-selectivity, i.e., it can recognize structural analogs of the imprinted molecule. This extends the applications of the produced MIP to also include enantiomer separations of closely resembling molecules. The selectivity of the MIP against a structurally related analog is, for obvious reasons, due to a less perfect fit in the imprint, less than for the imprinted analyte. Structural analogs to propranolol that can be separated on a *S*-propranolol MIP include prenalterol, atenolol, and pindolol.

References

1. Tang, Q. and Lee, M. L. (2000) Column technology for capillary electrochromatography. *Trends Anal. Chem.* **19,** 648–663.
2. Behnke, B., Johansson, J., Bayer, E., and Nilsson, S. (2000) Fluorescence imaging of frit effects in capillary separations. *Electrophoresis* **21,** 3102–3108.
3. Svec, F., Peters, E. C., Sykora, D., Yu, C., and Fréchet, J. M. J. (2000) Monolithic stationary phases for capillary electrochromatography based on synthetic polymers: designs and applications. *J. High. Resol. Chromatogr.* **23,** 3–18.
4. Jinno, K. and Sawada, H. (2000) Recent trends in open-tubular capillary electrochromatography. *Trends Anal. Chem.* **19,** 664–675.
5. Cikalo, M. G., Bartle, K. D., Robson, M. M., Myers, P. M., and Eubery, R. (1998) Capillary electrochromatography. *Analyst* **123,** 87R–102R.
6. Schweitz, L., Spégel, P., and Nilsson, S. (2001) Approaches to molecular imprinting based selectivity in capillary electrochromatography. *Electrophoresis* **22,** 4053–4063.
7. Svensson, J. and Nicholls, I. A. (2001) On the thermal and chemical stability of molecularly imprinted polymers. *Anal. Chim. Acta* **435,** 19–24.
8. Schweitz, L., Andersson, L. I., and Nilsson, S. (1997) Capillary electrochromatography with predetermined selectivity obtained through molecular imprinting. *Anal. Chem.* **69,** 1179–1183.
9. Lin, J.-M., Nakagama, T., Wu, X.-Z., Uchiyama, K., and Hobo, T. (1997) Capillary electrochromatographic separation of amino acid enantiomers with molecularly imprinted polymers as chiral recognition agents. *Fresenius J. Anal. Chem.* **357,** 130–132.
10. Schweitz, L., Andersson, L. I., and Nilsson, S. (1997) Capillary electrochromatography with molecular imprint-based selectivity for enantiomer separation of local anaesthetics. *J. Chromatogr. A* **792,** 401–409.
11. O'Shannessy, D., Ekberg, B., and Mosbach, K. (1989) Molecular imprinting of amino acid derivatives at low temperature (°C) using photolytic homolysis of azobisnitriles. *Anal. Biochem.* **177,** 144–149.
12. Schweitz, L., Andersson, L. I., and Nilsson, S. (2001) Rapid electrochromatographic enantiomer separations on short molecularly imprinted polymer monoliths. *Anal. Chim. Acta* **435,** 43–47.
13. Schweitz, L. (2002) Molecularly imprinted polymer coatings for open-tubular capillary electrochromatography prepared by surface initiation. *Anal. Chem.* **74,** 1192–1196.

14. Tan, J. Z. and Remcho, V. T. (1998) Molecular imprint polymers as highly selective stationary phases for open tubular liquid chromatography and capillary electrochromatography. *Electrophoresis* **19,** 2055–2060.
15. Spégel, P., Schweitz, L., and Nilsson, S. (2001) Molecularly imprinted microparticles for capillary electrochromatography: studies on microparticle synthesis and electrolyte composition. *Electrophoresis* **22,** 3833–3841.
16. Schweitz, L., Spégel, P., and Nilsson, S. (2000) Molecularly imprinted microparticles for capillary electrochromatographic enantiomer separation of propranolol. *Analyst* **125,** 1899–1901.
17. Spégel, P. and Nilsson, S. (2002) A new approach in capillary electrochromatography: disposable molecularly imprinted nanoparticles. *Am. Lab.* **34,** 29–33.

Index

A

Acetyl (Ac) amino acids, 325
Acidic compounds, 178
Affinity capillary electrophoresis,
      291–292, 295
   coated capillary, 296, 311
   diol coating, 296, 300
   linear acrylamide coating, 294, 297,
      300, 311
   partial filling technique, 294–295,
      309
   polyvinylalcohol coating, 311
Affinity electrochromatography,
      291–292
   α1- acid glycoprotein, 291, 294–297
   bovine serum albumin, 291,
      294–298, 300
   immobilization of proteins,
Alimemazine, 302
Aliphatic alcohols, 189
Alkaloids, 100, 185
N-Alkylamino acids, 37
Amines, 185,154
Amino acid derivatives, 88, 100, 149,
      324,325
Amino acid esters, 93
Amino acids, 37, 88, 149, 210, 211,
      213, 214, 317, 375, 376, 377,
      381, 382, 384
Amino alcohols, 376, 381, 309
Amino phosphonic acids, 88, 325
6-Aminoquinoylcarbamoyl-amino
      acids, 88
Amylose,
   tris(3,5-dimethylphenylcarbamate),
      173, 174, 185, 368
   tris (S-1-methylphenylcarbamate),
      174, 185

tris (S-1-phenylethylcarbamate), 173,
      174, 185, 368
Apparent electrophoretic mobility, 258
Aromatase inhibitors, 189
Aromatic amino acids, 100
Artificial receptor, 217
Aryl methyl esters, 185

B

Barbitals, 100
Basic compounds, 179
Benzoate, 175
Benzoin, 299
Benzoyl amino (Bz) acids, 325, 329
Benzyloxycarbonyl (Z) amino acids,
      149, 325
Biaryl compounds, 189
Biological samples, 12
β-Blockers, 32, 89, 100, 151, 185, 267,
      268, 309, 370, 375, 423
Butoxycarbonyl (Boc) amino acids,
      88, 150

C

Calcium channel blockers, 89,152
Calixarenes, 10
Capillary electrochromatography, 378
Capillary electrochromatography, 1,
      7, 266, 270, 387, 375, 378,
      401, 425
   continuous beds, 378, 381
   in situ polymerization, 378, 381
   monolithic phases, 378
Capillary electrophoresis, 1, 255, 265,
      275, 291, 307, 317, 323, 343,
      365, 375
Capillary zone electrophoresis, 343
Carbohydrates, 7, 173, 183, 343